WITHDRAWN

WITHDRAWN
Stafford Library
Columbia College
1001 Rogers Street
Columbia, MO 65216

Principles of Programming & Coding

✓

Principles of Programming & Coding

Editor
Donald R. Franceschetti, PhD
The University of Memphis

SALEM PRESS

A Division of EBSCO Information Services
Ipswich, Massachusetts

GREY HOUSE PUBLISHING

Cover image: By monsitj (iStock)

Copyright ©2018, by Salem Press, A Division of EBSCO Information Services, Inc., and Grey House Publishing, Inc.

All rights reserved. No part of this work may be used or reproduced in any manner whatsoever or transmitted in any form or by any means, electronic or mechanical, including photocopy, recording, or any information storage and retrieval system, without written permission from the copyright owner. For permissions requests, contact proprietarypublishing@ebsco.com.

∞ The paper used in these volumes conforms to the American National Standard for Permanence of Paper for Printed Library Materials, Z39.48 1992 (R2009).

Publisher's Cataloging-In-Publication Data
(Prepared by The Donohue Group, Inc.)

Names: Franceschetti, Donald R., 1947- editor.
Title: Principles of programming & coding / editor, Donald R. Franceschetti, PhD.
Other Titles: Principles of programming and coding
Description: [First edition]. | Ipswich, Massachusetts : Salem Press, a division of EBSCO Information Services, Inc. ; [Amenia, New York] : Grey House Publishing, [2018] | Series: Principles of | Includes bibliographical references and index.
Identifiers: ISBN 9781682176764 (hardcover)
Subjects: LCSH: Computer programming. | Coding theory.
Classification: LCC QA76.6 .P75 2018 | DDC 005.1—dc23

First Printing
Printed in the United States of America

CONTENTS

Publisher's Note vii
Editor's Introduction ix
Contributors xv

3D printing 1
Algorithms 4
American Standard Code for Information
Interchange (ASCII) 6
Android OS 8
Application 11
Autonomic computing 14
Avatars and simulation 16

Binary hexadecimal representations 19
Boolean operators 23
Branching logic 25

Characters and strings 28
Cloud computing 30
Coding and encryption 32
Color coding 35
Combinatorics 37
Comment programming 39
Comparison operators 41
Computer animation 44
Computer memory 46
Computer modeling 48
Computer security 51
Computer-aided design (CAD) 53
Computer-aided design and computer-aided
manufacturing software (CAD/CAM) 55
Computer-assisted instruction (CAI) 57
Conditional operators 59
Constraint programming 62
Control systems 64
Cowboy coding 66
CPU design 68
Crowdfunding 71
Crowdsourcing 72
Cryptography 74

Data mining 78
Data warehouse 80
Database design 82
Database structuring conventions 83
Debugging 85

Device drivers 87
Diffusion of innovations 89
Digital divide 91
Digital forensics 93
Digital libraries 95
Digital native 96
Digital photography 98
Digital signal processors (DSP) 99
Digital watermarking 101
Disk operating system (DOS) 103
Drone warfare 105
Drones 107

E-banking 110
E-learning 111
Electronic circuits 113
Electronic communication software 115
Encryption 117
Error handling 119
Event-driven marketing (EDM) 121
Expectancy theory 123
Experimenter's bias 124
Extreme programming 126

Firewalls 129
Firmware 131
Functional electrical stimulation (FES) 133

Game programming 135
Gamification 137
Graphical user interface (GUI) 138
Graphics formats 140
Guard clause 142

HTTP cookie 145

Imagined communities 147
Incremental development 149
Information technology (IT) 151
Information visualization 153
Internet Protocol (IP) 154
Inversion of control (Hollywood Principle) 155
iOS 158
Iterative constructs 160

Java programming language 163
JavaScript 164

v

Knowledge worker	166
Levels of processing theory	168
Logic synthesis	169
Logistics	170
Machine learning	173
Malware	174
Massive open online course (MOOC)	177
Meta-analysis	179
Metacomputing	181
Metadata	183
Microprocessors	184
Mixed methods research (MMR)	186
Mobile apps	188
Mobile technology	191
Motherboards	193
Multiprocessing operating systems (OS)	195
Multi-user operating system (OS)	197
Naming conventions	200
Net neutrality	203
Network security	205
Neuro-linguistic programming (NLP)	207
Neuromarketing	208
Neuromorphic chips	210
Objectivity	213
Object-oriented design (OOD)	215
Object-oriented programming (OOP)	217
Privacy rights	220
Programming languages	222
Prototyping	225
Quantum computing	228
Random access memory (RAM)	230
Rapid application development (RAD)	232
Rational choice theory	234
Search engine optimization (SEO)	237
Semantic memory	238
Semantics	239
Signal processing	240
Source code comments	243
Spiral development	245
Standard deviation	248
Standpoint theory	249
Statistical inference	251
String-oriented symbolic languages (SNOBOL)	252
Structural equation modeling (SEM)	254
Technology in education	256
Test doubles	259
Theory of multiple intelligences	261
Theory X and Theory Y	262
Transformation priority premise (TPP)	264
Tree structures	266
Turing test	269
Uncertainty reduction theory (URT)	272
Unicode	274
UNIX	276
Variables and values	279
Waterfall development	282
Web design	284
Web graphic design	287
Working memory	289
Worse-is-better	291
Time Line of Inventions and Advancements in Programming and Coding	295
Glossary	307
Bibliography	315
Index	345

Publisher's Note

Salem Press is pleased to add *Principles of Programming & Coding* as the tenth title in the *Principles of* series that includes *Biotechnology, Chemistry, Physics, Astronomy, Computer Science, Physical Science, Biology,* and *Scientific Research*. This new resource introduces students and researchers to the fundamentals of programming and coding using easy-to-understand language, giving readers a solid start and deeper understanding and appreciation of this complex subject.

The 138 articles in this volume explain basic principles of programming and coding, ranging from 3D printing and Algorithms to Waterfall development and Worse-is-better with attention paid to Information visualization, Uncertainty reduction theory, Technology in education, and more. All of the entries are arranged in an A to Z order, making it easy to find the topic of interest.

Entries related to basic principles and concepts include the following:

- Fields of study to illustrate the connections between the topic and the various branches of science and technology related to programming and coding;
- An Abstract that provides brief, concrete summary of the topic and how the entry is organized;
- Text that gives an explanation of the background and significance of the topic to programming and coding as well as describing the way a process works or how a procedure or technique is applied to achieve important goals related to writing effective programs and code.
- Illustrations that clarify difficult concepts via models, diagrams, and charts of such key topics as cloud computing, cryptography, e-banking, neuromarketing, and waterfall programming.
- Bibliographies that offer further reading that relates to the entry.

This reference work begins with a comprehensive introduction to the field, written by volume editor Donald Franceschetti, Professor Emeritus at University of Tennessee.

The book includes helpful appendixes as another valuable resource, including the following:

- Time Line of Inventions and Advancements in Programming and Coding;
- Glossary;
- Bibliography; and
- Subject Index.

Salem Press and Grey House Publishing extend their appreciation to all involved in the development and production of this work. The entries have been written by experts in the field. A list of heir names follows the Editor's Introduction.

Principles of Programming & Coding, as well as all Salem Press reference books, is available in print and as an e-book. Please visit www.salempress.com for more information.

Editor's Introduction

What Is a Program?

A program is a set of instructions, meant to be carried out by one of more individuals or intelligent machines. An Army drill sergeant uses a simple set of instructions to control his men. Once the meaning of the basic commands is understood, the sergeant could confine his men to a given area or control their behavior during a 20-mile hike. The set of commands he uses is similar to the commands in a computer program; it can be pre-recorded, so that no human programmer (or drill sergeant) is needed.

For the most part we are concerned with instructions that a computer programs gives to a digital computer. The program may by very simple, for instance, to input a number of statements and the then output them in alphabetical order, or it may be far more complex, involving branch points and convergence criteria to decide when it is done. Computer programs must be written in a computer language understandable to both the human programmer and the computer. The development of computer languages is itself a major subfield of linguistics—the study of languages and their grammars.

Any machine intended to perform its functions on its own, without guidance or interference, is called an *automaton*. The history of automata extends over several centuries. The Jacquard loom, invented in 1802 was an early example. The loom could be programmed to reproduce any pattern of threads in a fabric. Interestingly the loom was controlled by a sort of punch card, the precursor to punch cards used in computer programming. A variety of note-keeping systems were based on index cards, where the desired card would be the one that had all of its holes properly punched. A great step forward came as the result of the work of British mathematical logician, Alan M. Turing. Turing's main interest was in machines that could do elementary arithmetic. Turing began by breaking down each computation into elementary steps. The general framework introduced by Turing involves a machine with a single memory location and a potentially infinitely long tape. The Turing machine would then start from a designated square on the tape and, depending on the symbol written in that square and the symbol stored in memory, would write on the tape and then move the tape forward or backward by one space. It turned out that every effective procedure in arithmetic could be implemented as Turing machine and, further, that there were Turing machines that could read in description of another Turing machine and then emulate it. This in turn led to the notion of a universal Turing machine or programmable computer.

Now, a universal Turing machine was not needed for most applications. One could economize and use a reduced instruction set that would be useful for many purposes, particularly where speed or size was a consideration. A broad view of information technology shows the field growing out from two centers. One center might be labeled communication, starting with the telegraph and voice communication and then leading to television and, ultimately, virtual reality. The second center is computation which includes the several generations of computers. As time has gone by, the two "centers" have merged. Consider for instance the modern cell phone. It can be used for electronic mail, or as a camera, or as a computer terminal. This tendency to combine functions in one device is characteristic of most developments in information technology. While there will always be specialized applications the trend is clearly in the direction of greater generality.

What Is Coding?

If humans are to devise tasks that computers can execute, there must be a common language which is understood both by the human programmers and by computing machines. The history of computer science begins with the search for a common language. Now humans are already familiar with a number of simple codes. The color code by which an electronics student identifies a resistor with a particular resistance is one such. Various alphabetic codes have been developed for use in communication. The Morse code used by telegraphers is a noteworthy example as is semaphore code used for communication at sea. To make communications faster and more reliable various devices are added to the code, such as the parity bit.

When one looks at nature with the eyes of a coder, one finds that coding has evolved in a number of situations. One of the fundamental questions in biology is "Why do offspring look like their parents." The means of transmission is the genetic code with which virtually all organisms store the descriptions of the essential proteins in a sequence of triplets of DNA bases. Variations on the basic theme allow us to explain the immune response, mutation, and many other biological phenomena.

Coding may also occur in the dimension of time. The messages by which a sensation is encoded can be sampled by various means. The nervous system carries the message from one part of the body to another. It is also apparent that coding is used in the animal kingdom. The code used by bees to communicate the location of a food source is probably the best known example of coding in the insect world.

An Overview of Computer Evolution

It is difficult to predict the future evolution of computers. The physical sciences have been around since the time of Galileo and Newton. Electrical engineering is well over a hundred years old, though the field has grown markedly since the introduction of the transistor. Computational science has grown much more quickly as advances in computer use have had an exponential effect on the field and its growth. Subfields of computer science are constantly changing. The following is a generally accepted view of computer evolution and its effect on computational science.

Computer designers are in general agreement that electronic computers have passed through four generations of development and are starting on a fifth. While vacuum tube circuitry is obsolete, for the most part, the basic principles of electronic computation were established during the epoch of the first-generation electronic computers. The drawbacks of these computers were well known. They relied on vacuum tubes, which had limited lifetimes and were known to fail. To obtain a reliable solution, a preliminary program to test the functional soundness of every tube in the computer had to be run both before and after a program of interest to assure that no tubes had burned out. Computation schemes were developed that were meant to take into account the intrinsic variability of vacuum tubes. In addition to the inherent difficulties of working with vacuum tubes, the computers themselves were relatively large and generated a great deal of heat, which meant that housing the computers became a difficulty to overcome, as was the challenge of dissipating the wasted heat that the computers produced.

The invention of the transistor in 1947 gave rise to a second generation of electronic computers about a decade later. Transistor circuits were much smaller and far more and energy efficient than vacuum tubes. They quickly became far more uniform in their characteristics as well. The first compilers date back to this period, eliminating the need for tedious machine language programming. FORTRAN (formula translation) was invented Jim Backus of IBM in 1954. Fortran made it possible for scientists to do calculations using formulas without mastering the mysteries of machine language. Grace Hopper, an officer in the US Navy, developed the language known as COBOL (Common Business Oriented Language) in 1958. Incidentally the career of Grace Hopper who retired from active service as a Rear Admiral, is an inspiration to computer-minded boys and girls alike.

The third generation, considered as lasting from 1964 through 1971, was based on the integrated circuit. This was the time of large mainframe computers, like the IBM 360 and 370, and their counterparts. New languages like the Beginners' All-purpose Symbolic Instruction Code (BASIC) were developed during this period as well as the C language in multiple versions. Computers of this generation featured time-sharing, with a single central processing unit that could support the work of many peripheral devices. Also originating in this period are LISP, the list processing language, which was extensively used in early artificial intelligence research.

The fourth generation (1972–2010) was based on the microprocessor, now ubiquitous in all computers. This advance marked the beginning of the personal computer. While the need for very fast, very compact super computers remains, the development of the personal computer has had a large impact on computation in general. The semiconductor chip, on which personal computers were based, required several advances in materials science: A better understanding of crystal growth; ion implantation; epitaxy; and photolithography. When not needed by their owners, microcomputers can be networked together to join in such esoteric tasks as the search for extraterrestrial intelligence or the more practical modeling of protein folding. While an amateur could now assemble a working computer for a few hundred dollars, he was dependent on the chip manufacturers for the key component. Very few manufacturers would now have the high vacuum and ultra clean facilities to mass produce silicon chips.

At this point it is important to point out that not all physical problems have solutions that can be written out in closed form. In fact, the majority do not. As Stephen Wolfram, one of the guiding lights in the development of the Mathematica package, points out, computer simulation is possible in principle even in the most complex systems. If one writes the basic laws in local form, updating by small time increments, one can simulate numerically even the most complex systems. An interesting side effect was the development of chaos theory. It turned out that many mathematical procedures are extremely sensitive to initial conditions. With advances in computer graphics one could illustrate many special mathematical sets. Fractal geometry and the Mandelbrot set became popular computer related concepts.

The fifth generation is heralded as the advent of true artificial intelligence. Prominent in this area are IBM and several Japanese semiconductor manufacturers whose methods of organization and governmental-industrial collaboration are novel to the traditional economies of Western Europe and North America. It is noteworthy that IBM, which was very slow to get on the microcomputer bandwagon, has again established a leadership position with its Watson supercomputer. Following an initial launch, Watson was pitted against human opponents playing the televised game Jeopardy. Watson, a machine that can learn from its mistakes, demonstrated its power, and the machine was turned loose on the world's cancer literature. A variety of promising new therapies have their origin in Watson's work in the field.

What Is To Be Found in this Book?

This book presents over 140 articles on various aspects of coding and programming. The articles range from the strictly technical to general discussions of human psychology. Not all are free from controversy. Nor will reading of all the articles make one an expert on any branch of computer science. Nonetheless, the articles provide a snapshot of the current state of this rapidly changing field of human endeavor, one so rapidly changing that a bright high school student's contribution may have more impact than that of a Ph.D. Indeed, many of the most successful workers in this area have interrupted their formal education so that they could devote all their youthful energy to it. The only promise that can be made to the reader is that changes are taking place and that the pace of innovation is not slowing down.

The Role of Behavioral Science and Psychology

Should coders pay attention to their working environment, to what has been done in the past, and the obvious trade-off between hours spent on a project and the time needed to meet the immediate need for a solution? A coder working on a given project must decide which computer language to use and whether an existing program will work well enough, without giving away the advantage to a competitor. Copyright and proprietary information (trade secrets) may need protection. Articles on waterfall development or "worse is better" methodology provide a survey of different program writing approaches. The involvement of psychological factors should not be underestimated. While there are stories of a few classic examples of short programs which can pass the Turing test for a while, writing a program that displays anything that might be taken as intelligence is another matter. Currently a great many programs do not attempt intelligent behavior overall but rather take on just one aspect of human performance and use the computer's capacity for expanded search, communications, or reasoning to perform tasks without the limitations that humans bring to the problem.

An important open question is the extent to which computers should imitate the information processing found in nature. Should one write computer codes for a highly distributed processor like a neural network or for more deterministic computer, perhaps using random number generation to provide a measure of independence? Genetic algorithms, in which a trial solution is revised repeatedly by a process that resembles biological evolution, are but one example of a "biological" approach to computation.

Economic Transitions

One of the most significant changes involves the matter of who will pay for the research that will lead to new advances. Without a sense of history, one may be alarmed at cuts in government funding for research. The current situation is mainly a product of the Second World War and the nuclear arms race. Indeed, twentieth century history might have turned out differently had not one man, Albert Einstein sent the now famous telegram to President Franklin Roosevelt and the U. S. Army identified a charismatic leader, Dr. J. Robert Oppenheimer, to direct a massive, but secret operation. Interestingly Alan Turing was active in war research for the British government at the same time, trying to crack the German's Enigma code. Both Turing and Oppenheimer were somewhat suspect to the security officials of the time and needed friends in high places to maintain their position. Today the key figures may well be Bill Gates and Steve Jobs (who publicly who publicly acknowledged Oppenheimer as a role model) and the most intense battles are fought not using conventional or nuclear weapons but in the stock market indices.

The role of the American government is supporting marvels such as the Hubble Telescope and the human genome project is undeniable, but it is noteworthy that much of the research has been paid for by private companies. No longer is it a certainty that Nobel prizes will be awarded to academics or members of national research centers. The Nobel Prize for 2003 was awarded to Kary Mullis of Cetus Corporation for his work on the polymerase chain reaction, an essential tool in DNA fingerprinting and the Human Genome Project. Even in area traditionally left to governmental initiatives, like space exploration, one finds private companies and government-industry collaborations playing an increasing role.

The Internet

Upgrading the speed at which postal mail travels has been a goal of communications technology since the invention of the telegraph. The first advances meant that if one had a wired connection between two places, one could communicate by telegraph. The next step was to transmit voice, and then the telephone was invented. The discovery of radio transmission came next, followed within a couple of decades by television. The transmission of images was made possible by the facsimile machine. Now the process is very inexpensive and people can meet "face-to-face" over long distances, even around the globe, using Skype or similar group meeting software. The internet has to some extent replaced the need for microcomputers since it is accessible on portable cell phones and other small hand-held devices. With smaller hand-held devices a variety of computer languages are needed, so that the available memory is adequate for the system.

Following Biology

A significant amount of effort is now devoted to modeling mental processes with the computer. In 1943 Warren McCulloch, associated with the psychiatry department at the University of Illinois, published an article in the Bulletin of Mathematical Biophysics entitled "A logical calculus of the ideas Immanent in Nervous Activity. This paper proposed that the basic logical functions could be modeled by individual nerve cells with thresholds. This was the beginning of trying to understand what actually happens in the brain. Of course, neural processes involve many real neurons and the first neural nets were greatly oversimplified but investigations in this field continue.

Human Psychology, Perception, and Sensory Physics

Psychophysics Is the term used to describe the physics of human sensation, a field begun nearly a century ago by Ernst Mach. Every attribute of the visual or sonic output of a computer has its corresponding human sensation. Thus frequency, amplitude, and harmonic mixture describe the sensations of sound we perceive as pitch, loudness, and timbre, while a far more complex description is needed for vision. Even the chemical senses (taste and smell) are liable to computer analysis.

Computer Games

Last, but far from least, is the matter of computer games and alternative realities. While the advent of the personal computer brought the arcade into the home with video games such as Pac Man and Ping Pong, game designers took an interest in physics-like simulations, which had an appeal to slightly older players. While accurate simulations of spaceflight were much too slow to make for an exciting game, a number of approximate computational schemes made it possible to see believable results on a computer screen. Increasingly realistic simulations are now included in astronaut training. The frontier of gaming has attracted talented programmers responsible for both realistic simulations and pure fantasies that have gained legions of fans who are waiting for the next advance.

—*Donald R. Franceschetti, PhD*

For Further Reading:

Dyson, George. *Turing's Cathedral: The Origins of the Digital Universe.* London: Penguin Books, 2013. Print.

Feigenbaum, Edward Albert., and Julian Feldman. *Computers and Thought.* AAAI Press, 1995.

Franklin, Stan. *Artificial Minds.* MIT Press, 2001.

Khan, Jawad. "Five Generations of Computers." *Byte-Notes | Computer Science Learning Platform,* 31 Aug. 2013, www.byte-notes.com/five-generations-computers.

Whitson, G. M. "Artifical Intelligence." *Applied Science,* edited by Donald R. Franceschetti, Salem Press, a Division of EBSCO Information Services, Inc., 2012, pp. 121–127.

Winston, Patrick H. *Artificial Intelligence.* Reading, Mass. Addison-Wesley, 1999. Print.

Contributors

Andrew Farrell, MLIS
Andrew Hoelscher, MEng
Lindsay Rohland
Céleste Codington-Lacerte
Christopher Rager, MA
Daniel Horowitz
Daniel L. Berek, MA, MAT
Elizabeth Rholetter Purdy, MA, PhD
Gavin D. J. Harper, MSc, MIET
Hitaf R. Kady, MA
Isaiah Flair, MA
Jennifer A. Vadeboncoeur, MA, PhD
Jerry Johnson
Joel Christophel
Daniel Showalter, PhD
John Mark Froiland, PhD
John Vines
John Walsh, PhD
Joseph Dewey, PhD
Joshua Miller
Joy Crelin
Julia Gilstein
Kehley Coviello
Kenrick Vezina, MS
Laura L. Lundin, MA

Leon James Bynum, MA
Lisa U. Phillips, MFA
Luke E. A. Lockhart
Mariam Orkodashvili, PhD
Marjee Chmiel, MA, PhD
Mary Woodbury Hooper
Maura Valentino, MSLIS
Maura Valentino, MSLIS
Melvin O
Micah L. Issitt
Mitzi P. Trahan, MA, PhD
Narissra Maria Punyanunt-Carter, MA, PhD
Patricia Hoffman-Miller, MPA, PhD
Randa Tantawi, PhD
Richard De Veaux
Robert N. Stacy, CSS, MA
Robert N. Stacy, MA
Sarah E. Boslaugh, PhD, MPH
Scott Zimmer, JD
Shari Parsons Miller, MA
Steven Miller
Susan R. Adams, MA, PhD
Teresa E. Schmidt
Trevor Cunnington, MA, PhD

3D PRINTING

FIELDS OF STUDY
Computer Science; Digital Media

ABSTRACT
Additive manufacturing (AM), or 3D printing, comprises several automated processes for building three-dimensional objects from layers of plastic, paper, glass, or metal. AM creates strong, light 3D objects quickly and efficiently.

PRINCIPAL TERMS

- **binder jetting:** the use of a liquid binding agent to fuse layers of powder together.
- **directed energy deposition:** a process that deposits wire or powdered material onto an object and then melts it using a laser, electron beam, or plasma arc.
- **material extrusion:** a process in which heated filament is extruded through a nozzle and deposited in layers, usually around a removable support.
- **material jetting:** a process in which drops of liquid photopolymer are deposited through a printer head and heated to form a dry, stable solid.
- **powder bed fusion:** the use of a laser to heat layers of powdered material in a movable powder bed.
- **sheet lamination:** a process in which thin layered sheets of material are adhered or fused together and then extra material is removed with cutting implements or lasers.
- **vat photopolymerization:** a process in which a laser hardens layers of light-sensitive material in a vat.

Additive Manufacturing
3D printing, also called additive manufacturing (AM), builds three-dimensional objects by adding successive layers of material onto a platform. AM differs from traditional, or subtractive, manufacturing, also called machining. In machining, material is removed from a starting sample until the desired structure remains. Most AM processes use less raw material and are therefore less wasteful than machining.

The first AM process was developed in the 1980s, using liquid resin hardened by ultraviolet (UV) light. By the 2000s, several different AM processes had been developed. Most of these processes use liquid, powder, or extrusion techniques. Combined with complex computer modeling and robotics, AM could launch a new era in manufacture. Soon even complex mechanical objects could be created by AM.

Software and Modeling
3D printing begins with a computer-aided design (CAD) drawing or 3D scan of an object. These drawings or scans are usually saved in a digital file format known as STL, originally short for "stereolithography" but since given other meanings, such as "surface tessellation language." STL files "tessellate" the object—that is, cover its surface in a repeated pattern of shapes. Though any shape can be used, STL files use a series of non-overlapping triangles to model the curves and angles of a 3D object. Errors in the file may need repair. "Slices" of the STL file determine the number and thickness of the layers of material needed.

Liquid 3D Printing
The earliest AM technique was stereolithography (SLA), patented in 1986 by Chuck Hull. SLA uses liquid resin or polymer hardened by UV light to create a 3D object. A basic SLA printer consists of an elevator platform suspended in a tank filled with light-sensitive liquid polymer. A UV laser hardens a thin layer of resin. The platform is lowered, and the laser hardens the next layer, fusing it to the first. This process is repeated until the object is complete. The

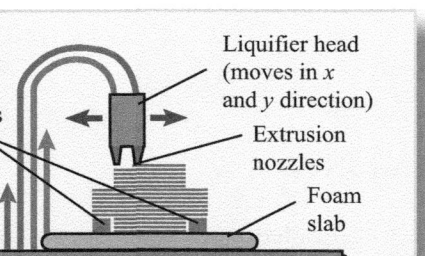
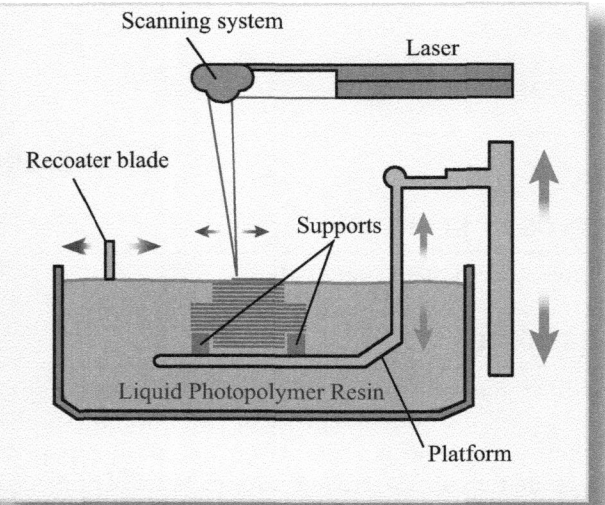
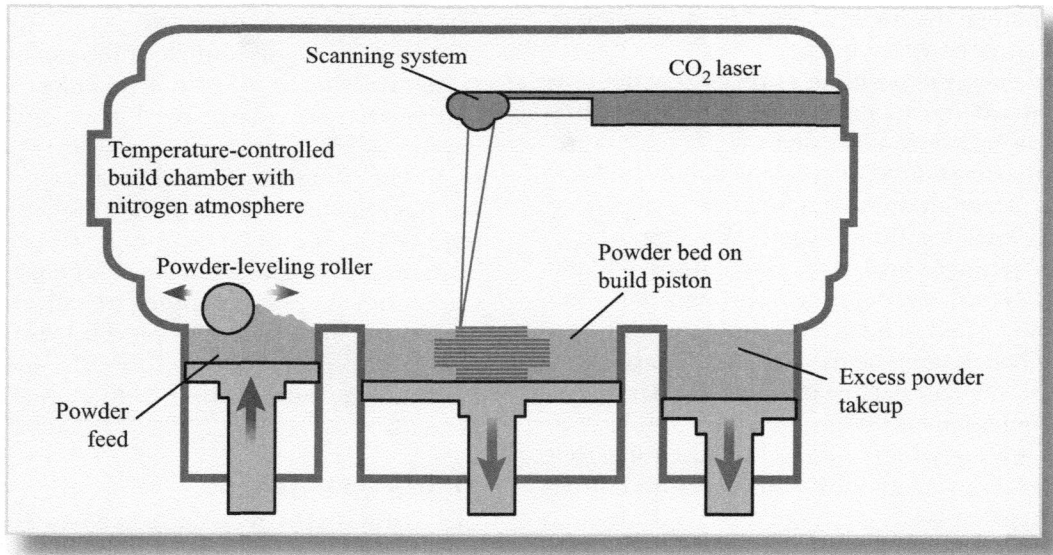

This presents a comparison of the three common 3-D printing processes: SLA (in which liquid polymer resin is solidified by a laser and support material is removed after completion), SLS (in which powder is fused by a CO2 laser and unfused powder acts as support), and FDM (in which liquid modeling material is extruded through extrusion nozzles and solidifies quickly, and a build material and a support material can be used in tandem, with the support material being removed after completion).

object is then cleaned and cured by UV. This AM technique is also called vat photopolymerization because it takes place within a vat of liquid resin. Various types of SLA printing processes have been given alternate names, such as "photofabrication" and "photo-solidification."

Powder-Based 3D Printing

In the 1980s, engineers at the University of Texas created an alternate process that uses powdered solids instead of liquid. Selective layer sintering (SLS), or powder bed fusion, heats powdered glass, metal, ceramic, or plastic in a powder bed until the material is "sintered." To sinter something is to cause its particles to fuse through heat or pressure without liquefying it. A laser is used to selectively sinter thin layers of the powder, with the unfused powder underneath giving structural support. The platform is lowered and the powder compacted as the laser passes over the object again.

Extrusion Printing

Material extrusion printing heats plastic or polymer filament and extrudes it through nozzles to deposit a layer of material on a platform. One example of this process is called fused deposition modeling (FDM). As the material cools, the platform is lowered and another layer is added atop the last layer. Creating extruded models often requires the use of a structural support to prevent the object from collapsing. Extrusion printing is the most affordable and commonly available 3D printing process.

Emerging and Alternative Methods

Several other 3D printing methods are also emerging. In material jetting, an inkjet printer head deposits liquefied plastic or other light-sensitive material onto a surface, which is then hardened with UV light. Another inkjet printing technique is binder jetting, which uses an inkjet printer head to deposit drops of glue-like liquid into a powdered medium. The liquid then soaks into and solidifies the medium. In directed energy deposition (DED), metal wire or powder is deposited in thin layers over a support before being melted with a laser or other heat source. Sheet lamination fuses together thin sheets of paper, metal, or plastic with adhesive. The resulting object is then cut with a laser or other cutting tool to refine the shape. This method is less costly but also less accurate than others.

The Future of 3D Printing

While AM techniques have been in use since the 1980s, engineers believe that the technology has not yet reached its full potential. Its primary use has been in rapid prototyping, in which a 3D printer is used to quickly create a 3D model that can be used to guide production. In many cases, 3D printing can create objects that are stronger, lighter, and more customizable than objects made through machining. Printed parts are already being used for planes, race cars, medical implants, and dental crowns, among other items. Because AM wastes far less material than subtractive manufacturing, it is of interest for conservation, waste management, and cost reduction. The technology could also democratize manufacturing, as small-scale 3D printers allow individuals and small businesses to create products that traditionally require industrial manufacturing facilities. However, intellectual property disputes could also occur more often as AM use becomes more widespread.

—*Micah L. Issitt*

Bibliography

"About Additive Manufacturing." *Additive Manufacturing Research Group*. Loughborough U, 2015. Web. 6 Jan. 2016.

Hutchinson, Lee. "Home 3D Printers Take Us on a Maddening Journey into Another Dimension." *Ars Technica*. Condé Nast, 27 Aug. 2013. Web. 6 Jan. 2016.

"Knowledge Base: Technologies in 3D Printing." *DesignTech*. DesignTech Systems, n.d. Web. 6 Jan. 2016.

Matulka, Rebecca. "How 3D Printers Work." *Energy.gov*. Dept. of Energy, 19 June 2014. Web. 6 Jan. 2016.

"The Printed World." *Economist*. Economist Newspaper, 10 Feb. 2011. Web. 6 Jan. 2016.

"3D Printing Processes: The Free Beginner's Guide." *3D Printing Industry*. 3D Printing Industry, 2015. Web. 6 Jan. 2016.

A

ALGORITHMS

FIELDS OF STUDY

Computer Science; Operating Systems; Software Engineering

ABSTRACT

An algorithm is set of precise, computable instructions that, when executed in the correct order, will provide a solution to a certain problem. Algorithms are widely used in mathematics and engineering, and understanding the design of algorithms is fundamental to computer science.

PRINCIPAL TERMS

- **deterministic algorithm:** an algorithm that when given a input will always produce the same output.
- **distributed algorithm:** an algorithm designed to run across multiple processing centers and so, capable of directing a concentrated action between several computer systems.
- **DRAKON chart:** a flowchart used to model algorithms and programmed in the hybrid DRAKON computer language.
- **function:** instructions read by a computer's processor to execute specific events or operations.
- **recursive:** describes a method for problem solving that involves solving multiple smaller instances of the central problem.
- **state:** a technical term for all the stored information, and the configuration thereof, that a program or circuit can access at a given time.

An Ancient Idea

The term "algorithm" is derived from the name al-Khwarizmi. Muhammad ibn Musa al-Khwarizmi was a ninth-century Persian mathematician who is credited with introducing the decimal system to the West. He has been celebrated around the world as a pioneer of mathematics and conceptual problem solving.

"Algorithm" has no precise definition. Broadly, it refers to a finite set of instructions, arranged in a specific order and described in a specific language, for solving a problem. In other words, an algorithm is like a plan or a map that tells a person or a machine what steps to take to complete a given task.

Algorithm Basics

In computer science, an algorithm is a series of instructions that tells a computer to perform a certain function, such as sorting, calculating, or finding data. Each step in the instructions causes the computer to transition from one state to another until it reaches the desired end state.

Any procedure that takes a certain set of inputs (a data list, numbers, information, etc.) and reaches a desired goal (finding a specific datum, sorting the list, etc.) is an algorithm. However, not all algorithms are equal. Algorithms can be evaluated for "elegance," which measures the simplicity of the coding. An elegant algorithm is one that takes a minimum number of steps to complete. Algorithms can also be evaluated in terms of "goodness," which measures the speed with which an algorithm reaches completion.

Algorithms can be described in a number of ways. Flowcharts are often used to visualize and map the steps of an algorithm. The DRAKON computer language, developed in the 1980s, allows users to program algorithms into a computer by creating a flowchart that shows the steps of each algorithm. Such a flowchart is sometimes called a DRAKON chart.

Algorithms often specify conditional processes that occur only when a certain condition has been met. For instance, an algorithm about eating lunch might begin with the question, "Are you hungry?"

If the answer is "yes," the algorithm will instruct the user to eat a sandwich. It will then ask again if the user is hungry. If the answer is still yes, the "eat a sandwich" instruction will be repeated. If the answer is "no," the algorithm will instruct the user to stop eating sandwiches. In this example, the algorithm repeats the "eat a sandwich" step until the condition "not hungry" is reached, at which point the algorithm ends.

TYPES OF ALGORITHMS
Various types of algorithms take different approaches to solving problems. An iterative algorithm is a simple form of algorithm that repeats the same exact steps in the same way until the desired result is obtained. A recursive algorithm attempts to solve a problem by completing smaller instances of the same problem. One example of a recursive algorithm is called a "divide and conquer" algorithm. This type of algorithm addresses a complex problem by solving less complex examples of the same problem and then combining the results to estimate the correct solution.

Algorithms can also be serialized, meaning that the algorithm tells the computer to execute one instruction at a time in a specific order. Other types of algorithms may specify that certain instructions should be executed simultaneously. Distributed algorithms are an example of this type. Different parts of the algorithm are executed in multiple computers or nodes at once and then combined.

An algorithm may have a single, predetermined output, or its output may vary based on factors other than the input. Deterministic algorithms use exact, specific calculations at every step to reach an answer to a problem. A computer running a deterministic algorithm will always proceed through the same sequence of states. Nondeterministic algorithms incorporate random data or "guessing" at some stage of the process. This allows such algorithms to be used as predictive or modeling tools, investigating problems for which specific data is lacking. In computational biology, for instance, evolutionary algorithms can be used to predict how populations will change over time, given estimations of population levels, breeding rates, and other environmental pressures.

ALGORITHM APPLICATIONS
One of the most famous applications of algorithms is the creation of "search" programs used to find

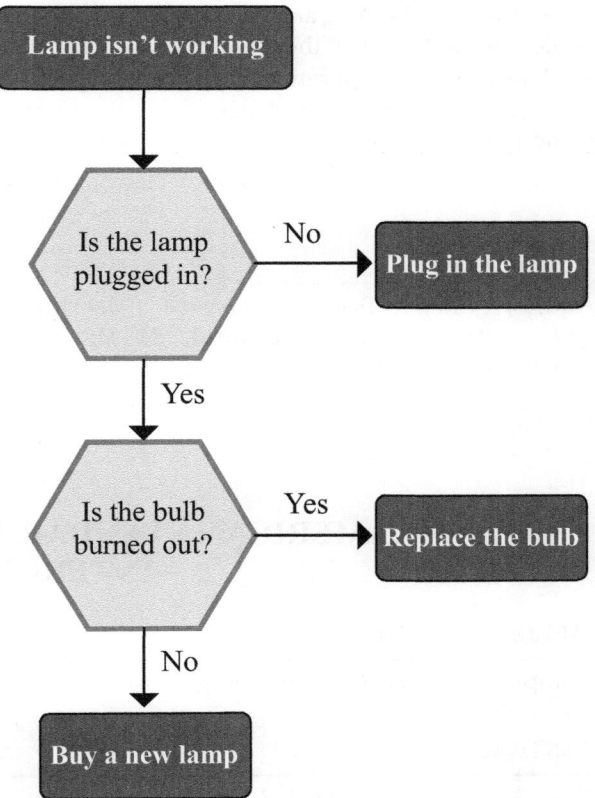

An algorithm is a set of operations or a procedure for solving a problem or processing data. Flowcharts are often used as a visualization of the process, showing the order in which steps are performed.

information on the Internet. The Google search engine can search through vast amounts of data and rank millions of search results in a specific order for different users. Sorting large lists of data was one of the earliest problems that computer scientists attempted to solve using algorithms. In the 1960s, the quicksort algorithm was the most successful sorting algorithm. Using a random element from the list as a "pivot," quicksort tells the computer to pick other elements from the list and compare them to the pivot. If the element is less than the pivot, it is placed above it; if it is greater, it is placed below. The process is repeated until each pivot is in its proper place and the data is sorted into a list. Computer scientists are still attempting to find search and sorting algorithms that are more "elegant" or "good" in terms of completing the function quickly and with the least demand on resources.

Searching and sorting are the most famous examples of algorithms. However, these are just two of the thousands of algorithm applications that computer scientists have developed. The study of algorithm design has become a thriving subfield within computer science.

—*Micah L. Issitt*

BIBLIOGRAPHY

Anthes, Gary. "Back to Basics: Algorithms." *Computerworld*. Computerworld, 24 Mar. 2008. Web. 19 Jan. 2016.

Bell, Tim, et al. "Algorithms." *Computer Science Field Guide*. U of Canterbury, 3 Feb. 2015. Web. 19 Jan. 2016.

Cormen, Thomas H. *Algorithms Unlocked*. Cambridge: MIT P, 2013. Print.

Cormen, Thomas H., et al. *Introduction to Algorithms*. 3rd ed. Cambridge: MIT P, 2009. Print.

"Intro to Algorithms." *Khan Academy*. Khan Acad., 2015. Web. 19 Jan. 2016.

Toal, Ray. "Algorithms and Data Structures." *Ray Toal*. Loyola Marymount U, n.d. Web. 19 Jan. 2016.

AMERICAN STANDARD CODE FOR INFORMATION INTERCHANGE (ASCII)

FIELDS OF STUDY

Computer Science; Computer Engineering

ABSTRACT

The American Standard Code for Information Interchange (ASCII) is a character encoding system. It enables computers and other electronic communication devices to store, process, transmit, print, and display text and other graphic characters. Initially published in 1963, ASCII formed the basis for several other character encoding systems developed for use with PCs and the Internet.

PRINCIPAL TERMS

- **bit width:** the number of bits used by a computer or other device to store integer values or other data.
- **character:** a unit of information that represents a single letter, number, punctuation mark, blank space, or other symbol used in written language.
- **control characters:** units of information used to control the manner in which computers and other devices process text and other characters.
- **hamming distance:** a measurement of the difference between two characters or control characters that effects character processing, error detection, and error correction.
- **printable characters:** characters that can be written, printed, or displayed in a manner that can be read by a human.

UNDERSTANDING CHARACTER ENCODING

Written language, or text, is composed of a variety of graphic symbols called characters. In many languages, these characters include letters, numbers, punctuation marks, and blank spaces. Such characters are also called printable characters because they can be printed or otherwise displayed in a form that can be read by humans. Another type of character is a control character. Control characters effect the processing of other characters. For example, a control character might instruct a printer to print the next character on a new line. Character encoding is the process of converting characters into a format that can be used by an electronic device such as a computer or telegraph.

Originally designed for use with Samuel Morse's telegraph system, Morse code was one of the first character encoding schemes adopted for widespread use. Telegraphs transmit information by sending electronic pulses over telegraph wires. Morse code assigns each character to a unique combination of short and long pulses. For example, the letter *A* was assigned to the combination of one short followed by one long pulse, while the letter *T* was assigned to a single long pulse. Using Morse code, a telegraph operator can

ASCII TABLE

Decimal	Hex	Char	Decimal	Hex	Char	Decimal	Hex	Char	Decimal	Hex	Char
0	0	[NULL]	32	20	[SPACE]	64	40	@	96	60	`
1	1	[START OF HEADING]	33	21	!	65	41	A	97	61	a
2	2	[START OF TEXT]	34	22	"	66	42	B	98	62	b
3	3	[END OF TEXT]	35	23	#	67	43	C	99	63	c
4	4	[END OF TRANSMISSION]	36	24	$	68	44	D	100	64	d
5	5	[ENQUIRY]	37	25	%	69	45	E	101	65	e
6	6	[ACKNOWLEDGE]	38	26	&	70	46	F	102	66	f
7	7	[BELL]	39	27	'	71	47	G	103	67	g
8	8	[BACKSPACE]	40	28	(	72	48	H	104	68	h
9	9	[HORIZONTAL TAB]	41	29	)	73	49	I	105	69	i
10	A	[LINE FEED]	42	2A	*	74	4A	J	106	6A	j
11	B	[VERTICAL TAB]	43	2B	+	75	4B	K	107	6B	k
12	C	[FORM FEED]	44	2C	,	76	4C	L	108	6C	l
13	D	[CARRIAGE RETURN]	45	2D	-	77	4D	M	109	6D	m
14	E	[SHIFT OUT]	46	2E	.	78	4E	N	110	6E	n
15	F	[SHIFT IN]	47	2F	/	79	4F	O	111	6F	o
16	10	[DATA LINK ESCAPE]	48	30	0	80	50	P	112	70	p
17	11	[DEVICE CONTROL 1]	49	31	1	81	51	Q	113	71	q
18	12	[DEVICE CONTROL 2]	50	32	2	82	52	R	114	72	r
19	13	[DEVICE CONTROL 3]	51	33	3	83	53	S	115	73	s
20	14	[DEVICE CONTROL 4]	52	34	4	84	54	T	116	74	t
21	15	[NEGATIVE ACKNOWLEDGE]	53	35	5	85	55	U	117	75	u
22	16	[SYNCHRONOUS IDLE]	54	36	6	86	56	V	118	76	v
23	17	[ENG OF TRANS. BLOCK]	55	37	7	87	57	W	119	77	w
24	18	[CANCEL]	56	38	8	88	58	X	120	78	x
25	19	[END OF MEDIUM]	57	39	9	89	59	Y	121	79	y
26	1A	[SUBSTITUTE]	58	3A	:	90	5A	Z	122	7A	z
27	1B	[ESCAPE]	59	3B	;	91	5B	[	123	7B	{
28	1C	[FILE SEPARATOR]	60	3C	<	92	5C	\	124	7C	|
29	1D	[GROUP SEPARATOR]	61	3D	=	93	5D	]	125	7D	}
30	1E	[RECORD SEPARATOR]	62	3E	>	94	5E	^	126	7E	~
31	1F	[UNIT SEPARATOR]	63	3F	?	95	5F	_	127	7F	[DEL]

This chart presents the decimal and hexidecimal ASCII codes for common characters on a keyboard.

send messages by transmitting a sequence of pulses. The sequence, or string, of pulses represents the characters that comprise the message text.

Other character encoding systems were created to meet the needs of new types of electronic devices including teleprinters and computers. By the early 1960s, the use of character encoding systems had become widespread. However, no standard character encoding system existed to ensure that systems from different manufacturers could communicate with each other. In fact, by 1963, over sixty different encoding systems were in use. Nine different systems were used by IBM alone. To address this issue, the American Standards Association (ASA) X3.4 Committee developed a standardized character encoding scheme called ASCII.

Understanding the ASCII Standard

The ASCII standard is based on English. It encodes 128 characters into integer values from 0 to 127. Thirty-three of the characters are control characters, and ninety-five are printable characters that include the upper- and lowercase letters from *A* to *Z*, the numbers zero to nine, punctuation marks, and a blank space. For example, the letter *A* is encoded as 65 and a comma as 44.

The encoded integers are then converted to bits, the smallest unit of data that can be stored by a computer system. A single bit can have a value of either zero or one. In order to store integers larger than one, additional bits must be used. The number of bits used to store a value is called the bit width. ASCII specifies a bit width of seven. For example, in ASCII, the integer value 65 is stored using seven bits, which can be represented as the bit string 1000001.

The ASCII seven-bit integer values for specific characters were not randomly assigned. Rather, the integer values of specific characters were selected to maximize the hamming distance between each value. Hamming distance is the number of bits set to different values when comparing two bit strings. For example, the bit strings 0000001 (decimal value 1) and 0000011 (decimal value 3) have a hamming distance of 1 as only the second to last bit differs between the two strings. The bit patterns 0000111 (decimal value 7) and 0000001 (decimal value 1) have a hamming distance of two as the bit in the third to last position also differs between the two strings. ASCII was designed to maximize hamming distance because larger hamming distances enable more efficient data processing as well as improved error detection and handling.

> **SAMPLE PROBLEM**
>
> ASCII defines the integer values for the first eleven lowercase letters of the alphabet as follows:
>
> $a = 97; b = 98; c = 99; d = 100; e = 101; f = 102;$
> $g = 103; h = 104; i = 105; j = 106; k = 107$
>
> Using this information, translate the word *hijack* to the correct ASCII integer values.
>
> **Answer:**
>
> The ASCII representation of the word hijack can be determined by comparing each character in the word to its defined decimal value as follows:
>
> | h | i | j | a | c | k |
> |---|---|---|---|---|---|
> | h (104) | i (105) | j (106) | a (97) | c (99) | k (107) |
> | 104 | 105 | 106 | 97 | 99 | 107 |
>
> The correct ASCII encoding for the word hijack is 104 105 106 97 99 107.

Beyond ASCII

Following its introduction in 1963, ASCII continued to be refined. It was gradually adopted for use on a wide range of computer systems including the first IBM PC. Other manufacturers soon followed IBM's lead. The ASCII standard was also widely adopted for use on the Internet. However, as the need for more characters to support languages other than English grew, other standards were developed to meet this need. One such standard, Unicode can encode more than 120,000 characters. ASCII remains an important technology, however. Many systems still use ASCII. Character encoding systems such as Unicode incorporate ASCII to promote compatibility with existing systems.

—*Maura Valentino, MSLIS*

Bibliography

Amer. Standards Assn. *American Standard Code for Information Interchange.* Amer. Standards Assn., 17 June 1963. Digital file.

Anderson, Deborah. "Global Linguistic Diversity for the Internet." *Communications of the ACM* Jan. 2005: 27. PDF file.

Fischer, Eric. *The Evolution of Character Codes, 1874–1968.* N.p.: Fischer, n.d. *Trafficways.org.* Web. 22 Feb. 2016.

Jennings, Tom. "An Annotated History of Some Character Codes." *World Power Systems.* Tom Jennings, 29 Oct. 2004. Web. 16 Feb. 2016.

McConnell, Robert, James Haynes, and Richard Warren. "Understanding ASCII Codes." *NADCOMM.* NADCOMM, 14 May 2011. Web. 16 Feb. 2016.

Silver, H. Ward. "Digital Code Basics." *Qst* 98.8 (2014): 58–59. PDF file.

"Timeline of Computer History: 1963." *Computer History Museum.* Computer History Museum, 1 May 2015. Web. 23 Feb. 2016.

ANDROID OS

FIELDS OF STUDY

Computer Science; Operating Systems; Mobile Platforms

ABSTRACT

This article briefly discusses the general issues involved with mobile computing and presents a history and analysis of Google's Android operating system. It concludes with a look at Android's future in the growing market for mobile technology.

PRINCIPAL TERMS

- **application program interface (API):** the code that defines how two pieces of software interact, particularly a software application and the operating system on which it runs.
- **immersive mode:** a full-screen mode in which the status and navigation bars are hidden from view when not in use.
- **Material Design:** a comprehensive guide for visual, motion, and interaction design across Google platforms and devices.

- **multitasking:** in the mobile phone environment, allowing different apps to run concurrently, much like the ability to work in multiple open windows on a PC.
- **multitouch gestures:** touch-screen technology that allows for different gestures to trigger the behavior of installed software.

A Force in Mobile Computing

Mobile computing is the fastest-growing segment of the tech market. As pricing has become more affordable, developing nations, particularly in Africa, are the largest growing market for smartphones. With smartphones, users shop, gather information, connect via social media such as Twitter and Facebook, and communicate—one of the uses more traditionally associated with phones.

By far the most popular operation system running on mobile phones is Android. It has outpaced Apple's iOS with nearly double the sales. As of 2014, more than a million Android devices were being activated daily. Since its launch in 2008, Android has far and away overtaken the competition.

Android Takes Off

Android came about amid a transformative moment in mobile technology. Prior to 2007, slide-out keyboards mimicked the typing experience of desktop PCs. In June of that year, Apple released its first iPhone, forever altering the landscape of mobile phones. Apple focused on multitouch gestures and touch-screen technology. Nearly concurrent with this, Google's Android released its first application program interface (API).

The original API of Google's new operating system (OS) first appeared in October 2008. The Android OS was first installed on the T-Mobile G1, also known as the HTC Dream. This prototype had a very small set of preinstalled apps, and as it had a slide-out QWERTY keyboard, there were no touch-screen capabilities. It did have native multitasking, which Apple's iOS did not yet have. Still, to compete with Apple, Google was forced to replace physical keyboards and access buttons with virtual onscreen controls. The next iteration of Android shipped with the HTC Magic and was accompanied by a virtual keyboard and a more robust app marketplace. Among the other early features that have stood the test of time are the pull-down notification list, home-screen

The swype keyboard, originally designed for Android operating systems, was developed to speed up typing capabilities by allowing the user to slide a finger over the keyboard from letter to letter without lifting their finger to choose each character. This standard software in the Android operating system allows for quick texting.

widgets, and strong integration with Google's Gmail service.

One later feature, the full-screen immersive mode, has become quite popular as it reduces distractions. First released with Android 4.4, "KitKat," in 2013, it hides the navigation and status bars while certain apps are in use. It was retained for the release of Android 5.0, "Lollipop," in 2015.

Android Changes and Grows

Both of Google's operating systems—Android and its cloud-based desktop OS, Chrome—are based on the free open-source OS Linux, created by engineer Linus Torvalds and first released in 1991. Open-source software is created using publicly available source code. The open-source development of

Android has allowed manufacturers to produce robust, affordable products that contribute to its widespread popularity in emerging and developing markets. This may be one reason why Android has had more than twice as many new users as its closest rival, Apple's iOS. This strategy has kept costs down and has also helped build Android's app marketplace, which offers more than one million native apps, many free of charge. By 2014 Android made up 54 percent of the global smartphone market.

This open-source development of Android has had one adverse effect: the phenomenon known as "forking," which occurs primarily in China. Forking is when a private company takes the OS and creates their own products apart from native Google services such as e-mail. Google seeks to prevent this loss of control (and revenue) by not supporting these companies or including their apps in its marketplace. Forked versions of Android made up nearly a quarter of the global market in early 2014.

Google's business model has always focused on a "rapid-iteration, web-style update cycle." By contrast, rivals such as Microsoft and Apple have had a far slower, more deliberate pace due to hardware issues. One benefit of Google's faster approach is the ability to address issues and problems in a timely manner. A drawback is the phenomenon known as "cloud rot." As the cloud-based OS grows older, servers that were once devoted to earlier versions are repurposed. Since changes to the OS initially came every few months, apps that worked a month prior would suddenly lose functionality or become completely unusable. Later Android updates have been released on a timescale of six months or more.

Android's Future

In 2014 alone, more than one billion devices using Android were activated. One of the biggest concerns about Android's future is the issue of forking. Making the code available to developers at no cost has made Android a desirable and cost-effective alternative to higher-end makers such as Microsoft and Apple, but it has also made Google a target of competitors.

Another consideration for Android's future is its inextricable link to the Chrome OS. Google plans to keep the two separate. Further, Google executives have made it clear that Chromebooks (laptops that run Chrome) and Android devices have distinct purposes. Android's focus has been on touch-screen technology, multitouch gesturing, and screen resolution, making it a purely mobile OS for phones, tablets, and more recently wearable devices and TVs. Meanwhile, Chrome has developed tools that are more useful in the PC and laptop environment, such as keyboard shortcuts. However, an effort to unify the appearance and functionality of Google's different platforms and devices called Material Design was introduced in 2014. Further, Google has ensured that Android apps can be executed on Chrome through Apps Runtime on Chrome (ARC). Such implementations suggest a slow merging of the Android and Chrome user experiences.

—*Andrew Farrell, MLIS*

Bibliography

Amadeo, Ron. "The History of Android." *Ars Technica*. Condé Nast, 15 June 2014. Web. 2 Jan. 2016.

"Android: A Visual History." *Verge*. Vox Media, 7 Dec. 2011. Web. 2 Jan. 2016.

Bajarin, Tim. "Google Is at a Major Crossroads with Android and Chrome OS." *PCMag*. Ziff Davis, 21 Dec. 2015. Web. 4 Jan. 2016.

Edwards, Jim. "Proof That Android Really Is for the Poor." *Business Insider*. Business Insider, 27 June 2014. Web. 4 Jan. 2016.

Goldsborough, Reid. "Android on the Rise." *Tech Directions* May 2014: 12. *Academic Search Complete*. Web. 2 Jan. 2016.

Manjoo, Farhad. "Planet Android's Shaky Orbit." *New York Times* 28 May 2015: B1. Print.

Newman, Jared. "Android Laptops: The $200 Price Is Right, but the OS May Not Be." *PCWorld*. IDG Consumer & SMB, 26 Apr. 2013. Web. 27 Jan. 2016.

Newman, Jared. "With Android Lollipop, Mobile Multitasking Takes a Great Leap Forward." *Fast Company*. Mansueto Ventures, 6 Nov. 2014. Web. 27 Jan. 2016.

APPLICATION

FIELDS OF STUDY

Applications; Software Engineering

ABSTRACT

In the field of information technology, an application is a piece of software created to perform a task, such as word processing, web browsing, or chess playing. Each application is designed to run on a particular platform, which is a type of system software that is installed on desktop computers, laptops, or mobile devices such as tablet computers or smartphones.

PRINCIPAL TERMS

- **app:** an abbreviation for "application," a program designed to perform a particular task on a computer or mobile device.
- **application suite:** a set of programs designed to work closely together, such as an office suite that includes a word processor, spreadsheet, presentation creator, and database application.
- **platform:** the specific hardware or software infrastructure that underlies a computer system; often refers to an operating system, such as Windows, Mac OS, or Linux.
- **system software:** the basic software that manages the computer's resources for use by hardware and other software.
- **utility program:** a type of system software that performs one or more routine functions, such as disk partitioning and maintenance, software installation and removal, or virus protection.
- **web application:** an application that is downloaded either wholly or in part from the Internet each time it is used.

APPLICATIONS IN CONTEXT

Applications are software programs that perform particular tasks, such as word processing or web browsing. They are designed to run on one or more specific platforms. The term "platform" can refer to any basic computer infrastructure, including the hardware itself and the operating system (OS) that manages it. An OS is a type of system software that manages a device's hardware and other software resources. Application designers may create different versions of an application to run on different platforms. A cross-platform application is one that can be run on more than one platform.

In the context of mobile devices such as tablets and smartphones, the term "application" is typically shortened to app. Since the introduction of the iPhone in 2007, apps have taken center stage in the realm of consumer electronics. Previously, consumers tended to be attracted more to a device's hardware or OS features. A consumer might have liked a certain phone for its solid design or fast processor, or they might have preferred the graphical interface of Microsoft Windows and Mac OS to the command-line interface of Linux. These features have since become much less of a concern for the average consumer. Instead, consumers tend to be more interested in finding a device that supports the specific apps they wish to use.

EVOLUTION OF APPLICATIONS

Over the years, apps have become more and more specialized. Even basic utility programs that were once included with OSs are now available for purchase as separate apps. In some cases, these apps are a more advanced version of the utility software that comes with the OS. For example, an OS may come with free antivirus software, but a user may choose to purchase a different program that offers better protection.

Some software companies offer application suites of interoperating programs. Adobe Creative Cloud is a cloud-based graphic design suite that includes popular design and editing programs such as Photoshop and InDesign. Microsoft Office is an office suite consisting of a word processor (Word), a spreadsheet program (Excel), and other applications commonly used in office settings. These programs are expensive and can take up large amounts of storage space on a user's computer. Before broadband Internet access became widely available, application suites were distributed on portable media such as floppy disks, CD-ROMs, or DVDs, because downloading them over a dial-up connection would have taken too long.

As high-speed Internet access has become much more common, application developers have taken a different approach. Instead of investing in bulky application suites, users often have the option of using web applications. These applications run partly or

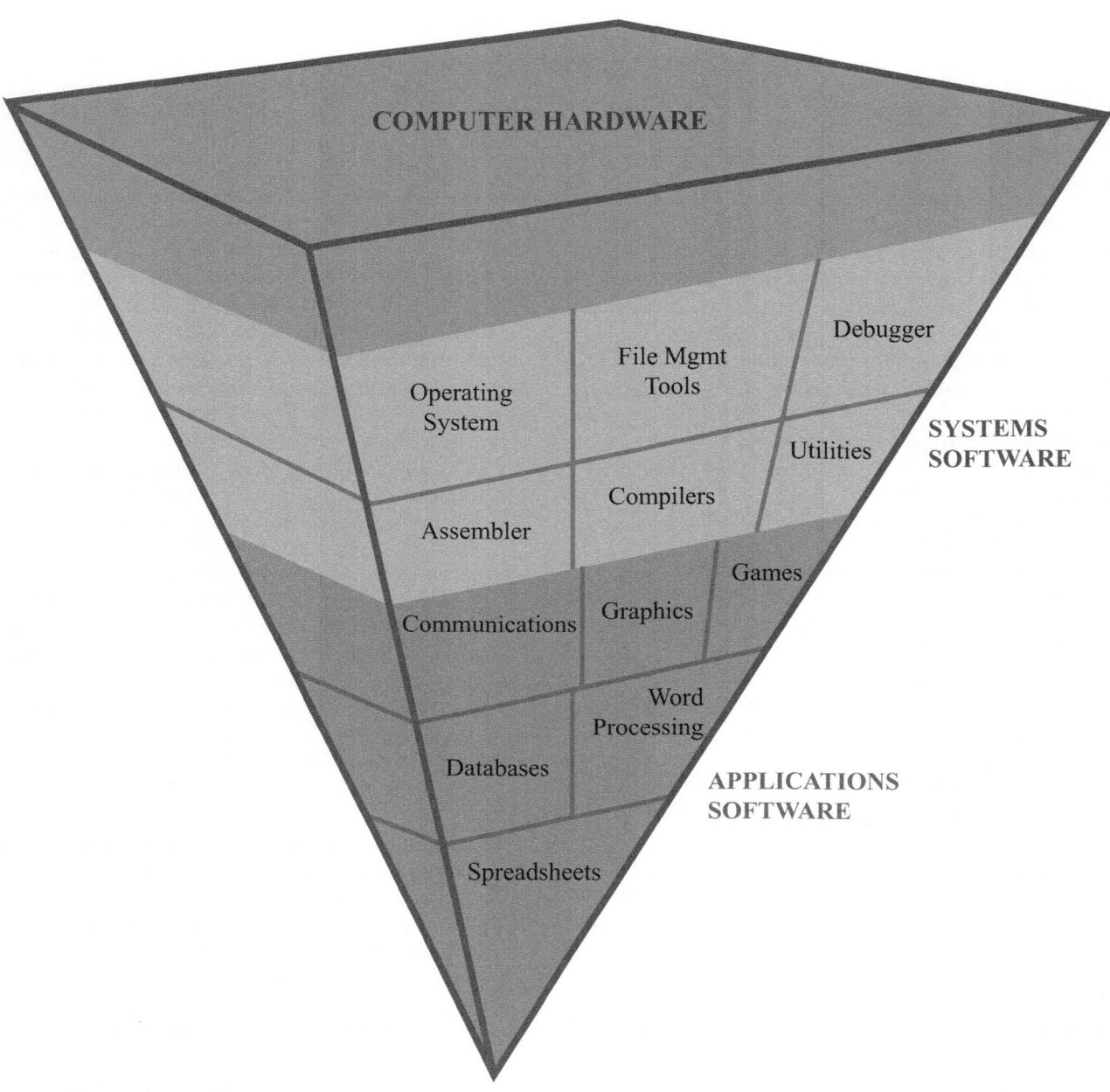

The variety and quantity of application software available is massive compared to the limited array of system software and hardware that support them.

entirely on remote servers, avoiding the need to install them on the computer's hard drive.

Types of Applications

Many different types of software fall under the broad heading of applications. A large segment of the application market is focused on office and productivity software. This category includes e-mail applications, word processors, spreadsheet software, presentation software, and database management systems. In an office environment, it is critical that users be able to create documents using these applications and share them with others. This often means that a business or organization will select a particular application suite and then require all employees to use it.

Other types of applications include games, audio-video editing and production software, and even software that helps programmers write new software. Due the complexity of software engineering, programmers have developed many applications to help them produce more polished, bug-free programs. Software developers may use multiple applications to code a single program. They might use a word processor or text editor to write the source code and explanatory comments, a debugging tool to check the code for errors, and a compiler to convert the code into machine language that a computer can execute. There is even a type of application that can emulate a virtual computer running inside another computer. These applications are often used by web-hosting companies. Instead of having to set up a new physical server for each customer that signs up, they can create another virtual server for the user to access.

Security Implications

Applications must have certain privileges in order to use the resources of the computer they are running on. As a result, they can sometimes be a point of weakness for attackers to exploit. A clever attacker can take over a vulnerable application and then use its privileges to make the computer behave in ways it should not. For example, the attacker could send spam e-mails, host illegally shared files, or even launch additional attacks against other computers on the same network.

Careers in Applications

Applications are the focus of a variety of career options for those interested in working with software. Apart from the obvious role of computer programmer, there are several other paths one might take. One option is quality assurance. Quality assurance staff are responsible for testing software under development to make sure it performs as it should. Technical support is another option. Technical support specialists assist users with operating the software and fixing errors it might cause. Yet another path is technical writing. Technical writers create software user manuals and training materials. Finally, some applications are so complex that using them can be a career in itself.

—*Scott Zimmer, JD*

Bibliography

Bell, Tom. *Programming: A Primer; Coding for Beginners.* London: Imperial Coll. P, 2016. Print.

Calude, Cristian S., ed. *The Human Face of Computing.* London: Imperial Coll. P, 2016. Print.

Dey, Pradip, and Manas Ghosh. *Computer Fundamentals and Programming in C.* 2nd ed. New Delhi: Oxford UP, 2013. Print.

Goriunova, Olga, ed. *Fun and Software: Exploring Pleasure, Paradox, and Pain in Computing.* New York: Bloomsbury, 2014. Print.

Neapolitan, Richard E. *Foundations of Algorithms.* 5th ed. Burlington: Jones, 2015. Print.

Talbot, James, and Justin McLean. *Learning Android Application Programming: A Hands-On Guide to Building Android Applications.* Upper Saddle River: Addison, 2014. Print.

AUTONOMIC COMPUTING

FIELDS OF STUDY
Computer Science; Embedded Systems; System-Level Programming

ABSTRACT
Autonomic computing is a subfield of computer science that focuses on enabling computers to operate independently of user input. First articulated by IBM in 2001, the concept has particular relevance to fields such as robotics, artificial intelligence (AI), and machine learning.

PRINCIPAL TERMS

- **autonomic components:** self-contained software or hardware modules with an embedded capacity for self-management, connected via input/outputs to other components in the system.
- **bootstrapping:** a self-starting process in a computer system, configured to automatically initiate other processes after the booting process has been initiated.
- **multi-agent system:** a system consisting of multiple separate agents, either software or hardware systems, that can cooperate and organize to solve problems.
- **resource distribution:** the locations of resources available to a computing system through various software or hardware components or networked computer systems.
- **self-star properties:** a list of component and system properties required for a computing system to be classified as an autonomic system.

SELF-MANAGING SYSTEMS
Autonomic computing is a branch of computer science aimed at developing computers capable of some autonomous operation. An autonomic system is one that is, in one or more respects, self-managing. Such systems are sometimes described as "self-*" or "self-star." The asterisk, or "star," represents different properties of autonomic systems (self-organization, self-maintenance). Autonomic computing aims to develop systems that require less outside input, allowing users to focus on other activities.

SELF-STAR SYSTEMS
The concept of autonomic computing is based on autonomic systems found in nature. Examples of such systems include the autonomic nervous system of humans and the self-regulation of colonial insects such as bees and ants. In an autonomic system, the behaviors of individual components lead to higher-order self-maintenance properties of the group as a whole.

The properties that a system needs to function autonomically are often called self-star properties. One is self-management, meaning that the system can manage itself without outside input after an initial setup. Computer scientists disagree about what other self-star properties a system must have to be considered autonomic. Proposed properties include:

> self-stabilization, the ability to return to a stable state after a change in configuration;
> self-healing, the ability to recover from external damage or internal errors;
> self-organization, the ability to organize component parts and processes toward a goal;
> self-protection, the ability to combat external threats to operation; and
> self-optimization, the ability to manage all resources and components to optimize operation.

Autonomic systems may also display self-awareness and self-learning. Self-awareness in a computer system differs from self-awareness in a biological system. In a computer system, self-awareness is better defined as the system's knowledge of its internal components and configuration. Self-learning is the ability to learn from experiences without a user programming new information into the system.

DESIGN OF AUTONOMIC SYSTEMS
An autonomic computer system is typically envisioned as having autonomic components (ACs), which are at least partly self-managing. An example of an AC is bootstrapping. Bootstrapping is the process by which a computer configures and initiates various processes during start-up. After a user turns on the computer, the bootstrapping process is self-managed. It proceeds through a self-diagnostic check and then activates various hardware and software components.

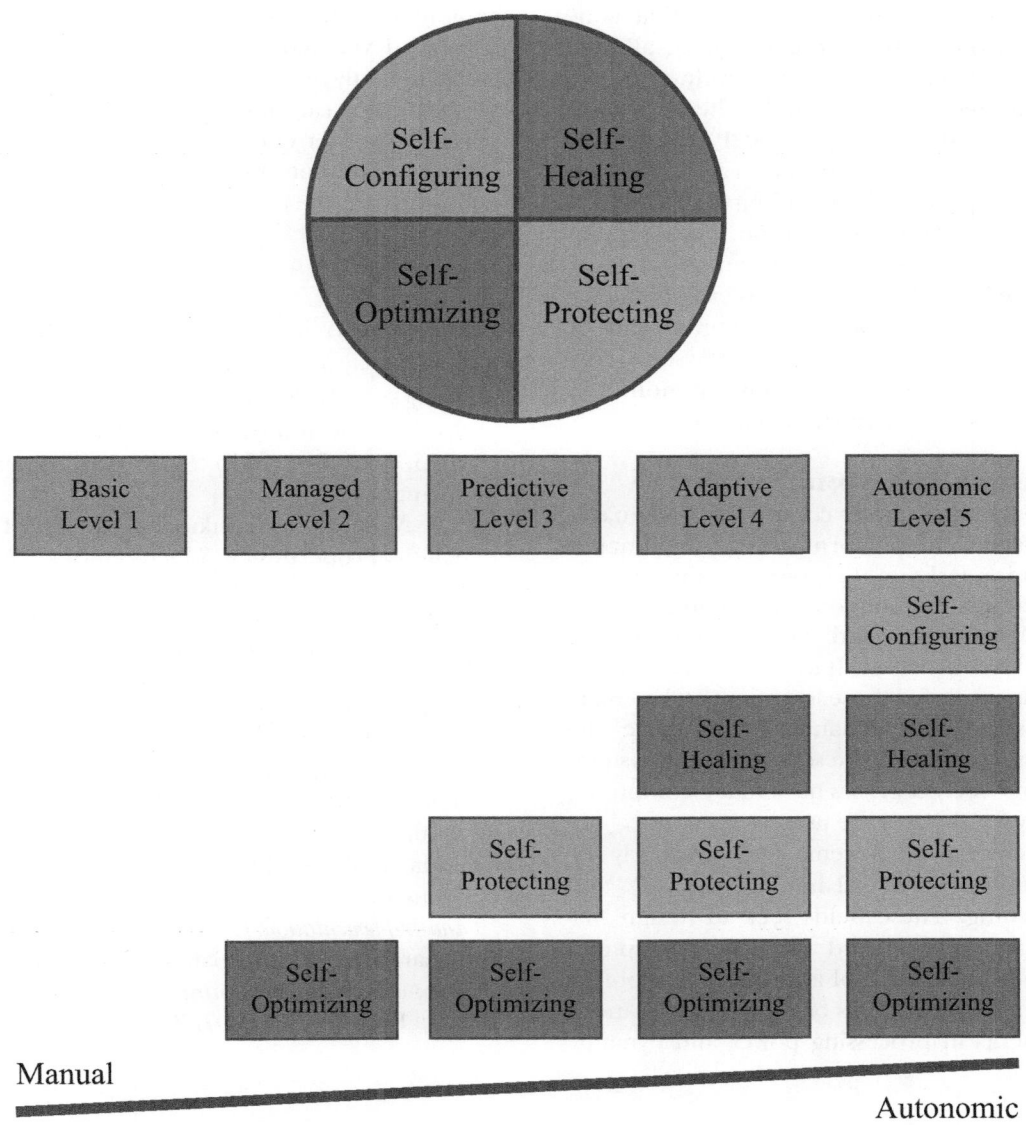

As computer systems have advanced from very basic technologies needing intense IT management toward autonomic systems that can self-manage, there have been four major stepping stones: self-optimizing, self-protecting, self-healing, and self-configuring. Each of these steps toward fully autonomic systems allows for more expansive computing while reducing the skill level required of the end users.

There are two basic models for autonomic computer design: a feedback control system and a multi-agent system. In a feedback control system, changing conditions provide feedback to the system that triggers changes in the system's function. Feedback control is often found in biological systems. In the autonomic nervous system, for example, levels of various neurotransmitters are linked to feedback systems that activate or deactivate ACs. A multi-agent system uses the collective functions of separate components to complete higher-order functions. For instance, groups of computers can be networked such that by performing individual functions, the components can collectively manage the system's functions and resources with reduced need for outside input. Linking multiple processors together changes the system's resource distribution, requiring that it be able to locate computing resources within all connected agents in to handle tasks effectively.

Semi-autonomic software and hardware systems are commonly used. Peer-to-peer (P2P) systems for social networking and communication generally have some autonomic properties, including self-configuration, self-tuning, and self-organization. These systems can determine a user's particular computer setup, tune themselves to function in various environments, and self-organize in response to changing data or configuration. Most modern computing systems contain ACs but are not considered fully autonomic, as they still require some external management.

THE PROMISE OF AUTONOMIC SYSTEMS

The main goal of autonomic computing is to enable computer systems to perform basic maintenance and optimization tasks on their own. Maximizing the number of automated functions that a computer can handle allows engineers and system administrators to focus on other activities. It also enhances ease of operation, especially for those less adept at data management or system maintenance. For instance, the bootstrapping system and the self-regulatory systems that detect and correct errors have made computing more friendly for the average user.

Autonomic computer systems are particularly important in robotics, artificial intelligence (AI), and machine learning. These fields seek to design machines that can work unaided after initial setup and programming. The science of autonomic computing is still in its infancy, but it has been greatly enhanced by advancements in processing power and dynamic computer networking. For instance, the AI system Amelia, developed by former IT specialist Chetan Dube, not only responds to verbal queries and answers questions but can also learn by listening to human operators answer questions that it cannot answer. To some, systems that can learn and alter their own programming are the ultimate goal of autonomic design.

—*Micah L. Issitt*

BIBLIOGRAPHY

Bajo, Javier, et al., eds. *Highlights of Practical Applications of Agents, Multi-Agent Systems, and Sustainability*. Proc. of the International Workshops of PAAMS 2015, June 3–4, 2015, Salamanca, Spain. Cham: Springer, 2015. Print.

Berns, Andrew, and Sukumar Ghosh. "Dissecting Self-* Properties." *SASO 2009: Third IEEE International Conference on Self-Adaptive and Self-Organizing Systems*. Los Alamitos: IEEE, 2009. 10–19. *Andrew Berns: Homepage*. Web. 20 Jan. 2016.

Follin, Steve. "Preparing for IT Infrastructure Autonomics." *IndustryWeek*. Penton, 19 Nov. 2015. Web. 20 Jan. 2016.

Gibbs, W. Wayt. "Autonomic Computing." *Scientific American*. Nature Amer., 6 May 2002. Web. 20 Jan. 2016.

Lalanda, Philippe, Julie A. McCann, and Ada Diaconescu, eds. *Autonomic Computing: Principles, Design and Implementation*. London: Springer, 2013. Print.

Parashar, Manish, and Salim Hariri, eds. *Autonomic Computing: Concepts, Infrastructure, and Applications*. Boca Raton: CRC, 2007. Print.

AVATARS AND SIMULATION

FIELDS OF STUDY

Digital Media; Graphic Design

ABSTRACT

Avatars and simulation are elements of virtual reality (VR), which attempts to create immersive worlds for computer users to enter. Simulation is the method by which the real world is imitated or approximated by the images and sounds of a computer. An avatar is the personal manifestation of a particular person. Simulation and VR are used for many applications, from entertainment to business.

PRINCIPAL TERMS

- **animation variables (avars):** defined variables used in computer animation to control the movement of an animated figure or object.

- **keyframing:** a part of the computer animation process that shows, usually in the form of a drawing, the position and appearance of an object at the beginning of a sequence and at the end.
- **modeling:** reproducing real-world objects, people, or other elements via computer simulation.
- **render farm:** a cluster of powerful computers that combine their efforts to render graphics for animation applications.
- **virtual reality:** the use of technology to create a simulated world into which a user may be immersed through visual and auditory input.

Virtual Worlds

Computer simulation and virtual reality (VR) have existed since the early 1960s. While simulation has been used in manufacturing since the 1980s, avatars and virtual worlds have yet to be widely embraced outside gaming and entertainment. VR uses computerized sounds, images, and even vibrations to model some or all of the sensory input that human beings constantly receive from their surroundings every day. Users can define the rules of how a VR world works in ways that are not possible in everyday life. In the real world, people cannot fly, drink fire, or punch through walls. In VR, however, all of these things are possible, because the rules are defined by human coders, and they can be changed or even deleted. This is why users' avatars can appear in these virtual worlds as almost anything one can imagine—a loaf of bread, a sports car, or a penguin, for example. Many users of virtual worlds are drawn to them because of this type of freedom.

Because a VR simulation does not occur in physical space, people can "meet" regardless of how far apart they are in the real world. Thus, in a company that uses a simulated world for conducting its meetings, staff from Hong Kong and New York can both occupy the same VR room via their avatars. Such virtual meeting spaces allow users to convey nonverbal cues as well as speech. This allows for a greater degree of authenticity than in telephone conferencing.

Mechanics of Animation

The animation of avatars in computer simulations often requires more computing power than a single workstation can provide. Studios that produce animated films use render farms to create the smooth and sophisticated effects audiences expect.

Before the rendering stage, a great deal of effort goes into designing how an animated character or avatar will look, how it will move, and how its textures will behave during that movement. For example, a fur-covered avatar that moves swiftly outdoors in the wind should have a furry or hairy texture, with fibers that appear to blow in the wind. All of this must be designed and coordinated by computer animators. Typically, one of the first steps is keyframing, in which animators decide what the starting and ending positions and appearance of the animated object will be. Then they design the movements between the beginning and end by assigning animation variables (avars) to different points on the object. This stage is called "in-betweening," or "tweening." Once avars are assigned, a computer algorithm can automatically change the avar values in coordination with one another. Alternatively, an animator can change "in-between" graphics by hand. When the program is run, the visual representation of the changing avars will appear as an animation.

In general, the more avars specified, the more detailed and realistic that animation will be in its movements. In an animated film, the main characters often have hundreds of avars associated with them. For instance, the 1995 film *Toy Story* used 712 avars for the cowboy Woody. This ensures that the characters' actions are lifelike, since the audience will focus attention on them most of the time. Coding standards for normal expressions and motions have been developed based on muscle movements. The MPEG-4 international standard includes 86 face parameters and 196 body parameters for animating human and humanoid movements. These parameters are encoded into an animation file and can affect the bit rate (data encoded per second) or size of the file.

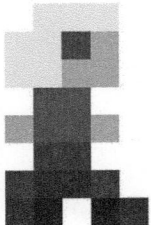

Avatars were around long before social media. As computers have become more powerful and the rendering capabilities more efficient, avatars have improved in detail and diversity.

Educational Applications

Simulation has long been a useful method of training in various occupations. Pilots are trained in flight simulators, and driving simulators are used to prepare for licensing exams. Newer applications have included training teachers for the classroom and improving counseling in the military. VR holds the promise of making such vocational simulations much more realistic. As more computing power is added, simulated environments can include stimuli that better approximate the many distractions and detailed surroundings of the typical driving or flying situation, for instance.

VR in 3D

Most instances of VR that people have experienced so far have been two-dimensional (2D), occurring on a computer or movie screen. While entertaining, such experiences do not really capture the concept of VR. Three-dimensional (3D) VR headsets such as the Oculus Rift may one day facilitate more lifelike business meetings and product planning. They may also offer richer vocational simulations for military and emergency personnel, among others.

—*Scott Zimmer, JD*

Bibliography

Chan, Melanie. *Virtual Reality: Representations in Contemporary Media.* New York: Bloomsbury, 2014. Print.

Gee, James Paul. *Unified Discourse Analysis: Language, Reality, Virtual Worlds, and Video Games.* New York: Routledge, 2015. Print.

Griffiths, Devin C. *Virtual Ascendance: Video Games and the Remaking of Reality.* Lanham: Rowman, 2013. Print.

Hart, Archibald D., and Sylvia Hart Frejd. *The Digital Invasion: How Technology Is Shaping You and Your Relationships.* Grand Rapids: Baker, 2013. Print.

Kizza, Joseph Migga. *Ethical and Social Issues in the Information Age.* 5th ed. London: Springer, 2013. Print.

Lien, Tracey. "Virtual Reality Isn't Just for Video Games." *Los Angeles Times.* Tribune, 8 Jan. 2015. Web. 23 Mar. 2016.

Parisi, Tony. *Learning Virtual Reality: Developing Immersive Experiences and Applications for Desktop, Web, and Mobile.* Sebastopol: O'Reilly, 2015. Print.

BINARY HEXADECIMAL REPRESENTATIONS

FIELDS OF STUDY
Computer Science; Computer Engineering; Software Engineering

ABSTRACT
The binary number system is a base-2 number system. It is used by digital devices to store data and perform mathematical operations. The hexadecimal number system is a base-16 number system. It enables humans to work efficiently with large numbers stored as binary data.

PRINCIPAL TERMS
- **base-16:** a number system using sixteen symbols, 0 through 9 and A through F.
- **base-2 system:** a number system using the digits 0 and 1.
- **bit:** a single binary digit that can have a value of either 0 or 1.
- **byte:** a group of eight bits.
- **nibble:** a group of four bits.

Understanding the Binary Number System

A mathematical number system is a way of representing numbers using a defined set of symbols. Number systems take their names from the number of symbols the system uses to represent numbers. For example, the most common mathematical number system is the decimal system, or base-10 system. *Deci-* means "ten." It uses the ten digits 0 through 9 as symbols for numbers. Number systems can be based on any number of unique symbols, however. For example, the number system based on the use of two digit symbols (0 and 1) is called the binary or base-2 system.

Both the decimal and binary number systems use the relative position of digits in a similar way when representing numbers. The value in the rightmost, or first, position is multiplied by the number of digits used in the system to the zero power. For the decimal system, this value is 10^0. For the binary system, this value is 2^0. Both 10^0 and 2^0 are equal to 1. Any number x raised to the zero power is equal to 1. The power used increases by one for the second position and so on.

Position	8	7	6	5	4	3	2	1
	Seventh Power	Sixth Power	Fifth Power	Fourth Power	Third Power	Second Power	First Power	Zero Power
Decimal	10,000,000 or 10^7	1,000,000 or 10^6	100,000 or 10^5	10,000 or 10^4	1,000 or 10^3	100 or 10^2	10 or 10^1	1 or 10^0
Binary	128 or 2^7	64 or 2^6	32 or 2^5	16 or 2^4	8 or 2^3	4 or 2^2	2 or 2^1	1 or 2^0

Decimal/Binary Table. Table organizes binary based values and decima based values by position, from 8 to 1, and by power, from seventh power to zero power.

Using the decimal number system, the integer 234 is represented by placing the symbols 2, 3, and 4 in positions 3, 2, and 1, respectively.

Position	3	2	1
Decimal	100 or 10^2	10 or 10^1	1 or 10^0
Digits	2	3	4

Decimal/Digits table. Table organizes decimal values and the number of digits by position, from 3 to 1.

In the decimal system, $234 = (2 \times 100) + (3 \times 10) + (4 \times 1)$, or $(2 \times 10^2) + (3 \times 10^1) + (4 \times 10^0)$. The binary system uses the relative position of the symbols 0 and 1 to express the integer 234 in a different manner.

Position	8	7	6	5	4	3	2	1
Binary	128 or 2^7	64 or 2^6	32 or 2^5	16 or 2^4	8 or 2^3	4 or 2^2	2 or 2^1	1 or 2^0
Bit	1	1	1	0	1	0	1	0

Binary/Bit Table. Table organizes binary based values and Bit values by position from 8 to 1.

In the binary system, $234 = (1 \times 128) + (1 \times 64) + (1 \times 32) + (0 \times 16) + (1 \times 8) + (0 \times 4) + (1 \times 2) + (0 \times 1)$, or $234 = (1 \times 2^7) + (1 \times 2^6) + (1 \times 2^5) + (0 \times 2^4) + (1 \times 2^3) + (0 \times 2^2) + (1 \times 2^1) + (0 \times 2^0)$.

The Importance of the Binary Number System

The binary number system is used to store numbers and perform mathematical operations in computers systems. Such devices store data using transistors, electronic parts that can each be switched between two states. One state represents the binary digit 0 and the other, the binary digit 1. These binary digits are bits, the smallest units of data that can be stored and manipulated. A single bit can be used to store the value 0 or 1. To store values larger than 1, groups of bits are used. A group of four bits is a nibble. A group of eight bits is a byte.

SAMPLE PROBLEM

To work with binary numbers in digital applications, it is important to be able to translate numbers from their binary values to their decimal values. Translate the following binary byte to its decimal value: 10111001

Answer:

The decimal value of the binary byte 10111001 is 185. The decimal value can be determined using a chart and then calculating.

8	7	6	5	4	3	2	1
128 or 2^7	64 or 2^6	32 or 2^5	16 or 2^4	8 or 2^3	4 or 2^2	2 or 2^1	1 or 2^0
1	1	1	0	1	0	1	0

$= (1 \times 2^7) + (0 \times 2^6) + (1 \times 2^5) + (1 \times 2^4) + (1 \times 2^3) + (0 \times 2^2) + (0 \times 2^1) + (1 \times 2^0)$
$= (1 \times 128) + (0 \times 64) + (1 \times 32) + (1 \times 16) + (1 \times 8) + (0 \times 4) + (0 \times 2) + (1 \times 1)$
$= 185$

DECIMAL	HEXADECIMAL	BINARY
0	00	00000000
1	01	00000001
2	02	00000010
3	03	00000011
4	04	00000100
5	05	00000101
6	06	00000110
7	07	00000111
8	08	00001000
9	09	00001001
10	0A	00001010
11	0B	00001011
12	0C	00001100
13	0D	00001101
14	0E	00001110
15	0F	00001111
16	10	00010000
17	11	00010001
18	12	00010010
19	13	00010011
20	14	00010100
21	15	00010101
22	16	00010110
23	17	00010111
24	18	00011000
25	19	00011001
26	1A	00011010
27	1B	00011011
28	1C	00011100
29	1D	00011101
30	1E	00011110
31	1F	00011111
32	20	00100000
33	21	00100001
34	22	00100010
35	23	00100011
36	24	00100100
37	25	00100101
38	26	00100110
39	27	00100111
40	28	00101000
41	29	00101001
42	2A	00101010
43	2B	00101011
44	2C	00101100
45	2D	00101101
46	2E	00101110
47	2F	00101111
48	30	00110000
49	31	00110001
50	32	00110010

The American Standard Code for Information Interchange (ASCII) was an early system used to translate basic characters into a numerical code readable by computers. The common characters on a keyboard are provided with decimal and hexadecimal codes.

Using Hexadecimal to Simplify Binary Numbers

The hexadecimal number system is a base-16 system. It uses the digits 0 through 9 and the letters A through F to represent numbers. The hexadecimal digit, or hex digit, A has a decimal value of 10. Hex digit B equals 11, C equals 12, D equals 13, E equals 14, and F equals 15. In hexadecimal, the value 10 is equal to 16 in the decimal system. Using hexadecimal, a binary nibble can be represented by a single symbol. For example, the hex digit F can be used instead of the binary nibble 1111 for the decimal value 15. Sixteen different combinations of bits are possible in a binary nibble. The hexadecimal system, with sixteen different symbols, is therefore ideal for working with nibbles.

One disadvantage of using binary is that large numbers of digits are needed to represent large integers. For example, 1,000,000 is shown in binary digits as 11110100001001000000. The same number is shown in hex digits as F4240, which is equal to $(15 \times 65{,}536) + (4 \times 4{,}096) + (2 \times 256) + (4 \times 16) + (0 \times 1)$.

Position	8	7	6	5	4	3	2	1
	Seventh Power	Sixth Power	Fifth Power	Fourth Power	Third Power	Second Power	First Power	Zero Power
Decimal	10,000,000 or 10^7	1,000,000 or 10^6	100,000 or 10^5	10,000 or 10^4	1,000 or 10^3	100 or 10^2	10 or 10^1	1 or 10^0
Hexadecimal	268,435,456 or 16^7	16,777,216 or 16^6	1,048,576 or 16^5	65,536 or 16^4	4,096 or 16^3	256 or 16^2	16 or 16^1	1 or 16^0

Decimal/Hexidecimal Table. table organizes decimal based values and hexidecimal based values by position from 8 to 1, and power from seventh to zero.

Computers can quickly and easily work with large numbers in binary. Humans have a harder time using binary to work with large numbers. Binary uses many more digits than hexadecimal does to represent large numbers. Hex digits are therefore easier for humans to use to write, read, and process than binary.

Position	5	4	3	2	1
	Fourth Power	Third Power	Second Power	First Powder	Zero Power
Hexadecimal	65,536 or 16^4	4,096 or 16^3	256 or 16^2	16 or 16^1	1 or 16^0
Hex digits	F or 15	4	2	4	0

Hexidecimal/Hex digits Table. Table organizes hexidecimal based values and corresponding hex digits by position from 5 to 1.

—Maura Valentino, MSLIS

Bibliography

Australian National University. *Binary Representation and Computer Arithmetic.* Australian National U, n.d. Digital file.

Cheever, Erik. "Representation of Numbers." *Swarthmore College.* Swarthmore College, n.d. Web. 20 Feb. 2016.

Govindjee, S. *Internal Representation of Numbers.* Dept. of Civil and Environmental Engineering, U of California Berkeley, Spring 2013. Digital File.

Glaser, Anton. *History of Binary and Other Nondecimal Numeration.* Rev. ed. Los Angeles: Tomash, 1981. Print.

Lande, Daniel R. "Development of the Binary Number System and the Foundations of Computer Science." *Mathematics Enthusiast* 1 Dec. 2014: 513–40. Print.

"A Tutorial on Data Representation: Integers, Floating-Point Numbers, and Characters." *NTU.edu.* Nanyang Technological U, Jan. 2014. Web. 20 Feb. 2016.

BOOLEAN OPERATORS

FIELD OF STUDY
Software Development; Coding Techniques; Computer Science

ABSTRACT
Boolean, or logical, operators are used to create Boolean expressions, which test whether a condition is true or false. Statements such as the if-statement use Boolean expressions to control the order in which a computer program processes commands and other statements.

PRINCIPAL TERMS

- **if-statement:** a command that instructs a computer program to execute a block of code if a certain condition is true.
- **operator:** a character that indicates a specific mathematical, relational, or logical operation.
- **string:** in computer programming, a data type consisting of characters arranged in a specific sequence.

UNDERSTANDING BOOLEAN EXPRESSIONS

In computer programming, an expression describes some value (much as a math equation does), often using integers, strings, constants, or variables. Boolean expressions are expressions that, when evaluated, return one of two values: "true/false," "yes/no," or "on/off." For example, the Boolean expression $x > 5$ would return a value of "false" if $x = 4$ and a value of "true" if $x = 6$. Boolean expressions can be used with different data types, including integers, strings, and variables. Complex Boolean expressions can be created using Boolean operators. Boolean operators combine two or more Boolean expressions into a single expression and can also reverse the value of a Boolean expression.

There are three main Boolean operators: AND, OR, and NOT. Different programming languages use different symbols to represent them, such as && for AND or || for OR in Java, but the uses remain the same. The AND operator combines two Boolean expressions into a single expression that evaluates to true if, and only if, both expressions are true. If either expression evaluates to false, then the combined expression is false. When two expressions are joined by the OR operator, the combined expression evaluates to true if either expression is true and only evaluates to false if both are false.

The NOT operator reverses the value of a Boolean expression. For example, the expression $x > 5$ evaluates to true if the value of x is greater than 5. However, the NOT operator can be used to reverse this result. The expression NOT $x > 5$ evaluates to false if the value of x is greater than 5.

The order in which operations are performed is called "precedence." Arithmetic is done before Boolean operations. Boolean operators with higher precedence are evaluated before those with lower precedence: NOT is evaluated before AND, which is evaluated before OR. The default precedence can be overridden using nested parentheses. Expressions within the innermost parentheses are evaluated first, followed by expressions contained within the parentheses that surround them, and so on. However, when using the AND operator, if any component expression evaluates to false, then any component expressions that follow the false expression are not evaluated. Likewise, when using the OR operator, if a component expression evaluates to true, then any component expressions that follow the true expression are not evaluated. This is called "short-circuiting."

The order in which statements are executed is called "control flow." Statements such as the if-statement and if-else statement alter a program's control flow. The if-statement tests for a condition, and if that condition is true, control flow passes to a designated block of code. If the condition is false, no action is performed. In an if-else statement, if the condition is false, control flow passes to a different block of code. Boolean expressions are used to create the conditions used by if-statements and if-else statements.

USING BOOLEAN OPERATORS EFFECTIVELY

Computer programs are often designed to help people accomplish their tasks more efficiently. As many tasks require making decisions based on whether different conditions are true or false, Boolean operators help computer programs complete these tasks effectively. In fact, simple conditional tests are easy to convert to code because simple Boolean expressions are coded much like words are

23

used in everyday language. Consider the following plain-language statement: "If the date for the concert is September 1, 2018, and the type of concert is folk music, then delete the concert from the master list." This statement could be expressed in code as follows:

```
IF (Date = "9/1/18") AND (ConcertType = "folk music") THEN
    Delete Concert
END IF
```

However, as Boolean expressions become more complicated, they become increasing difficult to understand correctly. For example, the expression NOT x < 5 AND y > 1 OR country = "UK" can only be executed correctly if the order of precedence for the operators is known. Programmers use parentheses to define the order precedence they desire to make such expressions easier to understand. For example, the above expression could be rewritten as follows:

NOT (x < 5 AND (y > 1 OR country = "UK"))

Conditional Statements in Practice

If statements and if-else statements could be used to assign a letter grade based on an exam score. First, a Boolean expression is used to check that the exam has been completed and scored. The AND operator is used because both conditions must be true for a letter grade to be assigned.

```
IF (ExamComplete = 1) AND (ExamScored = 1) THEN
    Retrieve ExamScore
END IF
Store ExamScore in a variable named Score
```

Next, the code uses a chain of if statements to assign a letter grade to the variable *Grade*. Note how the first statement does not use a Boolean operator because it does not need to evaluate more than one expression. The next three statements use the Boolean operator AND because they require both expressions to be true for the code to be executed. The final if-statement does not contain a Boolean expression because it executes only if all other if-statements fail to execute.

```
IF (Score >= 90) THEN
    Grade = "A"
ELSE IF (Score >= 80) AND (Score < 90) THEN
    Grade = "B"
ELSE IF (Score >= 70) AND (Score < 80) THEN
    Grade = "C"
ELSE IF (Score >= 60) AND (Score < 70) THEN
    Grade = "D"
ELSE
    Grade = "F"
END IF
```

Boolean Operators in the Real World

Boolean operators are commonly used when searching for information in a database. For instance, if a student needs to write a research paper on climate change, they might type in "global warming OR climate change." This would tell the library's database to return results that include either phrase and thus increase the overall number of results returned.

Boolean operators are also used to program software in which a sequential flow of control is not desired. The conditions that they enable allow programs not only to branch through selection, but also to loop (or repeat) sections of code. Thus, Boolean operators allow more complex software to be developed.

—*Maura Valentino, MSLIS*

Bibliography

Friedman, Daniel P., and Mitchell Wand. *Essentials of Programming Languages.* 3rd ed., MIT P, 2008.

Haverbeke, Marijn. *Eloquent JavaScript: A Modern Introduction to Programming.* 2nd ed., No Starch Press, 2014.

MacLennan, Bruce J. *Principles of Programming Languages: Design, Evaluation, and Implementation.* 3rd ed., Oxford UP, 1999.

Schneider, David I. *An Introduction to Programming using Visual Basic.* 10th ed., Pearson, 2016.

Scott, Michael L. *Programming Language Pragmatics.* 4th ed., Morgan Kaufmann Publishers, 2016.

Van Roy, Peter, and Seif Haridi. *Concepts, Techniques, and Models of Computer Programming.* MIT P, 2004.

BRANCHING LOGIC

FIELD OF STUDY

Coding Techniques; Software Development; Computer Science

ABSTRACT

Branching logic is used in computer programming when different sections of code need to be executed depending on one or more conditions. Each possible code pathway creates another branch. There is no limit to the number of branches that can be used to implement complex logic.

PRINCIPAL TERMS

- **Boolean logic:** a type of algebra in which variables have only two possible values, either "true" or "false."
- **if-then-else:** a conditional statement that evaluates a Boolean expression and then executes different sets of instructions depending on whether the expression is true or false.
- **parameter:** in computer programming, a variable with an assigned value that is passed into a function or subroutine within a larger program.

What Is Branching Logic?

Branching logic is a form of decision making in which a computer program follows different sets of instructions depending on whether or not certain conditions are met during the program's execution. Each set of instructions represents a different branch. Which branch of code is executed depends on the values assigned to the parameters of the branching procedure. The values of these parameters may be input by the user, or they may be generated by the output from a previous procedure.

A common form of branching logic is the if-then-else statement. Such statements generally take the following form:

```
IF condition 1 THEN
  outcome A
ELSE
  outcome B
END IF
```

In this statement, Boolean logic is used to test a given condition (condition 1), which can be found to be either true or false. If condition 1 evaluates to true, then code is executed to create outcome A. If it evaluates to false, then code is executed to create outcome B.

The exact syntax of the if-then-else statement varies among programming languages, but the logic remains the same. In some programming languages, the "else" statement is optional. In these cases, if the conditional statement evaluates to false, the program will continue with the instructions that come after the "end if" statement.

Multiple branches can be used to test multiple conditions. An "else if" statement can be added to the above example to test other conditions:

```
IF condition 1 THEN
  outcome A
ELSE IF condition 2 THEN
  outcome B
ELSE IF condition 3 THEN
  outcome C
ELSE
  outcome D
END IF
```

With the above structure, three conditions are tested and four outcomes are possible.

Most applications do not execute their code linearly, without variations. Rather, each time an application is run, different inputs can lead to different results. For example, in a word-processing application, pressing the same letter key on a computer keyboard will result in different outcomes depending on whether certain other keys are pressed at the same time. If a user presses the letter *a*, the application must evaluate whether or not that user also pressed either the shift key or the control key before it can determine if the letter should be displayed as a capital *A* or a lowercase *a*, or if some other code should be executed instead. The decision structure for this output might look something like the following:

```
IF shift key is also pressed THEN
  display A
ELSE IF control key is also pressed THEN
  select entire document
ELSE
  display a
END IF
```

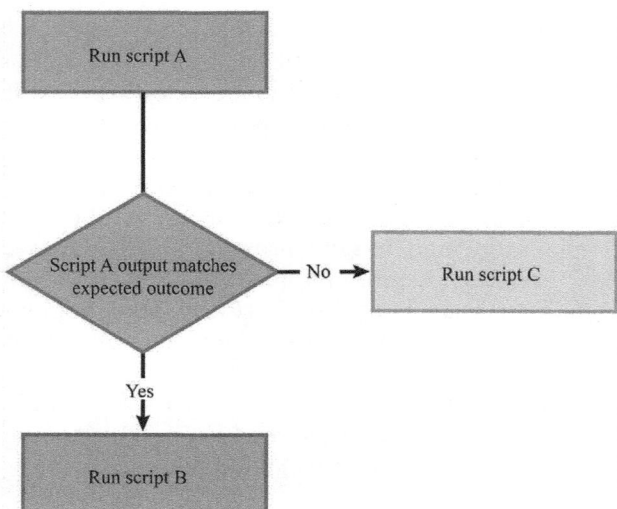

A model of branching logic will have multiple sequences of events. Which set of instructions is followed depends on the outcomes along specific nodes in the path. After script A is complete, if the output matches an expected outcome, then script B will run; if not, then script C will run.

In this example, the Boolean statements being evaluated are "shift key is also pressed" and "control key is also pressed." The keys pressed by the user are stored as parameters and then passed to the branching procedure, which uses those parameters to evaluate whether or not the statements are true. The user's actions can result in a new decision being created each time the same section of code is executed.

WHEN IS BRANCHING LOGIC USED?

It is imperative that computer programs and applications be able to execute different pieces of code depending on the input they receive. For example, branching logic is often used in surveys so that participants receive only questions that apply to them. If a survey contains questions that only apply to one sex or another, the first survey question may ask for the sex of the participant. The participant's response will be stored as a parameter that will then determine which additional questions they will be presented with. If every participant had to read through all of the possible questions to determine whether or not each one applied to them, the survey would be much more difficult and time consuming, and it would result in more errors.

Branching logic is also used in workflows in fields such as marketing. If a customer views one page of a website, certain marketing materials are sent to that customer. If a different page is viewed, different materials are sent.

The nesting structure of branching logic can be confusing to plan if there are many conditions to check. Programmers can use a script tool to create a form structure for users to input the parameters that will be used to evaluate the conditional statements. This technique reduces errors, because restrictions can be placed on what data is entered. The script tool can also be used to provide the user with information calculated from input parameters.

There are some minor drawbacks to using branching logic. For example, the nesting structure of the "else if" statement creates added difficulty for some compilers, and conditional statements can slow the processing of code. Faster structures have been developed to overcome these difficulties.

HOW IS BRANCHING LOGIC USED?

Branching logic is used by computer programs in many fields. In online retail stores, for example, some items only require the user to make one decision—to buy or not to buy—while other items require the user to choose a color and a size. If the color is chosen first, the store will only list the sizes available in that color, and vice versa. This is much more convenient, and much less frustrating to users, than it would be if the store presented users with choices that did not exist and did not inform them until much later in the process. A listing for an item that is available in three colors in sizes small and medium, four colors in size large, and one color in size extra large may follow the following decision structure:

```
IF size = small THEN
   colors = blue, green, red
ELSE IF size = medium THEN
   colors = blue, red, yellow
ELSE IF size = large THEN
   colors = blue, green, red, yellow
ELSE
   colors = blue
END IF
```

If the user chooses color before size, the ensuing decision structure would look similar. The two structures can be nested within a larger if-then-else statement that executes either one or the other depending on whether size or color is chosen first.

Why Is Branching Logic Important?

Many computer programs require decision making, and branching logic is one way of making decisions. It is so common in computer programming that many languages feature alternate structures that can execute branching procedures in much fewer lines of code. For example, conditional operators can produce the same results as if-then-else statements using just one line of code. A switch statement is another, more complicated form of branching logic that uses a table to check many values against the given condition. Branching logic is used extensively in modern programming and has led to many developments as programs have become more complex.

—*Maura Valentino, MSLIS*

Bibliography

Friedman, Daniel P., and Mitchell Wand. *Essentials of Programming Languages*. 3rd ed., MIT P, 2008.

Haverbeke, Marijn. *Eloquent JavaScript: A Modern Introduction to Programming*. 2nd ed., No Starch Press, 2015.

MacLennan, Bruce J. *Principles of Programming Languages: Design, Evaluation, and Implementation*. 3rd ed., Oxford UP, 1999.

Scott, Michael L. *Programming Language Pragmatics*. 4th ed., Elsevier, 2016.

Schneider, David I. *An Introduction to Programming Using Visual Basic*. 10th ed., Pearson, 2017.

Van Roy, Peter, and Seif Haridi. *Concepts, Techniques, and Models of Computer Programming*. MIT P, 2004.

C

CHARACTERS AND STRINGS

FIELDS OF STUDY
Software Development; Coding Techniques; Computer Science

ABSTRACT
Characters and strings are basic units of programming languages that direct computer programs to store, retrieve, and manipulate data. Characters represent the individual letters, numbers, punctuation marks, and other symbols in programming languages. Strings are groups of characters arranged in a specific sequence.

PRINCIPAL TERMS

- **command:** an instruction given by a user to a computer that directs the computer to perform a specific, predefined task.
- **interpolation:** a process by which a computer program evaluates variable expressions within a string.
- **sequence:** the specific order in which code is arranged, thereby affecting its output.
- **string literals:** a fixed sequence of characters that appear in source code enclosed by single or double quotation marks.
- **syntax:** in computer programming, the rules that govern how the elements of a programming language should be arranged.

WHAT ARE CHARACTERS AND STRINGS?
Characters are the most basic units of programming languages. Characters include letters, numbers, punctuation marks, blank spaces, and the other symbols. In computer programming, characters may be grouped together to form units of information called strings. A string can be created by placing one or more characters in sequential order. For example, the letters *O, P,* and *T* are characters. Those three characters can be combined to create the strings "TOP," "OPT," and "POT." Strings that use the same characters can have different meanings and outputs depending on the sequence in which the characters are arranged.

Computer programs use variables to store and manipulate data. Variables are referenced by unique symbolic names. The data stored in a variable is referred to as a value. Different types of data can be stored as variables, including numeric and textual information. For example, the variable used to hold an error message might be named *ErrorMessage*. If a computer program needs to notify the user that an error has occurred, it could accomplish this by storing the string "Error!" in the variable *ErrorMessage*. The program could then issue a command to instruct the computer to display the error message.

The rules that govern the arrangement and meanings of words and symbols in a programming language are the syntax. For example, one computer language might use the following syntax to display the error message.

```
ErrorMessage = "Error!"
echo ErrorMessage
```

In this example, *ErrorMessage* is the variable. "Error!" is the string. The command is named *echo*, and the statements are separated by line breaks. On the other hand, another language might use different syntax to accomplish the same task:

```
ErrorMessage = 'Error!';
printscreen ErrorMessage;
```

In this example, *ErrorMessage* is the variable and "Error!" is the string. However, *printscreen* is the command, and the statements are separated by semicolons and line breaks. In both examples, the string *Error!* is stored as a value for the variable *ErrorMessage* in the first statement. The string is then displayed on the user's monitor by a command in the second statement.

There are several ways in which string values may be created. One way is through the use of string literals. String literals are strings that represent a variable with a fixed value. String literals are enclosed in quotation marks in the source code. For example, the string "John Smith" might be stored in the variable *CustomerName* as follows:

CustomerName = "John Smith"

String literals can also be created by combining various placeholders. For example, the variable *CustomerName* might also be created as follows:

FirstName = "John"
LastName = "Smith"
CustomerName = FirstName + " " + LastName

The process of adding two string variables together in this manner is called concatenation.

Strings may also be created by combining variables, string literals, and constants. When a computer program evaluates various placeholders within a string, it is called interpolation. For example, the variables *FirstName* and *LastName*, along with the string literals "Dear" and blank spaces, might be used to store the string value "Dear John Smith" in the variable *Salutation*, as follows:

FirstName = "John"
LastName = "Smith"
Salutation = "Dear" + " " + FirstName + " " + LastName

In this example, the string value "John" is stored in the variable *FirstName*, and the string value "Smith" is stored in the variable *LastName*. The final line of code uses interpolation to combine the data stored in the variables with the string literals "Dear" and blank spaces to create a new variable named *Salutation*, which stores the string value "Dear John Smith."

Using Special Characters in Strings

As demonstrated in the previous examples, the syntax for a computer language might use double quotation marks to define a string as follows:

LastName = "Smith"

When a character, such as the double quotation mark, is used to define a string, it is called a special character. The use of a special character to define a string creates a problem when the special character itself needs to be used within the string. Programming languages solve this problem through the use of character combinations to escape the string. For example, a double quotation mark that is used in a string enclosed by double quotation marks might be represented by a backslash as follows:

MyQuotation = "/"Hello there, Mr. Jones!/""

This code would output the text "Hello there, Mr. Jones!" to the user's monitor. The backslash changes the meaning of the quotation mark that it precedes from marking the end of a string to representing an ordinary character within the string.

Strings and Characters in Real-World Data Processing

The use of characters and strings to store and manipulate the symbols that make up written languages is commonplace in computer programming. Strings are used to process textual data of all types. Computer programs use strings and characters to sort, search, and combine a vast array of textual information rapidly and accurately. The widespread use of textual data in computer programming has led to the development of industry standards that promote sharing of text across programming languages and software. For example, Unicode, the most widely used system for processing, displaying, and storing textual data, standardizes the encoding for more than 128,000 characters and 135 languages. Due to their widespread use, strings and characters form one of the cornerstones of computer and information science.

—*Maura Valentino, MSLIS*

Bibliography

Friedman, Daniel P., and Mitchell Wand. *Essentials of Programming Languages.* 3rd ed., MIT P, 2008.

Haverbeke, Marijn. *Eloquent JavaScript: A Modern Introduction to Programming.* 2nd ed., No Starch Press, 2014.

MacLennan, Bruce J. *Principles of Programming Languages: Design, Evaluation, and Implementation.* 3rd ed., Oxford UP, 1999.

Scott, Michael L. *Programming Language Pragmatics.* 4th ed., Morgan Kaufmann Publishers, 2016.

Schneider, David I. *Introduction to Programming using Visual Basic.* 10th ed., Pearson, 2016.

Van Roy, Peter, and Seif Haridi. *Concepts, Techniques, and Models of Computer Programming.* MIT P, 2004.

CLOUD COMPUTING

FIELDS OF STUDY
Information Technology; Computer Science; Software

ABSTRACT
Cloud computing is a networking model in which computer storage, processing, and program access are handled through a virtual network. Cloud computing is among the most profitable IT trends. A host of cloud-oriented consumer products are available through subscription.

PRINCIPAL TERMS
- **hybrid cloud:** a cloud computing model that combines public cloud services with a private cloud platform linked through an encrypted connection.
- **infrastructure as a service:** a cloud computing platform that provides additional computing resources by linking hardware systems through the Internet; also called "hardware as a service."
- **multitenancy:** a software program that allows multiple users to access and use the software from different locations.
- **platform as a service:** a category of cloud computing that provides a virtual machine for users to develop, run, and manage web applications.
- **software as a service:** a software service system in which software is stored at a provider's data center and accessed by subscribers.
- **third-party data center:** a data center service provided by a separate company that is responsible for maintaining its infrastructure.

CLOUD NETWORK DESIGN
Cloud computing is a networking model that allows users to remotely store or process data. Several major Internet service and content providers offer cloud-based storage for user data. Others provide virtual access to software programs or enhanced processing capabilities. Cloud computing is among the fastest-growing areas of the Internet services industry. It has also been adopted by government and research organizations.

TYPES OF CLOUD NETWORKS
Private clouds are virtual networks provided to a limited number of known users. These are often used in corporations and research organizations. Operating a private cloud requires infrastructure (software, servers, etc.), either on-site or through a third party. Public clouds are available to the public or to paying subscribers. The public-cloud service provider owns and manages the infrastructure. Unlike private clouds, public clouds provide access to an unknown pool of users, making them less secure. Public clouds tend to be based on open-source code, which is free and can be modified by any user.

The hybrid cloud lies somewhere between the two. It offers access to private cloud storage or software services, such as database servers, while keeping some services or components in a public cloud. Setup costs may be lower with hybrid cloud services. A group using a hybrid cloud outsources some aspects of infrastructure investment and maintenance but still enjoys greater security than with a public cloud. Hybrid clouds have become widespread in the health care, law, and investment fields, where sensitive data must be protected on-site.

CLOUD COMPUTING AS A SERVICE
The infrastructure as a service (IaaS) model offers access to virtual storage and processing capability through a linked network of servers. Cloud-based storage has become popular, with services such as Apple iCloud and Dropbox offering storage alternatives beyond the memory on users' physical

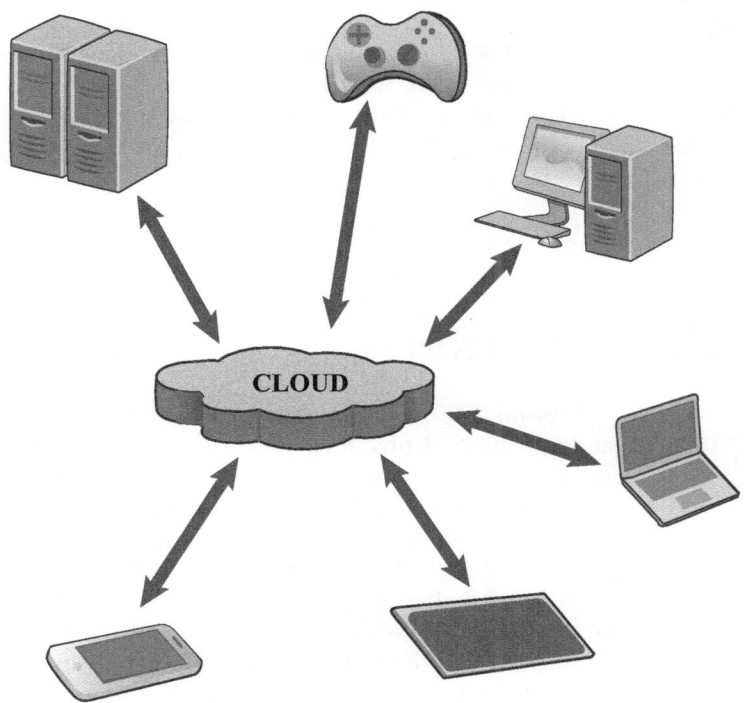

Cloud computing refers to the use of processors, memory, and other peripheral devices offsite, connected by a network to one's workstation. Use of the cloud protects data by storing it and duplicating it offsite and reduces infrastructure and personnel needs.

computers. IaaS can also give users greater computing power by allowing certain processes to run on virtual networks, rather than on the hardware of a single system. Using IaaS enables companies to create a corporate data center through third-party data centers. These third-party centers provide expert IT assistance and server resources, generally for subscription fees.

The platform as a service (PaaS) model mainly offers access to a specific platform that multiple users can use to develop software applications, or apps. Many apps require access to specific development programs. The Google App Engine and IBM's developerWorks Open provide an environment that stores, supports, and runs web apps. PaaS allows software developers to create apps without investing in infrastructure and data center support. Providers may also offer virtual storage, access to virtual networks, and other services.

The software as a service (SaaS) model offers users subscription-based or shared access to software programs through a virtual network. Adobe Systems' Creative Cloud provides access to programs such as Photoshop, Illustrator, and Lightroom for a monthly fee. Users pay a smaller amount over time rather than paying a higher cost up front to purchase the program. SaaS supports multitenancy, in which a single copy of a program is available to multiple clients. This allows software providers to earn revenue from multiple clients through a single instance of a software program.

Advantages and Disadvantages of the Cloud

Cloud networking allows small companies and individuals access to development tools, digital storage, and software that once were prohibitively expensive or required significant management and administration. By paying subscription fees, users can gain monthly, yearly, or as-used access to software or other computing tools with outsourced administration. For service providers, cloud computing is cost effective because it eliminates the cost of packaging and selling individual programs and other products.

Data security is the chief concern among those considering cloud computing. The private and hybrid cloud models provide a secure way for companies to reap the benefits of cloud computing. Firewalls and encryption are common means of securing data in these systems. Providers are working to increase the security of public clouds, thus reducing the need for private or hybrid systems.

—*Micah L. Issitt*

Bibliography

Beattie, Andrew. "Cloud Computing: Why the Buzz?" *Techopedia*. Techopedia, 30 Nov. 2011. Web. 21 Jan. 2016.

Huth, Alexa, and James Cebula. *The Basics of Cloud Computing*. N.p.: Carnegie Mellon U and US Computer Emergency Readiness Team, 2011. PDF file.

Kale, Vivek. *Guide to Cloud Computing for Business and Technology Managers*. Boca Raton: CRC, 2015. Print.

Kruk, Robert. "Public, Private and Hybrid Clouds: What's the Difference?" *Techopedia*. Techopedia, 18 May 2012. Web. 21 Jan. 2016.

Rountree, Derrick, and Ileana Castrillo. *The Basics of Cloud Computing*. Waltham: Elsevier, 2014. Print.

Ryan, Janel. "Five Basic Things You Should Know about Cloud Computing." *Forbes*. Forbes.com, 30 Oct. 2013. Web. 30 Oct. 2013.

Sanders, James. "Hybrid Cloud: What It Is, Why It Matters." *ZDNet*. CBS Interactive, 1 July 2014. Web. 10 Jan. 2016.

CODING AND ENCRYPTION

ABSTRACT

Mathematical algorithms are used in modern encryption and decryption.

OVERVIEW

Human beings have a propensity to preserve and share secret information. Cryptography, from the Greek *kryptos* (hidden) and *graphein* (to write), is the art and science of coding and decoding messages containing secret information. Encryption is the algorithmic process that converts plain-text into cipher-text (looks like a collection of unintelligible symbols), while decryption is the reverse process that converts the cipher-text back to the original plain-text. A cipher algorithm and its associated key control both directions of the sequence, with the code's security level directly related to the algorithm's complexity. The two fundamental types of cryptography are symmetric (or secret keys) or asymmetric (or public-key), with multiple variations. Claude Shannon, an American mathematician and electronic engineer, is known as the father of information theory and cryptography. Some claim that his master's thesis, which demonstrates that electrical applications of Boolean algebra can construct and resolve any logical numerical relationship, is the most important master's thesis of all time.

Around 2000 bce, Egyptian scribes included nonstandard hieroglyphs in carved inscriptions. During war campaigns, Julius Caesar sent coded information to Roman generals. Paul Revere's signal from a Boston bell tower in 1775 is even a simple example of a coded message. Success of the Allies in both World Wars depended on their breaking of the German's Enigma code. With the world-wide need for more sophisticated coding algorithms to transmit secure messages for military forces, businesses, and governments, people began capitalizing on the combined powers of mathematics, computer technology, and engineering.

The simplest examples of ciphers involve either transpositions or substitutions. In 450 bce, the Spartans used transposition ciphers when they wound a narrow belt spirally around a thick staff and wrote a plain-text (or message) along the length of the rod. Once unwound, the belt appeared to be a meaningless sequence of symbols. To decipher the cipher-text, the receiver wound the belt around a similar staff. Variations of transposition ciphers are the route cipher and the Cardan grill.

Julius Caesar used substitution ciphers, where each letter of the plain-text is replaced by some other letter or symbol, using a substitution dictionary. For example, suppose:

Original Alphabet:

| A | B | C | D | E | F | G | H | I | J | K | L | M | N | O | P | Q | R | S | T | U | V | W | X | Y | Z |

Key Dictionary:

| K | L | M | N | O | P | Q | R | S | T | U | V | W | X | Y | Z | A | B | C | D | E | F | G | H | I | J |

A man writing codes on his laptop

where the key dictionary is made by starting with "code" letter K and then writing the alphabet as if on a loop. To encode the plain-text, "The World Is Round," each letter is substituted by its companion letter, producing the cipher-text "CRO FXAUN SB AXDWN." To disguise word lengths and to add complexity, the cipher-text was sometimes blocked into fixed-length groups of letters such as "CROF XAUN SBAX DWN." To decipher the cipher-text, one needed to know only the "code" letter. Though simple and initially confusing, substitution ciphers now are easily broken using frequency patterns of letters and words. Variations of the substitution cipher involve the suppression of letter frequencies, syllabic substitutions, or polyalphabetic substitutions such as the Vigenère or Beaufort ciphers.

The Playfair Square cipher used by Great Britain in World War I is a substitution cipher, but its encryption of letter pairs in place of single letters is more powerful yet easy to use. The cipher-key is a 5 × 5 table initiated by a key word, such as "mathematics."

Exchange Table Model

M	A	T	H	E
I	C	S	B	D
F	G	K	L	N
O	P	Q	R	U
V	W	X	Y	Z

The table is built by moving left to right and from top to bottom (or other visual pattern as in a spiral) by first filling in the table's cells with the keyword's letters—avoiding duplicate letters. Then, the subsequent cells are filled with the remaining letters of the alphabet, using the "I" to represent the "J" to reduce the alphabet to 25 letters (instead of 26). Both the coder and the decoder need to know the both the keyword and the conventions used to construct the common cipher-key.

The coder first breaks the plain-text into two-letter pairs and uses the cipher-key via a system of rules:

If double letters occur in the plain-text, insert an X between them.

Rewrite the plain-text as a sequence of two-letter pairs, using an X as a final filler for last letter-pair.

If the two letters lie in the same row, replace each letter by the letter to its right (for example, CS becomes SB).

If the two letters lie in the same column, replace each letter by the letter below it (TS becomes SK and PW becomes WA).

If the two letters lie at corners of a rectangle embedded in the square, replace them by their counterpart in the same rectangle (TB becomes SH and CR becomes PB).

Using this cipher-key, the plain-text "The World Is Round" becomes first

33

TH EW OR LD IS RO UN DX

which when encoded, becomes

HE ZA PU BN CB UP ZU ZS.

The same cipher-key is used to decode this message, but the rules are interpreted in reverse. It is quite difficult to decode this cipher-text without access to both the keyword and the conventions to construct the common cipher-key, though very possible.

The problem with all substitution and transposition encryption systems is their dependence on shared secrecy between the coders and the intended decoders. To transmit plain-text via cipher-text and then decode it back to public-text successfully, both parties would have to know and use common systems, common key-words, and common visual arrangements. In turn, privacy is required, since these systems are of no value if the user learns the key-word or is able to use frequency techniques of word/letter patterns to break the code. A more complicated and secure encryption process was needed, but it was not invented until the 1970s.

The revolutionary idea in encryption was the idea of a public key system, where the encryption key is known by everyone (that is, the public). However, the twist was that this knowledge was not useful in figuring out the decryption key, which was not made public. The RSA public-key cipher, invented in 1977 by Ronald Rivest, Adi Shamir, and Leonard Adleman ("RSA" stands for the names of the inventors), all of whom have bachelor's degrees in mathematics and advanced degrees in computer science, is still used today thanks to powerful mathematics and powerful computer systems.

In a RSA system, the "receiver" of the intended message is the driver of the process. In lieu of the "sender," the receiver chooses both the encryption key and the matching decryption key. In fact, the "receiver" can make the encryption key public in a directory so any "sender" can use it to send secure messages, which only the "receiver" knows how to decrypt. Again, the latter decryption process is not even known by the "sender."

Because the problem is quite complex and uses both congruence relationships and modular arithmetic, only a sense of the process can be described as follows:

As the "receiver," start with the product n equal to two very large prime numbers p and q.

Choose a number e relatively prime to $(p-1)(q-1)$.

The published encryption key is the pair (n,e).

Change plain-text letters to equivalent number forms using a conversion such as $A=2$, $B=3$, $C=4,\ldots$, $Z=27$.

Using the published encryption key, the "sender" encrypts each number z using the formula $m \equiv z^e \mod(n)$, with the new number sequence being the cipher-text.

To decode the text, the "receiver" not only knows both e and the factors of n but also the large primes p and q as prime factors of n.

Then, the decryption key d is private but can be computed by the "receiver" using an inverse relationship $ed \equiv 1 \mod(p-1)(q-1)$, which allows the decoding of the encrypted number into a set of numbers that can be converted back into the plain-text.

The RSA public key system works well, but the required primes p and q have to be very large and often involve more than 300 digits. If they are not large, powerful computers can determine the decryption key d from the given encryption key (n,e) by factoring the number n. This decryption is possible because of the fact that, while computers can easily multiply large numbers, it is much more difficult to factor large numbers on a computer.

Regardless of its type, a cryptographic system must meet multiple characteristics. First, it must reflect the user's abilities and physical context, avoiding extreme complexity and extraneous physical apparatus. Second, it must include some form of error checking, so that small errors in composition or transmission do not render the message into meaningless gibberish. Third, it must ensure that the decoder of the cipher-text will produce a single, meaningful plain-text. There are many mathematicians working for government agencies like the National Security Agency (NSA), as well as for private companies that are developing improved security for storage and transmission of digital information using these principles. In fact, the NSA is the largest employer of mathematicians in the United States.

—*Jerry Johnson*

Bibliography

Churchhouse, Robert. *Codes and Ciphers*. Cambridge, England: Cambridge University Press, 2002.

Kahn, David. *The Codebreakers: The Story of Secret Writing*. New York: Macmillan, 1967.

Lewand, Robert. *Cryptological Mathematics*. Washington, DC: Mathematical Association of America, 2000.

Smith, Laurence. *Cryptography: The Science of Secret Writing*. New York: Dover Publications, 1971.

COLOR CODING

FIELDS OF STUDY

Programming Methodologies; Software Development

ABSTRACT

Many text editors display different elements of a computer programming or markup language in different colors. The goal of color coding is to make it easier for programmers to write, read, and understand source code. Syntax highlighting color-codes parts of the syntax of a programming language, such as keywords. Semantic highlighting color-codes other aspects, such as variables, based on the code content.

PRINCIPAL TERMS

- **annotation:** in computer programming, a comment or other documentation added to a program's source code that is not essential to the execution of the program but provides extra information for anybody reading the code.
- **class:** in object-oriented programming, a category of related objects that share common variables and methods.
- **command:** an instruction given by a user to a computer that directs the computer to perform a specific, predefined task.
- **method:** in object-oriented programming, a procedure or function specific to objects of a particular class.
- **syntax:** in computer programming, the rules that govern how the elements of a programming language should be arranged.
- **variable:** in computer programming, a symbolic name that refers to data stored in a specific location in a computer's memory, the value of which can be changed.

Coloring Code for Clarity

Computer code is, first and foremost, a set of commands intended for a computer. In practice, however, the process of writing software requires programmers to spend time reading their code and their colleagues' code. Reading and understanding code can be daunting and time-consuming.

In order to facilitate the process of reading code, text editors and integrated development environments (IDEs) began incorporating visual aids in the late 1980s. Among these aids is syntax highlighting, a technique that assigns various colors (and sometimes fonts and styles) to important elements of the code's syntax. For example, numbers could be represented in green and keywords in blue.

The goal of color coding is to reduce the amount of time it takes programmers to understand the purpose of a segment of code. When a program's source code is displayed in one color, programmers must read each line carefully for full comprehension. Color coding aims to allow programmers to skim like-colored words and patterns and to help identify errors.

Studies examining the effects of color coding on comprehension have had mixed findings. A 2006 Finnish experiment found no benefit to color coding when programmers searched code by sight. Such searches are crucial to comprehension and debugging. By contrast, two small Cambridge University experiments from 2015 found that colored code significantly reduced code-writing time among novice programmers but less so among experienced ones. Interestingly, the Finnish subjects favored colored code even when its effectiveness was not shown to be significant.

Approaches to Color Coding

For a text editor or IDE to use color coding, it must know which language a given segment of code was

```
<!--ClickTail Top part-->
<div id="content">...</div>
<script type="text/javascript">
var WRInitTime=(new Date()).getTime();
</script>
<!--ClickTail end of Top part-->
<!--end wrapperA-->
<!--ClickTail Bottom part-->
<div id="ClickTailDiv" style="display: none;"> </div>
<script type="text/javascript">...</script>
<script src="http://s.clicktail.net/WRc9.is" type="text/javascript"> </script>
<script type="text/javascript">...</script>
<!--ClickTail end of Bottom part-->
<span style="height: 20px; width: 40px; position: absolute; opacity: 1; z-index: 8
```

Using colored font in code does not change the function of the code itself, but it does make the code easier to read. This block of code has separate colors for commands, variables, functions, strings, and comments; a coder can use sight to quickly find particular elements.

```
public class Foo { // Program by Joe
   public static void main(String[] args){
      String helloText = "Hello world";
      String goodbyeText = "Goodbye world";
      System.out.println(helloText);
      System.out.println(goodbyeText);
   }
}
```

written in and must understand the syntax rules for that language. Some editors support only a single programming language. Such editors assume that all code they encounter is in the language they support. More versatile editors examine file extensions and the content of source code files to determine which language is being used.

Some editors rely entirely upon built-in knowledge about the syntax of certain languages for color coding. More versatile editors allow for the installation of language extensions or plug-ins. This makes it possible to use highlighting with languages about which the editor had no prior knowledge.

Some text editors allow programmers to customize the details of their color coding. This may include choices about which elements to color and which colors to use.

COMMONLY COLORED ELEMENTS

Standard approaches to syntax highlighting assign colors to a range of syntactical elements, often including the following:

> **SAMPLE PROBLEM**
>
> Consider the following code segment, written in the Java programming language. Even those who are unfamiliar with Java can appreciate how the coloring might make it easier to read the code.
>
> Identify which colors in the segment correspond to the language elements listed below. For each element, list as many colors as apply.
>
> Class names
> Comments
> Language keywords
> Variable names
> Literal values
> Method names
>
> **Answer:**
> Dark blue
> Dark green
> Purple
> Orange, grey, and black
> Red
> Bright green

- Keywords (such as for, true, or if)
- Literal values and constants (such as "Hello world," 3.14)
- Matching paired brackets (if the cursor is adjacent to such a character)
- Comments and annotations (such as //fix this)

A powerful extension of syntax highlighting, called "semantic highlighting," attempts to make further coloring decisions based on the meaning of the code. Semantic highlighting may expand color coding to the names of variables (sometimes using a different color for each), classes, and methods.

COLOR CODING IN THE REAL WORLD

Color coding has become widely used in computer programming. Not only is it supported by most major code editors, but code-sharing websites have adopted the technique as well. Almost anywhere code is displayed, it is in color. Although the effectiveness of color coding remains under debate, colorful code appears to have a firm foothold in the world of programming.

Each new language attempts to fix inefficiencies of prior languages, and with each IDE update comes new features aimed at further streamlining the task of programming. At each step, programmers spend less time puzzling over tools such as visual aids and more time imagining elegant solutions to difficult problems.

—*Joel Christophel and Daniel Showalter, PhD*

BIBLIOGRAPHY

Beelders, Tanya R., and Jean-Pierre L. Du Plessis. "Syntax Highlighting as an Influencing Factor When Reading and Comprehending Source Code." *Journal of Eye Movement Research*, vol. 9, no. 1, 2016, pp. 2207–19.

Deitel, Paul J., et al. Preface. *Android: How to Program, Global Edition*. 2nd ed., Pearson, 2015, pp. 19–30.

Dimitri, Giovanna Maria. "The Impact of Syntax Highlighting in Sonic Pi." *Psychology of Programming Interest Group*, 2015, www.ppig.org/sites/default/files/2015-PPIG-26th-Dimitri.pdf. Accessed 22 Feb. 2017.

Everitt, Paul. "Make Sense of Your Variables at a Glance with Semantic Highlighting." *PyCharm Blog*, JetBrains, 19 Jan. 2017, blog.jetbrains.com/pycharm/2017/01/make-sense-of-your-variables-at-a-glance-with-semantic-highlighting/. Accessed 17 Feb. 2017.

Hakala, Tuomas, et al. "An Experiment on the Effects of Program Code Highlighting on Visual Search for Local Patterns." *18th Workshop of the Psychology of Programming Interest Group*, U of Sussex, Sept. 2006, www.ppig.org/papers/18th-hakala.pdf. Accessed 7 Mar. 2017.

Kingsley-Hughes, Adrian, and Kathie Kingsley-Hughes. *Beginning Programming*. Wiley Publishing, 2005.

Sarkar, Advait. "The Impact of Syntax Colouring on Program Comprehension." *Psychology of Programming Interest Group*, 2015, www.ppig.org/sites/default/files/2015-PPIG-26th-Sarkar.pdf. Accessed 17 Feb. 2017.

COMBINATORICS

FIELDS OF STUDY

Information Technology; Algorithms; System Analysis

ABSTRACT

Combinatorics is a branch of mathematics that is concerned with sets of objects that meet certain conditions. In computer science, combinatorics is used to study algorithms, which are sets of steps, or rules, devised to address a certain problem.

PRINCIPAL TERMS

- **analytic combinatorics:** a method for creating precise quantitative predictions about large sets of objects.
- **coding theory:** the study of codes and their use in certain situations for various applications.
- **combinatorial design:** the study of the creation and properties of finite sets in certain types of designs.
- **enumerative combinatorics:** a branch of combinatorics that studies the number of ways that certain patterns can be formed using a set of objects.
- **graph theory:** the study of graphs, which are diagrams used to model relationships between objects.

BASICS OF COMBINATORICS

Combinatorics is a branch of mathematics that studies counting methods and combinations, permutations, and arrangements of sets of objects. For instance, given a set of fifteen different objects, combinatorics studies equations that determine how many different sets of five can be created from the original set of fifteen. The study of combinatorics is crucial to the study of algorithms. Algorithms are sets

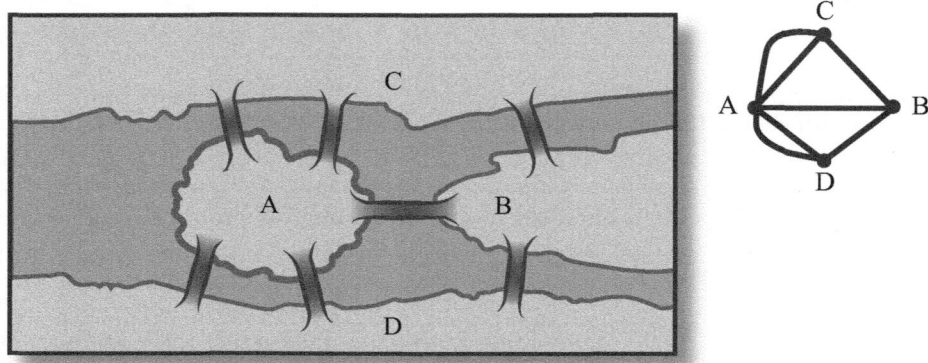

The Konigsberg bridge problem is a common example of combinatorial structures used to identify and quantify possible combinations of values. Each landmass becomes a vertex or node, and each bridge becomes an arc or edge. In computer science, these graphs are used to represent networks or the flow of computation.

of rules, steps, or processes that are linked together to address a certain problem.

THE SCIENCE OF COUNTING AND COMBINATIONS

Combinatorics is often called the "science of counting." It focuses on the properties of finite sets of objects, which do not have infinite numbers of objects and so are theoretically countable. The process of describing or counting all of the items in a specific set is called "enumeration." Combinatorics also includes the study of combinations, a process of selecting items from a set when the order of selection does not matter. Finally, combinatorics also studies permutations. Permutations involve selecting or arranging items in a list, when the order of arrangement is important. Combinatorics also studies the relationships between objects organized into sets in various ways.

There are numerous subfields of combinatorics used to study sets of objects in different ways. Enumerative combinatorics is the most basic branch of the field. It can be described as the study of counting methods used to derive the number of objects in a given set. By contrast, analytic combinatorics is a subfield of enumerative combinatorics. It deals with predicting the properties of large sets of objects, using quantitative analysis. All combinatorics analysis requires detailed knowledge of calculus. Many subfields make extensive use of probability theory and predictive analysis.

COMBINATORICS APPLICATIONS

There are many different applications for combinatorics in analytic mathematics, engineering, physics, and computer science. Among the most familiar basic examples of combinatorics is the popular game sudoku. The game challenges players to fill in the blanks in a "magic square" diagram with specific column and row values. Sudoku puzzles are an example of combinatorial design, which is a branch of combinatorics that studies arrangements of objects that have symmetry or mathematical/geometric balance between the elements.

Combinatorics is also crucial to graph theory. Graph theory is a field of mathematics that deals with graphs, or representations of objects in space. Graphs are used in geometry, computer science, and other fields to model relationships between objects. In computer science, for instance, graphs are typically used to model computer networks, computational flow, and the structure of links within websites. Combinatorics is used to study the enumeration of graphs. This can be seen as counting the number of different possible graphs that can be used for a certain application or model.

COMBINATORICS IN ALGORITHM DESIGN

In computer science, combinatorics is used in the creation and analysis of algorithms. Algorithms are sets of instructions, computing steps, or other processes that are linked together to address a certain computational problem. As algorithms are essentially sets, the steps within algorithms can be studied using combinatorial analysis, such as enumeration or permutation. As combinatorics can help researchers to find more efficient arrangements of objects, or sets, combinatorial analysis is often used to test and assess the efficiency of computer algorithms. Combinatorics is also key to the design of sorting algorithms. Sorting algorithms allow computers to sort objects (often pieces of data, web pages, or other informational elements), which is a necessary step in creating effective algorithms for web searching.

COMBINATORICS IN CODING

Combinatorics is also used in coding theory, the study of codes and their associated properties and characteristics. Codes are used for applications, including cryptography, compressing or translating data, and correcting errors in mathematical, electrical, and information systems. Coding theory emerged from combinatorial analysis. These two branches of mathematics are distinct but share theories and techniques.

Combinatorics is an advanced field of study. The arrangement, organization, and study of relationships between objects provides analytical information applicable to many academic and practical fields. Combinatorics influences many aspects of computer design and programming including the development of codes and the precise study of information and arrangement of data.

—*Micah L. Issitt*

BIBLIOGRAPHY

Beeler, Robert A., *How to Count: An Introduction to Combinatorics*. New York: Springer, 2015. Print.

"Combinatorics." *Mathigon*. Mathigon, 2015. Web. 10 Feb. 2016.

Faticoni, Theodore G., *Combinatorics: An Introduction*. New York: Wiley, 2014. Digital file.

Guichard, David. "An Introduction to Combinatorics and Graph Theory." *Whitman*. Whitman Coll., 4 Jan 2016. Web. 10 Feb. 2016.

Roberts, Fred S., and Barry Tesman. *Applied Combinatorics*. 2nd ed. Boca Raton: Chapman, 2012. Print.

Wagner, Carl. "Choice, Chance, and Inference." *Math.UTK.edu*. U of Tennessee, Knoxville, 2015. Web. 10 Feb. 2016.

COMMENT PROGRAMMING

FIELDS OF STUDY

Software Development; Coding Techniques; Computer Science

ABSTRACT

Comments are annotations that are used to explain the purpose and structure of the code used to create computer programs. The use of appropriate comments makes it more efficient to create, debug, and maintain computer programs and enables programmers to work together more effectively.

PRINCIPAL TERMS

- **annotation:** in computer programming, a comment or other documentation added to a program's source code that is not essential to the execution of the program but provides extra information for anybody reading the code.
- **class:** in object-oriented programming, a category of related objects that share common variables and methods.
- **command:** an instruction given by a user to a computer that directs the computer to perform a specific, predefined task.
- **method:** in object-oriented programming, a procedure or function specific to objects of a particular class.
- **syntax:** in computer programming, the rules that govern how the elements of a programming language should be arranged.
- **variable:** in computer programming, a symbolic name that refers to data stored in a specific location in a computer's memory, the value of which can be changed.

Using Comments to Annotate Code

Anyone familiar with the syntax of a computer programming language can read the code itself to gain a basic understanding what a statement or series of statements does. However, the code itself cannot reveal to a programmer why a statement or series of statements was used or the overall purpose of an entire program or a section of code. This can make computer programs difficult to understand, debug, and maintain. To address this problem, programmers use comments to annotate the code that they create.

Comments can be used to annotate many types of computer code. For example, an explanation of why a certain command was used can be provided in a comment. Comments can also be used to describe functions, subroutines, classes, and methods, and to explain why variables are used.

Comments are written in plain language, not in the syntax of a programming language. However, each programming language provides its own rules that allow the program to distinguish properly marked comments from executable lines of code and to ignore them. For example, a comment might be distinguished by using two forward slashes before the text that makes up the comment, as follows:

// This is a comment.

How to Create Useful Comments

There is no good substitute for annotation. Although this takes time, the time spent annotating code can greatly reduce the time spent on debugging and maintenance. Comments may be added before, while, or after the code is written. Updating comments to reflect changes in the code is just as critical as adding them at the start.

Because annotation takes valuable programming time, learning to create short but effective and useful comments is an important skill. Comments should not simply restate what a line or block of code does, but rather focus on explaining why the code does what it does and how it relates to the overall structure and function of the program. Consider the following comment associated with the declaration for the variable *age*.

// Declare an integer variable named age.

This comment does not convey any meaning or information that could not be determined from the code statement itself. A more effective comment could be written as follows.

// Variable stores applicant age for use in determining eligibility status for
// Social Security benefits.

This comment explains succinctly why the variable is being declared and how the variable functions within the program.

Comments are expected in several places. An entire program typically has a comment at the beginning. Called a "header comment," this states the author, the date of creation, the date of last update, the purpose of the program, and its function. Each function should be called out, as should any single line of code that is not intuitive. Too many in-line comments may indicate that the code is overly complex. Or, it may mean that the names for variables, constants, functions, and the like are not descriptive enough.

Using Comments to Improve Software Development

Creating effective and useful comments is extremely important in real-world programming environments, where teams of programmers often work together on complex applications. For example, an application used to track client data for a major insurer might consist of tens of thousands of lines of code containing numerous variables, methods, classes, commands, and other statements. Such projects are usually developed over many weeks, months, or even years by dozens of programmers. Thus, proper use of comments allows programmers to quickly understand the purpose of and interactions among lines of code written by other programmers and at other times. This greatly increases the efficiency with which applications can be developed and updated. Consider the following code example in the Visual Basic language:

```
Dim y as integer
Dim t as integer
y = GetYesterdaysTotal()
t = GetTodaysTotal()
n = y + t
echo z
```

While that example may be syntactically correct, the statements themselves do little to explain the purpose of the code. However, comments and clearer names can be used to make the purpose of the code clearer, as follows.

```
// Declare a variable eggsYesterday to hold the
number of eggs laid yesterday.
Dim eggsYesterday as integer
// Declare a variable eggsToday to hold the
number of eggs laid today.
Dim eggsToday as integer
// Retrieve total number of eggs laid yesterday
from Egg1 database.
eggsYesterday = GetYesterdaysTotal()
// Retrieve total number of eggs laid today from
Egg1 database.
eggsToday = GetTodaysTotal()
// Calculate total number of eggs laid in past two
days.
eggsNow = eggsYesterday + eggsToday
// Display total number of eggs laid in past two days.
echo eggsNow
```

Along with comments, adding line breaks and indenting nested blocks of code make the code easier to read and understand.

Comments Make Good Code More Useful

Programmers focus on writing effective code that has correct syntax. However, the inherent limitations of programming languages make it difficult to describe code fully without using comments, even if the program has few functions and a single author. Larger projects that provide complex functionality and involve multiple programmers working together are exponentially more difficult to manage. The consistent use of appropriate comments provides significant benefits to programmers and results in greater efficiency, lower costs, and more robust and effective code.

—*Maura Valentino, MSLIS*

Bibliography

Friedman, Daniel P., and Mitchell Wand. *Essentials of Programming Languages*. 3rd ed., MIT P, 2008.

Haverbeke, Marijn. *Eloquent JavaScript: A Modern Introduction to Programming*. 2nd ed., No Starch Press, 2014.

MacLennan, Bruce J. *Principles of Programming Languages: Design, Evaluation, and Implementation*. 3rd ed., Oxford UP, 1999.

Schneider, David I. *An Introduction to Programming using Visual Basic*. 10th ed., Pearson, 2016.

Scott, Michael L. *Programming Language Pragmatics*. 4th ed., Morgan Kaufmann Publishers, 2016.

Van Roy, Peter, and Seif Haridi. *Concepts, Techniques, and Models of Computer Programming*. MIT P, 2004.

COMPARISON OPERATORS

FIELDS OF STUDY

Software Development; Coding Techniques; Computer Science

ABSTRACT

Comparison operators compare two values in an expression that resolves to a value of true or false. The main comparison operators are equal to, not equal to, greater than, greater than or equal to, less than, and less than or equal to. Comparison operators are used in conditional expressions to determine if one block of code or another executes, thus controlling flow in a computer program. They thereby support complex decision making in computer programs.

PRINCIPAL TERMS

- **array:** a data structure holding a fixed sized collection of values or variables of the same type.
- **Boolean operator:** one of a predefined set of words—typically AND, OR, NOT, and variations—that is used to combine expressions that resolve to either-or values.
- **conditional expression:** a statement executed based on whether a condition is found to be true or false.

True Example	False Example	Statement	Comparison Operator	Description
int $x = 2$	int $x = 3$	$x == 2$	==	is equal to
int $x = 3$	int $x = 2$	$x != 2$	!=	is not equal to
int $x = 2$	int $x = 13$	$x < 13$	<	is less than
int $x = 2$	int $x = 3$	$x <= 2$	<=	is less than or equal to
int $x = 13$	int $x = 2$	$x > 2$	>	is greater than
int $x = 13$	int $x = 2$	$x >= 3$	>=	is greater than or equal to

Operators used to compare the value of a variable (x) to a given value are called "comparison operators." The table provides multiple comparison operators, definitions of the operators, and statements using them. True and false examples of values for the variable are provided for each statement.

- **object:** in object-oriented programming, a self-contained module of data belonging to a particular class that shares the variables and methods assigned to that class.
- **string:** in computer programming, a data type consisting of characters arranged in a specific sequence.

UNDERSTANDING COMPARISON OPERATORS

In computer programming, comparison operators are used in conditional expressions to determine which block of code executes, thus controlling the program flow. Comparison operators compare two values in an expression that resolves to a value of true or false. In if-then-else statements, comparison operators enable different blocks of code to execute based on whether the result of the comparison is true or false. Boolean operators can be used to combine either-or expressions to create complex Boolean expressions that also resolve to true or false. This allows more complex scenarios to be tested to determine which block of code should execute.

There are six main comparison operators: equal to, not equal to, greater than, greater than or equal to, less than, and less than or equal to. Different programming languages use different syntax to express these operators, but the meanings are the same. Some languages include additional operators, such as the equal value and equal type operator (===) and the not equal value or not equal type operator (!==) in JavaScript.

Comparison operators can compare a variety of data types, including integers and strings. For example, if x is an integer variable, x is less than 5 evaluates to true if x equals 4 or less but evaluates to false if x equals 5 or more. If x and y are string variables, x equals y evaluates to true only if the strings are the same length and are composed of the same characters in the same order. Therefore, "abc" is equal to "abc" but "abc" is not equal to "bac." For some operations, only data of the same type can be compared. For example, if x is an integer and y is a string, then expression x is greater than y would return an error. Some languages, such as JavaScript and MySQL, use numeric values for the characters in a string to make the comparison.

Arrays are not typically compared using comparison operators. Different languages return different results when comparing arrays for equivalence. For example, the C language returns a pointer to the both array's location in memory and then compares the memory locations for equivalence. Thus, the contents of the arrays, their stored values, are not directly compared.

Similarly, when objects are compared with the "equal to" comparison operator in languages like JavaScript or C#, the comparison operator checks whether the objects are of the same class and does not compare the data stored within the objects. In the language PHP, however, the values are also assessed. It is thus important to be aware of the rules of the programming language in question.

USING COMPARISON OPERATORS

Comparison operators are used in computer programs in many ways. For example, personal information, stored in a database, could be compared to eligibility requirements for a program or benefit. Using the Boolean operator AND allows the program to check two conditions, both of which must be true to return a value of true.

```
IF candidateAge > 65 AND candidateAnnualIncome < 250,000 THEN
    SocialSecurityEligible()
END IF
```

They are also used to compare strings. Here another Boolean operator, OR, has been used to check two comparisons, and if either is true, the entire statement returns a value of true and the function is called:

```
IF userState = "NY" OR userState = "New York"
THEN
    NewYorkResident()
END IF
```

Comparison operators are an important development tool because of their support for the if-else conditional logic. With the added power of Boolean logic, complex decision-making structures are possible. For instance, "branching," or "selection," allows for various courses of action to be taken (that is, different else-if statements to execute) depending on the condition.

However, using comparison operators incorrectly can lead to errors that are difficult to correct. Programmers must ensure they are aware of how comparison operators behave when used to compare different types of data in the using different computer languages, paying particular attention to how arrays, objects, and strings are handled.

Comparison Operators in Practice

Comparison operators are often used in process codes. In the following example, an error code is returned from a function that updates a customer's name in a database. Comparison operators are then used to determine which error has occurred and then to transfer control to the appropriate block of code. First, the function is called and the value it returns is stored in a variable named *code*.

```
code = updateDatabase("Jane Smith")
```

Next, an if statement is used to execute different blocks of code depending on which value is returned by the function.

```
IF code = 1 THEN
    Display a message that the update was successful.
IF code = 2 THEN
    Display a message that the database was not found.
IF code = 3 THEN
    Display a message that the user was not found.
ELSE
    Display a message that an unknown error has occurred.
END IF
```

The Power of Comparison

Comparison operators are used in a wide variety of development scenarios. They are used in authentication routines, where exact comparison is crucial. For example, a user's credentials may be checked against a database of authorized users before access to the database is permitted. Comparison operators may also be used for more mundane tasks such as processing discount codes. When a user types in a discount code, it is compared to valid codes, and if the code is valid, the applicable discount is applied.

—*Maura Valentino, MSLIS*

Bibliography

Friedman, Daniel P., and Mitchell Wand. *Essentials of Programming Languages.* 3rd ed., MIT P, 2008.

Haverbeke, Marijn. *Eloquent JavaScript: A Modern Introduction to Programming.* 2nd ed., No Starch Press, 2014.

MacLennan, Bruce J. *Principles of Programming Languages: Design, Evaluation, and Implementation.* 3rd ed., Oxford UP, 1999.

Nixon, Robin. *Learning PHP, MySQL, JavaScript, CSS & HTML5: A Step-by-Step Guide to Creating Dynamic Websites.* 3rd ed., O'Reilly, 2014.

Scott, Michael L. *Programming Language Pragmatics.* 4th ed., Morgan Kaufmann Publishers, 2016.

Van Roy, Peter, and Seif Haridi. *Concepts, Techniques, and Models of Computer Programming.* MIT P, 2004.

COMPUTER ANIMATION

FIELDS OF STUDY
Digital Media; Graphic Design

ABSTRACT
Computer animation is the creation of animated projects for film, television, or other media using specialized computer programs. As animation projects may range from short, simple clips to detailed and vibrant feature-length films, a wide variety of animation software is available, each addressing the needs of animators. The computer animation process includes several key steps, including modeling, keyframing, and rendering. These stages are typically carried out by a team of animators.

PRINCIPAL TERMS

- **animation variables (avars):** defined variables that control the movement of an animated character or object.
- **keyframing:** the process of defining the first and last—or key—frames in an animated transition.
- **render farms:** large computer systems dedicated to rendering animated content.
- **3D rendering:** the process of creating a 2D animation using 3D models.
- **virtual reality:** a form of technology that enables the user to view and interact with a simulated environment.

History of Computer Animation

Since the early twentieth century, the field of animation has been marked by frequent, rapid change. Innovation in the field has been far reaching, filtering into film, television, advertising, video games, and other media. It was initially an experimental method and took decades to develop. Computer animation revitalized the film and television industries during the late twentieth and early twenty-first centuries, in many ways echoing the cultural influence that animation had decades before.

Prior to the advent of computer animation, most animated projects were created using a process that later became known as "traditional," or "cel," animation. In cel animation, the movement of characters, objects, and backgrounds was created frame by frame. Each frame was drawn by hand. This time-consuming and difficult process necessitated the creation of dozens of individual frames for each second of film.

As computer technology developed, computer researchers and animators began to experiment with creating short animations using computers. Throughout the 1960s, computers were used to create 2D images. Ed Catmull, who later founded the studio Pixar in 1986, created a 3D animation of his hand using a computer in 1972. This was the first 3D computer graphic to be used in a feature film when it appeared in *Futureworld* (1976). Early attempts at computer animation were found in live-action films. The 1986 film *Labyrinth*, for instance, notably features a computer-animated owl flying through its opening credits. As technology improved, computer animation became a major component of special effects in live-action media. While cel animation continued to be used in animated feature films, filmmakers began to include some computer-generated elements in such works. The 1991 Walt Disney Studios film *Beauty and the Beast*, for instance, featured a ballroom in one scene that was largely created using a computer.

In 1995, the release of the first feature-length computer-animated film marked a turning point in the field of animation. That film, *Toy Story*, was created by Pixar, a pioneer in computer animation. Over the following decades, Pixar and other studios, including Disney (which acquired Pixar in 2006) and DreamWorks, produced numerous computer-animated films. Computer animation became a common process for creating animated television shows as well as video games, advertisements, music videos, and other media.

In the early twenty-first century, computer animation also began to be used to create simulated environments accessed through virtual reality equipment such as the head-mounted display Oculus Rift. Much of the computer-animated content created during this time featured characters and surroundings that appeared 3D. However, some animators opted to create 2D animations that more closely resemble traditionally animated works in style.

From designing the original animation model to creating algorithms that control the movement of fluids, hair, and other complex systems, computer software has drastically changed the art of animation. Through software that can manipulate polygons, a face can be rendered and further manipulated to create a number of images much more efficiently than with hand-drawn illustrations. Thus, the detail of the imaging is increased, while the time needed to develop a full animation is reduced.

Three-Dimensional Computer Animation

Creating a feature-length computer-animated production is a complex and time-intensive process that is carried out by a large team of animators, working with other film-industry professionals. When creating a 3D computer-animated project, the animation team typically begins by drawing storyboards. Storyboards are small sketches that serve as a rough draft of the proposed scenes.

Next, animators transform 2D character designs into 3D models using animation software. They use animation variables (avars) to control the ways in which the 3D characters move, assigning possible directions of movement to various points on the characters' bodies. The number of avars used and the areas they control can vary widely. The 2006 Pixar film *Cars* reportedly used several hundred avars to control the characters' mouths alone. Using such variables gives animated characters a greater range of motion and often more realistic expressions and gestures. After the characters and objects are modeled and animated, they are combined with backgrounds as well as lighting and special effects. All of the elements are then combined to transform the 3D models into a 2D image or film. This process is known as 3D rendering.

Two-Dimensional Computer Animation

Animating a 2D computer-animated work is somewhat different from its 3D counterpart, in that it does not rely on 3D modeling. Instead, it typically features the use of multiple layers, each of which contains different individual elements. This method of animating typically features keyframing. In this procedure, animators define the first and last frames in an animated sequence and allow the computer to fill in the movement in between. This process, which in traditionally animated films was a laborious task done by hand, is often known as "inbetweening," or "tweening."

Tools

Various animation programs are available to animators, each with its own strengths and weaknesses. Some animation software, such as Maya and Cinema 4D, are geared toward 3D animation. Others, such as Adobe Flash, are better suited to 2D animation. Adobe Flash has commonly been used to produce 2D cartoons for television, as it is considered a quick and low-cost means of creating such content. Animation studios such as Pixar typically use proprietary animation software, thus ensuring that their specific needs are met.

In addition to animation software, the process of computer animation relies heavily on hardware, as many steps in the process can be taxing for the systems in use. Rendering, for example, often demands a sizable amount of processing power. As such, many studios make use of render farms, large, powerful computer systems devoted to that task.

—*Joy Crelin*

Bibliography

Carlson, Wayne. "A Critical History of Computer Graphics and Animation." *Ohio State University.* Ohio State U, 2003. Web. 31 Jan. 2016.

Highfield, Roger. "Fast Forward to Cartoon Reality." *Telegraph.* Telegraph Media Group, 13 June 2006. Web. 31 Jan. 2016.

Parent, Rick. *Computer Animation: Algorithms and Techniques.* Waltham: Elsevier, 2012. Print.

"Our Story." *Pixar.* Pixar, 2016. Web. 31 Jan. 2016.

Sito, Tom. *Moving Innovation: A History of Computer Animation.* Cambridge: MIT P, 2013. Print.

Winder, Catherine, and Zahra Dowlatabadi. *Producing Animation.* Waltham: Focal, 2011. Print.

COMPUTER MEMORY

Fields of Study

Computer Science; Computer Engineering; Information Technology

Abstract

Computer memory is the part of a computer used for storing information that the computer is currently working on. It is different from computer storage space on a hard drive, disk drive, or storage medium such as CD-ROM or DVD. Computer memory is one of the determining factors in how fast a computer can operate and how many tasks it can undertake at a time.

Principal Terms

- **flash memory:** nonvolatile computer memory that can be erased or overwritten solely through electronic signals, i.e. without physical manipulation of the device.
- **nonvolatile memory:** memory that stores information regardless whether power is flowing to the computer.
- **random access memory (RAM):** memory that the computer can access very quickly, without regard to where in the storage media the relevant information is located.
- **virtual memory:** memory used when a computer configures part of its physical storage (on a hard drive, for example) to be available for use as additional RAM. Information is copied from RAM and moved into virtual memory whenever memory resources are running low.
- **volatile memory:** memory that stores information in a computer only while the computer has power; when the computer shuts down or power is cut, the information is lost.

Overview of Computer Memory

Computer memory is an extremely important part of configuring and using computers. Many users find

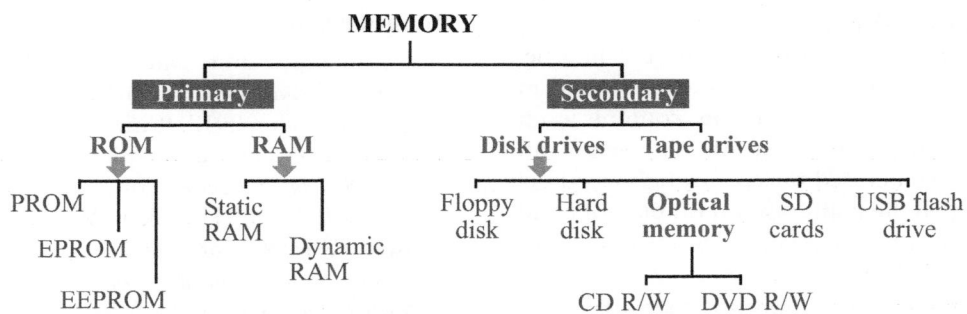

Computer memory comes in a number of formats. Some are the primary CPU memory, such as RAM and ROM; others are secondary memory, typically in an external form. External memory formats have changed over the years and have included hard drives, floppy drives, optical memory, flash memory, and secure digital (SD) memory. Adapted from the ItsAllAboutEmbedded blog.

themselves confused by the concepts of computer memory and computer storage. After all, both store information. Storage serves a very different purpose from memory, however. Storage is slower, because it uses a series of spinning platters of magnetic disks and a moving read/write head. These create a tiny magnetic field that can modify the polarity of tiny sections of the platters to record or read information. The benefit of storage is that it is nonvolatile memory, meaning that its contents remain even when the power is turned off. Computer memory, by contrast, is volatile memory because it only stores information while power is flowing through the computer.

This can be a drawback, especially when a user has vital information, such as a newly created but unsaved document, that is lost when a power surge shuts down the system. However, memory is also incredibly useful because it allows the computer to store information that it is currently working on. For example, if a user wished to update a résumé file saved on their hard drive, they would first tell the computer to access its storage to find a copy of the file. The computer would locate the file and then copy the file's contents into its volatile memory and open the document. As the user makes changes to the file, these changes are reflected only in the memory, until the user saves them to storage. Some have compared storage and memory to a librarian. Storage is like the thousands of books the librarian has in the library, which they can consult if given enough time. Memory is like the knowledge the librarian carries around in their head. It is accessible very quickly but holds a smaller amount of information than the books.

Random and Virtual Memory

One feature of memory that makes it much faster than other types of information storage is that it is a type of random access memory (RAM). Any address in the memory block can be accessed directly, without having to sort through the other entries in the memory space. This contrasts with other types of memory, such as magnetic tape, which are sequential access devices. To get to a certain part of the tape, one must move forward through all the other parts of the tape that come first. This adds to the time it takes to access that type of memory.

Memory, more than almost any other factor, determines how fast a computer responds to requests and completes tasks. This means that more memory is constantly in demand, as consumers want systems that are faster and can on more tasks at the same time. When a computer is running low on memory because too many operations are going on at once, it may use virtual memory to try to compensate. Virtual memory is a technique in which the computer supplements its memory space by using some of its storage space. If the computer is almost out of memory and another task comes in, the computer copies some contents of its memory onto the hard drive. Then it can remove this information from memory and make space for the new task. Once the new task is managed, the computer pulls the information that was copied to the hard drive and loads it back into memory.

Flash and Solid-State Memory

Some newer types of memory straddle the line between memory and storage. Flash memory, used in many mobile devices, can retain its contents even when power to the system is cut off. However, it can be accessed or even erased purely through electrical signals. Wiping all the contents of flash memory and replacing them with a different version is sometimes called "flashing" a device.

Many newer computers have incorporated solid state disks (SSDs), which are similar in many respects to flash memory. Like flash memory, SSDs have no moving parts and can replace hard drives because they retain their contents after system power has shut down. Many advanced users of computers have adopted SSDs because they are much faster than traditional computer configurations. A system can be powered on and ready to use in a matter of seconds rather than minutes.

—*Scott Zimmer, JD*

Bibliography

Biere, Armin, Amir Nahir, and Tanja Vos, eds. *Hardware and Software: Verification and Testing*. New York: Springer, 2013. Print.

Englander, Irv. *The Architecture of Computer Hardware and System Software: An Information Technology Approach*. 5th ed. Hoboken: Wiley, 2014. Print.

Kulisch, Ulrich. *Computer Arithmetic and Validity: Theory, Implementation, and Applications*. 2nd ed. Boston: De Gruyter, 2013. Print.

Pandolfi, Luciano. *Distributed Systems with Persistent Memory: Control and Moment Problems*. New York: Springer, 2014. Print.

Patterson, David A., and John L. Hennessy. *Computer Organization and Design: The Hardware/Software Interface*. 5th ed. Waltham: Morgan, 2013. Print.

Soto, María, André Rossi, Marc Sevaux, and Johann Laurent. *Memory Allocation Problems in Embedded Systems: Optimization Methods*. Hoboken: Wiley, 2013. Print.

COMPUTER MODELING

Fields of Study

Computer Science; Computer Engineering; Software Engineering

Abstract

Computer modeling is the process of designing a representation of a particular system of interacting or interdependent parts in order to study its behavior. Models that have been implemented and executed as computer programs are called computer simulations.

Principal Terms

- **algorithm:** a set of step-by-step instructions for performing computations.
- **data source:** the origin of the information used in a computer model or simulation, such as a database or spreadsheet.
- **parameter:** a measurable element of a system that affects the relationships between variables in the system.
- **simulation:** a computer model executed by a computer system.
- **system:** a set of interacting or interdependent component parts that form a complex whole.
- **variable:** a symbol representing a quantity with no fixed value.

Understanding Computer Models

A computer model is a programmed representation of a system that is meant to mimic the behavior of the system. A wide range of disciplines, including meteorology, physics, astronomy, biology, and economics, use computer models to analyze different types of systems. When the program representing the system is executed by a computer, it is called a simulation.

One of the first large-scale computer models was developed during the Manhattan Project by scientists designing and building the first atomic bomb. Early computer models produced output in the form of tables or matrices that were difficult to analyze. It was later discovered that humans can see data trends

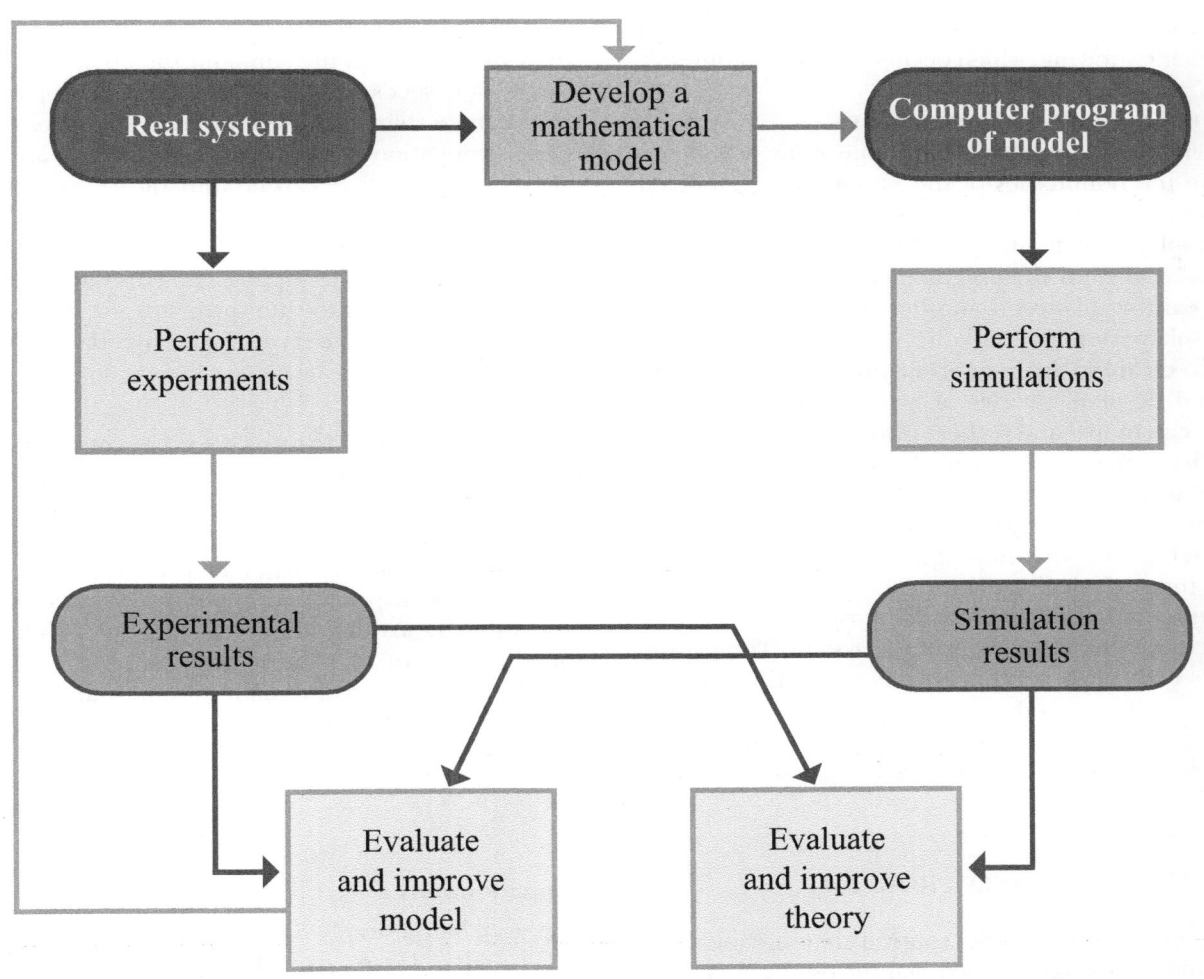

Mathematical models are used to identify and understand the details influencing a real system. Computer programs allow one to evaluate many variations in a mathematical model quickly and accurately, thus efficiently evaluating and improving models. (Adapted from Allen and Tildesly, "Computer Simulation of Liquids.")

more easily if the data is presented visually. For example, humans find it easier to analyze the output of a storm-system simulation if it is presented as graphic symbols on a map rather than as a table of meteorological data. Thus, simulations that produced graphic outputs were developed.

Computer models are used when a system is too complex or hard to study using a physical model. For example, it would be difficult to create a physical model representing the gravitational effects of planets and moons on each other and on other objects in space, although the National Aeronautics and Space Administration (NASA) has done exactly that for the Voyager space probes now leaving the solar system.

There are several different types of models. Static models simulate a system at rest, such as a building design. Dynamic models simulate a system that changes over time. A dynamic model could be used to simulate the effects of changing ocean temperatures on the speed of ocean currents throughout the year. A continuous model simulates a system that changes constantly, while a discrete model simulates a system that changes only at specific times. Some models contain both discrete and continuous elements. A farming model might simulate the effects of both weather patterns, which constantly change, and pesticide spraying, which occurs at specified times.

How Computer Models Work

To create a computer model, one must first determine the boundaries of the system being modeled and what aspect of the system is being studied. For example, if the model is of the solar system, it might be used to study the potential effect on the orbits of the existing planets if another planet were to enter the solar system.

To create such a model, a computer programmer would develop a series of algorithms that contain the equations and other instructions needed to replicate the operation of the system. Variables are used to represent the input data needed. Examples of variables that might be used for the solar system model include the mass, diameter, and trajectory of the theoretical new planet. The values that define the system, and thus how the variables affect each other, are the parameters of the system. The parameters control the outputs of the simulation when it is run. Different values can be used to test different scenarios related to the system and problem being studied. Example parameters for the solar system model might include the orbits of the known planets, their distance from the sun, and the equations that relate an object's mass to its gravity. Certain parameters can be changed to test different scenarios each time a simulation is run. Because parameters are not always constant, they can be difficult to distinguish from variables at times.

The model must also have a data source from which it will draw the input data. This data may be directly entered into the program or imported from an external source, such as a file, database, or spreadsheet.

Why Computer Models Are Important

Computer models have provided great benefits to society. They help scientists explore the universe, understand the earth, cure diseases, and discover and test new theories. They help engineers design buildings, transportation systems, power systems, and other items that affect everyday life. With the development of more powerful computer systems, computer models will remain an important mechanism for understanding the world and improving the human condition.

—Maura Valentino, MSLIS

SAMPLE PROBLEM

It is important to select appropriate variables and parameters when designing a computer model. List two variables and two parameters that might be used to determine the speed of a baseball after it has traveled three feet from the pitcher's throwing hand.

Answer:

Variables needed to determine the ball's speed might include the mass and size of the ball, the angle at which the ball is released, and the force with which it is thrown.

Parameters that define the system, which may be changed to test different scenarios, include wind speed, aerodynamic drag on the ball, and whether or not it is raining when the ball is thrown.

Bibliography

Agrawal, Manindra, S. Barry Cooper, and Angsheng Li, eds. *Theory and Applications of Models of Computation: 9th Annual Conference, TAMC 2012, Beijing, China, May 16–21, 2012*. Berlin: Springer, 2012. Print.

Edwards, Paul N. *A Vast Machine: Computer Models, Climate Data, and the Politics of Global Warming*. Cambridge: MIT P, 2010. Print.

Kojić, Miloš, et al. *Computer Modeling in Bioengineering: Theoretical Background, Examples and Software*. Hoboken: Wiley, 2008. Print.

Law, Averill M. *Simulation Modeling and Analysis*. 5th ed. New York: McGraw, 2015. Print.

Morrison, Foster. *The Art of Modeling Dynamic Systems: Forecasting for Chaos, Randomness, and Determinism*. 1991. Mineola: Dover, 2008. Print.

Seidl, Martina, et al. *UML@Classroom: An Introduction to Object-Oriented Modeling*. Cham: Springer, 2015. Print.

COMPUTER SECURITY

FIELDS OF STUDY
Information Technology; Security

ABSTRACT
The goal of computer security is to prevent computer and network systems from being accessed by those without proper authorization. It encompasses different aspects of information technology, from hardware design and deployment to software engineering and testing. It even includes user training and workflow analysis. Computer security experts update software with the latest security patches, ensure that hardware is designed appropriately and stored safely, and train users to help protect sensitive information from unauthorized access.

PRINCIPAL TERMS

- **backdoor:** a hidden method of accessing a computer system that is placed there without the knowledge of the system's regular user in order to make it easier to access the system secretly.
- **device fingerprinting:** information that uniquely identifies a particular computer, component, or piece of software installed on the computer. This can be used to find out precisely which device accessed a particular online resource.
- **intrusion detection system:** a system that uses hardware, software, or both to monitor a computer or network in to determine when someone attempts to access the system without authorization.
- **phishing:** the use of online communications to trick a person into sharing sensitive personal information, such as credit card numbers or social security numbers.
- **principle of least privilege:** a philosophy of computer security that mandates users of a computer or network be given, by default, the lowest level of privileges that will allow them to perform their jobs. This way, if a user's account is compromised, only a limited amount of data will be vulnerable.
- **trusted platform module (TPM):** a standard used for designing cryptoprocessors, which are special chips that enable devices to translate plain text into cipher text and vice versa.

HARDWARE SECURITY
The first line of defense in the field of computer security concerns the computer hardware itself. At a basic level, computer hardware must be stored in secure locations where it can only be accessed by authorized personnel. Thus, in many organizations, access to areas containing employee workstations is restricted. It may require a badge or other identification to gain access. Sensitive equipment such as an enterprise server is even less accessible, locked away in a climate-controlled vault. It is also possible to add hardware security measures to existing computer systems to make them more secure. One example of this is using biometric devices, such as fingerprint scanners, as part of the user login. The computer will only allow logins from people whose fingerprints are authorized. Similar restrictions can be linked to voice authentication, retina scans, and other types of biometrics.

Inside a computer, a special type of processor based on a trusted platform module (TPM) can manage encrypted connections between devices. This ensures that even if one device is compromised, the whole system may still be protected. Another type of security, device fingerprinting, can make it possible to identify which device or application was used to access a system. For example, if a coffee shop's wireless access point was attacked, the access point's logs could be examined to find the machine address of the device used to launch the attack. One highly sophisticated piece of security hardware is an intrusion detection system. These systems can take different forms but generally consist of a device through which all traffic into and out of a network or host is filtered and analyzed. The intrusion detection system examines the flow of data to pinpoint any attempt at hacking into the network. The system can then block the attack before it can cause any damage.

NETWORK SECURITY
Network security is another important aspect of computer security. In theory, any computer connected to the Internet is vulnerable to attack. Attackers can try to break into systems by exploiting weak points in the software's design or by tricking users into giving away

Computer security

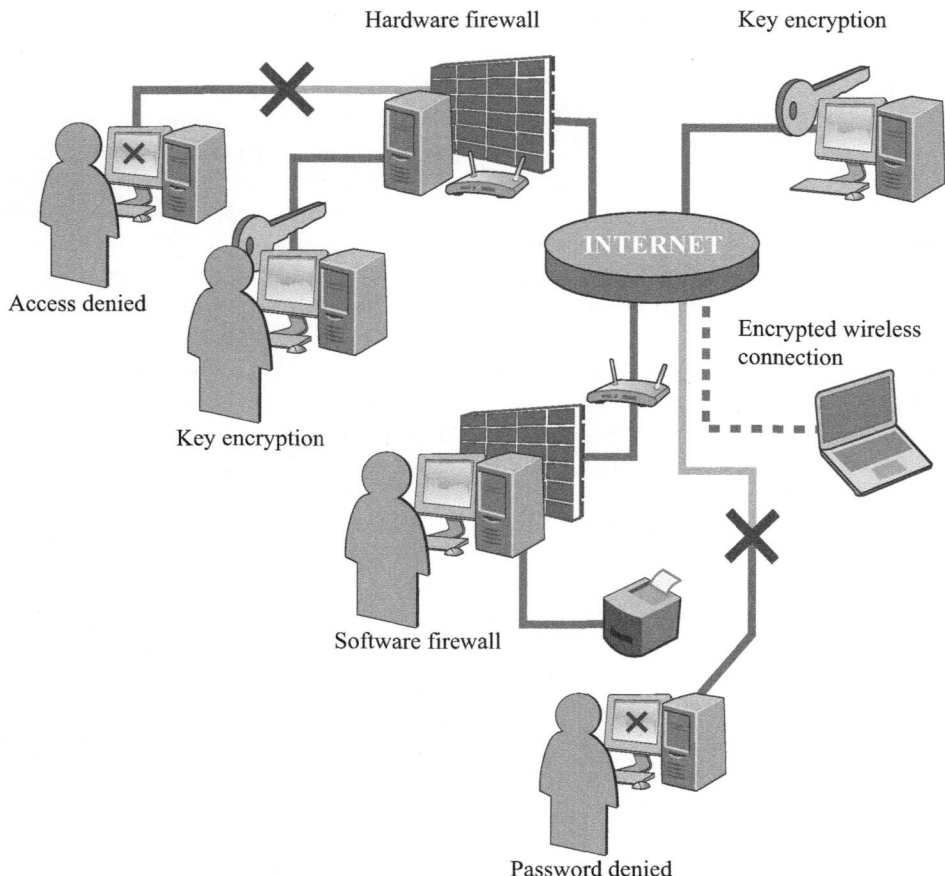

Computer security can come in many forms to ensure data and programs are protected. Passwords limit who can access a computer, key encryption limits who can read transmitted data, and firewalls can limit intrusion from the Internet.

their usernames and passwords. The latter method is called phishing, because it involves "fishing" for information. Both methods can be time consuming, however. So once a hacker gains access, they may install a backdoor. Backdoors allow easy, undetected access to a system in future.

One way of preventing attackers from tricking authorized users into granting access is to follow the principle of least privilege. According to this principle, user accounts are given the minimum amount of access rights required for each user to perform their duties. For instance, a receptionist's account would be limited to e-mail, scheduling, and switchboard functions. This way, a hacker who acquired the receptionist's username and password could not do things such as set their own salary or transfer company assets to their own bank account. Keeping privileges contained thus allows an organization to minimize the damage an intruder may try to inflict.

Software Security
Software represents another vulnerable point of computer systems. This is because software running on a computer must be granted certain access privileges to function. If the software is not written in a secure fashion, then hackers may be able to enhance the software's privileges. Hackers can then use these enhanced privileges to perform unintended functions or even take over the computer running the software. In the vernacular of hackers, this is known as "owning" a system.

A Moving Target

Computer security professionals have an unenviable task. They must interfere with the way users wish to use their computers, to make sure that hardware and software vulnerabilities are avoided as much as possible. Often, the same users whom they are trying to protect attempt to circumvent those protective measures, finding them inconvenient or downright burdensome. Computer security in these cases can become a balancing act between safety and functionality.

—*Scott Zimmer, JD*

Bibliography

Boyle, Randall, and Raymond R. Panko. *Corporate Computer Security*. 4th ed. Boston: Pearson, 2015. Print.

Brooks, R. R. *Introduction to Computer and Network Security: Navigating Shades of Gray*. Boca Raton: CRC, 2014. Digital file.

Jacobson, Douglas, and Joseph Idziorek. *Computer Security Literacy: Staying Safe in a Digital World*. Boca Raton: CRC, 2013. Print.

Schou, Corey, and Steven Hernandez. *Information Assurance Handbook: Effective Computer Security and Risk Management Strategies*. New York: McGraw, 2015. Print.

Vacca, John R. *Computer and Information Security Handbook*. Amsterdam: Kaufmann, 2013. Print.

Williams, Richard N. *Internet Security Made Easy: Take Control of Your Computer*. London: Flame Tree, 2015. Print.

Computer-Aided Design (CAD)

Computer-aided design, more commonly referred to as CAD, is the use of computer software, programs, and systems to create two-dimensional (2D) vector-based graphics and three-dimensional (3D) models that are used extensively in the architectural, design, drafting, and civil, electrical, and mechanical engineering industries. The introduction of CAD programs and software beginning in the 1980s significantly reduced manufacturing costs and expanding design and concepts capabilities within these and other fields.

Overview

CAD programs and software encompass a wide variety of applications that enable users to design not only machine parts, automobiles, buildings, and prosthetics, but also to create visual effects and computer animation, video games, and maps, and geospatial applications.

CAD was first introduced in the early 1980s and began to transform the profession of drafting by replacing the need for many manual drafting techniques used in architecture and engineering. However, many in the field credit Ivan Sutherland's 1963 MIT PhD dissertation, the computer program Sketchpad, for propelling graphics technology into the digital age. The capabilities and uses of subsequent programs grew to incorporate other design applications as computer technology improved and personal computers became more affordable. This, in turn, allowed businesses in various design fields the opportunity to reduce overall production costs by enabling fewer people to accomplish the work of many in a fraction of the time compared to traditional drafting, design, and engineering techniques.

The value of this technology to the industry goes beyond cost efficiency and productivity. The advanced technology associated with these systems has improved manufacturing and construction techniques, increasing the stability and safety of products and buildings, with increased accuracy and with fewer errors introduced into the process. This improved accuracy has also allowed designers, engineers, and architects to look at improving the materials used in a project, because they can determine how the material and structure will act in a 3D model prior to construction. CAD has also expanded the possibilities of construction with new shapes and complex geometrical designs that would have been cost-prohibitive to carry out via traditional techniques, such as hand-drawing and models. Many professionals credit

Located at Chambersburg, Pennsylvania. A view of the Computer Aided Design/Computer Aided Manufacturing (CAD/CAM) system central processing facility and graphic plotter at Letterkenny Army Depot.

CAD for enhancing the design process by encouraging architects, designers, and engineers to conceptualize nontraditional shapes that can be applied to their products and structures while simultaneously improving the accuracy needed to manufacture or build the resulting design.

CAD software comes in many varieties, including many that are available as open source and free software. One such program is SketchUp, which was released in the summer of 2000 and won the Community Choice Award that same year. A free version of SketchUp that combines 3D modeling techniques with a geolocation component allows multiple users to download others' designs and contribute to large multigroup projects associated with Google Earth. Widely used CAD software packages include CADKEY, CATIA, IEAS-MS, Inventor, Pro-E, SolidEdge, SolidWorks, and Unigraphics.

—*Laura L. Lundin, MA*

Bibliography

Bethune, James. *Engineering Graphics with AutoCAD 2014*. San Francisco: Peachpit, 2013. Print.

Farin, G. E. *Curves and Surfaces for Computer-Aided Geometric Design: A Practical Guide*. 4th ed. San Diego: Academic, 1997. Print.

Innovation in Produce Design: From CAD to Virtual Prototyping. Eds. Monica Bordegoni and Caterina Rizze. London: Springer, 2001. Print.

Rao, P. N. *CAD/CAM: Principles and Applications*. 3rd Ed. India: Tata McGraw, 2010. Print.

Stojkovic, Zlatar. *Computer-Aided Design in Power Engineering: Application of Software Tools*. New York: Springer, 2012. Print.

Sutherland, Ivan. Sketchpad: A Man-Machine Graphical Communication System. Diss. MIT. Cambridge: MIT, 1963. Print.

Thakur, A. A. G. Banerjee, and S. K. Gupta. "A Survey of CAD Model Simplification Techniques for Physics-Based Simulation Applications." *CAD Computer Aided Design* 41.2 (2009): 65–80. Print.

Yu, Cheng, and Song Jia. *Computer Aided Design: Technology, Types, and Practical Applications*. New York: Nova, 2012. *eBook Collection (EBSCOhost)*. Web. 5 June 2015.

COMPUTER-AIDED DESIGN AND COMPUTER-AIDED MANUFACTURING SOFTWARE (CAD/CAM)

FIELDS OF STUDY

Applications; Graphic Design

ABSTRACT

Computer-aided design (CAD) and computer-aided manufacturing (CAM) are software that enable users to design products and, through the use of computer-guided machinery, manufacture them according to the necessary specifications. CAD/CAM programs are used in a wide range of industries and play a key role in rapid prototyping, a process that allows companies to manufacture and test iterations of a product.

PRINCIPAL TERMS

- **four-dimensional building information modeling (4D BIM):** the process of creating a 3D model that incorporates time-related information to guide the manufacturing process.
- **rapid prototyping:** the process of creating physical prototype models that are then tested and evaluated.
- **raster:** a means of storing, displaying, and editing image data based on the use of individual pixels.
- **solid modeling:** the process of creating a 3D representation of a solid object.
- **vector:** a means of storing, displaying, and editing image data based on the use of defined points and lines.

APPLICATIONS OF CAD/CAM

The term "CAD/CAM" is an acronym for "computer-aided design" and "computer-aided manufacturing." CAD/CAM refers collectively to a wide range of computer software products. Although CAD software and CAM software are considered two different types of programs, they are frequently used in concert and thus associated strongly with each other. Used primarily in manufacturing, CAD/CAM software enables users to design, model, and produce various objects—from prototypes to usable parts—with the assistance of computers.

CAD/CAM software originated in the 1960s, when researchers developed computer programs to assist professionals with design and modeling. Prior to that point, designing objects and creating 3D models was a time-consuming process. Computer programs designed to aid with such tasks represented a significant time savings. By the late 1960s, CAD programs began to be used alongside early CAM software.

Computer-Aided Design

Using CAD/CAM software is a two-part process that begins with design. In some cases, the user begins designing an object by using CAD software to create 2D line drawings of the object. This process is known as "drafting." He or she may then use tools within the CAD software to transform those 2D plans into a 3D model. As CAD/CAM is used to create physical objects, the modeling stage is the most essential stage in the design process. In that stage, the user creates a 3D representation of the item. This item may be a part for a machine, a semiconductor component, or a prototype of a new product, among other possibilities.

In some cases, the user may create what is known as a "wire-frame model," a 3D model that resembles the outline of an object. However, such models do not include the solid surfaces or interior details of the object. Thus, they are not well suited for CAM, the goal of which is to manufacture a solid object. As such, those using CAD software in a CAD/CAM context often focus more on solid modeling. Solid modeling is the process of creating a 3D model of an object that includes the object's edges as well as its internal structure. CAD software typically allows the user to rotate or otherwise manipulate the created model. With CAD, designers can ensure that all the separate parts of a product will fit together as intended. CAD also enables users to modify the digital model. This is less time-consuming and produces less waste than modifying a physical model.

When designing models with the intention of manufacturing them through CAM technology, users must be particularly mindful of their key measurements. Precision and accurate scaling are crucial. As such, users must be sure to use vector images when designing their models. Unlike raster images, which are based on the use of individual pixels, vector images are based on lines and points that have defined relationships to one another. No matter how much a user shrinks or enlarges a vector image, the image will retain the correct proportions in terms of the relative placement of points and lines.

Computer-Aided Manufacturing

After designing an object using CAD software, a user may use a CAM program to manufacture it. CAM programs typically operate through computer numerical control (CNC). In CNC, instructions are transmitted to the manufacturing machine as a series

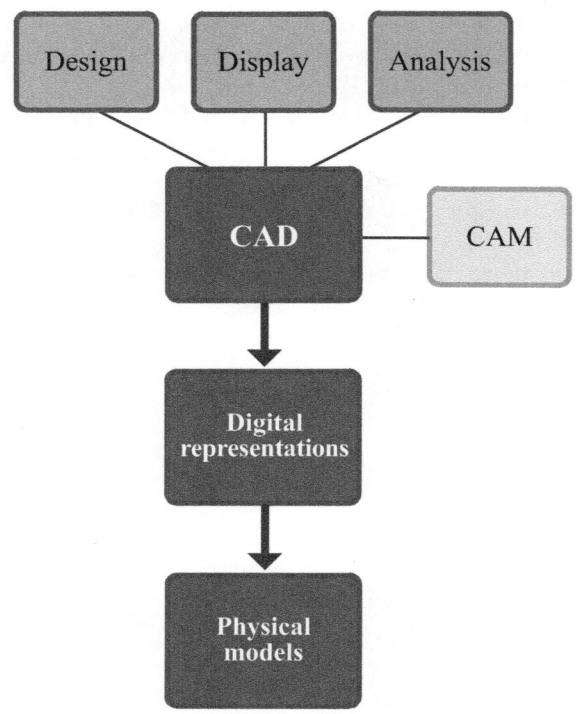

Computer-aided design (CAD) and computer-aided manufacturing (CAM) are used in many industries to fulfill the same basic goals. 3-D items are scanned and analyzed, new items are designed, and those designs can then be translated into manufactured items through CAM, which can develop the program necessary for machines to properly create the new item.

CAM enabled users to instruct computer-compatible machinery to manufacture various objects according to digital designs. The use of CAD/CAM software became widespread over the following decades.

CAD/CAM is now a key part of the manufacturing process for numerous companies, from large corporations to small start-ups. Industries in which CAD/CAM proved particularly useful include the automotive and computer technology industries. However, CAM software has also been widely used in less obvious fields, including dentistry and textile manufacturing.

In addition to its use alongside CAM software, CAD software functions alone in a number of fields. CAD software allows users to create 3D models of objects or structures that do not need to be manufactured by machine. For instance, specialized CAD software are used in architecture to design floor plans and 3D models of buildings.

of numbers. Those instructions tell the machine how to move and what actions to perform in order to construct the object. The types of machines used in that process vary and may include milling machines, drills, and lathes.

In the early twenty-first century, 3D printers, devices that manufacture objects out of thin layers of plastic or other materials, began to be used in CAM. Unlike traditional CNC machinery, 3D printers are typically used by individuals or small companies for whom larger-scale manufacturing technology is excessive.

SPECIALIZED APPLICATIONS

As CAD/CAM technology has evolved, it has come to be used for a number of specialized applications. Some CAD software, for instance, is used to perform four-dimensional building information modeling (4D BIM). This process enables a user to incorporate information related to time. For instance, the schedule for a particular project can be accounted for in the modeling process with 4D BIM.

Another common CAD/CAM application is rapid prototyping. In that process, a company or individual can design and manufacture physical prototypes of an object. This allows the designers to make changes in response to testing and evaluation and to test different iterations of the product. The resulting prototypes are often manufactured using 3D printers. Rapid prototyping results in improved quality control and a reduced time to bring a product to market.

—*Joy Crelin*

BIBLIOGRAPHY

Bryden, Douglas. *CAD and Rapid Prototyping for Product Design*. London: King, 2014. Print.

Chua, Chee Kai, Kah Fai Leong, and Chu Sing Lim. *Rapid Prototyping: Principles and Applications*. Hackensack: World Scientific, 2010. Print.

"Computer-Aided Design (CAD) and Computer-Aided Manufacturing (CAM)." *Inc*. Mansueto Ventures, n.d. Web. 31 Jan. 2016.

"Design and Technology: Manufacturing Processes." *GCSE Bitesize*. BBC, 2014. Web. 31 Jan. 2016.

Herrman, John. "How to Get Started: 3D Modeling and Printing." *Popular Mechanics*. Hearst Digital Media, 15 Mar. 2012. Web. 31 Jan. 2016.

Krar, Steve, Arthur Gill, and Peter Smid. *Computer Numerical Control Simplified*. New York: Industrial, 2001. Print.

Sarkar, Jayanta. *Computer Aided Design: A Conceptual Approach*. Boca Raton: CRC, 2015. Print.

COMPUTER-ASSISTED INSTRUCTION (CAI)

FIELDS OF STUDY

Computer Science; Information Systems

ABSTRACT

Computer-assisted instruction is the use of computer technology as a means of instruction or an aid to classroom teaching. The instructional content may or may not pertain to technology. Computer-assisted instruction often bridges distances between instructor and student and allows for the instruction of large numbers of students by a few educators.

PRINCIPAL TERMS

- **learner-controlled program:** software that allows a student to set the pace of instruction, choose which content areas to focus on, decide which areas to explore when, or determine the medium or difficulty level of instruction; also known as a "student-controlled program."
- **learning strategy:** a specific method for acquiring and retaining a particular type of knowledge, such as memorizing a list of concepts by setting the list to music.
- **learning style:** an individual's preferred approach to acquiring knowledge, such as by viewing visual stimuli, reading, listening, or using one's hands to practice what is being taught.
- **pedagogy:** a philosophy of teaching that addresses the purpose of instruction and the methods by which it can be achieved.
- **word prediction:** a software feature that recognizes words that the user has typed previously and offers to automatically complete them each time the user begins typing.

Computer-Assisted vs. Traditional Instruction

In a traditional classroom, a teacher presents information to students using basic tools such as pencils, paper, chalk, and a chalkboard. Most lessons consist of a lecture, group and individual work by students, and the occasional hands-on activity, such as a field trip or a lab experiment. For the most part, students must adapt their learning preferences to the teacher's own pedagogy, because it would be impractical for one teacher to try to teach to multiple learning styles at once.

Computer-assisted instruction (CAI) supplements this model with technology, namely a computing device that students work on for some or all of a lesson. Some CAI programs offer limited options for how the material is presented and what learning strategies are supported. Others, known as learner-controlled programs, are more flexible. A typical CAI lesson focuses on one specific concept, such as long division or the history of Asia. The program may present information through audio, video, text, images, or a combination of these. It then quizzes the student to make sure that they have paid attention and understood the material. Such instruction has several benefits. First, students often receive information through several mediums, so they do not have to adapt to the teacher's preferred style. Second, teachers can better support students as they move through the lessons without having to focus on presenting information to a group.

Advantages and Disadvantages

CAI has both benefits and drawbacks. Certain software features make it easier to navigate the learning environment, such as word prediction to make typing easier and spell-checking to help avoid spelling mistakes. Copy-and-paste features save users time that they would otherwise spend reentering the same information over and over. Speech recognition can assist students who are blind, have physical disabilities, or have a learning disability that affects writing. Other

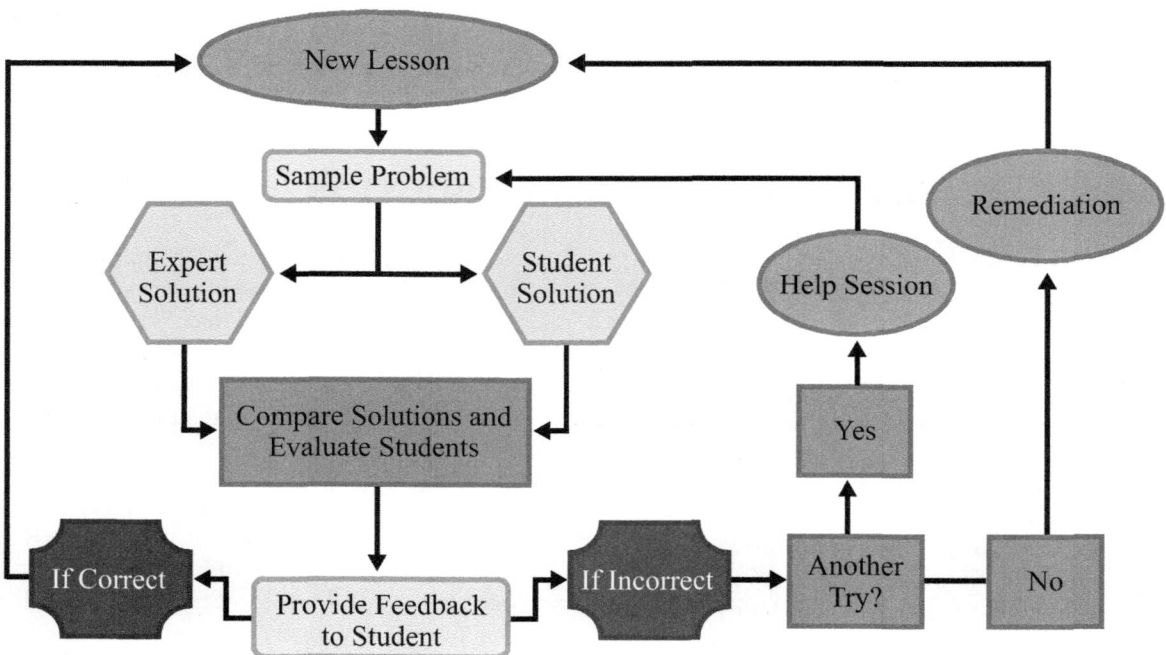

Computer-assisted instruction uses programming to determine whether a student understands the lesson (correctly answers sample problems) or needs more help or remediation (incorrectly answers sample problems). This flowchart indicates a general path of computer-assisted instruction. (Adapted from S. Egarievwe, A. O. Ajiboye, and G. Biswas, "Internet Application of LabVIEW in Computer Based Learning," 2000.)

helpful features are present despite not having been intended to benefit students. For example, CAI video lessons include the option to pause playback, skip ahead or back, or restart. These functions can be vital to a student who is struggling to grasp an especially difficult lesson or is practicing note-taking. They can stop the video at any time and restart it after catching up with the content that has been presented. By contrast, in a regular classroom, the lecturer often continues at the same pace regardless how many students may be struggling to understand and keep up.

Learning is rarely a "one size fits all" affair. Different topics pose greater or lesser challenges to different students, depending on their natural abilities and study habits. In regular classrooms, teachers can often sense when some students are not benefiting from a lesson and adapt it accordingly. With some CAI, this is not an option because the lesson is only presented in one way.

Adaptive Instruction

Fortunately, some forms of CAI address different learning rates by using adaptive methods to present material. These programs test students' knowledge and then adapt to those parts of the lesson with which they have more difficulty. For instance, if a math program notices that a student often makes mistakes when multiplying fractions, it might give the student extra practice in that topic. Adaptive programs give teachers the means to better assess students' individual needs and track their progress. As the technology improves, more detailed and specific results may bolster teachers' efforts to tailor instruction further.

Distance Education

CAI is especially important to the growing field of online education. Online instructors often use elements of CAI to supplement their curricula. For example, an online course might require students to watch a streaming video about doing library research so that they will know how to complete their own research paper for the course. Online education also enables just a few instructors to teach large numbers of students across vast distances. Tens of thousands of students may enroll in a single massive open online course (MOOC).

—*Scott Zimmer, JD*

Bibliography

Abramovich, Sergei, ed. *Computers in Education*. 2 vols. New York: Nova, 2012. Print.

Erben, Tony, Ruth Ban, and Martha E. Castañeda. *Teaching English Language Learners through Technology*. New York: Routledge, 2009. Print.

Miller, Michelle D. *Minds Online: Teaching Effectively with Technology*. Cambridge: Harvard UP, 2014. Print.

Roblyer, M. D., and Aaron H. Doering. *Integrating Educational Technology into Teaching*. 6th ed. Boston: Pearson, 2013. Print.

Tatnall, Arthur, and Bill Davey, eds. *Reflections on the History of Computers in Education: Early Use of Computers and Teaching about Computing in Schools*. Heidelberg: Springer, 2014. Print.

Tomei, Lawrence A., ed. *Encyclopedia of Information Technology Curriculum Integration*. 2 vols. Hershey: Information Science Reference, 2008. Print.

CONDITIONAL OPERATORS

FIELDS OF STUDY

Coding Techniques; Software Development; Computer Science

ABSTRACT

Conditional operators perform a similar function in code to other conditional expressions such as the if-then-else construct. Conditional operators can be used to reduce the number of lines of code needed for branching logic. Such refactoring makes code easier to understand, debug, and maintain.

PRINCIPAL TERMS

- **bool:** in many programming languages, a data type consisting of either-or values, such as "true" or "false," "yes" or "no," and "on" or "off."

- **operand:** in computer programming, the part of a computer instruction that specifies the data on which a given mathematical, relational, or logical operation is to be performed; also refers to the data itself.
- **ternary operator:** an operator that takes three operands; most often refers to a conditional operator.

What Is a Conditional Operator?

In computer programming, an operand represents the data on which a specific mathematical, relational, or logical operation is to be performed. A conditional operator is an operator that takes three operands. The first operand is a conditional statement, the second operand is the result to return if the condition is true, and the third operand is the result to return if the condition is false. In many programming languages, the conditional operator is the only ternary operator that is commonly used. Because of this, the terms "conditional operator" and "ternary operator" are often used interchangeably.

The conditional operator is a type of conditional expression. Conditional expressions result in different actions being taken depending on whether a Boolean condition evaluates to true or false. A conditional operator executes different blocks of code based on the bool value returned by a Boolean condition. The condition can be a simple Boolean condition, such as $x = 5$. Alternatively, it can be a complex Boolean expression constructed with Boolean operators, such as $x = 5$ AND $y = 6$ AND $z = 7$.

True Example	False Example	Statement	Conditional Operator	Description
int x = 2, int y = 8	int x = 4, int y = 6	(x < 3) && (y > 6)	&&	If one condition AND another condition are met, then the statement is TRUE
int x = 1, int y = 8	int x = 1, int y = 13	(x < 1) || (y <= 10)	||	If either one condition OR another condition is met, then the statement is TRUE
int x = 2, int y = 8	int x = 6, int y = 8	!(x = 6)	!	If condition is NOT met, then the statement is TRUE

Along with a comparison operator, conditional operators can be used to combine more than one condition into a single test. True and false examples of integers are provided for the following conditional statements: (x < 3) AND (y > 6), (x < 1) OR (y <= 10), and the statement NOT (x = 6).

In programming code, conditional operators perform a similar function to that of other conditional expressions, such as the if-then-else construct. They can control program flow and implement conditional logic as needed by the application.

In many programming languages, the conditional operator uses the following syntax:

[condition] ? [result if condition is true] : [result if condition is false]

Conditional operators can be nested in parentheses in order to test multiple conditions, much as "if" statements can be nested:

[condition 1] ? ([condition 2] ? [result A] : [result B]) : [result C]

Here, if condition 1 evaluates to false, the program will simply return result C. If condition 1 evaluates to true, however, the program will then evaluate condition 2. If condition 2 evaluates to true, the program will return result A; if it evaluates to false, the program will return result B.

Considerations with Conditional Operators

Conditional operators can be used to reduce the number of lines of code needed to create branching logic. Such refactoring makes code easier to understand, debug, and maintain. If many conditions need to be tested, however, an if-then-else construct might be easier to understand.

It is crucial that developers be aware of how conditional operators are implemented in a specific programming language. For example, the conditional operator is right associative in many languages. However, it is left associative in PHP. Associativity refers to the direction in which the operation is performed. To avoid errors of this type, parentheses should be used to specify operator precedence.

Conditional Operators in Practice

A developer is writing code to add functionality to an e-commerce application. The retailer wants to assign a special holiday discount based on how much a customer has spent that year. The developer could use the following pseudocode if-then-else construct to accomplish this task:

```
IF yearlyPurchases > 1000 THEN
  Discount = .1
ELSE
  Discount = .05
END IF
```

This code sets *Discount* to 0.1, or 10 percent, if a customer has spent more than $1,000 during the year and 0.05, or 5 percent, if the customer has spent $1,000 or less. A more efficient way to write this instruction would be to use a conditional operator, as below:

```
yearlyPurchases > 1000 ? Discount = .1 :
Discount = .05
```

Note how the Boolean expression from the if-then-else statement is the first operand of the conditional operator, the code from the "if" block is the second operand, and the code from the "else" block is the third operand.

Suppose the retailer wanted a more refined discount system in which customers who spent more than $1,000 during the year received a discount of 10 percent, those who spent more than $5,000 received 15 percent, and all others received 5 percent. A nested if-then-else statement could be used to accomplish the task, as follows:

```
IF yearlyPurchases > 1000 THEN
  IF yearlyPurchases < = 5000 THEN
  Discount = .1
  ELSE
  Discount = .15
  END IF
ELSE
  Discount = .05
END IF
```

However, using a conditional operator could reduce the lines of code required from nine to one:

```
yearlyPurchases > 1000 ? (yearlyPurchases < =
5000 ? Discount = .1 : Discount = .15) :
Discount = .05
```

The Boolean expression used in the outermost "if" statement becomes the first operand of the outer conditional operator. Likewise, the Boolean expression used in the innermost "if" statement becomes the first operand of the inner conditional operator.

The Benefits of Simplified Code

Conditional operators provide an alternative to the if-then-else construct that can reduce the number of lines of code needed to create branching logic, particularly in complicated nesting scenarios. This reduces the time needed to write, debug, and maintain code and makes the code easier to understand. The time saved can be significant, as conditional expressions are among the most common constructs used in computer programming. Branching logic is at the core of many algorithms, making it a core capability required of any computer language. Thus, conditional operators will continue to be relevant as new computer languages are developed and existing languages are improved.

—*Maura Valentino, MSLIS*

Bibliography

Friedman, Daniel P., and Mitchell Wand. *Essentials of Programming Languages*. 3rd ed., MIT P, 2008.

Haverbeke, Marijn. *Eloquent JavaScript: A Modern Introduction to Programming*. 2nd ed., No Starch Press, 2015.

MacLennan, Bruce J. *Principles of Programming Languages: Design, Evaluation, and Implementation*. 3rd ed., Oxford UP, 1999.

Schneider, David I. *An Introduction to Programming Using Visual Basic*. 10th ed., Pearson, 2017.

Scott, Michael L. *Programming Language Pragmatics*. 4th ed., Elsevier, 2016.

Van Roy, Peter, and Seif Haridi. *Concepts, Techniques, and Models of Computer Programming*. MIT P, 2004.

CONSTRAINT PROGRAMMING

FIELDS OF STUDY
Software Engineering; System-Level Programming

ABSTRACT
Constraint programming typically describes the process of embedding constraints into another language, referred to as the "host" language since it hosts the constraints. Not all programs are suitable for constraint programming. Some problems are better addressed by different approaches, such as logic programming. Constraint programming tends to be the preferred method when one can envision a state in which multiple constraints are simultaneously satisfied, and then search for values that fit that state.

PRINCIPAL TERMS

- **constraints:** limitations on values in computer programming that collectively identify the solutions to be produced by a programming problem.
- **domain:** the range of values that a variable may take on, such as any even number or all values less than −23.7.
- **functional programming:** a theoretical approach to programming in which the emphasis is on applying mathematics to evaluate functional relationships that are, for the most part, static.
- **imperative programming:** programming that produces code that consists largely of commands issued to the computer, instructing it to perform specific actions.

Models of Constraint Programming

Constraint programming tends to be used in situations where the "world" being programmed has multiple constraints, and the goal is to have as many of them as possible satisfied at once. The aim is to find values for each of the variables that collectively describe the world, such that the values fall within that variable's constraints.

To achieve this state, programmers use one of two main approaches. The first approach is the perturbation model. Under this model, some of the variables are given initial values. Then, at different times, variables are changed ("perturbed") in some way. Each change then moves through the system of interrelated variables like ripples across the surface of a pond: other values change in ways that follow their constraints and are consistent with the relationships between values. The perturbation model is much like a spreadsheet. Changing one cell's value causes changes in other cells that contain formulas that refer to the value stored in the original cell. A single change to a cell can propagate throughout many other cells, changing their values as well.

A contrasting approach is the refinement model. Whereas the perturbation model assigns particular values to variables, under the refinement model, each variable can assume any value within its domain. Then, as time passes, some values of one variable will inevitably be ruled out by the values assumed by other variables. Over time, each variable's possible values are refined down to fewer and fewer options. The refinement model is sometimes considered more flexible, because it does not confine a variable to one possible value. Some variables will occasionally have multiple solutions.

A Different Approach to Programming

Constraint programming is a major departure from more traditional approaches to writing code. Many programmers are more familiar with imperative programming, where commands are issued to the computer to be executed, or functional programming, where the program receives certain values and then performs various mathematical functions using those values. Constraint programming, in contrast, can be less predictable and more flexible.

A classic example of a problem that is especially well suited to constraint programming is that of map coloring. In this problem, the user is given a map of a country composed of different states, each sharing one or more borders with other states. The user is also given a palette of colors. The user must find a way to assign colors to each state, such that no adjacent states (i.e., states sharing a border) are the same color. Map makers often try to accomplish this in real life so that their maps are easier to read.

Those experienced at constraint programming can immediately recognize some elements of this problem. The most obvious is the constraint, which is the restriction that adjacent states may not be the

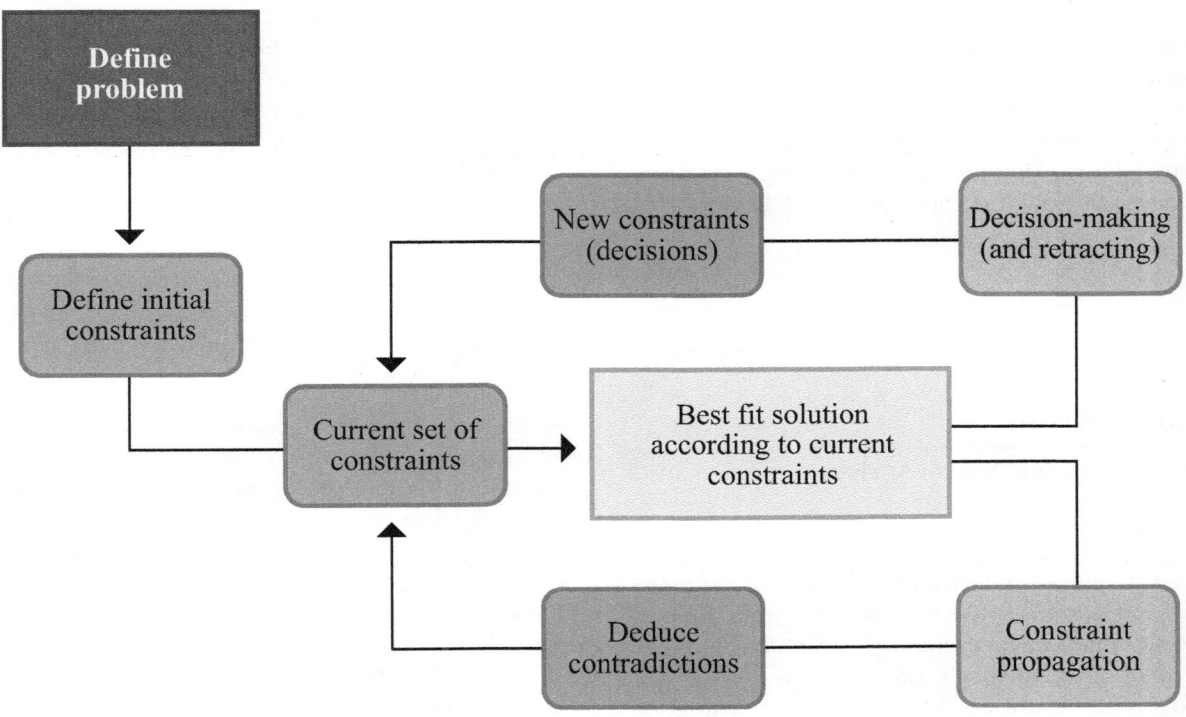

In constraint programming, the design of the program is dictated by specific constraints determined prior to development. After a problem is identified, the initial constraints are defined and condensed into the current program constraints. As solutions are developed and tested, decisions are made as to whether the constraints must be modified or retracted to remove duplications or whether new ones must be developed. The process cycles through current restraints and solutions until a solution is developed that meets all the constraints and solves the initial problem. Adapted from Philippe Baptiste, "Combining Operations Research and Constraint Programming to Solve Real-Life Scheduling Problems."

same color. Another element is the domain, which is the list of colors that may be assigned to states. The fewer the colors included in the domain, the more challenging the problem becomes. While this map-coloring problem may seem simplistic, it is an excellent introduction to the concept of constraint programming. It provides a useful situation for student programmers to try to translate into code.

Feasibility vs. Optimization

Constraint programming is an approach to problem solving and coding that looks only for a solution that works. It is not concerned with finding the optimal solution to a problem, a process known as "optimization." Instead, it seeks values for the variables that fit all of the existing constraints. This may seem like a limitation of constraint programming. However, its flexibility can mean that it solves a problem faster than expected.

Another example of a problem for which constraint programming is well suited is that of creating a work schedule. The department or team contains multiple variables (the employees), each with their own constraints. Mary can work any day except Friday, Thomas can work mornings on Monday through Thursday but only evenings on Friday, and so forth. The goal is to simply find a schedule that fits all of the constraints. It does not matter whether it is the best schedule, and in fact, there likely is no "best" schedule.

—*Scott Zimmer, JD*

Bibliography

Baptiste, Philippe, Claude Le Pape, and Wim Nuijten. *Constraint-Based Scheduling: Applying Constraint Programming to Scheduling Problems.* New York: Springer, 2013. Print.

Ceberio, Martine, and Vladik Kreinovich. *Constraint Programming and Decision Making*. New York: Springer, 2014. Print.

Henz, Martin. *Objects for Concurrent Constraint Programming*. New York: Springer, 1998. Print.

Hofstedt, Petra. *Multiparadigm Constraint Programming Languages*. New York: Springer, 2013. Print.

Pelleau, Marie, and Narendra Jussien. *Abstract Domains in Constraint Programming*. London: ISTE, 2015. Print.

Solnon, Christine. *Ant Colony Optimization and Constraint Programming*. Hoboken: Wiley, 2010. Print.

CONTROL SYSTEMS

FIELDS OF STUDY

Embedded Systems; System Analysis

ABSTRACT

A control system is a device that exists to control multiple other systems or devices. For example, the control system in a factory would coordinate the operation of all of the factory's interconnected machines. Control systems are used because the coordination of these functions needs to be continuous and nearly instantaneous, which would be difficult and tedious for human beings to manage.

PRINCIPAL TERMS

- **actuator:** a motor designed to move or control another object in a particular way.
- **automatic sequential control system:** a mechanism that performs a multistep task by triggering a series of actuators in a particular sequence.
- **autonomous agent:** a system that acts on behalf of another entity without being directly controlled by that entity.
- **fault detection:** the monitoring of a system to identify when a fault occurs in its operation.
- **system agility:** the ability of a system to respond to changes in its environment or inputs without failing altogether.
- **system identification:** the study of a system's inputs and outputs to develop a working model of it.

TYPES OF CONTROL SYSTEMS

Different types of control systems are used in different situations. The nature of the task being performed usually determines the design of the system. At the most basic level, control systems are classified as either open-loop or closed-loop systems. In an open-loop system, the output is based solely on the input fed into the system. In a closed-loop system, the output generated is used as feedback to adjust the system as necessary. Closed-loop systems can make fault detection more difficult, as they are designed to minimize the deviations created by faults.

One of the first steps in designing a control system is system identification. This is the process of modeling the system to be controlled. It involves studying the inputs and outputs of the system and determining how they need to be manipulated in order to produce the desired outcome. Some control systems require a certain degree of human interaction or guidance during their operation. However, control-system designers generally prefer systems that can function as autonomous agents. The purpose of a control system is to reduce human involvement in the process as much as possible. This is partly so personnel can focus on other, more important tasks and partly because problems are more likely to arise from human error.

The downside of an autonomous agent is that it requires greater system agility, so that when problems are encountered, the control system can either continue to perform its role or "fail gracefully" rather than failing catastrophically or ceasing to function altogether. This could mean the difference between a jammed conveyor belt being automatically corrected and that same belt forcing an entire assembly line to shut down.

CONTROL SYSTEM PERFORMANCE

Control systems are more than just computers that monitor other processes. Often, they also control

Feedback

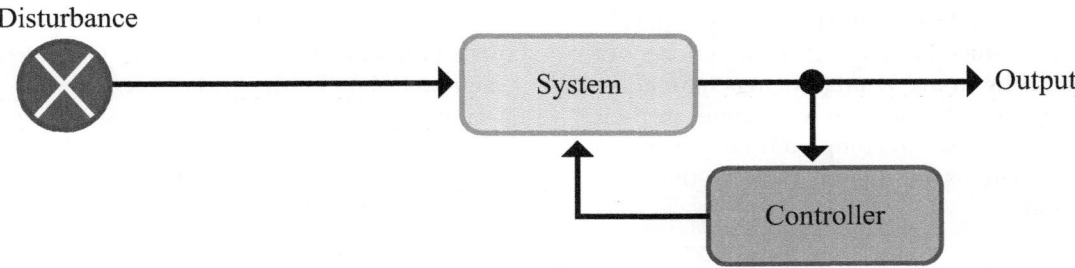

Feed Forward/Anticipation/Planning

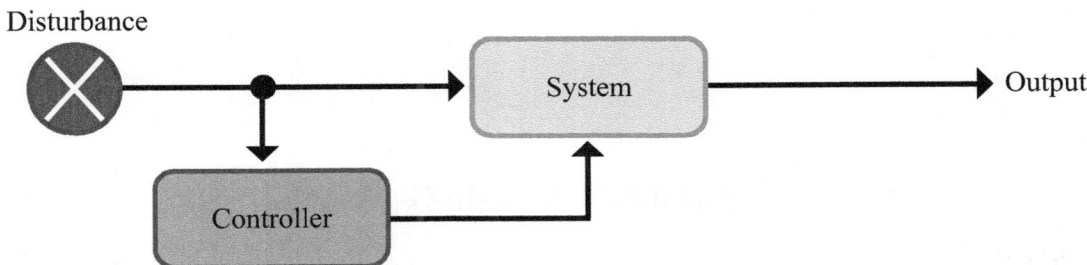

Control systems are used to alter a process pathway to reach a desired output. A controller senses some type of information along the pathway and causes a change in the system. Making changes to the system prior to system input is called a "feed-forward control"; making changes to the system based on the output is called "feedback control." (Adapted from http://www.csci.csusb.edu/dick/cs372/a1.html.)

the physical movements of system components. A control system may cause an assembly-line robot to pivot an automobile door so that another machine can attach its handle, or it may cause a device to fold cardboard into boxes to be filled with candy. Such tasks are accomplished via actuators, which are motors that the control system manipulates to execute the necessary physical movements. Sometimes only a single movement is required from a device. At other times, a device may have to perform several different movements in a precisely timed sequence. If this is the case, an automatic sequential control system is used to tell the machine what to do and when.

The development of very small and inexpensive microprocessors and sensors has made it possible for them to be incorporated into control systems. Similar to the closed-loop approach, these tiny computers help a control system monitor its performance and provide detailed information about potential issues. For example, if a control system were using a new type of raw material that produced increased heat due to friction and caused the system to operate less efficiently, microsensors inside the machinery could detect this and alert those in charge of supervising the system. Such information could help avoid costly breakdowns of equipment and delays in production by allowing supervisors to address problems before they become severe.

Linear Control Systems

Linear control systems are a type of closed-loop system. They receive linear negative feedback from their outputs and adjust their operating parameters in response. This allows the system to keep relevant variables within acceptable limits. If the sensors in a linear control system detect excess heat, as in the example above, this might cause the system to initiate additional cooling. An example of this is when a computer's fan speeds up after an intensive application causes the system to begin generating heat.

Behind the Scenes

Control systems are a vital part of everyday life in the industrialized world. Most of the products people use every day are produced or packaged using dozens of different control systems. Human beings have come to rely heavily on technology to assist them in their work, and in some cases to completely take over that work. Control systems are the mechanisms that help make this happen.

—Scott Zimmer, JD

Bibliography

Ao, Sio-Iong, and Len Gelman, eds. *Electrical Engineering and Intelligent Systems.* New York: Springer, 2013. Print.

Chen, Yufeng, and Zhiwu Li. *Optimal Supervisory Control of Automated Manufacturing Systems.* Boca Raton: CRC, 2013. Print.

Janert, Philipp K. *Feedback Control for Computer Systems.* Sebastopol: O'Reilly, 2014. Print.

Li, Han-Xiong, and XinJiang Lu. *System Design and Control Integration for Advanced Manufacturing.* Hoboken: Wiley, 2015. Print.

Song, Dong-Ping. *Optimal Control and Optimization of Stochastic Supply Chain Systems.* London: Springer, 2013. Print.

Van Schuppen, Jan H., and Tiziano Villa, eds. *Coordination Control of Distributed Systems.* Cham: Springer, 2015. Print.

COWBOY CODING

FIELDS OF STUDY

Software Development; Programming Methodologies; Coding Techniques

ABSTRACT

Cowboy coding is a style of software development that frees the programmer from the constraints of formal software-development methodologies. Proponents of cowboy coding maintain it promotes creativity, experimentation, and "outside the box" thinking, while critics maintain it is inefficient and leads to the development of lower-quality software.

PRINCIPAL TERMS

- **agile development:** an approach to software development that focuses on improving traditional software-development methodologies and adapting to changes during the software life cycle.
- **requirement:** a necessary characteristic of a software system or other product, such as a service it must provide or a constraint under which it must operate.
- **software architecture:** the overall structure of a software system, including the components that make up the system, the relationships among them, and the properties of both those components and their relationships.
- **waterfall development:** a traditional, linear software-development methodology based on a series of ordered phases.

What Is Cowboy Coding?

Cowboy coding is a style of software development that gives individual programmers complete control of all aspects of the development process. Unlike traditional software-development methodologies such as waterfall development, cowboy coding does not follow a defined approach. Rather, it allows programmers to use any approach when developing software. Cowboy coders may use elements of formal software methodologies in any combination or order, or they may forgo formal methodologies entirely. Cowboy coding is typically used by students, hobbyists, and developers working alone. It can be practiced by individuals within larger development teams on projects using formal methodologies, but cowboy coding practices are typically considered inappropriate and disruptive in such environments.

Cowboy coding is often confused with agile development. Like cowboy coding, agile software development aims to free programmers from constraints imposed by traditional development methodologies.

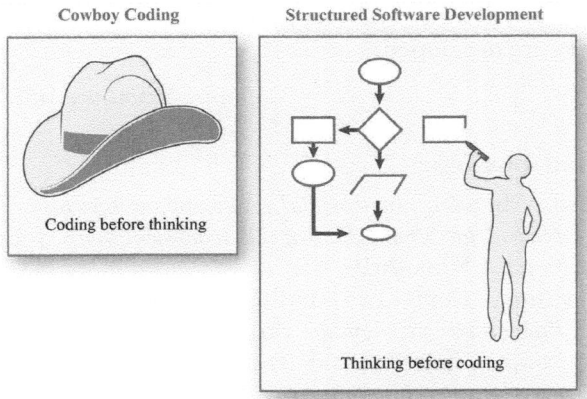

The approach to creating programming where one writes code without thinking out a plan first is called cowboy coding. Alternatively, one would think up a strategy for development and the user expectations of that software before writing code.

However, unlike cowboy coding, agile development does not dispense with formal methodologies when designing and implementing software architecture. Instead, it focuses on improving and optimizing the use of existing methodologies. For example, rather than abandoning the comprehensive software testing included in many development methodologies, agile development encourages a focus on useful testing that adds value to the development process. In other words, a test should be conducted to achieve a specific purpose, not simply because it is part of a given methodology.

Agile also retains many traditional software-development elements, including a focus on proper documentation and requirements identification. These elements are frequently abandoned by cowboy coders. In addition, agile development encourages programmers to work in well-integrated development teams, in contrast to the typically individual focus of cowboy coding.

Cowboy Coding Is Controversial

Cowboy coding encourages programmers to use whatever process, model, or technology works best for them in any given situation. Proponents of this approach believe it promotes creativity, experimentation, and thinking outside the box. They note that its focus on results rather than process often produces innovative solutions in an efficient manner. They stress that cowboy coding does not prohibit the programmer from using established methodologies; it simply frees them to use their own experience and judgment to select which methodologies to use.

Critics of cowboy coding believe that the approach does not work well in a professional software-development environment. They argue that it tends to lead to solutions that are poorly designed and implemented. Lack of a formal planning and design process can lead to project delays, incorrect scheduling, wasted effort, and solutions that are incomplete or do not meet the purpose for which they are designed. Critics also maintain that teamwork is a key component of all but the smallest development projects. If developers work with a high degree of independence, it can be difficult to effectively coordinate their efforts with those of other stakeholders on the types of large, complex projects often faced by professional developers.

When Is It Good to Be a Cowboy?

While cowboy coding is not typically used in professional software development, it is often used by individual software developers who work in isolation. For example, this approach is frequently used by independent game developers who design and code their own games. Such a scenario minimizes cowboy coding's weaknesses while emphasizing its strengths. The lack of a development team reduces the need for the planning and coordination required with larger projects. The lack of outside customers and other stakeholders reduces the need for agreed-upon benchmark deliverables and dates, detailed project specification documents, scheduled releases, and timely updates.

The principal resource being devoted to a small-scale cowboy coding project is the developer's time. As such, cost overruns and other financial and resource-related risks addressed by formal development methodologies are not as critical, as they only affect the developer. An experienced developer can mitigate the negative effects of cowboy coding by employing elements of standard methodologies and other professional practices. For this reason, cowboy coding is best used by experienced developers who are familiar with such techniques.

Rein Them In or Let Them Roam?

Cowboy coding remains controversial among software developers. Opponents believe that it undermines the success of software-development projects and that it has no place in the world of professional software development. They maintain that the principles of creativity and flexibility embraced by cowboy coding can be encouraged within the framework of defined software methodologies. They argue that defined software methodologies were developed to address real-world needs identified by programmers and have been proved to produce effective software solutions in a timely and efficient manner. According to this view, this effectiveness is threatened by cowboy coders who never bother to learn best practices.

Cowboy coders maintain that they produce the best solutions when they work independently and free of any restraints on their process or creativity. They reject the use of "cowboy coding" as a derogatory term to imply sloppiness or lack of a comprehensive outlook. In their view, the skill of the individual programmer matters more than the methods used. Defined software development methodologies remain integral to professional programming, but many programmers1 begin by working alone and have an independent nature. Because of this, conflict between cowboy coders and traditionalists is likely to persist as new methodologies and technologies are developed.

—*Maura Valentino, MSLIS*

BIBLIOGRAPHY

Bell, Michael. *Incremental Software Architecture: A Method for Saving Failing IT Implementations.* John Wiley & Sons, 2016.

Friedman, Daniel P., and Mitchell Wand. *Essentials of Programming Languages.* 3rd ed., MIT P, 2008.

Jayaswal, Bijay K., and Peter C. Patton. *Design for Trustworthy Software: Tools, Techniques, and Methodology of Developing Robust Software.* Prentice Hall, 2007.

MacLennan, Bruce J. *Principles of Programming Languages: Design, Evaluation, and Implementation.* 3rd ed., Oxford UP, 1999.

Scott, Michael L. *Programming Language Pragmatics.* 4th ed., Elsevier, 2016.

Van Roy, Peter, and Seif Haridi. *Concepts, Techniques, and Models of Computer Programming.* MIT P, 2004.

Wysocki, Robert K. *Effective Project Management: Traditional, Agile, Extreme.* 7th ed., John Wiley & Sons, 2014.

CPU DESIGN

FIELDS OF STUDY

Computer Engineering; Information Technology

ABSTRACT

CPU design is an area of engineering that focuses on the design of a computer's central processing unit (CPU). The CPU acts as the "brain" of the machine, controlling the operations carried out by the computer. Its basic task is to execute the instructions contained in the programming code used to write software. Different CPU designs can be more or less efficient than one another. Some designs are better at addressing certain types of problems.

PRINCIPAL TERMS

- **control unit design:** describes the part of the CPU that tells the computer how to perform the instructions sent to it by a program.
- **datapath design:** describes how data flows through the CPU and at what points instructions will be decoded and executed.
- **logic implementation:** the way in which a CPU is designed to use the open or closed state of combinations of circuits to represent information.
- **microcontroller:** a tiny computer in which all the essential parts of a computer are united on a single microchip—input and output channels, memory, and a processor.
- **peripherals:** devices connected to a computer but not part of the computer itself, such as scanners, external storage devices, and so forth.
- **protocol processor:** a processor that acts in a secondary capacity to the CPU, relieving it from some of the work of managing communication protocols that are used to encode messages on the network.

CPU Design Goals

The design of a CPU is a complex undertaking. The main goal of CPU design is to produce an architecture that can execute instructions in the fastest, most efficient way possible. Both speed and efficiency are relevant factors. There are times when having an instruction that is fast to execute is adequate, but there are also situations where it would not make sense to have to execute that simple instruction hundreds of times to accomplish a task.

Often the work begins by designers considering what the CPU will be expected to do and where it will be used. A microcontroller inside an airplane performs quite different tasks than one inside a kitchen appliance, for instance. The CPU's intended function tells designers what types of programs the CPU will most often run. This, in turn, helps determine what type of instruction-set architecture to use in the microchip that contains the CPU. Knowing what types of programs will be used most frequently also allows CPU designers to develop the most efficient logic implementation. Once this has been done, the control unit design can be defined. Defining the datapath design is usually the next step, as the CPU's handling of instructions is given physical form.

Often a CPU will be designed with additional supports to handle the processing load, so that the CPU itself does not become overloaded. Protocol processors, for example, may assist with communications

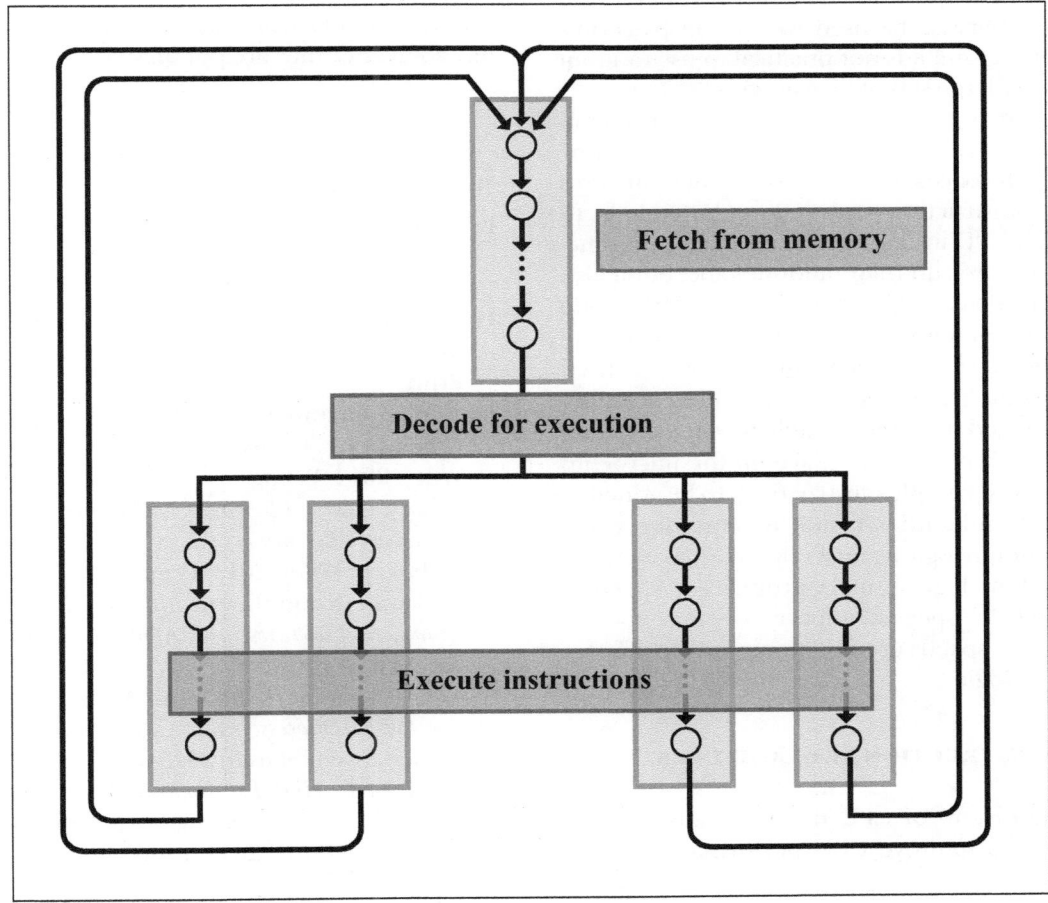

A generic state diagram shows the simple processing loop: fetch instructions from memory, decode instructions to determine the proper execute cycle, execute instructions, and then fetch next instructions from memory and continue the cycle. A state diagram is the initial design upon which data paths and logic controls can be designed.

protocol translation involving the transmission of data between the computer and its peripherals or over the Internet. Internet communication protocols are quite complex. They involve seven different, nested layers of protocols, and each layer must be negotiated before the next can be addressed. Protocol processors take this workload off the CPU.

INSTRUCTION SETS

CPU design is heavily influenced by the type of instruction sets being used. In general, there are two approaches to instructions. The first is random logic. Sometimes this is referred to as "hardwired instructions." Random logic uses logic devices, such as decoders and counters, to transport data and to perform calculations. Random logic can make it possible to design faster chips. The logic itself takes up space that might otherwise be used to store instructions, however. Therefore, it is not practical to use random logic with very large sets of instructions.

The second approach to instruction sets is microcode. Microcode is sometimes called "emulation" because it references an operations table and uses sets of microinstructions indexed by the table in to execute each CPU instruction. Microcode can sometimes be slower to run than random logic, but it also has advantages that offset this weakness. Microcode breaks down complex instructions into sets of microinstructions. These microinstructions are used in several complex instructions. A CPU executing microcode would therefore be able to reuse microinstructions. Such reuse saves space on the microchip and allows more complex instructions to be added.

The most influential factor to consider when weighing random logic against microcode is memory speed. Random logic usually produces a speedier CPU when CPU speeds outpace memory speeds. When memory speeds are faster, microcode is faster than random logic.

REDUCED INSTRUCTION SET COMPUTER (RISC)

Early in the history of CPU design, it was felt that the best way to improve CPU performance was to continuously expand instruction sets to give programmers more options. Eventually, studies began to show that adding more complex instructions did not always improve performance, however. In response, CPU manufacturers produced reduced instruction set computer (RISC) chips. RISC chips could use less complex instructions, even though this meant that a larger number of instructions were required.

MOORE'S LAW

Moore's law is named after Gordon Moore, a co-founder of the computer manufacturer Intel. In 1975, Moore observed that the computing power of an integrated circuit or microchip doubles, on average, every two years. This pace of improvement has been responsible for the rapid development in technological capability and the relatively short lifespan of consumer electronics, which tend to become obsolete soon after they are purchased.

—*Scott Zimmer, JD*

BIBLIOGRAPHY

Englander, Irv. *The Architecture of Computer Hardware, Systems Software, & Networking: An Information Technology Approach.* Hoboken: Wiley, 2014. Print.

Hyde, Randall. *Write Great Code: Understanding the Machine.* Vol. 1. San Francisco: No Starch, 2005. Print.

Jeannot, Emmanuel, and J. Žilinskas. *High Performance Computing on Complex Environments.* Hoboken: Wiley, 2014. Print.

Lipiansky, Ed. *Electrical, Electronics, and Digital Hardware Essentials for Scientists and Engineers.* Hoboken: Wiley, 2013. Print.

Rajasekaran, Sanguthevar. *Multicore Computing: Algorithms, Architectures, and Applications.* Boca Raton: CRC, 2013. Print.

Stokes, Jon. *Inside the Machine: An Illustrated Introduction to Microprocessors and Computer Architecture.* San Francisco: No Starch, 2015. Print.

Wolf, Marilyn. *High Performance Embedded Computing: Architectures, Applications, and Methodologies.* Amsterdam: Elsevier, 2014. Print.

CROWDFUNDING

Crowdfunding is a method by which an individual or group raises money for a project, cause, or business by receiving donations from a large group of individuals, often by means of many small donations from people who have no relationship to each other beyond interest in the project. Crowdfunding began largely as a way for musicians, artists, and other creative people to gain support for their projects and relies heavily on the Internet to reach potential funders. Crowdfunding is also used to raise money for philanthropic causes and to raise investment capital for new businesses.

OVERVIEW

There are two primary models for crowdfunding: donation-based crowdfunding and investment crowdfunding. In donation-based crowdfunding, donors contribute money for a project, such as the creation of a film or book, and in turn receive gifts such as a personal note from the author, a coffee mug or t-shirt advertising the project, or a copy of the product created (e.g., a copy of the book or film). This model is similar to the fundraising method used on National Public Radio (NPR) and PBS television, where people make a charitable contribution to their local station or to a particular program they enjoy and may elect to receive some branded item in return. In investment crowdfunding, businesses sell debt or equity stakes to investors, and those investors will share in the company's profits, if there are any. Indiegogo, founded in 2008, allows fundraising for any purpose other than business investment, including charity, and AngelList is specifically for tech startup companies looking for business investors.

A photo of crowd funding expert Pim Betist

GoFundMe, founded in 2010, allows for any kind of project, creative, business, charitable, or otherwise, as well as allowing people to fundraise for personal needs such as medical expenses.

Despite its burgeoning popularity in the Internet era, crowdfunding is not a new concept. For centuries artists and publishers have sought subscriptions to finance creative projects, and in the United States notable examples include the public solicitation of contributions to pay for the pedestal for the Statue of Liberty in 1884, and the 1933 public contributions to build a swimming pool in the White House so that President Franklin Roosevelt, who was disabled as the result of polio, could enjoy water therapy. In the 2010s, crowdfunding is primarily conducted over the Internet, often through websites such as Kickstarter and the aforementioned Indiegogo and AngelList. There are over 400 such sites in existence. Each site has its own set of rules; for instance, Kickstarter, launched in 2009, only hosts campaigns for creative projects, and requires that project creators set a goal (amount of money to be raised) and a deadline; if the goal is not met by the deadline, no money is collected from the donors who pledged it. Other sites, such as Indiegogo, allow campaign creators to keep the donations from a partially funded campaign.

The field of crowdfunding is growing and changing rapidly. According to Chance Barnett, writing for Forbes.com in August 2013, in 2012 the crowdfunding industry raised $2.7 billion across more than one million individual campaigns. In May 2014, a report by the Crowdfunding Centre estimated that, worldwide, more than $60,000 were being raised via crowdfunding every hour, and 442 crowdfunding campaigns were being initiated each day.

—Sarah E. Boslaugh, PhD, MPH

BIBLIOGRAPHY

Barnett, Chance. "Top 10 Crowdfunding Sites for Fundraising." *Forbes.com.* Forbes.com LLC, 8 May 2013. Web. 7 Aug. 2013.

Clifford, Catherine. "Crowdfunding Generates More Than $60,000 an Hour." *Entrepreneur.* Entrepreneur Media, 19 May 2014. Web. 22 May 2015.

Cunningham, William Michael. *The JOBS Act: Crowdfunding for Small Businesses and Startups.* New York: Springer, 2012. Print.

Dredge, Stuart. "Kickstarter's Biggest Hits: Why Crowdfunding Now Sets the Trends." *Guardian.* Guardian News and Media, 17 Apr. 2014. Web. 22 May 2015.

Dushnitsky, Gary. "What If Crowdfunding Becomes the Leading Source of Finance for Entrepreneurs or Growing Companies?" *Forbes.* Forbes, 20 May 2015. Web. 22 May 2015.

Jacobs, Deborah L. "The Trouble with Crowdfunding." *Forbes.com.* Forbes.com LLC, 17 Apr. 2013. Web. 7 Aug. 2013.

Lawton, Kevin, and Dan Marom. *The Crowdfunding Revolution: How to Raise Venture Capital Using Social Media.* New York: McGraw. 2012. Print.

Steinberg, Don. *The Kickstart Handbook: Real-life Crowdfunding Success Stories.* Philadelphia: Quirk, 2012. Print.

Steinberg, Scott. "Amanda Palmer on Crowdfunding and the Rebirth of the Working Musician." *Rollingstone.com.* Rolling Stone, 29 Aug. 2013. Web. 7 Aug. 2013.

Trigonis, John T. *Crowdfunding for Filmmakers: The Way to a Successful Film Campaign.* Studio City: Michael Wiese Productions, 2013. Print.

Young, Thomas Elliott. *The Everything Guide to Crowdfunding: Learn How to Use Social Media for Small-Business Funding.* Avon: F+W Media, 2013. Print.

CROWDSOURCING

The term *crowdsourcing* made its first print appearance in the June 2006 issue of *Wired* magazine, and by 2013 it was an established way of conducting business on the Internet. Crowdsourcing can be thought of a specialized variant of outsourcing, one where specific content is solicited from a large, and often anonymous, online community. Individuals respond to a call for information by providing small amounts of data. The entities that sponsor crowdsourcing include corporations, academic institutions, governments, and individuals. Contributors to crowdsourcing projects may be paid professionals or volunteers merely interested in sharing their knowledge.

Overview

The appeal of crowdsourcing is based on its relatively inexpensive cost and its ability to undertake problems too complex to solve with an organization's existing staff, as well as the idea that harnessing the knowledge of multiple individuals, who possess different knowledge and experiences, produces better, more accurate data. While individuals who participate in crowdsourcing are typically are demographically diverse, many have academic training or significant professional experience relevant to the project. Critics decry the quality and accuracy of crowdsourcing projects, but by the 2010s, the practice had become commonplace across a wide array of industries and endeavors.

Specialized forms of crowdsourcing allow the efforts of multiple users to achieve numerous types of distinct goals. Crowd creation is used frequently to solve complicated scientific problems. Crowdfunding involves collecting small sums of money from numerous contributors to meet fundraising goals for a start-up company or an artistic endeavor. Crowd searching leverages large networks of Internet and smartphone users to ascertain the location of a missing item or individual. Crowd voting makes use of communal judgment to grade, sort, and categorize Web-based content; it is the most common type of crowdsourcing (such as used on Reddit).

Although crowdsourcing is a creation of the Internet era, information-gathering projects with considerable numbers of participants have existed for decades. Crowdsourcing served as the means for the creation of the Oxford English Dictionary, was responsible for new methods for determining a ship's location in the Longitude Prize contest in Great Britain, and became a way to collect genealogical data for members of the Church of Jesus Christ of Latter-day Saints.

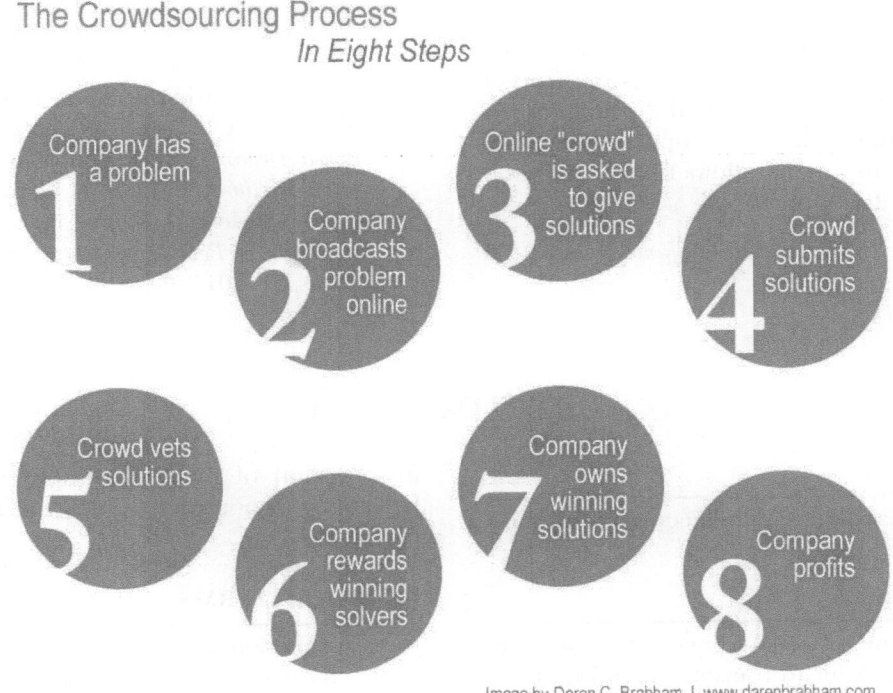

The crowdsourcing process in eight steps.

The Internet Movie Database and Wikipedia are two of the most successful examples of crowd-sourced Internet resources. Another well-received crowdsourcing project was the New York Public Library's "What's on the Menu," which began in 2011 and sought to transcribe more than 45,000 historical restaurant menus from the 1840s to the 2010s. In 2011 the California Digital Newspaper Collection utilized crowdsourcing to correct the text of digitized newspapers, some of which pre-date California's founding in 1850. The Katrina PeopleFinder Project was developed in the wake of Hurricane Katrina in 2005; more than ninety thousand entries were inputted in an effort to keep track of the victims of the natural disaster. Carnegie Mellon University's CAPTCHA system, established in 2000, is a type of implicit crowdsourcing, where users identify characters as part of a logon sequence, which, in turn, is used to determinate text that optical character recognition (OCR) software cannot, thereby aiding the digitization of historical text documents.

Cloud-Based Software Crowdsourcing, (2015) edited by Wei Li, Michael N. Huhns, Wei-Tek Tsai, and Wenjun Wu, discusses the practice of using cloud computing to crowdsource computer software development. By moving every stage of the software development process to the cloud, it becomes a process of cocreation, in which the traditional distinctions between end-users and software developers are blurred. Cloud-based crowdsourcing allows for efficient and scalable software development.

—*Leon James Bynum, MA*

BIBLIOGRAPHY

Brabham, Daren C. *Crowdsourcing*. Cambridge: MIT Press, 2013. Print.

Dawson, Ross, and Steve Bynghall. *Getting Results from Crowds: The Definitive Guide to Using Crowdsourcing to Grow Your Business*. San Francisco: Advanced Human Technologies, 2011. Print.

Estelles-Miguel, Sofia, Ignacio Gil-Pechuán, and Fernando J. Garrigos-Simon. *Advances in Crowdsourcing*. Cham: Springer, 2015. eBook Collection (EBSCOhost). Web. 19 June 2015.

Howe, Jeff. *Crowdsourcing: Why the Power of the Crowd Is Driving the Future of Business*. New York: Crown, 2009. Print.

Powell, Juliette. *33 Million People in the Room: How to Create, Influence, and Run a Successful Business with Social Networking*. Upper Saddle River: FT, 2009. Print.

Sloane, Paul, ed. *A Guide to Open Innovation and Crowdsourcing: Advice from Leading Experts*. London: Kogan, 2012. Print.

Suie, Daniel, Sara Elwood, and Michael Goodchild, eds. *Crowdsourcing Geographic Knowledge: Volunteered Geographic Information (VGI) in Theory and Practice*. New York: Spring, 2013. Print.

Surowiecki, James. *The Wisdom of Crowds: Why the Many Are Smarter than the Few and How Collective Wisdom Shapes Business, Economies, Societies, and Nations*. New York: Doubleday, 2004. Print.

Tsai, Wei-Tek, et al., eds. *Crowdsourcing : Cloud-Based Software Development*. Heidelberg: Springer, 2015. eBook Collection (EBSCOhost). Web. 19 June 2015.

Winograd, Morley. *Millennial Momentum: How a New Generation Is Remaking America*. New Brunswick: Rutgers UP, 2011. Print.

CRYPTOGRAPHY

FIELDS OF STUDY

Computer Science; Computer Engineering; Algorithms

ABSTRACT

Cryptography is the process of encrypting messages and other data in order to transmit them in a form that can only be accessed by the intended recipients. It was initially applied to written messages. With the introduction of modern computers, cryptography became an important tool for securing many types of digital data.

PRINCIPAL TERMS

- **hash function:** an algorithm that converts a string of characters into a different, usually smaller, fixed-length string of characters that is ideally impossible either to invert or to replicate.

- **public-key cryptography:** a system of encryption that uses two keys, one public and one private, to encrypt and decrypt data.
- **substitution cipher:** a cipher that encodes a message by substituting one character for another.
- **symmetric-key cryptography:** a system of encryption that uses the same private key to encrypt and decrypt data.
- **transposition cipher:** a cipher that encodes a message by changing the order of the characters within the message.

What Is Cryptography?

The word "cryptography" comes from the Greek words *kryptos* ("hidden," "secret") and *graphein* ("writing"). Early cryptography focused on ensuring that written messages could be sent to their intended recipients without being intercepted and read by other parties. This was achieved through various encryption techniques. Encryption is based on a simple principle: the message is transformed in such a way that it becomes unreadable. The encrypted message is then transmitted to the recipient, who reads it by transforming (decrypting) it back into its original form.

Early forms of encryption were based on ciphers. A cipher encrypts a message by altering the characters that comprise the message. The original message is called the "plaintext," while the encrypted message is the "ciphertext." Anyone who knows the rules of the cipher can decrypt the ciphertext, but it remains unreadable to anyone else. Early ciphers were relatively easy to break, given enough time and a working knowledge of statistics. By the 1920s, electromechanical cipher machines called "rotor machines" were creating complex ciphers that posed a greater challenge. The best-known example of a rotor machine was the Enigma machine, used by the German military in World War II. Soon after, the development of modern computer systems in the 1950s would change the world of cryptography in major ways.

Cryptography in the Computer Age

With the introduction of digital computers, the focus of cryptography shifted from just written language to

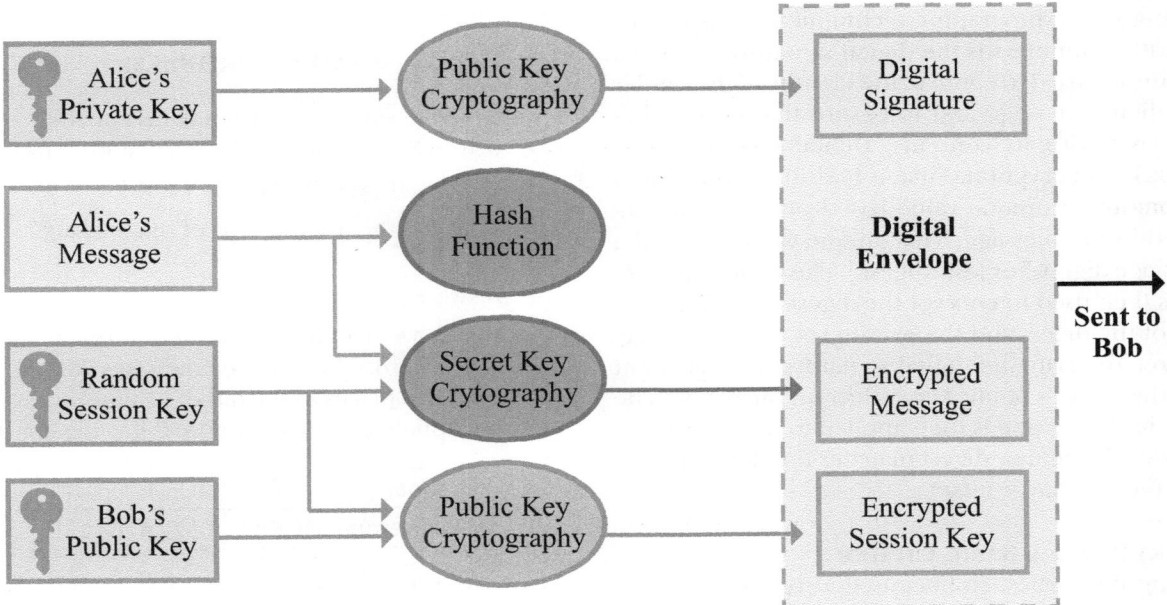

This is a diagram of cryptography techniques: public (asymmetric) key, private (symmetric) key, and hash functions. Each secures data in a different way and requires specific types of keys to encrypt and decrypt the data. Adapted from Gary C. Kessler, "An Overview of Cryptography."

Original	A	B	C	D	E	F	G	H	I	J	K	L	M	N	O	P	Q	R	S	T	U	V	W	X	Y	Z
Replacement	X	Y	Z	A	B	C	D	E	F	G	H	I	J	K	L	M	N	O	P	Q	R	S	T	U	V	W

Caesar's cipher table. Table organized original letters and replacement letters by Caesar's cipher.

any data that could be expressed in binary format. The encryption of binary data is accomplished through the use of keys. A key is a string of data that determines the output of a cryptographic algorithm. While there are many different types of cryptographic algorithms, they are usually divided into two categories. Symmetric-key cryptography uses a single key to both encrypt and decrypt the data. Public-key cryptography, also called "asymmetric-key cryptography," uses two keys, one public and one private. Usually, the public key is used to encrypt the data, and the private key is used to decrypt it.

When using symmetric-key cryptography, both the sender and the recipient of the encrypted message must have access to the same key. This key must be exchanged between parties using a secure channel, or else it may be compromised. Public-key cryptography does not require such an exchange. This is one reason that public-key cryptography is considered more secure.

Another cryptographic technique developed for use with computers is the digital signature. A digital signature is used to confirm the identity of the sender of a digital message and to ensure that no one has tampered with its contents. Digital signatures use public-key encryption. First, a hash function is used to compute a unique value based on the data contained in the message. This unique value is called a "message digest," or just "digest." The signer's private key is then used to encrypt the digest. The combination of the digest and the private key creates the signature. To verify the digital signature, the recipient uses the signer's public key to decrypt the digest. The same hash function is then applied to the data in the message. If the new digest matches the decrypted digest, the message is intact.

Decrypting a Basic Cipher

Among the earliest ciphers used were the transposition cipher and the substitution cipher. A transposition cipher encrypts messages by changing the order of the letters in the message using a well-defined scheme. One of the simplest transposition ciphers involves reversing the order of letters in each word. When encrypted in this fashion, the message "MEET ME IN THE PARK" becomes "TEEM EM NI EHT KRAP." More complicated transposition ciphers might involve writing out the message in a particular orientation (such as in stacked rows) and then reading the individual letters in a different orientation (such as successive columns).

A substitution cipher encodes messages by substituting certain letters in the message for other letters. One well-known early substitution cipher is the Caesar cipher, named after Julius Caesar. This cipher encodes messages by replacing each letter with a letter that is a specified number of positions to its right or left in the alphabet. For example, Caesar is reported to have used a left shift of three places when encrypting his messages.

SAMPLE PROBLEM

The following message has been encoded with a Caesar cipher using a left shift of five:

JSHWDUYNTS NX KZS

What was the original text of the message?

Answer:

The answer can be determined by replacing each letter in the encoded message with the letter five places to the left of its position in the alphabet, as shown in the following chart:

Original	A	B	C	D	E	F	G	H	I	J
Replacement	V	W	X	Y	Z	A	B	C	D	E

Cipher table. Table organizes original letters and replacement letters using a left shift of five cipher.

The original text read "ENCRYPTION IS FUN."

Using this cipher, Julius Caesar's famous message "I came, I saw, I conquered" becomes "F ZXJB F PXT F ZLKNRBOBA" when encrypted.

Why Is Cryptography Important?

The ability to secure communications against interception and decryption has long been an important part of military and international affairs. In the modern age, the development of new computer-based methods of encryption has had a major impact on many areas of society, including law enforcement, international affairs, military strategy, and business. It has also led to widespread debate over how to balance the privacy rights of organizations and individuals with the needs of law enforcement and government agencies. Businesses, governments, and consumers must deal with the challenges of securing digital communications for commerce and banking on a daily basis. The impact of cryptography on society is likely to increase as computers grow more powerful, cryptographic techniques improve, and digital technologies become ever more important.

—*Maura Valentino, MSLIS*

Bibliography

Esslinger, Bernhard, et al. *The CrypTool Script: Cryptography, Mathematics, and More*. 11th ed. Frankfurt: CrypTool, 2013. *CrypTool Portal*. Web. 2 Mar. 2016.

Hoffstein, Jeffrey, Jill Pipher, and Joseph H. Silverman. *An Introduction to Mathematical Cryptography*. 2nd ed. New York: Springer, 2014. Print.

Katz, Jonathan, and Yehuda Lindell. *Introduction to Modern Cryptography*. 2nd ed. Boca Raton: CRC, 2015. Print.

Menezes, Alfred J., Paul C. van Oorschot, and Scott A. Vanstone. *Handbook of Applied Cryptography*. Boca Raton: CRC, 1996. Print.

Neiderreiter, Harald, and Chaoping Xing. *Algebraic Geometry in Coding Theory and Cryptography*. Princeton: Princeton UP, 2009. Print.

Paar, Christof, and Jan Pelzi. *Understanding Cryptography: A Textbook for Students and Practitioners*. Heidelberg: Springer, 2010. Print.

D

DATA MINING

ABSTRACT
Data mining is the relatively recent practice of using algorithms to distill patterns, summaries, and other specific forms of information from databases.

CATEGORY
Business, Economics, and Marketing.

FIELDS OF STUDY
Fields of Study: Data Analysis and Probability; Measurement; Number and Operations.

Advances in technology in the latter half of the twentieth century led to the accumulation of massive data sets in government, business, industry, and various sciences. Extracting useful information from these large-scale data sets required new mathematical and statistical methods to model data, account for error, and handle issues like missing data values and different variable scales or measures. Data mining uses tools from statistics, machine learning, computer science, and mathematics to extract information from data, especially from large databases. The concepts involved in data mining are drawn from many mathematical fields such as fuzzy sets, developed by mathematician and computer scientist Lotfi Zadeh, and genetic algorithms, based on the work of mathematicians such as Nils Barricelli. Because of the massive amounts of data processed, data mining relies heavily on computers, and mathematicians contribute to the development of new algorithms and hardware systems. For example, the Gfarm Grid File System was developed in the early twenty-first century to facilitate high-performance petascale-level computing and data mining.

HISTORY
Data mining has roots in three areas: classical statistics, artificial intelligence, and machine learning. In the late 1980s and early 1990s, companies that owned large databases of customer information, in particular credit card banks, wanted to explore the potential for learning more about their customers through their transactions. The term "data mining" had been used by statisticians since the 1960s as a pejorative term to describe the undisciplined exploration of data. It was also called "data dredging" and "fishing." However, in the 1990s, researchers and practitioners from the field of machine learning began successfully applying their algorithms to these large databases in order to discover patterns that enable businesses to make better decisions and to develop hypotheses for future investigations.

Partly to avoid the negative connotations of the term "data mining," researchers coined the term "knowledge discovery in databases" (KDD) to describe the entire process of finding useful patterns in databases, from the collection and preparation of the data, to the end product of communicating the results of the analyses to others. This term gained popularity in the machine learning and AI fields, but the term "data mining" is still used by statisticians. Those who use the term "KDD" refer to data mining as only the specific part of the KDD process where algorithms are applied to the data. The broader interpretation will be used in this discussion.

Software programs to implement data mining emerged in the 1990s and continue to evolve today. There are open-source programs (such as WEKA, http://www.cs.waikato.ac.nz/ml/weka and packages in R, http://www.r-project.org) and many commercial programs that offer easy-to-use graphical user interfaces (GUIs), which can facilitate the spread of data mining practice throughout an organization.

Types of Problems

The specific types of tasks that data mining addresses are typically broken into four types:

1. Predictive Modeling (classification, regression)
2. Segmentation (data clustering)
3. Summarization
4. Visualization

Predictive modeling is the building of models for a response variable for the main purpose of predicting the value of that response under new—or future—values of the predictor variables. Predictive modeling problems, in turn, are further broken into classification problems or regression problems, depending on the nature of the response variable being predicted. If the response variable is categorical (for example, whether a customer will switch telephone providers at the end of a subscription period or will stay with his or her current company), the problem is called a "classification." If the response is quantitative (for example, the amount a customer will spend with the company in the next year), the problem is a "regression problem." The term "regression" is used for these problems even when techniques other than regression are used to produce the predictions. Because there is a clear response variable, predictive modeling problems are also called "supervised problems" in machine learning. Sometimes there is no response variable to predict, but an analyst may want to divide customers into segments based on a variety of variables. These segments may be meaningful to the analyst, but there is no response variable to predict in order to evaluate the accuracy of the segmentation. Such problems with no specified response variable are known as "unsupervised learning problems."

Summarization describes any numerical summaries of variables that are not necessarily used to model a response. For example, an analyst may want to examine the average age, income, and credit scores of a large batch of potential new customers without wanting to predict other behaviors. Any use of graphical displays for this purpose, especially those involving many variables at the same time, is called "visualization."

Algorithms

Data mining uses a variety of algorithms (computer code) based on mathematical equations to build models that describe the relationship between the response variable and a set of predictor variables. The algorithms are taken from statistics and machine learning literature, including such classical statistical techniques as linear regression and logistic regression and time series analysis, as well as more recently developed techniques like classification and regression trees (ID3 or C4.5 in machine learning), neural networks, naïve Bayes, K-nearest neighbor techniques, and support vector machines.

One of the challenges of data mining is to choose which algorithm to use in a particular application. Unlike the practice in classical statistics, the data miner often builds multiple models on the same data set, using a new set of data (called the "test set") to evaluate which model performs best.

Recent advances in data mining combine models into ensembles in an effort to collect the benefits of the constituent models. The two main ensemble methods are known as "bootstrap aggregation" (bagging) and "boosting." Both methods build many (possibly hundreds or even thousands of) models on resampled versions of the same data set and take a (usually weighted) average (in the case of regression) or a majority vote (in the case of classification) to combine the models. The claim is that ensemble methods produce models with both less variance and less bias than individual models in a wide variety of applications. This is a current area of research in data mining.

Applications

Data mining techniques are being applied everywhere there are large data sets. A number of important application areas include the following:

1. *Customer relationship management (CRM).* Credit card banks formed one of the first groups of companies to use large transactional databases in an attempt to predict and understand patterns of customer behavior. Models help banks understand acquisition, retention, and cross-selling opportunities.

2. *Risk and collection analytics.* Predicting both who is most likely to default on loans and which type of collection strategy is likely to be successful is crucial to banks.
3. *Direct marketing.* Knowing which customers are most likely to respond to direct marketing could save companies billions of dollars a year in junk mail and other related costs.
4. *Fraud detection.* Models to identify fraudulent transactions are used by banks and a variety of government agencies including state comptroller's offices and the Internal Revenue Service (IRS).
5. *Terrorist detection.* Data mining has been used by various government agencies in an attempt to help identify terrorist activity—although concerns of confidentiality have accompanied these uses.
6. *Genomics and proteomics.* Researchers use data mining techniques in an attempt to associate specific genes and proteins with diseases and other biological activity. This field is also known as "bioinformatics."
7. *Healthcare.* Data mining is increasingly used to study efficiencies in physician decisions, pharmaceutical prescriptions, diagnostic results, and other healthcare outcomes.

Concerns and Controversies

Privacy issues are some of the main concerns of the public with respect to data mining. In fact, some kinds of data mining and discovery are illegal. There are federal and state privacy laws that protect the information of individuals. Nearly every Web site, credit card company, and other information collecting organization has a publicly available privacy policy. Social networking sites, such as Facebook, have been criticized for sharing and selling information about subscribers for data mining purposes. In healthcare, the Health Insurance Portability and Accountability Act of 1996 (HIPAA) was enacted to help protect individuals' health information from being shared without their knowledge.

—*Richard De Veaux*

Bibliography

Berry, M. A. J., and G. Linoff. *Data Mining Techniques For Marketing, Sales and Customer Support.* Hoboken, NJ: Wiley, 1997.

De Veaux, R. D. "Data Mining: A View From Down in the Pit." Stats 34 (2002).

———, and H. Edelstein. "Reducing Junk Mail Using Data Mining Techniques." In *Statistics: A Guide to the Unknown.* 4th ed. Belmont, CA: Thomson, Brooks-Cole, 2006.

Piatetsky-Shapiro, Gregory. "Knowledge Discovery in Real Databases: A Workshop Report." *AI Magazine* 11, no. 5 (January 1991).

DATA WAREHOUSE

A data warehouse is a database that collects information from various disparate sources and stores it in a central location for easy access and analysis. The data stored in a data warehouse is a static record, or a snapshot, of what each item looked like at a specific point in time. This data is not updated; rather, if the information has changed over time, later snapshots of the same item are simply added to the warehouse. Data warehouses are commonly used in business to aid management in tracking a company's progress and making decisions.

Overview

Computer scientist William H. Inmon, is largely credited with having codified and popularized the concept of data warehousing. Inmon began discussing the underlying principles in the 1970s and published the first book on the subject, *Building the Data Warehouse*, in 1992. The term itself was coined in 1988 when Barry A. Devlin and Paul T. Murphy published their paper "An Architecture for a Business and Information System," which discusses a new software architecture they describe as a "business data warehouse."

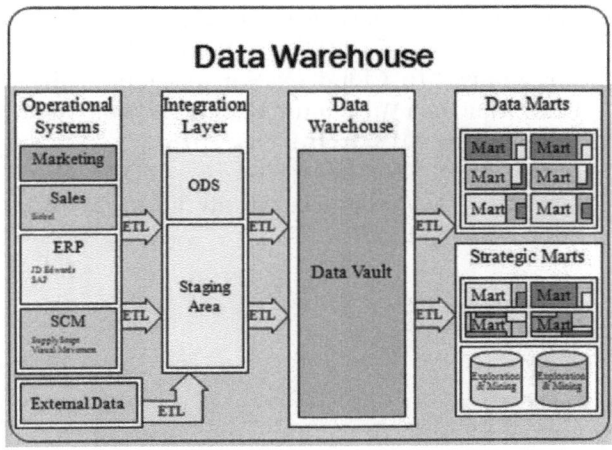

Data warehouse overview.

Simply put, a data warehouse is a central database that contains copies of information from several different sources, such as other, smaller databases and company records, stored in a read-only format so that nothing in the warehouse can be changed or overwritten. Since its inception, data warehousing has been associated with the field of business intelligence, which deals with analyzing a company's accumulated data for the purpose of developing business strategies. However, data warehouses are used in nonbusiness contexts as well, such as scientific research or analysis of crime statistics.

There are two main approaches to designing a data warehouse: from the top down and from the bottom up. Inmon champions the top-down model, in which the warehouse is designed as a central information repository. Data is entered into the warehouse in its most basic form, after which the warehouse can be subdivided into data marts, which group together related data for easy access by the team or department to which that information pertains. The advantages of creating data marts in a top-down data warehouse include increased performance for the user, as they can be accessed with greater efficiency than the much larger warehouse, as well as the ability to institute individual security protocols on different subsets of data. However, data marts are not a necessary component of top-down warehouse design, and they have their drawbacks, such as a propensity toward duplicated or inconsistent data.

Alternatively, the bottom-up model of data warehousing, developed by Ralph Kimball, views a data warehouse as simply a union of the various data marts. In this method, the data marts—separate databases used and maintained by individual departments—are created first and then linked to create a data warehouse. The merits of one approach over the other are often debated; the cheaper, faster, and more flexible bottom-up design is generally regarded as better in small or medium-sized businesses but too unwieldy for large corporations, where the sheer number of different people entering new data into different data marts would make any efforts to maintain consistency and avoid duplication extremely time consuming.

Cloud computing is the use of a remote server network on the Internet rather than a local server or hard drive to store, manage, and process data. Cloud computing can be used for data warehousing; warehousing data in the cloud offers both benefits and challenges. The benefits of moving a data warehouse to the cloud include the potential to lower costs and increase flexibility and access. Challenges include maintaining data security and privacy, network performance, and management of intellectual property.

—*Randa Tantawi, PhD*

BIBLIOGRAPHY

Chandran, Ravi. "DW-on-Demand: The Data Warehouse Redefined in the Cloud." *Business Intelligence Journal* 20.1 (2015): 8–13. *Business Source Complete*. Web. 8 June 2015.

Devlin, Barry A., and Paul T. Murphy. "An Architecture for a Business and Information System." *IBM Systems Journal* 27.1 (1988): 60–80. Print.

Greenfield, Larry. *Data Warehousing Information Center*. LGI Systems, 1995. Web. 8 Oct. 2013.

Griffith, Eric. "What Is Cloud Computing?" *PC Magazine*. Ziff Davis, PCMag Digital Group, 17 Apr. 2015. Web. 8 June 2015.

Henschen, Doug, Ben Werther, and Scott Gnau. "Big Data Debate: End Near for Data Warehousing?" *InformationWeek*. UBM Tech, 19 Nov. 2012. Web. 8 Oct. 2013.

Inmon, William H. *Building the Data Warehouse*. 4th ed. Indianapolis: Wiley, 2011. Print.

Laberge, Robert. *The Data Warehouse Mentor: Practical Data Warehouse and Business Intelligence Insights*. New York: McGraw, 2011. Print.

Mohamed, Arif. "A History of Cloud Computing." *Computer Weekly*. TechTarget, 2000–2015. Web. 8 June 2015.

Singh, Ajit, D. C. Upadhyay, and Hemant Yadav. "The Analytical Data Warehouse: A Sustainable Approach for Empowering Institutional Decision Making." *International Journal of Engineering Science and Technology* 3.7 (2011): 6049–57. PDF file.

Steier, Sandy. "To Cloud or Not to Cloud: Where Does Your Data Warehouse Belong?" *Wired*. Condé Nast, 29 May 2013. Web. 9 Oct. 2013.

Williams, Paul. "A Short History of Data Warehousing." *Dataversity*. Dataversity Educ., 23 Aug. 2012. Web. 8 Oct. 2013.

DATABASE DESIGN

Database design comprises the plan, models, specifications, and instructions for creating a structure (database) where data can be stored in a permanent (persistent) manner. Users can extract that data in combinations that answer questions in a particular area (domain). Successful database design requires complete and current understanding of database technologies and methods. Further, designers must understand how people work and the types of questions a database will answer. Because of the wide variety of technical and intellectual skills that come into play, there is an ongoing debate as to whether database design is an art or a science.

Overview

Databases store data so users can access it to provide information and insights into how an organization is functioning. Electronic databases, as we currently use the term, date back only to the late 1950s when they were developed for the US Department of Defense. By the 1960s databases were being designed and created for business and academic use.

In the intervening years database design has changed in response to increased computing capabilities and, just as importantly, to the changing needs of organizations and the increasingly sophisticated information expertise of potential users.

Several database design methods have come into favor over the years, resulting from an evolution in how users and designers have looked at data and how it can serve their needs. Organizations' information requirements have become larger and more complex; how people look at information and its use has also grown more sophisticated.

Variations in database design exist but methodologies generally follow a sequence of steps starting with gathering information and user requirements through choosing the appropriate software and hardware technologies and then iterative development and testing. A six-step process defined by the University of Liverpool's Computer Science Department is typical.

The first step in the Liverpool methodology is the requirements analysis where an organization's information needs are documented. What does the organization need to know? What pieces of data are required to form the picture that the organization requires? How are these data to be organized and relationships defined? The second step, conceptual database design, identifies all of the pieces of data and puts them into a model in which their place and relationships are also defined. At the same time, the business processes that create the data are also modeled. This second step is extremely critical because it requires that the database designers thoroughly understand the customers' needs and way of working. Often a problem arises at this stage because while database designers are experts at databases they do not always have similar expertise in the organization's business area.

The third step is the choice of hardware and software, including the choice of security programs to protect the database. In some instances, the available options are wide, allowing potential users to get the capabilities they need on an affordable scale.

The fourth step is logical design. Business processes (transactions) are documented and diagrammed followed by a mapping of these workflows to the capabilities of the selected (or newly developed) database. The fifth step (physical design) determines where data will be stored and how it is moved in the database.

The final step is planning how the database is actually developed with continual (iterative) testing at every stage until installation and use when it is then usually managed by a company's database administrator.

—*Robert N. Stacy, CSS, MA*

BIBLIOGRAPHY

Badia, Antonio, and Daniel Lemire. "A Call to Arms: Revisiting Database Design." *ACM SIGMOD Record* 40.3 (2009): 61–69. Print.

Buxton, Stephen. *Database Design: Know it All.* Boston: Morgan Kaufmann, 2009. Print.

Churcher, Clare. *Beginning Database Design: From Novice to Professional.* 2nd ed. Berkeley: Apress, 2012. Print.

Creswell, John. W. *Research Design: Qualitative, Quantitative, and Mixed Methods Approaches.* Thousand Oaks: Sage, 2013. Print.

Currim, Sabah, et al. Using a Knowledge Learning Framework to Predict Errors in Database Design." *Elsevier* 40 (Mar. 2014): 11–31. Print.

Grad, Burton, and Thomas J. Bergin. "History of Database Management Systems." *IEEE Annals of the History of Computing* 31.4 (2009): 3–5. Print.

Hernandez, Michael J. *Database Design for Mere Mortals: A Hands-On Guide to Relational Database Design.* Upper Saddle River: Addison-Wesley, 2013. Print.

Stephens, Rod. *Beginning Database Design Solutions.* Indianapolis: John Wiley & Sons, 2009. Print.

DATABASE STRUCTURING CONVENTIONS

FIELDS OF STUDY

Software Development; Coding Techniques; Computer Science

ABSTRACT

Database structuring conventions are used to ensure that a database design provides access to the required data quickly and accurately while making efficient use of system resources such as network bandwidth and database storage. Standard conventions are used in defining entities, attributes, keys, views, and triggers.

PRINCIPAL TERMS

- **attribute:** in computer science, a property or characteristic of an object or entity.
- **entity:** in a database, an item of interest whose attributes constitute the data being stored.
- **foreign key:** an attribute of an entity in one table of a database that is the primary key of an entity in another table of the same database and serves to establish a relationship between the two entities.
- **primary key:** an attribute that uniquely defines each entity stored in a database.
- **trigger:** a procedure that executes when data is changed in a database.

ENTITIES AND THEIR ATTRIBUTES

Databases are built on entities. An entity is simply a single item of interest, such as a person, thing, event, or place, about which data is stored. For example, entities in a database might represent employees (people), products (things), concerts (events), or towns (places). Each entity has an associated set of attributes, which constitute the information stored in the database. For example, attributes of employees might include position, salary, and years of employment. Databases can be modeled as tables, with each entity represented by a row in the table and the various attributes represented by columns. Entities of

Entities	Attributes
employee	id, first_name, last_name, department_id
department	id, name, campus_id
campus	id, name, location

different types are listed in separate tables. If a database contained records of both a company's employees and the different departments within that company, for example, one table would list the employees and their attributes, and the other would list the departments and their attributes.

Relationships can be constructed between entities of different types. These relationships take one of the following main forms: one-to-one, one-to-many, or many-to-many. For example, multiple employees (many entities) might each have a relationship with the same department (one entity). To construct relationships, each entity must first be assigned an attribute called a primary key. This is a unique identifier whose value must not be repeated within the same table. An employee number might be used to uniquely identify each employee, and a unique department identification (ID) number could be created for each department. Next, the relationship between entities is defined by making the primary key of an entity in one table an attribute of an entity in the other table. This new attribute is called a foreign key. For example, the foreign key column of the employee table would hold the department ID of each employee's department, establishing a relationship between the employee and their department.

When data is updated in one table, it is often necessary to update data in a related table. For example, if the department ID changes, that information must also be changed in the employee table for all employees of that department. Triggers and constraints can be used to ensure that the correct updates occur.

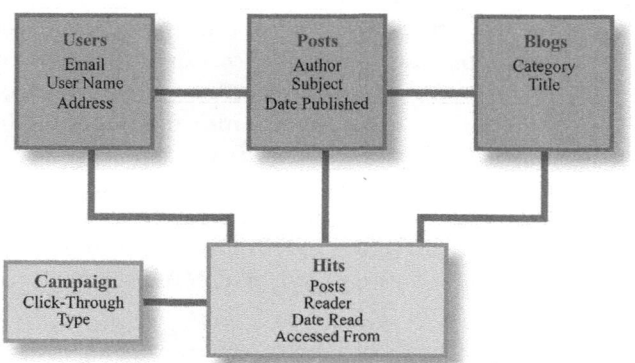

Common conventions for building a database can be visualized in an entity relationship model. This model shows the relationships between users, blog posts, blogs, hits, and campaigns and lists some of the attributes of each of those entities.

Establishing a consistent naming convention is an important part of database design. When naming objects in a database, keep names as short, clear, and descriptive as possible. If abbreviations are needed, use standard ones. Avoid using terms that might change over time, as database objects can be difficult to rename. An organization's naming conventions, such as those used for file naming, should inform database naming. Various database systems also impose their own limits on object naming. In general, underscores, not spaces, are used to separate words. A prefix or suffix may identify views.

Using Normalization to Optimize Databases

Once the necessary entities and attributes are determined, the database structure can be optimized through normalization. Normalization maximizes the use of keys to relate tables to reduce duplication. For example, if the database of a US-based company had one table for employees and another for customers, both tables would likely have an attribute for each entity's state of residence. As a result, data would be duplicated between the two tables, taking up unnecessary space. In addition, if the state names were abbreviated differently in the different tables—for example, if Texas were abbreviated as "Tex." in one table and "TX" in the other—this could cause errors. During normalization, a state table would be created to store the name and abbreviation of each state. This table would then be linked to the employee and customer tables using keys, thus preventing data duplication.

Normalization also ensures that the correct attributes are stored in the correct tables. For instance, it is more logical for an employee's work hours to be stored in the employee table than in the department table. This makes searches, or queries, of the data easier to perform.

Normalization reduces the size of tables, which offers performance advantages. Updates and insertions are faster because there is no need to update or add data in multiple places. Deletions do not risk the loss of other, needed data. In general, normalized databases perform well when data is written to the database or changed more often than it is read.

In contrast, denormalized databases offer performance advantages when data is read from the database more often than it is written. This is because the data required can be retrieved from a single

> **SAMPLE PROBLEM**
>
> A developer is designing a database to store information used by a university's payroll department. The database will contain at least three tables: one for the university's employees, one for its departments, and one for the different campus locations. The following is a partial list of the names that have been assigned to various database objects.
>
> In addition, two views have been named: vw_humanities_departments and vw_downtown_maintenance_employees. Given these names, describe the naming conventions used by the developer.
>
> **Answer:**
>
> All names are in lowercase, and multiple words in names are separated by underscores. Few standard abbreviations have been used (e.g., in department_id, "id" is an abbreviation, but "department" is spelled out rather than abbreviated as "dept"). Entity and attribute names are singular. The primary keys for each table are assigned a short name (id), and foreign key names combine the name of the entities in the referenced table with the name of the primary key (e.g., department_id). Views are identified with a prefix (vw_).

Normalization and denormalization must therefore balance the read and write performance for a particular database.

GOOD STRUCTURE MAXIMIZES PERFORMANCE

In order for a database to offer the maximum possible performance while using system resources efficiently, it must be well designed. The foundation of a successful design is a sound database structure. Established database structuring conventions help developers create the most efficient database based on the individual requirements of the end user. Following these conventions has proved to result in improved database design.

—*Maura Valentino, MSLIS*

BIBLIOGRAPHY

Churcher, Clare. *Beginning Database Design: From Novice to Professional.* 2nd ed., Apress, 2012.

Harrington, Jan L. *Relational Database Design and Implementation.* 4th ed., Elsevier, 2016.

Hernandez, Michael J. *Database Design for Mere Mortals: A Hands-On Guide to Relational Database Design.* 3rd ed., Addison-Wesley, 2013.

MacLennan, Bruce J. *Principles of Programming Languages: Design, Evaluation, and Implementation.* 3rd ed., Oxford UP, 1999.

Mullins, Craig S. *Database Administration: The Complete Guide to DBA Practices and Procedures.* 2nd ed., Addison-Wesley, 2013.

Scott, Michael L. *Programming Language Pragmatics.* 4th ed., Elsevier, 2016.

Van Roy, Peter, and Seif Haridi. *Concepts, Techniques, and Models of Computer Programming.* MIT P, 2004.

table, which is faster than accessing multiple tables, as is needed in a normalized database. Index usage is also more efficient if a database is denormalized.

DEBUGGING

FIELDS OF STUDY

Computer Science; Software Engineering

ABSTRACT

Debugging is the process of identifying and addressing errors, known as "bugs," in computer systems. It is an essential step in the development of all kinds of programs, from consumer programs such as web browsers and video games to the complex systems used in transportation and infrastructure. Debugging can be carried out through a number of methods depending on the nature of the computer system in question.

PRINCIPAL TERMS

- **delta debugging:** an automated method of debugging intended to identify a bug's root cause while eliminating irrelevant information.
- **in-circuit emulator:** a device that enables the debugging of a computer system embedded within a larger system.
- **integration testing:** a process in which multiple units are tested individually and when working in concert.
- **memory dumps:** computer memory records from when a program crashed, used to pinpoint and address the bug that caused the crash.
- **software patches:** updates to software that correct bugs or make other improvements.

UNDERSTANDING DEBUGGING

Debugging is the process of testing software or other computer systems, noting any errors that occur, and finding the cause of those errors. Errors, or "bugs," in a computer program can seriously affect the program's operations or even prevent it from functioning altogether. The goal of debugging is to get rid of the bugs that have been identified. This should ensure the smooth and error-free operation of the computer program or system.

Computer programs consist of long strings of specialized code that tell the computer what to do and how to do it. Computer code must use specific vocabulary and structures to function properly. As such code is written by human programmers, there is always the possibility of human error, which is the cause of many common bugs. Perhaps the most common bugs are syntax errors. These are the result of small mistakes, such as typos, in a program's code. In some cases, a bug may occur because the programmer neglected to include a key element in the code or structured it incorrectly. For example, the code could include a command instructing the computer to begin a specific process but lack the corresponding command to end it.

Bugs fall into one of several categories, based on when and how they affect the program. Compilation errors prevent the program from running. Run-time errors, meanwhile, occur as the program is running. Logic errors, in which flaws in the program's logic produce unintended results, are a particularly common form of bug. Such errors come about when a program's code is syntactically correct but does not make logical sense. For instance, a string of code with flawed logic may cause the program to become caught in an unintended loop. This can cause it to become completely unresponsive, or freeze. In other cases, a logic error might result when a program's code instructs the computer to divide a numerical value by zero, a mathematically impossible task.

WHY DEBUG?

Bugs may interfere with a program's ability to perform its core functions or even to run. Not all bugs are related to a program's core functions, and some programs may be usable despite the errors they contain. However, ease of use is an important factor that many people consider when deciding which program to use. It is therefore in the best interest of software creators to ensure that their programs are as free of errors as possible. In addition to testing a program or other computer system in house prior to releasing them to the public, many software companies collect reports of bugs from users following its release. This is often done through transfers of collected data commonly referred to as memory dumps. They can then address such errors through updates known as software patches.

While bugs are an inconvenience in consumer computer programs, in more specialized computer systems, they can have far more serious consequences. In areas such as transportation, infrastructure, and finance, errors in syntax and logic can place lives and livelihoods at risk. Perhaps the most prominent example of such a bug was the so-called Y2K bug. This bug was projected to affect numerous computer systems beginning on January 1, 2000. The problem would have resulted from existing practices related to the way dates were written in computer programs. However, it was largely averted through the work of programmers who updated the affected programs to prevent that issue. As the example of the far-reaching Y2K bug shows, the world's growing reliance on computers in all areas of society has made thorough debugging even more important.

IDENTIFYING AND ADDRESSING BUGS

The means of debugging vary based on the nature of the computer program or system in question. However, in most cases bugs may be identified and addressed through the same general process. When a bug first appears, the programmer or tester must first attempt to reproduce the bug in order to identify the results of the error and the conditions under which they occur. Next, the programmer must attempt to determine which part of the program's code is causing the error to occur. As programs can be quite complex, the programmer must simplify this process by eliminating as much irrelevant data as possible. Once the faulty segment of code has been found, the programmer must identify and correct the specific problem that is causing the bug. If the cause is a typo or a syntax error, the programmer may simply need to make a small correction. If there is a logic error, the programmer may need to rewrite a portion of the code so that the program operates logically.

Debugging in Practice

Programmers use a wide range of tools to debug programs. As programs are frequently complex and lengthy, automating portions of the debugging process is often essential. Automated debugging programs, or "debuggers," search through code line by line for syntax errors or faulty logic that could cause bugs. A technique known as delta debugging provides an automated means of filtering out irrelevant information when the programmer is looking for the root cause of a bug.

Different types of programs or systems often require different debugging tools. An in-circuit emulator is used when the computer system being tested is an embedded system (that is, one located within a larger system) and cannot otherwise be accessed. A form of debugging known as integration testing is often used when a program consists of numerous components. After each component is tested and debugged on its own, they are linked together and tested as a unit. This ensures that the different components function correctly when working together.

Historical Note

The first computer bug was an actual bug (insect) that had flown into one of the old vacuum tube computers and fried itself. The name stuck even after the technology changed.

—*Joy Crelin*

Bibliography

Foote, Steven. *Learning to Program.* Upper Saddle River: Pearson, 2015. Print.

McCauley, Renée, et al. "Debugging: A Review of the Literature from an Educational Perspective." *Computer Science Education* 18.2 (2008): 67–92. Print.

Myers, Glenford J., Tom Badgett, and Corey Sandler. *The Art of Software Testing.* Hoboken: Wiley, 2012. Print.

St. Germain, H. James de. "Debugging Programs." *University of Utah.* U of Utah, n.d. Web. 31 Jan. 2016.

"What Went Wrong? Finding and Fixing Errors through Debugging." *Microsoft Developer Network Library.* Microsoft, 2016. Web. 31 Jan. 2016.

Zeller, Andreas. *Why Programs Fail: A Guide to Systematic Debugging.* Burlington: Kaufmann, 2009. Print.

DEVICE DRIVERS

FIELDS OF STUDY

Computer Engineering; Software Engineering

ABSTRACT

Device drivers are software interfaces that allow a computer's central processing unit (CPU) to communicate with peripherals such as disk drives, printers, and scanners. Without device drivers, the computer's operating system (OS) would have to come preinstalled with information about all of the devices it could ever need to communicate with. OSs contain some device drivers, but these can also be installed when new devices are added to a computer.

PRINCIPAL TERMS

- **device manager:** an application that allows users of a computer to manipulate the device drivers installed on the computer, as well as adding and removing drivers.
- **input/output instructions:** instructions used by the central processing unit (CPU) of a computer

when information is transferred between the CPU and a device such as a hard disk.
- **interface:** the function performed by the device driver, which mediates between the hardware of the peripheral and the hardware of the computer.
- **peripheral:** a device that is connected to a computer and used by the computer but is not part of the computer, such as a printer.
- **virtual device driver:** a type of device driver used by the Windows operating system that handles communications between emulated hardware and other devices.

How Device Drivers Work

The main strength of device drivers is that they enable programmers to write software that will run on a computer regardless of the type of devices that are connected to that computer. Using device drivers allows the program to simply command the computer to save data to a file on the hard drive. It needs no specific information about what type of hard drive is installed in the computer or connections the hard drive has to other hardware in the computer. The device driver acts as an interface between computer components.

When a program needs to send commands to a peripheral connected to the computer, the program communicates with the device driver. The device driver receives the information about the action that the device is being asked to perform. It translates this information into a format that can be input into the device. The device then performs the task or tasks requested. When it finishes, it may generate output that is communicated to the driver, either as a message or as a simple indication that the task has been completed. The driver then translates this information into a form that the original program can understand. The device driver acts as a kind of translator between the computer and its peripherals, conveying input/output instructions between the two. Thus, the computer program does not need to include all the low-level commands needed to make the device function. The program only needs to be able to tell the device driver what it wants the device to do. The device driver takes care of translating this into concrete steps.

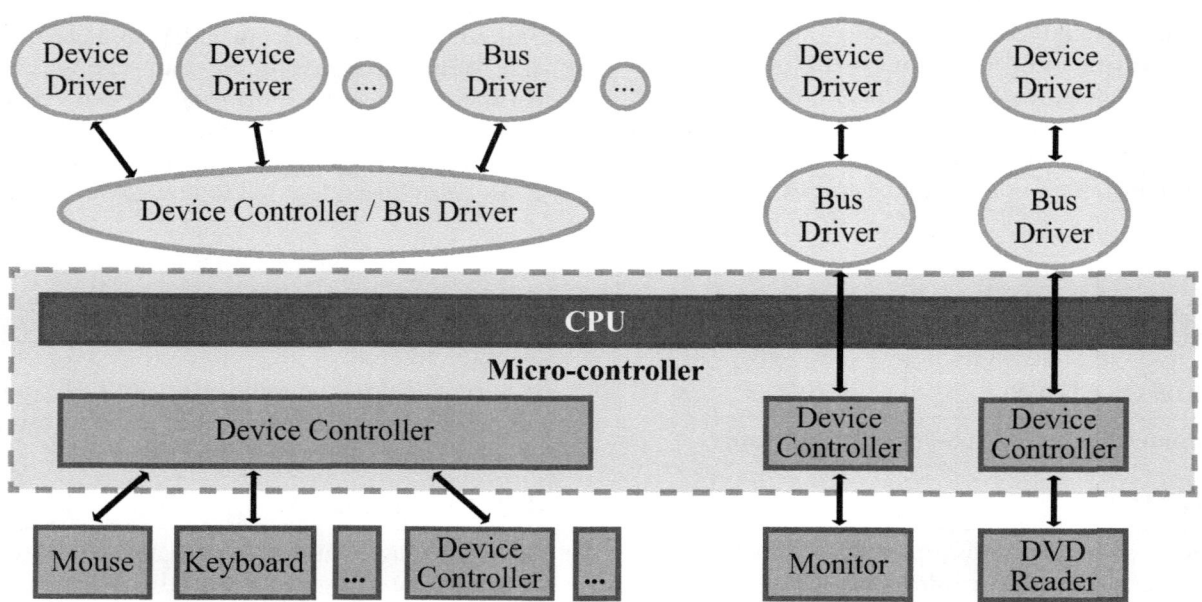

Each device connected to a CPU is controlled by a device driver, software that controls, manages, and monitors a specific device (e.g., keyboard, mouse, monitor, DVD reader). Device drivers may also drive other software that drives a device (e.g., system management bus, universal serial bus controller).

How Device Drivers Are Made

Writing device drivers is a highly technical undertaking. It is made more challenging by the fact that device drivers can be unforgiving when a mistake is made in their creation. This is because higher-level applications do not often have unlimited access to all of the computer's functionality. Issuing the wrong command with unrestricted privileges can cause serious damage to the computer's operating system (OS) and, in some cases, to the hardware. This is a real possibility with device drivers, which usually need to have unrestricted access to the computer.

Because writing a device driver requires a lot of specialized information, most device drivers are made by software engineers who specialize in driver development and work for hardware manufacturers. Usually the device manufacturer has the most information about the device and what it needs to function properly. The exception to this trend is the impressive amount of driver development accomplished by the open-source movement. Programmers all over the world have volunteered their own time and talent to write drivers for the Linux OS.

Often development is separated into logical and physical device driver development. Logical device driver development tends to be done by the creator of the OS that the computer will use. Physical device driver development, meanwhile, is handled by the device manufacturer. This division of labor makes sense, but it does require coordination and a willingness to share standards and practices among the various parties.

Virtual Device Drivers

Virtual device drivers are a variation on traditional device drivers. They are used when a computer needs to emulate a piece of hardware. This often occurs when an OS runs a program that was created for a different OS by emulating that operating environment. One example would be a Windows OS running a DOS program. If the DOS program needed to interface with an attached printer, the computer would use a virtual device driver.

Device Managers

Most OSs now include device managers that make it easier for the user to manage device drivers. They allow the user to diagnose problems with devices, troubleshoot issues, and update or install drivers. Using the graphical interface of a device manager is less intimidating than typing in text commands to perform driver-related tasks.

—*Scott Zimmer, JD*

Bibliography

Corbet, Jonathan, Alessandro Rubini, and Greg Kroah-Hartman. *Linux Device Drivers*. 3rd ed. Cambridge: O'Reilly, 2005. Print.

McFedries, Paul. *Fixing Your Computer: Absolute Beginner's Guide.* Indianapolis: Que, 2014. Print.

Mueller, Scott. *Upgrading and Repairing PCs*. 22nd ed. Indianapolis: Que, 2015. Print.

Noergaard, Tammy. *Embedded Systems Architecture: A Comprehensive Guide for Engineers and Programmers*. 2nd ed. Boston: Elsevier, 2012. Print.

Orwick, Penny, and Guy Smith. *Developing Drivers with the Windows Driver Foundation*. Redmond: Microsoft P, 2007. Print.

"What Is a Driver?" *Microsoft Developer Network*. Microsoft, n.d. Web. 10 Mar. 2016.

DIFFUSION OF INNOVATIONS

Diffusion of innovations (DOI) theory offers a framework for studying the processes of adopting an innovation through the lens of change; it tells the story of why and how quickly change occurs. Everett Rogers's 1962 seminal book, *Diffusion of Innovations*, defines innovations as ideas, practices, or objects perceived as new by an individual or culture. Diffusion is the process whereby an innovation is communicated over time among the members of a social system resulting in individual or social change. Diffusion of innovations applies primarily to technology integration and paves the way to study adoption in other disciplines.

Overview

In the fifth edition of *Diffusions of Innovations* Rogers describes adoption as the "full use of an innovation as the best course of action available" and rejection as the decision "not to adopt an innovation." Four elements of DOI theory include innovation, time, communication channels, and social systems. The acceptance of an innovation is influenced by uncertainty. Change can often be a slow, uncomfortable process, which takes time before diffusion is accomplished.

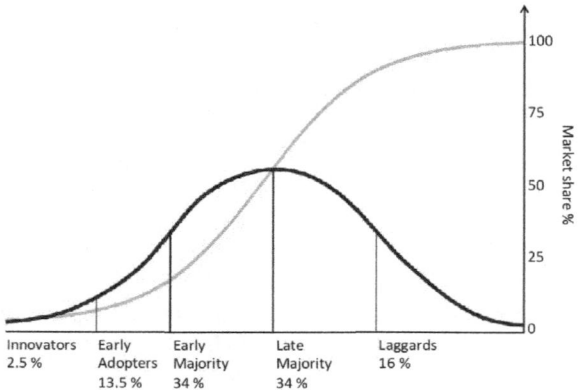

The diffusion of innovations according to Rogers (1962). With successive groups of consumers adopting the new technology (shown in blue), its market share (yellow) will eventually reach the saturation level.

Specialized interpersonal communication channels are necessary for diffusion. Social systems ultimately decide to modify an innovation to fit their culture.

Rogers proposes that the rate of adoption is influenced by multiple perceptions of relative advantage over a previous technology, compatibility with existing needs, complexity and perceived difficulty of use, and available trialability and observability to experiment and see the results of the innovation. The rate of adoption has been found to be positively correlated to these characteristics.

Diffusion of an innovation occurs in stages, beginning with knowledge and followed by persuasion, decision, implementation, and finally confirmation. Knowledge relates to the upfront understanding of the new technology. Persuasion is a positive attitudinal disposition. Decision is closely aligned to commitment followed by implementation. Confirmation is determined by the positive outcomes that reinforce use.

A final component offered by the DOI theory is a classification structure of adopters. Innovators tend to rapidly embrace a technology, followed closely by early adopters who readily accept change. Alternately, early majority adopters typically need more time, whereas, late majority adopters openly express skepticism but will eventually buy-in to the innovation once the majority has accepted the change. Individuals more comfortable with the status quo are termed laggards.

Adoption research has kindled the development of two popular instruments designed to measure change. In 1973 Gene Hall and colleagues developed the concerns-based adoption model (CBAM) that measures change along the continuum of skill development. A second dimension of CBAM identifies stages of concern (SoC), in which concern is conceptualized as the intensity of motivations, thoughts, and feelings moving an individual past internal acceptance toward a more global perspective.

Rogers's theoretical framework remains the most widely used theory leading to a deeper understanding of the process and rate of adoption of an innovation. The characteristics and classifications described are the pivotal factors influencing individual change and acceptance of an innovation into a social system.

—*Mitzi P. Trahan, MA, PhD*

Bibliography

Gollakota, Kamala, and Kokila Doshi. "Diffusion of Technological Innovations in Rural Areas." *Journal of Corporate Citizenship* 41 (2011): 69–82. Print.

Hall, Gene E., R. C. Wallace, and W. A. Dossett. *A Developmental Conceptualization of the Adoption Process within Educational Institutions*. Austin: U of Texas, 1973. Print.

Hebert, Dustin M. "Innovation Diffusion Factors Affecting Electronic Assessment System Adoption in Teacher Education." *National Teacher Educational Journal* 5.2 (2012): 35–44. Print.

Magsamen-Conrad, Kate, et al. "Bridging the Divide: Using UTAUT to Predict Multigenerational Tablet Adoption Practices." *Computers in Human Behavior* 50 (2015): 186–196. Print.

Neo, Emily, and Philip J. Calvert. "Facebook and the Diffusion of Innovation in New Zealand Public Libraries." *Journal of Librarianship and Information Science* 44.4 (2012): 227–37. Print.

Parrinello, Michael. C. "Prevention of Metabolic Syndrome from Atypical Antipsychotic Medications: Applying Rogers' Diffusion of Innovations Model in Clinical Practice." *Journal of Psychosocial Nursing & Mental Health Services* 50.12 (2012): 36–44. Print.

Rogers, Everett M. *Diffusion of Innovations*. 5th ed. New York: Free Press, 2003. Print.

Sahin, Ismail I. "Detailed Review of Rogers' Diffusion of Innovations Theory and Educational Technology-related Studies Based on Rogers' Theory." *The Turkish Online Journal of Educational Technology* 5.2 (2006): 14–23. PDF file.

Shinohara, Kazunori, and Hiroshi Okuda. "Dynamic Innovation Diffusion Modeling." *Computational Economics* 35.1 (2010): 51–62. Print.

Soffer, Tal, Rafi Nachmias, and Judith Ram. "Diffusion of Web Supported Instruction in Higher Education—the Case of Tel-Aviv University." *Journal of Educational Technology & Society* 13.3 (2010): 212. Print.

Stummer, Christian, et al. "Innovation Diffusion of Repeat Purchase Products in a Competitive Market: An Agent-Based Simulation Approach." *European Journal of Operational Research* 245.1 (2015): 157–167. Print.

Zhang, Xiaojun, et al. "Using Diffusion of Innovation Theory to Understand the Factors Impacting Patient Acceptance and Use of Consumer E-Health Innovations: A Case Study in a Primary Care Clinic." *BMC Health Services Research* 15.1 (2015): 1–15. PDF file.

DIGITAL DIVIDE

The advent of Information and Communication Technology (ICT), including a range of digital devices that allow connection to the Internet, has changed the way people communicate, interact, learn, produce, and circulate information. While ICTs have been lauded for their potential to democratize access to information, a serious gap exists in access to and use of ICTs. This gap is often referred to as the *digital divide*. The digital divide surfaces between countries based on economic and technological infrastructure. It also emerges within a country, between individuals who differ in terms of social class, education, location (urban or rural), age, and racial and ethnic backgrounds. Over time, the definition of the digital divide has expanded to include access to skill development, as well as access and use of ICTs.

OVERVIEW

Initially, the digital divide was defined as the gap between those with access to ICTs, the "haves," and those without access, the "have-nots." The absence of or limited access to affordable ICT infrastructure and services is viewed as a contributing factor to the growing digital divide between and within countries. Within countries, differences along the lines of income, education, location, age, and race and ethnicity are factors that differentiate ICT and Internet use. For example, in the United States, 2013 data from the Pew Research Center shows that children and youth from American Indian, Latino, and African-American backgrounds have significantly lower access to technology than children and youth from white and Asian-American backgrounds; residents in rural areas less likely to have access to computers in their households; and there is a generational gap between young people, who have had access to and experience with ICTs, and older adults, for whom opportunities to access and learn with ICTs has been limited. People with disabilities and people who speak Spanish as their primary language are also significantly less likely to use the Internet at all, and the gap is even larger when looking only at those who can access the Internet at home. The factors that appear to be most significant for higher levels of access and use, however, are higher levels of income and education.

The digital divide identifies a new landscape of inequity. People who do not have access to ICTs early in life and/or frequently are at a disadvantage. They will have limited access to information, including media reports, medical information, and opportunities to learn, as well as limited access to current methods of communication. Cost, inadequate computer skills, and lack of interest are some factors relating to people's disconnection from the Internet. In response, scholars are studying a second kind of divide that

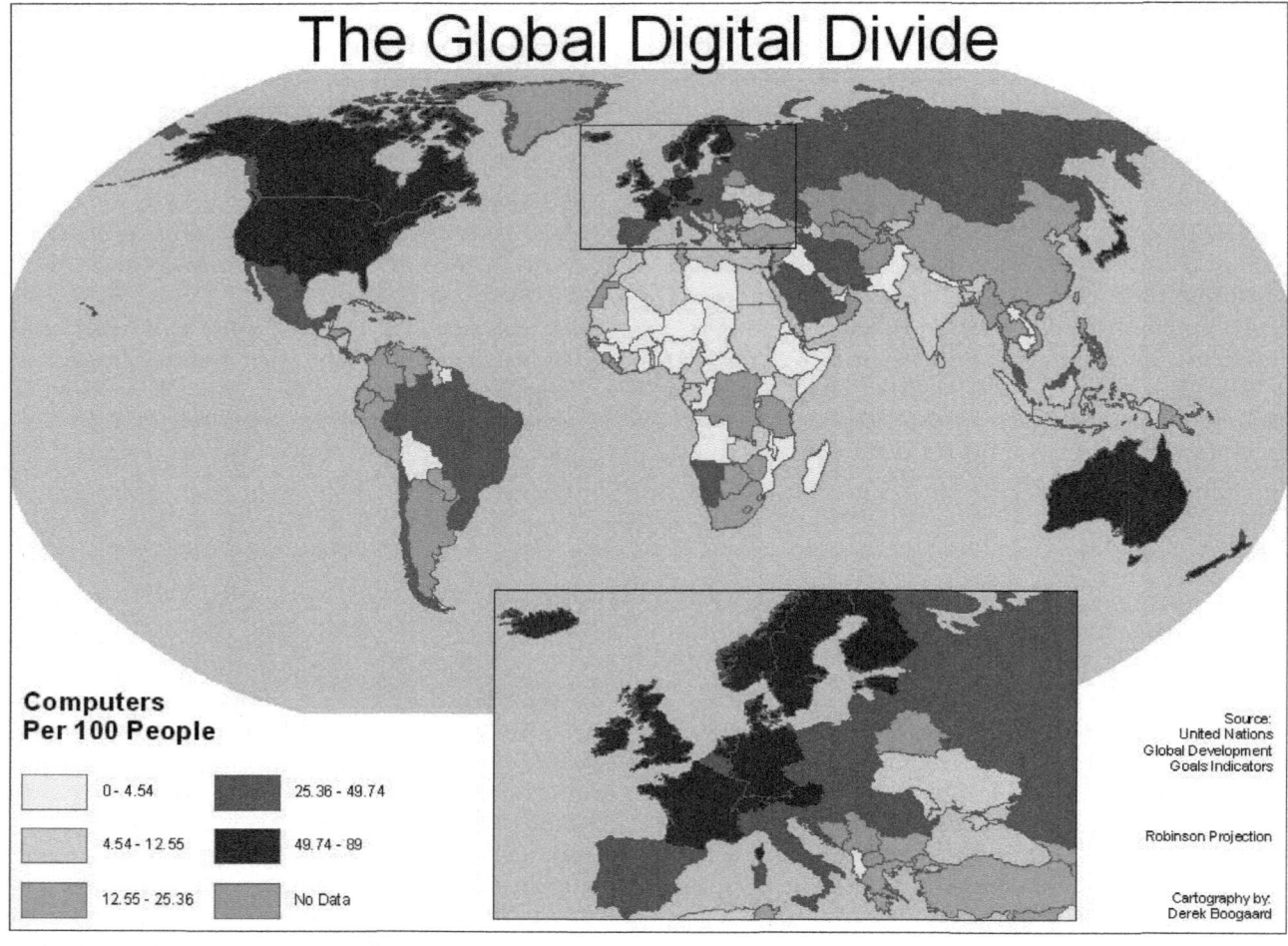

Der Digitale Graben (vgl. 1. Abbildung); Karte ohne Jahr.

includes differences in ICT skills and skill development as attributes of a widening digital gap.

Access to, use of, and skills for engaging with ICTs are not developing equally across groups, communities, and countries. A lack of availability, access, and experience with ICTs, along with the thinking and practical skills that develop as a result, is hampered by insufficient public services and scarcity in technology resources. Educational environments are positioned to either: (1) contribute to the divide, for example, when educators lack preparation and support to teach with ICTs; or (2) enable children and youth to overcome it, for example, when the use of ICTs is integrated throughout classroom practices as a method for both teaching and learning. Educational policies and practices are changing in ways that support the effective use of ICTs in classrooms and enable students to benefit from, and grow with, the digitized world. However, until availability, access, and skills are made equally available to everyone, inside of and outside of schools, the digital divide is likely to exist.

—*Hitaf R. Kady, MA, and Jennifer A. Vadeboncoeur, MA, PhD*

Bibliography

Creeber, Glen, and Royston Martin, eds. *Digital Cultures: Understanding New Media*. Berkshire: Open UP, 2009. Print.

File, Thom. "Computer and Internet Use in the United States: Population Characteristics." *Census.gov*. US Dept. of Commerce, 2013. PDF file.

Graham, Mark. "The Knowledge Based Economy and Digital Divisions of Labour." *The Companion to Development Studies.* Ed. V. Desai and R. Potter. 3rd ed. London: Routledge, 2014. 189–95. Print.

Kellner, Douglas. "New Media and New Literacies: Reconstructing Education for the New Millennium." The Handbook of New Media: Social Shaping and Consequences of ICTs. Eds. Leah Lievrouw and Sonia Livingstone. Thousand Oaks: Sage, 2002. Print.

Light, Jennifer. "Rethinking the Digital Divide." Harvard Educational Review 71.4 (2001): 709–33. Print.

Looker, Dianne, and Victor Thiessen. "The Digital Divide in Canadian Schools: Factors Affecting Student Access to and Use of Information Technology." Statistics Canada. Statistics Canada, 2003. PDF file.

Min, Seong-Jae. "From the Digital Divide to the Democratic Divide: Internet Skills, Political Interest, and the Second-Level Digital Divide in Political Internet Use." Journal of Information Technology & Politics 7 (2010): 22–35. *Taylor & Francis Online.* PDF file.

Nair, Mahendhiran, Mudiarasan Kuppusamy, and Ron Davison. "A Longitudinal Study on the Global Digital Divide Problem: Strategies to Close Cross-Country Digital Gap." *The Business Review Cambridge* 4.1 (2005): 315–26. PDF file.

National Telecommunications and Information Administration. "Digital Nation: Expanding Internet Usage." *National Telecommunications and Information Administration.* US Dept. of Commerce, 2011. PDF file.

Rainie, Lee. "The State of Digital Divides." *Pew Research Internet Project.* Pew Research Center, 5 Nov. 2013. Web. 11 Nov. 2014.

Tatalović, Mićo. "How Mobile Phones Increased the Digital Divide." *SciDev.Net.* SciDev.Net, 26 Feb. 2014. Web. 11 Nov. 2014.

van Deursen, Alexander, and Jan van Dijk. "Internet Skills and the Digital Divide." *New Media and Society* 13.6 (2010): 893–911. PDF file.

DIGITAL FORENSICS

FIELDS OF STUDY

Information Technology; System Analysis; Privacy

ABSTRACT

Digital forensics is a branch of science that studies stored digital data. The field emerged in the 1990s but did not develop national standards until the 2000s. Digital forensics techniques are changing rapidly due to the advances in digital technology.

PRINCIPAL TERMS

- **cybercrime:** crime that involves targeting a computer or using a computer or computer network to commit a crime, such as computer hacking, digital piracy, and the use of malware or spyware.
- **Electronic Communications Privacy Act:** a 1986 law that extended restrictions on wiretapping to cover the retrieval or interception of information transmitted electronically between computers or through computer networks.
- **logical copy:** a copy of a hard drive or disk that captures active data and files in a different configuration from the original, usually excluding free space and artifacts such as file remnants; contrasts with a physical copy, which is an exact copy with the same size and configuration as the original.
- **metadata:** data that contains information about other data, such as author information, organizational information, or how and when the data was created.
- **Scientific Working Group on Digital Evidence (SWGDE):** an American association of various academic and professional organizations interested in the development of digital forensics systems, guidelines, techniques, and standards.

An Evolving Science

Digital forensics is the science of recovering and studying digital data, typically in criminal investigations. Digital forensic science is used to investigate

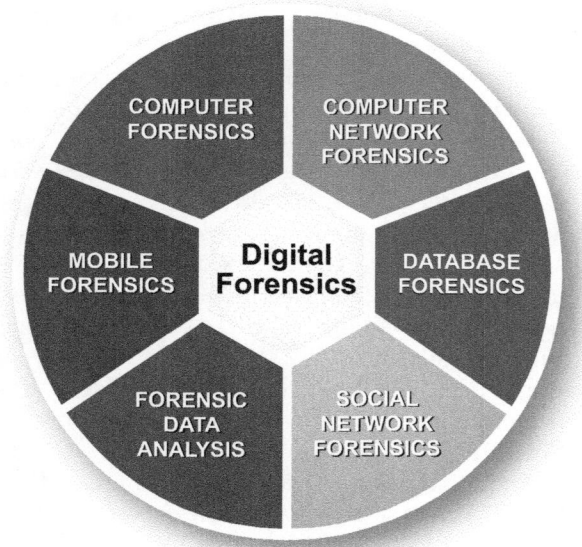

Digital forensics encompasses computer forensics, mobile forensics, computer network forensics, social networking forensics, database forensics, and forensic data analysis or the forensic analysis of large-scale data.

cybercrimes. These crimes target or involve the use of computer systems. Examples include identity theft, digital piracy, hacking, data theft, and cyberattacks. The Scientific Working Group on Digital Evidence (SWGDE), formed in 1998, develops industry guidelines, techniques, and standards.

Digital Forensics Policy

Digital forensics emerged in the mid-1980s in response to the growing importance of digital data in criminal investigations. The first cybercrimes occurred in the early 1970s. This era saw the emergence of "hacking," or gaining unauthorized access to computer systems. Some of the first documented uses of digital forensics data were in hacking investigations.

Prior to the Electronic Communications Privacy Act (ECPA) of 1986, digital data or communications were not protected by law and could be collected or intercepted by law enforcement. The ECPA was amended several times in the 1990s and 2000s to address the growing importance of digital data for private communication. In 2014, the Supreme Court ruled that police must obtain a warrant before searching the cell phone of a suspect arrested for a crime.

Digital Forensics Techniques

Once forensic investigators have access to equipment that has been seized or otherwise legally obtained, they can begin forensic imaging. This process involves making an unaltered copy, or forensic image, of the device's hard drive. A forensic image records the drive's structures, all of its contents, and metadata about the original files.

A forensic image is also known as a "physical copy." There are two main methods of copying computer data, physical copying and logical copying. A physical copy duplicates all of the data on a specific drive, including empty, deleted, or fragmented data, and stores it in its original configuration. A logical copy, by contrast, copies active data but ignores deleted files, fragments, and empty space. This makes the data easier to read and analyze. However, it may not provide a complete picture of the relevant data.

After imaging, forensics examiners analyze the imaged data. They may use specialized tools to recover deleted files using fragments or backup data, which is stored on many digital devices to prevent accidental data loss. Automated programs can be used to search and sort through imaged data to find useful information. (Because searching and sorting are crucial to the forensic process, digital forensics organizations invest in research into better search and sort algorithms). Information of interest to examiners may include e-mails, text messages, chat records, financial files, and various types of computer code. The tools and techniques used for analysis depend largely on the crime. These specialists may also be tasked with interpreting any data collected during an investigation. For instance, they may be called on to explain their findings to police or during a trial.

Challenges for the Future

Digital forensics is an emerging field that lags behind fast-changing digital technology. For instance, cloud computing is a fairly new technology in which data storage and processing is distributed across multiple computers or servers. In 2014, the National Institute of Standards and Technology identified sixty-five challenges that must be addressed regarding cloud computing. These challenges include both technical problems and legal issues.

The SWGDE works to create tools and standards that will allow investigators to effectively retrieve and analyze data while keeping pace with changing technology. It must also work with legal rights organizations to ensure that investigations remain within boundaries set to protect personal rights and privacy. Each forensic investigation may involve accessing personal communications and data that might be protected under laws that guarantee free speech and expression or prohibit unlawful search and seizure. The SWGDE and law enforcement agencies are debating changes to existing privacy and surveillance laws to address these issues while enabling digital forensic science to continue developing.

—*Micah L. Issitt*

BIBLIOGRAPHY

"Digital Evidence and Forensics." *National Institute of Justice*. Office of Justice Programs, 28 Oct. 2015. Web. 12 Feb. 2016.

Gogolin, Greg. *Digital Forensics Explained*. Boca Raton: CRC, 2013. Print.

Holt, Thomas J., Adam M. Bossler, and Kathryn C. Seigfried-Spellar. *Cybercrime and Digital Forensics: An Introduction*. New York: Routledge, 2015. Print.

Pollitt, Mark. "A History of Digital Forensics." *Advances in Digital Forensics VI*. Ed. Kam-Pui Chow and Sujeet Shenoi. Berlin: Springer, 2010. 3–15. Print.

Sammons, John. *The Basics of Digital Forensics: The Primer for Getting Started in Digital Forensics*. Waltham: Syngress, 2012. Print.

Shinder, Deb. "So You Want to Be a Computer Forensics Expert." *TechRepublic*. CBS Interactive, 27 Dec. 2010. Web. 2 Feb. 2016.

DIGITAL LIBRARIES

In the early 1990s, when traditional library systems were first galvanized by the sweeping possibilities of digitalizing centuries of paper artifacts and documents into computer databases, library scientists believed a global network of digitalized materials was just around the corner. This grand digital library would be accessible to individual users anytime on any computer, thus rendering obsolete the notion of a library as a public building servicing a given community. However, that bold vision proved to be a significant challenge to bring to fruition. Computer software experts and library scientists charged with devising the systems, system links, and databases and amassing the archives to create digital libraries realized that reaching their goal would involve overcoming a myriad of complexities that still remained unresolved in the early twenty-first century.

OVERVIEW

Not every available archive of information, documents, specialized publications, and data is a digital library; by that definition, any Internet search engine would qualify. Instead, much like traditional libraries, digital libraries stress organization of the materials in addition to the traditional functions of long-term storage and preservation.

Digital libraries offer an important strategy for extending the reach of traditional libraries, and public and university libraries routinely subscribe to global databases for patron use in addition to participating in ongoing projects to convert centuries of print material into digital format. Although library science is well on its way to catching up to the possibilities of digital collections, distinctions have been established between materials that have to be digitalized—a process that is relatively quick, cheap, and applicable to basically to any publication from before 1990—and materials that are "born digital." Those charged with developing the templates for digital libraries stress the virtual system's need to organize the ever-growing body of materials to permit efficient and transparent access to users worldwide, given that there will be no central digital library, simply links that connect databases and archives around the world. Such a system

The World Digital Library logo.

would require well-trained professionals to move users smoothly through an often-intimidating network of information.

The development of digital libraries poses a number of significant challenges, including designing the superstructure of an architecturally sound interrelated network of systems, rewriting existing international copyright laws for reproduction and distribution of materials, and generating metadata—that is, data that describes and can be used to catalog primary materials, much like traditional card catalogs or indexes. In addition, libraries must keep up with the ever-expanding body of data that includes not only traditional print materials, such as books and periodicals, but also films, music, government records, and scientific and research data. Coordinating a theoretically unlimited number of digital libraries, each storing and organizing a specific area of available materials, and putting that network within reach of users ably assisted by digital librarians is the challenge that remains for library scientists and computer engineers.

—*Joseph Dewey, MA, PhD*

BIBLIOGRAPHY

Candela, Leonardo, Donatella Castelli, and Pasquale Pagano. "History, Evolution, and Impact of Digital Libraries." *E-Publishing and Digital Libraries: Legal and Organizational Issues.* Ed. Ioannis Iglezakis, Tatiana-Eleni Synodinou, and Sarantos Kapidakis. Hershey: IGI Global, 2011. 1–30. Print.

Chowdhury, G. G., and Schubert Foo, eds. *Digital Libraries and Information Access: Research Perspectives.* New York: Neal, 2012. Print.

Goodchild, Michael F. "The Alexandria Digital Library Project: Review, Assessment, and Prospects." *Trends in Information Management* 1.1 (2005): 20–25. Web. 21 Aug. 2013.

Jeng, Judy. "What Is Usability in the Context of the Digital Library and How Can It Be Measured?" *Information Technology and Libraries* 24.2 (2005): 47–56. Web. 21 Aug. 2013.

Lesk, Michael. *Understanding Digital Libraries.* 2nd ed. San Francisco: Morgan, 2004. Print.

McMurdo, Thomas, and Birdie MacLennan. "The Vermont Digital Newspaper Project and the National Digital Newspaper Program: Cooperative Efforts in Long-Term Digital Newspaper Access and Preservation." *Library Resources & Technical Services* 57.3 (2013): 148–63. Web. 21 Aug. 2013.

Reese, Terry, Jr., and Kyle Banerjee. *Building Digital Libraries: A How-to-Do-It Manual.* New York: Neal, 2008. Print.

Yi, Esther. "Inside the Quest to Put the World's Libraries Online." *Atlantic.* Atlantic Monthly Group, 26 July 2012. Web. 21 Aug. 2013.

DIGITAL NATIVE

A digital native is someone who was born during or after the information age, a period of emerging digital technology. The term "digital native" most commonly refers to those born at the end of the twentieth century and into the twenty-first century. Digital natives began interacting with digital technology at a young age and are unfamiliar with life without tools such as the Internet, cell phones, and computers. Digital natives (often identified as millennials, generation Y, and generation Z) stand in contrast to digital immigrants, who were born before the prevalence of digital technology and had to adopt it later in life.

OVERVIEW

The term "digital native" was coined by writer Marc Prensky in 2001 in an article for the journal *On the Horizon* about an emerging issue in education; he also created the term "digital immigrant" in the same article. In the article, he was writing about the changing nature of teaching—students have grown up around digital technology, while their teachers did not—and how it is affecting the educational system. He argued that the discontinuity has changed how students think critically and process information, which has widened the gap between them and their teachers, who must adapt to the new environment.

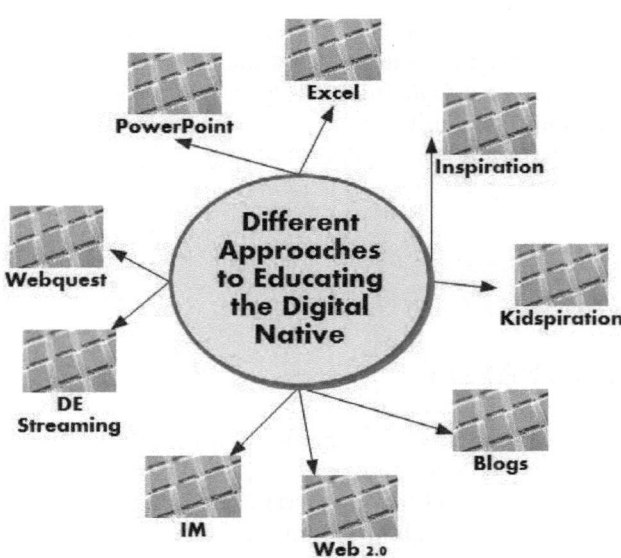

Different approaches to educate the digital native (Wikipedia).

Prensky further noted that digital immigrants often retain an "accent" that blends outdated language and behavior, highlighting their struggle to reconcile the new and the old.

The primary trait of digital natives is the ability to receive information rapidly, often through parallel processes and multitasking, taking on many things at once. Digital natives, who became used to these skills through their formative years, have entirely different frameworks for learning from those of the digital immigrants who are their teachers. This can be as simple as a student being able to read and watch television at the same time or a much wider divide related to communication, networking, and connectivity in instruction. Another related issue is the disconnect between the attention of natives and the methodology of immigrants.

Beyond education, the concept of digital natives has become, in the broader sense, related to demographic and generational differences. The differences between the two groups—occasionally referred to as "analog versus digital"—are increasingly important as technology development continues to accelerate. Digital natives adapt to new technology almost instinctively, utilizing prior experience to intuit how to use something new; digital immigrants often struggle to keep up and often fail when they attempt to apply more traditional styles of thinking to new contexts.

The gap between digital natives and digital immigrants, known as the digital divide, has further implications when examined on a global scale. Poverty in developing nations is a potent example of this gap between the connected and disconnected, making it clear that it is not just a generational concern and instead runs the risk of being an ever-widening issue. The concept of digital natives is not without some controversy, however, with some critics noting that learning the use of digital technology is more simply a matter of access and cultural capital.

In their 2015 book, *The New Digital Natives: Cutting the Chord*, Dylan Seychell and Alexiei Dingli draw a distinction between first generation digital natives and second generation digital natives (2DNs). First generation digital natives grew up before the advent of smartphones, tablets, social networking, and widespread wifi and laptop use. Their web access started only when they were able to use a mouse or read, unlike their children, the 2DNs, many of whom first accessed the web via touchscreens as toddlers.

—*Kehley Coviello*

Bibliography

Bennett, Sue, and K. Maton. "Beyond the 'Digital Natives' Debate: Towards a More Nuanced Understanding of Students' Technology Experiences." *Journal of Computer Assisted Learning* 26.5 (2010): 321–31. Print.

Bittman, Michael, Leonie Rutherford, Jude Brown, and Lens Unsworth. "Digital Natives? New and Old Media and Children's Outcomes." *Australian Journal of Education* 55.2 (2011): 161–75. Print.

Dingli, Alexiei, and Dylan Seychell. *The New Digital Natives: Cutting the Chord*. Heidelberg: Springer, 2015. eBook Collection (EBSCOhost). Web. 19 June 2015.

Joiner, Richard, et al. "Comparing First and Second Generation Digital Natives' Internet Use, Internet Anxiety, and Internet Identification." *CyberPsychology, Behavior & Social Networking* 16.7 (2013): 549–52. Print.

Jones, Chris. "A New Generation of Learners? The Net Generation and Digital Natives." *Learning, Media & Technology* 35.4 (2010): 365–68. Print.

Pentland, Alex. "The Data-Driven Society." *Scientific American* 309.4 (2013): 78–83. Print.

Ransdell, Sarah, Brianna Kent, Sandrine Gaillard-Kenney, and John Long. "Digital Immigrants Fare Better than Digital Natives Due to Social Reliance." *British Journal of Educational Technology* 42.6 (2011): 931–38. Print.

Teo, Timothy. "An Initial Development and Validation of a Digital Natives Assessment Scale (DNAS)." *Computers & Education* 67 (2013): 51–57. Print.

Thompson, Penny. "The Digital Natives as Learners: Technology Use Patterns and Approaches to Learning." *Computers & Education* 65 (2013): 12–33. Print.

DIGITAL PHOTOGRAPHY

Long before the written word, people recorded stories in pictures in the form of rock paintings and engravings, which go back tens of thousands of years. Throughout most of history, recording a scene or producing a detailed image was a laborious, time-intensive process. Artists created paintings and etchings from sketches. The invention of the camera allowed people without these artistic talents to record images.

Background

The first cameras were large, cumbersome tools, requiring glass or metal plates, chemicals, and other equipment. In 1884, George Eastman introduced the technology of film to replace the large plates. Sixteen years later, in 1900, the company he founded, Kodak, would allow everyone to take pictures with a simple, yet effective little camera, the Brownie.

Digital photography was the next step in allowing the general public to produce photographic images with ease. In 1975, Cromemco released the Cyclops, a digital camera available commercially; however, much of the technology used was parlayed into other products. Though development had begun much earlier, with prototypes surfacing in the 1970s, the first digital camera widely available on the market was the Dycam Model 1, released in 1990 by Swiss tech company Logitech.

Overview

Digital cameras use electronic photodetectors to capture images. Although the science of light and optics (lenses) is largely the same as with traditional film cameras, the image is stored electronically, without the use of light-sensitive film. Electronic images are stored either as RGB color space—where tints of red, green, and blue can be combined to various degrees to form any of all possible colors—or raw data, sometimes also called digital negatives, which store all photographic metadata, including light, color, color saturation, white balance, contrast, and sharpness. The advantages of a camera that can store raw images is that the digital negatives contain significantly more information to produce higher quality and more accurate photographs during photo editing. In either case, the images are stored on solid-state digital memory devices.

The technical (rather than aesthetic) quality of a digital image depends on several factors. One is pixel count: the number of red, green, or blue image points within a given space. Modern cameras record images that contain millions of pixels (1 megapixel = 1,000,000 pixels). Camera manufacturers and vendors like to quote megapixels. Although these numbers can help the consumer

Cromemco Cyclops: First all-digital camera with solid-state area image sensor.

compare cameras, the number of pixels along either axis of the photograph (rather than the entire area) is a much more accurate factor in determining the quality of a photograph, as each increase in linear resolution results in the square of total resolution.

Digital cameras have many advantages over traditional film cameras. First, they allow the photographer to take many photos easily, without relying on the limitations of the number of exposures on a roll of film. Digital cameras also allow the photographer to see an image immediately and decide whether it is necessary to reshoot a scene. Finally, digital photography allows the photographer to edit photographs in a photo-editing program such as Adobe Photoshop or Lightshop, bringing out the dynamic range of colors and shades in the photograph to reveal greater detail. Some photographers also use photo editing to make changes in the light, hue, and texture of a photograph for esthetic purposes.

In the world of photography, there are purists who prefer the warmth of traditional film (and traditional photo developing), just as in the music world there are audiophiles who prefer the warmth and dynamic range of vinyl recordings. Digital photography will never replace traditional analog photography, but the two can easily coexist. The camera industry continues to evolve, as advances in camera technologies continue.

—*Daniel L. Berek, MA, MAT*

BIBLIOGRAPHY

Ang, Tom. *Digital Photography Essentials*. New York: DK, 2011. Print.

_____. *Digital Photography Masterclass*. New York: DK, 2013. Print.

_____. *How to Photograph Absolutely Everything: Successful Pictures from Your Digital Camera*. New York: DK, 2009. Print.

Cotton, Charlotte. *The Photograph as Contemporary Art*. London: Thames, 2009. Print.

Davis, Harold. *The Way of the Digital Photographer: Walking the Photoshop Post-Production Path to More Creative Photography*. San Francisco: Peachpit, 2013. Print.

Horenstein, Henry. *Digital Photography: A Basic Manual*. Boston: Little, 2011. Print.

Kelby, Scott. *The Digital Photography Book: Part One*. 2nd ed. San Francisco: Peachpit, 2013. Print.

Schewe, Jeff. *The Digital Negative: Raw Image Processing in Lightroom, Camera Raw, and Photoshop*. San Francisco: Peachpit, 2012. Print.

_____. *The Digital Print: Preparing Images in Lightroom and Photoshop for Printing*. San Francisco: Peachpit, 2013. Print.

Sheppard, Rob. *Digital Photography: Top One Hundred Simplified Tips and Tricks*. Indianapolis, IN: Wiley, 2010. Print.

DIGITAL SIGNAL PROCESSORS (DSP)

FIELDS OF STUDY

Computer Science; Information Technology; Network Design

ABSTRACT

Digital signal processors (DSPs) are microprocessors designed for a special function. DSPs are used with analog signals to continuously monitor their output, often performing additional functions such as filtering or measuring the signal. One of the strengths of DSPs is that they can process more than one instruction or piece of data at a time.

PRINCIPAL TERMS

- **fixed-point arithmetic:** a calculation involving numbers that have a defined number of digits before and after the decimal point.
- **floating-point arithmetic:** a calculation involving numbers with a decimal point, and an exponent, as is done in scientific notation.
- **Harvard architecture:** a computer design that has physically distinct storage locations and signal routes for data and for instructions.
- **multiplier-accumulator:** a piece of computer hardware that performs the mathematical operation of

This is a block diagram for an analog-to-digital processing system. Digital signal processors are responsible for performing specified operations on digital signals. Signals that are initially analog must first be converted to digital signals in order for the programs in the digital processor to work correctly. The output signal may or may not have to be converted back to analog.

multiplying two numbers and then adding the result to an accumulator.
- **pipelined architecture:** a computer design where different processing elements are connected in a series, with the output of one operation being the input of the next.
- **semiconductor intellectual property (SIP) block:** a quantity of microchip layout design that is owned by a person or group; also known as an "IP core."

DIGITAL SIGNAL PROCESSING

Digital signal processors (DSPs) are microprocessors designed for a special function. Semiconductor intellectual property (SIP) blocks designed for use as DSPs generally have to work in a very low latency environment. They are constantly processing streams of video or audio. They also need to keep power consumption to a minimum, particularly in the case of mobile devices, which rely heavily on DSPs. To make this possible, DSPs are designed to work efficiently on both fixed-point arithmetic and the more computationally intensive floating-point arithmetic. The latter, however, is not needed in most DSP applications. DSPs tend to use chip architectures that allow them to fetch multiple instructions at once, such as the Harvard architecture. Many DSPs are required to accept analog data as input, convert this to digital data, perform some operation, and then convert the digital signals back to analog for output. This gives DSPs a pipelined architecture. They use a multistep process in which the output of one step is the input needed by the next step.

An example of the type of work performed by a DSP can be seen in a multiplier-accumulator. This is a piece of hardware that performs a two-step operation. First, it receives two values as inputs and multiplies one value by the other. Next, the multiplier-accumulator takes the result of the first step and adds it to the value stored in the accumulator. At the end, the accumulator's value can be passed along as output. Because DSPs rely heavily on multiplier-accumulator operations, these are part of the instruction set hardwired into such chips. DSPs must be able to carry out these operations quickly in order to keep up with the continuous stream of data that they receive.

The processing performed by DSPs can sometimes seem mysterious. However, in reality it often amounts to the performance of fairly straightforward mathematical operations on each value in the stream of data. Each piece of analog input is converted to a digital value. This digital value is then added to, multiplied by, subtracted from, or divided by another value. The result is a modified data stream that can then be passed to another process or generated as output.

APPLICATIONS OF DIGITAL SIGNAL PROCESSORS

There are many applications in which DSPs have become an integral part of daily life. The basic purpose of a DSP is to accept as input some form of analog information from the real world. This could include anything from an audible bird call to a live video-feed broadcast from the scene of a news event.

DSPs are also heavily relied upon in the field of medical imaging. Medical imaging uses ultrasound or other technologies to produce live imagery of what is occurring inside the human body. Ultrasound is often used to examine fetal developmental, from what position the fetus is in to how it is moving to how its heart is beating, and so on. DSPs receive the ultrasonic signals from the ultrasound equipment and convert it into digital data. This digital data can then be used to produce analog output in the form of a video display showing the fetus.

Digital signal processing is also critical to many military applications, particularly those that rely on the use of sonar or radar. As with ultrasound devices, radar and sonar send out analog signals in the form of energy waves. These waves bounce off features in the environment and back to the radar- or sonar-generating device. The device uses DSPs to receive this analog information and convert it into digital data. The data can then be analyzed and converted into graphical displays that humans can easily interpret. DSPs must be able to minimize delays in processing, because a submarine using sonar to navigate underwater cannot afford to wait to find out whether obstacles are in its path.

Biometric Scanning

A type of digital signal processing that many people have encountered at one time or another in their lives is the fingerprint scanner used in many security situations to verify one's identity. These scanners allow a person to place his or her finger on the scanner, and the scanner receives the analog input of the person's fingerprint. This input is then converted to a digital format and compared to the digital data on file for the person to see whether they match. As biometric data becomes increasingly important for security applications, the importance of digital signal processing will likely grow.

—Scott Zimmer, JD

Bibliography

Binh, Le Nguyen. *Digital Processing: Optical Transmission and Coherent Receiving Techniques.* Boca Raton: CRC, 2013. Print.

Iniewski, Krzysztof. *Embedded Systems: Hardware, Design, and Implementation.* Hoboken: Wiley, 2013. Print.

Kuo, Sen M., Bob H. Lee, and Wenshun Tian. *Real-Time Digital Signal Processing: Fundamentals, Implementations and Applications.* 3rd ed. Hoboken: Wiley, 2013. Print.

Snoke, David W. *Electronics: A Physical Approach.* Boston: Addison, 2014. Print.

Sozański, Krzysztof. *Digital Signal Processing in Power Electronics Control Circuits.* New York: Springer, 2013. Print.

Tan, Li, and Jean Jiang. *Digital Signal Processing: Fundamentals and Applications.* 2nd ed. Boston: Academic, 2013. Print.

DIGITAL WATERMARKING

FIELDS OF STUDY

Computer Science; Digital Media; Security

ABSTRACT

Digital watermarking protects shared or distributed intellectual property by placing an additional signal within the file. This signal can be used to inform users of the copyright owner's identity and to authenticate the source of digital data. Digital watermarks may be visible or hidden.

PRINCIPAL TERMS

- **carrier signal:** an electromagnetic frequency that has been modulated to carry analog or digital information.
- **crippleware:** software programs in which key features have been disabled and can only be activated after registration or with the use of a product key.
- **multibit watermarking:** a watermarking process that embeds multiple bits of data in the signal to be transmitted.
- **noise-tolerant signal:** a signal that can be easily distinguished from unwanted signal interruptions or fluctuations (i.e., noise).
- **1-bit watermarking:** a type of digital watermark that embeds one bit of binary data in the signal to be transmitted; also called "0-bit watermarking."
- **reversible data hiding:** techniques used to conceal data that allow the original data to be recovered in its exact form with no loss of quality.

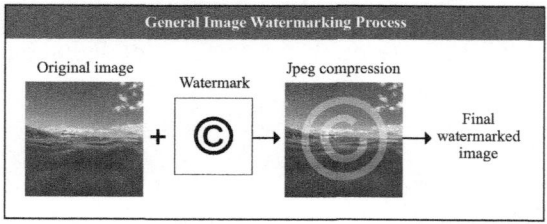
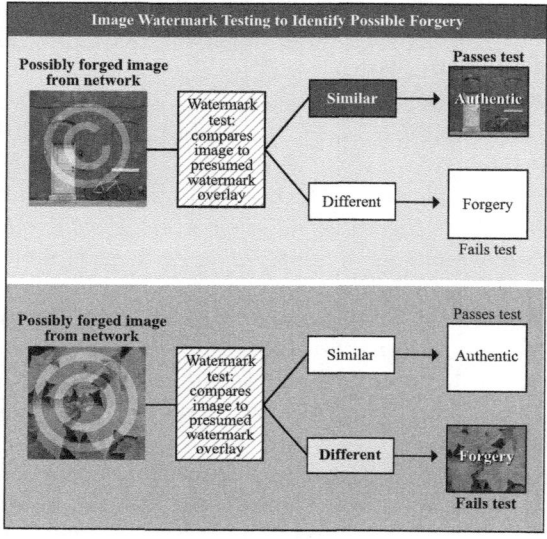

Diagrams of the watermarking process and a process for testing watermarks.

Protecting Ownership and Security of Digital Data

Digital watermarking is a technique that embeds digital media files with a hidden digital code. It was first developed in the late twentieth century. These hidden codes can be used to record copyright data, track copying or alteration of a file, or prevent alteration or unauthorized efforts to copy a copyrighted file. Digital watermarking is therefore commonly used for copyright-protected music, video, and software downloads. Governments and banks also rely on it to ensure that sensitive documents and currency are protected from counterfeiting and fraud.

Basics of Watermarking

A paper watermark is an image embedded within another image or a piece of paper. It can be seen by shining light on the image. Watermarks are used on banknotes, passports, and other types of paper documents to verify their authenticity. Similarly, digital watermarking involves embedding data within a digital signal in order to verify the authenticity of the signal or identify its owners. Digital watermarking was invented in the late 1980s or early 1990s. It uses techniques that are also used in steganography (the concealment of messages, files, or other types of data within images, audio, video, or even text).

Most digital watermarks are not detectable without an algorithm that can search for the signal embedded in the carrier signal. For a carrier signal to be watermarked, it must be tolerant of noise. Noise-tolerant signals are generally strong signals that resist degradation or unwanted modulation. Typically, digital watermarks are embedded in data by using an algorithm to encode the original signal with a hidden signal. The embedding may be performed using either public- or private-key encryption, depending on the level of security required.

Qualities of Digital Watermarks

One way to classify digital watermarks is by capacity, which measures how long and complex a watermarking signal is. The simplest type is 1-bit watermarking. This is used to encode a simple message that is meant only to be detected or not (a binary result of 1 or 0). In contrast, multibit watermarking embeds multiple bits of data in the original signal. Multibit systems may be more resistant to attack, as an attacker will not know how or where the watermark has been inserted.

Watermarks may also be classified as either robust or fragile. Robust watermarks resist most types of modification and therefore remain within the signal after any alterations, such as compression or cropping. These watermarks are often used to embed copyright information, as any copies of the file will also carry the watermark. Fragile watermarks cannot be detected if the signal is modified and are therefore used to determine if data has been altered.

In some cases, a digital watermark is designed so that users can easily detect it in the file. For instance, a video watermark may be a visible logo or text hovering onscreen during playback. In most cases, however, digital watermarks are hidden signals that can only be detected using an algorithm to retrieve the watermarking code. Reversible data hiding refers to cases in which the embedding of a watermark can be reversed by an algorithm to recover the original file.

APPLICATIONS OF DIGITAL WATERMARKING

A primary function of digital watermarking is to protect copyrighted digital content. Audio and video players may search for a digital watermark contained in a copyrighted file and only play or copy the file if it contains the watermark. This essentially verifies that the content is legally owned.

Certain types of programs, known colloquially as crippleware, use visible digital watermarks to ensure that they are legally purchased after an initial free evaluation period. Programs used to produce digital media files, such as image- or video-editing software, can often be downloaded for free so users can try them out first. To encourage users to purchase the full program, these trial versions will output images or videos containing a visible watermark. Only when the program has been registered or a product key has been entered will this watermark be removed.

Some creators of digital content use digital watermarking to embed their content with their identity and copyright information. They use robust watermarks so that even altered copies of the file will retain them. This allows a content owner to claim their work even if it has been altered by another user. In some cases, watermarked data can be configured so that any copies can be traced back to individual users or distributors. This function can be useful for tracing illegal distribution of copyrighted material. It can also help investigations into the unauthorized leaking of sensitive or proprietary files.

More recently, digital watermarking has been used to create hidden watermarks on product packaging. This is intended to make it easier for point-of-sale equipment to find and scan tracking codes on a product. Digital watermarking is also increasingly being used alongside or instead of regular watermarking to help prevent the counterfeiting of important identification papers, such as driver's licenses and passports.

—*Micah L. Issitt*

BIBLIOGRAPHY

Chao, Loretta. "Tech Partnership Looks beyond the Bar Code with Digital Watermarks." *Wall Street Journal.* Dow Jones, 12 Jan. 2016. Web. 14 Mar. 2016.

"Frequently Asked Questions." *Digital Watermarking Alliance.* DWA, n.d. Web. 11 Mar. 2016.

Gupta, Siddarth, and Vagesh Porwal. "Recent Digital Watermarking Approaches, Protecting Multimedia Data Ownership." *Advances in Computer Science* 4.2 (2015): 21–30. Web. 14 Mar. 2016.

Patel, Ruchika, and Parth Bhatt. "A Review Paper on Digital Watermarking and Its Techniques." *International Journal of Computer Applications* 110.1 (2015): 10–13. Web. 14 Mar. 2016.

Savage, Terry Michael, and Karla E. Vogel. *An Introduction to Digital Multimedia.* 2nd ed. Burlington: Jones, 2014. 256–58. Print.

"Unretouched by Human Hand." *Economist.* Economist Newspaper, 12 Dec. 2002. Web. 14 Mar. 2016.

DISK OPERATING SYSTEM (DOS)

FIELDS OF STUDY

Computer Science; Information Technology; Operating Systems

ABSTRACT

The term DOS is an acronym for "disk operating system." DOS refers to any of several text-based operating systems that share many similarities. Perhaps the best-known form of DOS is MS-DOS. MS-DOS is an operating system developed by the technology company Microsoft. It is based heavily on an earlier operating system known as QDOS. DOS was largely replaced by operating systems featuring graphical user interfaces by the end of the twentieth century. It continues to be used in certain specialized contexts in the twenty-first century, however.

PRINCIPAL TERMS

- **command line:** a text-based computer interface that allows the user to input simple commands via a keyboard.
- **graphical user interface (GUI):** an interface that allows users to control a computer or other device by interacting with graphical elements such as icons and windows.
- **nongraphical:** not featuring graphical elements.
- **shell:** an interface that allows a user to operate a computer or other device.

BACKGROUND ON DOS

"Disk operating system" (DOS) is a catchall term for a variety of early operating systems designed for personal computers (PCs). Operating systems have existed since the early days of computers. They became more important in the late 1970s and early 1980s. At that time PCs became available to the public. An operating system allows users unfamiliar with computer programming to interface directly with these devices. Among the first companies to offer PCs for sale was the technology company IBM, a long-time leader in the field of computer technology.

In 1980 IBM sought to license an operating system for its new device, the IBM PC. IBM approached the company Digital Research to license its Control Program for Microcomputers (CP/M). However, the two companies were unable to come to an agreement. IBM made a deal with the software company Microsoft instead. Microsoft had been founded in 1975. It initially specialized in creating and licensing computer programs and programming languages. In order to supply IBM with an operating system, Microsoft licensed the Quick and Dirty Operating System (QDOS) from the company Seattle Computer Products. This system, later renamed 86-DOS, was based in part on CP/M and used similar commands but differed in several key ways. Notably, the operating system was designed to be used with computers, such as the IBM PC, that featured sixteen-bit microprocessors, which could process sixteen bits, or pieces of binary information, at a time. Microsoft also hired QDOS creator Tim Patterson. Microsoft asked Patterson to create a new version of QDOS to be licensed to IBM under the name PC-DOS. Microsoft went on to develop an essentially identical version called MS-DOS, which the company sold and licensed to computer manufacturers itself.

The FreeDOS command line interface is based on the original DOS (disk operating system) command line interface to provide individuals with an alternative to the more prevalent graphical user interface available with most operating systems.

As IBM's PCs became increasingly popular, competing hardware manufacturers created similar computers. Many of these computers, commonly known as "IBM clones," used MS-DOS. In addition to MS-DOS, PC-DOS, and 86-DOS, other DOS or DOS-compatible systems entered the market over the decades. However, MS-DOS dominated the market and was often referred to simply as DOS.

UNDERSTANDING DOS

MS-DOS is the most popular and best-known form of DOS. MS-DOS consists of three key parts: the input/output (I/O) handler, the command processor, and the auxiliary utility programs. The I/O handler manages data input and output and consists of two programs, IO.SYS and MSDOS.SYS. The command processor enables the computer to take commands from the user and carry them out. The most commonly used commands and associated routines are stored in the command processor. Others are stored on the system disk and loaded as needed. Those routines are known as "auxiliary utility programs."

Like all operating systems, DOS functions as a shell. A shell is an interface that allows a user to operate a computer without needing knowledge of computer programming. A DOS or DOS-compatible system does require users to enter text-based commands. These are relatively simple and limited in number, however. As such, the public found PCs featuring the early forms of DOS fairly easy to use.

Using DOS

MS-DOS and later DOS and DOS-compatible systems are in many ways quite similar to those systems that came before, such as QDOS and CP/M. In general, MS-DOS and similar systems are nongraphical systems. Thus, they do not contain graphical elements found in later graphical user interfaces (GUIs), such as clickable icons and windows. Instead, nongraphical systems allow the user to operate the computer by entering commands into the command line. By entering basic commands, the user can instruct the computer to perform a wide range of functions, including running programs and opening or copying files.

Impact of DOS

Although DOS and other nongraphical systems were largely phased out by the mid-1990s as GUIs became more popular, they remained an influential part of the development of PC operating systems. Some graphical operating systems, such as Microsoft's Windows 95, were based in part on DOS, and had the ability to run DOS programs when opened in a specialized mode. Despite advances in computer technology, MS-DOS and similar systems are still used in the twenty-first century for applications. In some cases, companies or institutions continue to use the operating system in order to maintain access to software compatible only with DOS. Some individuals use DOS and DOS-compatible systems to play early games originally designed for those systems. Companies such as Microsoft no longer sell DOS to the public. However, a number of companies and organizations are devoted to providing DOS-compatible systems to users, often in the form of open-source freeware.

—*Joy Crelin*

Bibliography

Doeppner, Thomas W. *Operating Systems in Depth*. Hoboken: Wiley, 2011. Print.

Gallagher, Sean. "Though 'Barely an Operating System,' DOS Still Matters (to Some People)." *Ars Technica*. Condé Nast, 14 July 2014. Web. 31 Jan. 2016.

McCracken, Harry. "Ten Momentous Moments in DOS History." *PCWorld*. IDG Consumer, n.d. Web. 31 Jan. 2016.

Miller, Michael J. "The Rise of DOS: How Microsoft Got the IBM PC OS Contract." *PCMag.com*. PCMag Digital Group, 10 Aug. 2011. Web. 31 Jan. 2016.

"MS-DOS: A Brief Introduction." *Linux Information Project*. Linux Information Project, 30 Sept. 2006. Web. 31 Jan. 2016.

"Part Two: Communicating with Computers—The Operating System." *Computer Programming for Scientists*. Oregon State U, 2006. Web. 31 Jan. 2016.

Shustek, Len. "Microsoft MS-DOS Early Source Code." *Computer History Museum*. Computer History Museum, 2013. Web. 31 Jan. 2016.

DRONE WARFARE

Drone warfare represents one of the most dramatic developments in the United States' continual effort to maintain military superiority by implementing new technology. Unmanned aerial vehicles (UAVs), controlled from half a world away, can perform extended surveillance and intelligence gathering missions. When required, they can conduct air strikes against enemy targets. These capabilities can be achieved without placing a pilot in danger or by relying upon a single individual to perform flight duty over extended periods. The technical capabilities have raised major ethical and moral issues, not the least of which is the possibility that UAVs may violate international and domestic law.

Overview

Almost from its invention in 1903 the powered aircraft has been used as a tool allowing military commanders to see and strike the enemy from a somewhat safe distance. By 1911 the Italian Army used aircraft as part of its efforts to crush a native rebellion in Libya. Through both world wars, aircraft progressed in their range, speed, ability to gather intelligence, and lethality.

Despite these growing capabilities, aircraft were still limited due to the pilot's physical condition and concern over pilot safety. As an effort to surmount those limitations, drones made their first appearance in the 1950s and 1960s. They were first used

An MQ-1B Predator unmanned aircraft from the 361st Expeditionary Reconnaissance Squadron.

as flying targets, allowing antiaircraft crews to track and destroy targets without killing a pilot. During the Vietnam War drones performed aerial reconnaissance. While these drones provided a valuable service, they could not return and safely land at their base but crashed after performing their mission.

In the early 1970s Israeli-developed drones performed military reconnaissance missions and air strikes. An Israeli drone designer, Abraham Karem, moved to the United States in the 1980s and began developing drones, an activity funded at first by the US Defense Advanced Research Projects Agency (DARPA) and later by the Central Intelligence Agency (CIA). By the early 1990s drones were being used by the CIA in Bosnia to provide information for convoys and to spot Serbian military positions.

In 2000 drones were used extensively for intelligence purposes, and their use increased substantially after the 9/11 terrorist attacks. Drones have been widely used in both Afghanistan and Pakistan since 2004 for reconnaissance and as a means to kill selected individuals. After Barack Obama became president in 2009, his administration began defending the strategy and ramped up the number of drone strikes in Yemen, Somalia, and especially Pakistan. A drone strike in Pakistan in 2012 successfully killed Abu Yahya al-Libi, the man considered to be al-Qaeda's number-two leader. However, criticism of the tactic continued to rise on behalf of Pakistanis and Americans alike, especially concerning the loss of civilian lives. That following year, Obama gave a speech in which he promised tighter restrictions and rules regarding the use of drone strikes. Yet in early 2015, it was announced that two al-Qaeda hostages, an American and an Italian man, were killed during another attack by drones in Pakistan. President Obama made a public address to apologize to the families.

As a technological solution to military and foreign policy, drones allow their users to find and destroy targets without endangering the pilot or requiring a single individual to fly an aircraft for more hours than a normal pilot could. However, several issues mitigate the effectiveness of UAVs. First, they are not perfect; they have often hit the wrong target and have also caused civilian casualties. Their use as a selective means to assassinate individuals has drawn a great deal of criticism and is seen by many as a breach of international law. In the effort to win over populations, their remoteness and lethality can push public opinion against their users. Finally, in the United States, at least, their use within US borders to perform surveillance (a program acknowledged by the head of the Federal Bureau of Intelligence) has created significant concern, particularly with regard to whether they violate privacy rights as well as their being used as a means of attacking US citizens.

—*Robert N. Stacy, MA*

BIBLIOGRAPHY

Alkire, Brien. *Applications for Navy Unmanned Aircraft Systems.* Santa Monica: Rand, 2010. Print.

Barnhardt, Richard K., and Eric Shappee. *Introduction to Unmanned Aircraft Systems.* New York: CRC, 2012. Print.

Benjamin, Medea. *Drone Warfare: Killing by Remote Control.* London: Verso. 2013. Print.

Bergen, Peter, and Jennifer Rowland. "Did Obama Keep His Drone Promises?" *CNN.* Cable News Network, 25 Oct. 2013. Web. 22 May 2015.

Coker, Christopher. *Warrior Geeks: How 21st Century Technology Is Changing the Way We Fight and Think about War.* London: Hurst, 2013. Print.

Coll, Steve. "The Unblinking Stare." *New Yorker.* Condé Nast, 24 Nov. 2014. Web. 22 May 2015.

Hansen, V. "Predator Drone Attacks." *New England Law Review* 46 (2011): 27–36. Print.

Metz, J. R. *The Drone Wars: Uncovering the Dynamics and Scope of United States Drone Strikes.* Diss. Wesleyan U, 2013. Print.

Pugliese, Joseph. *State Violence and the Execution of Law: Biopolitical Caesurae of Torture, Black Sites, Drones.* New York: Routledge, 2013. Print.

Turse, Nick, and Tom Engelhardt. *Terminator Planet: The First History of Drone Warfare, 2001–2050.* TomDispatch, 2012. Kindle file.

United States Congress. House Committee on Oversight and Government Reform. Subcommittee on National Security and Foreign Affairs. *Rise of the Drones: Hearing before the Subcommittee on National Security and Foreign Affairs of the Committee on Oversight and Government Reform, House of Representatives, One Hundred Eleventh Congress, second session.* Washington: GPO, 2011. Print.

DRONES

ABSTRACT

A drone is a kind of aircraft. A drone is also known as an unmanned aerial vehicle (UAV). A drone is "unmanned" because it does not need a pilot on board to fly it. A person on the ground flies a drone. Some drones are guided by a remote control. Others are guided by computers. Some are flown by people thousands of miles away. Drones are available in different sizes and shapes. Many have cameras that take pictures or record video. The U.S. military has used drones since the mid-1990s. Since early 2013, many people have started flying drones for fun. So many people own drones that the government had to create rules for flying them.

HISTORY OF DRONES

Militaries have used unmanned aircraft for many years. During the Civil War (1861–1865), armies tried using balloons to drop explosives into enemy camps. The British Royal Navy created a drone-like vehicle in the 1930s. Pilots used it to practice their shooting skills.

The U.S. military began using modern drones in 1995. The military first used drones to collect information. For example, a drone could be used to find an enemy's hiding place. Unlike an airplane or a helicopter, a drone does not need a pilot on board. The people who fly drones can be thousands of miles away. For this reason, drones are safer to fly over dangerous areas.

The U.S. military later added weapons to drones. They began to use drones to perform drone strikes. During a drone strike, a drone fires its weapons at an enemy.

A few companies introduced drones for ordinary citizens in early 2013. People could buy their own drones for a few hundred dollars. Most of these drones are small—about the size of a textbook. They are not very heavy. These drones do not have weapons. However, some have cameras that take pictures or record video. People buy drones for many reasons.

USES FOR DRONES

Drones can be helpful. Farmers use them to check on fields and crops. Firefighters fly drones over forests to check for wildfires. Filmmakers may use them to record scenes in movies. Scientists may fly drones into big storms to gather information such as temperature and wind speed.

Some companies think drones could help grow their business. For example, Amazon and Walmart are interested in using drones to deliver goods to

Unmanned aerial vehicles, more commonly known as drones, come in sizes ranging from a small toy to a large plane. They are used by the military for surveillance as well as by civilians for recreation.

people's homes. Drones can deliver goods in minutes. People would not have to wait for goods to arrive in the mail. Other companies want to use drones to deliver food.

People do not always use drones to help others. Some use drones to break the law. For example, people have tried to deliver items to prisoners in jail using drones. People have flown drones over baseball and football stadiums, which is not allowed. Someone even crashed a drone onto the lawn of the White House. The White House is where the president lives.

Problems with Drones

Drones have become very popular. However, they do cause some problems. Many people do not trust drones. They think drones will record what they do and say. They do not want drones to invade their privacy. When a person's privacy is invaded, others know about the person's private life.

Drones can cause property damage. Property is something owned by someone. Houses and cars are examples of property. Many people lose control of their drones while flying them. Drones may crash into people's houses or cars and cause damage. Out-of-control drones also can hit people and hurt them.

Another problem is that drones could collide with other aircraft. They could get sucked into a jet's engine. Such events could have serious effects, especially if a helicopter or an airplane crashes as a result. The government has created rules to prevent some of these problems.

Rules for Drones

The Federal Aviation Administration (FAA) is the part of the U.S. government that makes rules about flying. The FAA introduced rules for drones in 2015. First, people must register their drones. To register is to add one's name to an official list. The FAA then gives the drone owner an identification number. This number must be displayed on the drone. The following are more FAA rules for drones:

Do not fly above 400 feet.
Keep the drone in sight at all times.
Stay away from other aircraft.
Do not fly near airports.
Do not fly near people or stadiums.

Do not fly a drone that weighs more than 55 pounds. If you put others in danger, you may have to pay a fine.

—By Lindsay Rohland

Bibliography

Anderson, Chris. "How I Accidentally Kickstarted the Domestic Drone Boom." *Wired* (June, 2012). http://www.wired.com/2012/06/ff_drones/. This article describes a number of drone uses as well as the author's recollection of building his own drone.

Bowden, Mark. "How the Predator Drone Changed the Character of War." *Smithsonian* (November, 2013). http://www.smithsonianmag.com/history/how-the-predator-drone-changed-the-character-of-war-3794671/?no-ist. This article discusses the use of drones, particularly Predator drones, in war.

Eadicicco, Lisa. "Here's Why Drone Delivery Won't Be a Reality Any Time Soon." *Time* (November, 2015). http://time.com/4098369/amazon-google-drone-delivery/. This article identifies a number of potential issues related to drone delivery services.

New York Times Editorial Board. "Ruling Drones, Before They Rule Us." *New York Times* (January 10, 2016): SR10. http://www.nytimes.com/2016/01/10/opinion/sunday/drone-regulations-should-focus-on-safety-and-privacy.html?_r=0. This article discusses the need for regulation of drones.

Ripley, Amanda. "Playing Defense Against the Drones." *Atlantic* (November, 2015). http://www.theatlantic.com/magazine/archive/2015/11/playing-defense-against-the-drones/407851/. This article discusses interactions with and reactions to drones in various places.

Suebsaeng, Asawin. "Drones: Everything You Ever Wanted to Know but Were Always Afraid to Ask." *Mother Jones* (March, 2013). http://www.motherjones.com/politics/2013/03/drones-explained. This question-and-answer article provides an overview of drones.

Federal Aviation Administration. https://www.faa.gov/uas/model_aircraft/. "Model Aircraft Operations." Accessed February, 2016. This site identifies the rules and regulations that the Federal Aviation Administration has established for people who fly drones as a hobby.

Federal Aviation Administration. https://www.faa.gov/uas/model_aircraft/. "Model Aircraft Operations." Accessed February, 2016. This site identifies the rules and regulations that the Federal Aviation Administration has established for people who fly drones as a hobby.

E

E-BANKING

E-banking, also referred to as Internet banking or online banking, allows customers of a financial institution an opportunity to conduct his or her financial transactions, such as day-to-day account statements to loan applications, in an online, virtual environment. Most traditional banks, such as Bank of America or Chase, offer extensive online services that allow customers to avoid brick-and-mortar locations altogether. These services include deposits, transfers, payments, account management, budgeting tools, and loan applications. In fact, there are several banks, such as Ally, that operate completely online. E-banking is built on a foundation of several electronic and mobile technologies, such as online data transfers and automated data collection systems. In the world's global market, e-banking has played a significant role in the rise of e-commerce, especially in how people and businesses interact in the buying and selling of goods and services. Due in part to a rise in online banking, e-commerce has gained a significant foothold in the consumer environment and has increased the consumer base of many businesses exponentially because e-banking and e-commerce are not limited to specific geographic locations.

Overview

Modern banking no longer relies on geographic locations; instead, the industry has come to embrace consumers who prefer an online, virtual environment. In fact, some banks operate solely online. One benefit of e-banking is a reduction in overhead costs usually associated with a brick-and-mortar location that are passed on to the customer. Other benefits include convenience and instant access for the customer to manage financial accounts and holdings. E-banking also plays a large role in the success of the electronic commerce environment in a global market, allowing financial transactions to happen almost instantaneously, regardless of geographic location.

However, online security of account information is a concern for both the online banking industry and its customers. This has spawned the rise of the information security industry, or infosec, which is the practice of protecting information from unauthorized use, disclosure, access, modification, or destruction. Infosec applies to all information regardless of the form it may take and is comprised of two major categories: information assurance, which is ability to ensure data is not lost to a breakdown in system security, due to theft, natural disasters, or technological malfunction; and IT security, which is the security applied to computer networks where information resides. Improvements in information technology and related security elements are integral to the e-banking industry and foster reliable and effective security practices to protect proprietary and confidential information. Overall, consumers are operating more and more in an online environment and are demanding online capabilities for their smartphones and other devices. Businesses and banks will continue to adapt toward this trend by providing secure applications and environments.

—*L. L. Lundin, MA*

Bibliography

Cavusgil, S. Tamer, et al. *International Business: The New Realities.* Frenchs Forest, 2015. Print.

Chau, S. C., and M. T. Lu. "Understanding Internet Banking Adoption and Use Behavior: A Hong Kong Perspective." *Journal of Global Information Management* 12.3 (2009): 21–43. Print.

Dahlberg, T., et al. "Past, Present, and Future of Mobile Payments Research: A Literature Review."

Electronic Commerce Research and Applications 7.2 (2008): 165–81. Print.
Demirkan, H., and R. J. Kauffman. "Service-Oriented Technology and Management: Perspectives on Research and Practice for the Coming Decade." *Electronic Commerce Research and Applications* 7.4 (2008): 356–76. Print.
Fox, Susannah. "51% of U.S. Adults Bank Online." *Pew Internet: Pew Internet & American Life Project*. Pew Research Center. 7 Aug. 2013. Web. 20 Aug. 2013.
Gkoutzinis, Apostolos. *Internet Banking and the Law in Europe: Regulation, Financial Integration and Electronic Commerce*. Cambridge UP, 2010. Print.
Hoehle, Hartmut, Eusebio Scornavacca, and Sid Huff. "Three Decades of Research on Consumer Adoption and Utilization of Electronic Banking Channels: A Literature Analysis." *Decision Support Systems* 54.1 (2012): 122–32. Print.
Lee, M. C. "Factors Influencing the Adoption of Internet Banking: An Integration of TAM and TPB with Perceived Risk and Perceived Benefit." *Electronic Commerce Research and Applications* 8.3 (2009): 130–41. Print.
Santora, Marc. "In Hours, Thieves Took $45 Million in A.T.M. Scheme." *New York Times* 9 May 2013, New York ed.: A1. Print.

E-LEARNING

E-learning leverages communications technology to broaden the learning environment for students. The Internet has had a significant impact the ways in which students learn and teachers teach; e-learning, or electronic learning, extends the learning environment into this virtual realm. E-learning takes place in an online, computer-based environment and covers a broad range of teaching techniques and practices. These include online instructional presentations, interactive lessons, and computer-supported in-class presentations.

Overview

Also known as online learning or virtual education, e-learning is commonly used in public high schools or at college and universities. It may use of a variety of electronic media, including, but not limited to, text, streaming video, instant messaging, document sharing software, Blackboard learning environments, webcams, blogging, and streaming video. E-learning's key benefits include broader access to education by a wide range of students; collaborative interaction between students, peers, and teachers; development of technological skills and knowledge; and independent study.

E-learning is not synonymous with distance or mobile learning. Instead, e-learning refers solely to the use of online, computer-based technologies to complete an area of instruction. E-learning can take place in or out of a classroom and can be led either by a teacher in real time or completed at a pace set by the student, although many e-learning courses include some hybrid version of those two scenarios. The virtual learning environment is typically collaborative and often incorporates a blog or wiki entries to facilitate this relationship.

Researchers suggest that the success of e-learning is compounded when the information presented is coherent in design and does not include extraneous material or information. Educators suggest that when using an e-learning environment with their students,

Cpl. Chad Tucker, a financial management resource analyst with Marine Expeditionary Brigade–Afghanistan, studies at the end of his day here, Feb. 20. Like Tucker, many Marines have taken advantage of online education while deployed.

instructors should still employ good pedagogical practices and techniques as a foundation upon which to build. Additionally, many educators cite the interactions between a student and teacher and between students and their peers as a crucial for the successful completion of an e-course. In a 2013 study, however, Fain reported that while 97 percent of community colleges offered online courses, only 3 percent of students attending those institutions were enrolled in entirely online degree programs. For the most part, students reported feeling that they learned better in face-to-face instructional settings, especially in science and foreign language classes.

E-learning is also at the forefront of college and university education structures, with a significant number of university curriculums offering courses that take place solely in an online, e-learning environment. These courses, called massive open online courses, or MOOCs, are often available to anyone in the world without charge. Additionally, other universities have taken e-learning a step further to build entirely mobile programs of study, establishing online higher-education programs where students can complete a bachelor's or master's degree program entirely in an e-learning format. The US Department of Education reported that 62 percent of the postsecondary institutions surveyed offered online education courses in the 2011–12 academic year.

In one 2012 study, Groux reported that while the "income, race, and ethnicity" is the same among online students as their peers in traditional courses, students who take online course tend to be older than traditional students, with about 40 percent of online students being under the age of thirty with only 20 percent under the age of twenty-five. Kolowich reported in his 2012 study that sixty percent of online students are white, 20 percent African American, and 8 percent identifying as Hispanic.

E-learning is also being implemented by many companies and organizations to keep their employees up-to-date on various topics and to ensure compliance with the organization's policies and practices, offering training seminars in a completely online environment. This frees up resources by placing the burden of learning on the student, alleviating some oversight and human resources costs associated with training workforces.

—*Laura L. Lundin, MA*

BIBLIOGRAPHY

Berge, Zane L., and Lin Muilenburg. *Handbook of Mobile Learning.* New York: Routledge, 2013. Print.

Dikkers, Seann, John Martin, and Bob Coulter. *Mobile Media Learning: Amazing Uses of Mobile Devices for Learning.* Pittsburgh: ETC P, Carnegie Mellon U, 2011. Print.

Fain, Paul. "Only Sometimes for Online." *Inside Higher Ed.* Inside Higher Ed., 26 Apr. 2013. Web. 14 Sept. 2014.

Groux, Catherine. "Study Analyzes Characteristics of Online Learners." *Learning House.* Learning House, 27 July 2012. Web. 14 Sept. 2014.

Hill, Janette R., Llyan Song, and Richard E. West. "Social Learning Theory and Web-based Learning Environments: A Review of Research and Discussion Implications." American Journal of Distance Education 23.2 (2009): 88–103. Print.

Kolowich, Steve. "The Online Student." *Inside Higher Ed.* Inside Higher Ed., 25 July 2012. Web. 14 Sept. 2014.

Mayer, Richard E. *The Cambridge Handbook for Multimedia Learning.* New York: Cambridge UP, 2005. Print.

Rosenberg, Marc J. *E-Learning: Strategies for Delivering Knowledge in the Digital Age.* New York: McGraw, 2001. Print.

Salmon, Gilly. *E-Moderating: The Key to Teaching and Learning Online.* London: Kogan. 2003. Print.

Tu, Chih-Hsiung, and Marina McIsaac. "The Relationship of Social Presence and Interaction in Online Classes." *American Journal of Distance Education* 16.3 (2002): 131–50. Print.

Wang, C., D. M. Shannon, and M. E. Ross. "Students' Characteristics, Self-Regulated Learning, Technology Self-Efficacy, and Course Outcomes in Online Learning." *Distance Education* 34.3 (2013): 302–23. Print.

Woodill, Gary. *The Mobile Learning Edge: Tools and Technologies for Developing Your Teams.* New York: McGraw, 2011. Print.

ELECTRONIC CIRCUITS

FIELDS OF STUDY
Computer Engineering

ABSTRACT
Electronic circuits actively manipulate electric currents. For many years these circuits have been part of computer systems and important home appliances such as televisions. In information processing, the most important innovation has been the integrated circuit, which is power efficient, small, and powerfully capable. Integrated circuits form the basic operational units of countless everyday gadgets.

PRINCIPAL TERMS

- **BCD-to-seven-segment decoder/driver:** a logic gate that converts a four-bit binary-coded decimal (BCD) input to decimal numerals that can be output to a seven-segment digital display.
- **counter:** a digital sequential logic gate that records how many times a certain event occurs in a given interval of time.
- **inverter:** a logic gate whose output is the inverse of the input; also called a NOT gate.
- **negative-AND (NAND) gate:** a logic gate that produces a false output only when both inputs are true
- **programmable oscillator:** an electronic device that fluctuates between two states that allows user modifications to determine mode of operation.
- **retriggerable single shot:** a monostable multivibrator (MMV) electronic circuit that outputs a single pulse when triggered but can identify a new trigger during an output pulse, thus restarting its pulse time and extending its output.

Electric versus Electronic Circuits
Electrical circuits have developed over the years since the discovery of the Leyden jar in 1745. An electrical circuit is simply a path through which electric current can travel. Its primary components include resistors, inductors, and capacitors. The resistor controls the amount of current that flows through the circuit. It is so called because it provides electrical resistance. The inductor and the capacitor both store energy. Inductors store energy in a magnetic field, while capacitors store it in the electric field. The Leyden jar was the first capacitor, designed to store static electricity.

Electronic circuits are a type of electrical circuit. However, they are distinct from basic electrical circuits in one important respect. Electrical circuits passively conduct electric current, while electronic circuits actively manipulate it. In addition to the passive components of an electrical circuit, electronic circuits also contain active components such as transistors and diodes. A transistor is a semiconductor that works like a switch. It can amplify an electronic signal or switch it on or off. A diode is a conductor with very low electrical resistance in one direction and very high resistance in the other. It is used to direct the flow of current.

Integrated Circuits
The most important advance in electronic circuits was the development of the integrated circuit (IC) in the mid-twentieth century. An IC is simply a semiconductor chip containing multiple electronic circuits. Its development was enabled by the invention of the transistor in 1947. Previously, electric current was switched or amplified through a vacuum tube. Vacuum tubes are much slower, bulkier, and less efficient than transistors. The first digital computer, ENIAC (Electronic Numerical Integrator and Computer), contained about eighteen thousand vacuum tubes and weighed more than thirty tons. Once the transistor replaced the vacuum tube, electronic circuits could be made much smaller.

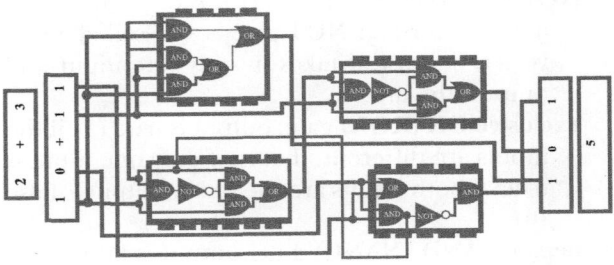

Electronic circuits are designed to use a series of logic gates to send a charge through the circuit in a particular manner. These logic gates control the charge output and thus the output of the circuits. In this example, the circuit is designed to add two binary numbers together using a series of AND, OR, and NOT commands to determine the route of the charge and the resulting output from each circuit component.

By 1958, several scientists had already proposed ideas for constructing an IC. That year, Jack Kilby, a scientist at Texas Instruments, was the first to put the idea into practice. He designed a chip constructed entirely from a single block of semiconductor material. Because there were no individual components, the circuits did not have to be large enough to assemble manually. The number of circuits in the chip was limited only by the number of transistors that could fit in it. Early ICs contained only a few transistors each. By the twenty-first century, the maximum possible number of transistors per IC was in the billions.

Logic Gates

Active manipulation of electric current is accomplished through logic gates. A logic gate is an electronic circuit that implements a Boolean function. Broadly speaking, a Boolean function is a function that produces one of two potential outputs—either 0 or 1—based on a given rule. Because transistors work as switches, which can take one of two values (e.g., "on" or "off"), they are ideal for implementing logic gates. Most logic gates take in two inputs and produce a single output.

There are seven basic types of logic gates: AND, OR, NOT, XOR, NAND, NOR, and XNOR. These logic gates only accept two input values, 0 and 1, which represent "false" and "true" respectively. They are distinguished from one another based on what output is produced by each combination of inputs:

- **AND** gate: output is only true (1) if both inputs are true; otherwise, output is false (0).
- **OR** gate: output is false (0) only if both outputs are false; otherwise, output is true (1).
- **NOT** gate: output is true (1) if input is false (0), and vice versa. A **NOT** gate is also called an inverter, because it takes in only one input and outputs the inverse.
- **exclusive-OR** (**XOR**) gate: output is true (1) if the inputs are different, that is, if only one input is true; if both inputs are the same, output is false (0).
- **negative-AND** (**NAND**) gate: output is false (0) if all inputs are true (1); otherwise output is true. A **NAND** gate is essentially an **AND** gate followed by an inverter
- **NOR** gate: output is true (1) only if both inputs are false (0); otherwise, output is false. A **NOR** gate is an **OR** gate followed by an inverter.
- **exclusive-NOR** (**XNOR**) gate: output is true (1) if both inputs are the same and false (0) if they are different. An **XNOR** gate is a **XOR** gate followed by an inverter.

Electronic circuits transmit binary data in the form of electric pulses, where, for example, 0 and 1 are represented by pulses of different voltages. These seven gates can be combined in different ways to complete more complex operations. For example, a BCD-to-seven-segment decoder/driver is a logic gate that converts binary data from a counter to a decimal number display. The "seven segment" refers to the number display system common in digital clocks and other devices, where each numeral is represented by up to seven short parallel or perpendicular line segments. Another complex circuit is a retriggerable single shot. This is a type of time-delay relay circuit that can generate an output pulse of a predetermined length and then extend the output indefinitely if the input is repeated. The purpose of this is to change the timing of another circuit, such as a programmable oscillator.

Life without Integrated Circuits

Whether in home appliances, computer systems, or mobile devices, electronic circuits make modern life possible. Without advanced electronic circuits such

---

SAMPLE PROBLEM

Determine the output of a NAND logic gate for all possible combinations of two input values (0 and 1).

Answer:

The combination of an AND gate and a NOT gate forms a NAND logic gate. The output of each input combination is the inverse of the AND gate output. The NAND gate accepts four possible combinations of two inputs and produces outputs as shown:

0,0 = 1
0,1 = 1
1,0 = 1
1,1 = 0

as ICs, personal computers and small, portable electronic devices could not exist. Despite improvements over the years, ICs have maintained their silicon-based design. Scientists predict that the only thing to replace ICs would be a new kind of biologically based circuit technology.

—Melvin O

Bibliography

Frenzel, Louis E., Jr. *Electronics Explained: The New Systems Approach to Learning Electronics.* Burlington: Elsevier, 2010. Print.

Harris, David Money, and Sarah L. Harris. *Digital Design and Computer Architecture.* 2nd ed. Waltham: Morgan, 2013. Print.

"The History of the Integrated Circuit." *Nobelprize.org.* Nobel Media, 2014. Web. 31 Mar. 2016.

Kosky, Philip, et al. *Exploring Engineering: An Introduction to Engineering and Design.* 4th ed. Waltham: Academic, 2016. Print.

Tooley, Mike. *Electronic Circuits: Fundamentals and Applications.* 4th ed. New York: Routledge, 2015. Print.

Wilson, Peter. *The Circuit Designer's Companion.* 3rd ed. Waltham: Newnes, 2012. Print.

ELECTRONIC COMMUNICATION SOFTWARE

FIELDS OF STUDY

Information Systems; Information Technology

ABSTRACT

Electronic communication software is used to transfer information via the Internet or other transmission-and-reception technology. As technology has evolved, electronic communication software has taken on many new forms, from text-based instant messaging using computers to SMS messages sent between cell phones on opposite sides of the world. Electronic communication software allows people to communicate in real time using audio and video and to exchange digital files containing text, photos, and other data.

PRINCIPAL TERMS

- **Electronic Communications Privacy Act (ECPA):** a regulation enacted in 1986 to limit the ability of the US government to intrude upon private communications between computers.
- **multicast:** a network communications protocol in which a transmission is broadcast to multiple recipients rather than to a single receiver.
- **push technology:** a communication protocol in which a messaging server notifies the recipient as soon as the server receives a message, instead of waiting for the user to check for new messages.
- **Short Message Service (SMS):** the technology underlying text messaging used on cell phones.
- **Voice over Internet Protocol (VoIP):** a set of parameters that make it possible for telephone calls to be transmitted digitally over the Internet, rather than as analog signals through telephone wires.

Asynchronous Communication

Many types of electronic communication software are asynchronous. This means that the message sender and the recipient communicate with one another at different times. The classic example of this type of electronic communication software is e-mail. E-mail is asynchronous because when a person sends a message, it travels first to the server and then to the recipient. The server may use push technology to notify the recipient that a message is waiting. It is then up to the recipient to decide when to retrieve the message from the server and read it.

E-mail evolved from an earlier form of asynchronous electronic communication: the bulletin-board system. In the 1980s and earlier, before the Internet was widely available to users in their homes, most people went online using a dial-up modem. A dial-up modem is a device that allows a computer to connect to another computer through a telephone line. To connect, the first computer dials the phone number assigned to the other computer's phone line. Connecting in this way was slow and cumbersome compared to the broadband Internet access

Electronic communication software has many forms to satisfy many uses. The most popular social media and communication companies implement programming that provides users with attributes they deem important for electronic communication, such as private and/or public sharing and posting to a network, saving, contributing, subscribing, and commenting across a number of formats.

common today. This was in part because phone lines could only be used for one purpose at a time, so a user could not receive phone calls while online. Because users tended to be online only in short bursts, they would leave messages for each other on online bulletin-board systems (BBSs). Like e-mail, the message would stay on the BBS until its recipient logged on and saw that it was waiting.

Another popular method of asynchronous communication is text messaging. Text messaging allows short messages to be sent from one mobile phone to another. The communications protocol technology behind text messages is called Short Message Service (SMS). SMS messages are limited to 160 characters. They are widely used because they can be sent and received using any kind of cell phone.

Synchronous Communication

Other forms of electronic communication software allow for synchronous communication. This means that both the recipient and the sender interact through a communications medium at the same time. The most familiar example of synchronous communication is the telephone, and more recently the cell phone. Using either analog protocols or voice over Internet Protocol (VoIP), users speak into a device. Their speech is translated into electronic signals by the device's communication software and then transmitted to the recipient. There are sometimes minor delays due to network latency. However, most of the conversation happens in the same way it would if the parties were face to face.

Another form of electronic synchronous communication is instant messaging or chat. Instant messaging occurs when multiple users use computers or mobile devices to type messages to one another. Each time a user sends a message, the recipient or recipients see it pop up on their screens. Chat can occur between two users, or it can take the form of a multicast in which one person types a message and multiple others receive it.

Multicast can also be delivered asynchronously. An example of this type of electronic communication is a performer who records a video of themselves using a digital camera and then posts the video on an online platform such as YouTube. The video would then stay online, available for others to watch at any time, until its creator decided to take it down. This type of electronic communication is extremely popular because viewers do not have to be online at a prearranged time in order to view the performance, as was the case with television broadcasts in the past.

Privacy Concerns

To some extent, the rapid growth of electronic communication software caught regulators off guard. There were many protections in place to prevent the government from eavesdropping on private communications using the telephone. However similar protections for electronic communications were lacking until the passage of the Electronic Communications Privacy Act (ECPA) in the late 1980s. This act extended many traditional communication protections to VoIP calls, e-mails, SMS messages, chat and instant messaging logs, and other types of communications.

Cutting Edge

Some of the newest forms of electronic communication software are pushing the boundaries of what is possible. One example of this is video calling using cell phones. This technology is available in many consumer devices, but its utility is often limited by the amount of bandwidth available in some locations. This causes poor video quality and noticeable delays in responses between users.

—*Scott Zimmer, JD*

Bibliography

Bucchi, Massimiano, and Brian Trench, eds. *Routledge Handbook of Public Communication of Science and Technology*. 2nd ed. New York: Routledge, 2014. Print.

Cline, Hugh F. *Information Communication Technology and Social Transformation: A Social and Historical Perspective*. New York: Routledge, 2014. Print.

Gibson, Jerry D., ed. *Mobile Communications Handbook*. 3rd ed. Boca Raton: CRC, 2012. Print.

Gillespie, Tarleton, Pablo J. Boczkowski, and Kirsten A. Foot, eds. *Media Technologies: Essays on Communication, Materiality, and Society*. Cambridge: MIT P, 2014. Print.

Hart, Archibald D., and Sylvia Hart Frejd. *The Digital Invasion: How Technology Is Shaping You and Your Relationships*. Grand Rapids: Baker, 2013. Print.

Livingston, Steven, and Gregor Walter-Drop, eds. *Bits and Atoms: Information and Communication Technology in Areas of Limited Statehood*. New York: Oxford UP, 2014. Print.

ENCRYPTION

FIELDS OF STUDY

Security; Privacy; Algorithms

ABSTRACT

Encryption is the encoding of information so that only those who have access to a password or encryption key can access it. Encryption protects data content, rather than preventing unauthorized interception of or access to data transmissions. It is used by intelligence and security organizations and in personal security software designed to protect user data.

PRINCIPAL TERMS

- **asymmetric-key encryption:** a process in which data is encrypted using a public encryption key but can only be decrypted using a different, private key.
- **authentication:** the process by which the receiver of encrypted data can verify the identity of the sender or the authenticity of the data.
- **hashing algorithm:** a computing function that converts a string of characters into a different, usually smaller string of characters of a given length, which is ideally impossible to replicate without knowing both the original data and the algorithm used.
- **Pretty Good Privacy:** a data encryption program created in 1991 that provides both encryption and authentication.

CRYPTOGRAPHY AND ENCRYPTION

Encryption is a process in which data is translated into code that can only by read by a person with the correct encryption key. It focuses on protecting data content rather than preventing unauthorized interception. Encryption is essential in intelligence and national security and is also common in commercial applications. Various software programs are available that allow users to encrypt personal data and digital messages.

The study of different encryption techniques is called "cryptography." The original, unencrypted data is called the "plaintext." Encryption uses an algorithm called a "cipher" to convert plaintext into ciphertext. The ciphertext can then be deciphered by using another algorithm known as the "decryption key" or "cipher key."

TYPES OF ENCRYPTION

A key is a string of characters applied to the plaintext to convert it to ciphertext, or vice versa. Depending on the keys used, encryption may be either symmetric or asymmetric. Symmetric-key encryption uses the same key for both encoding and decoding. The key used to encode and decode the data must be kept secret, as anyone with access to the key can translate the ciphertext into plaintext. The oldest known cryptography systems used alphanumeric substitution algorithms, which are a type of symmetric encryption. Symmetric-key algorithms are simple to create but vulnerable to interception.

In asymmetric-key encryption, the sender and receiver use different but related keys. First, the receiver

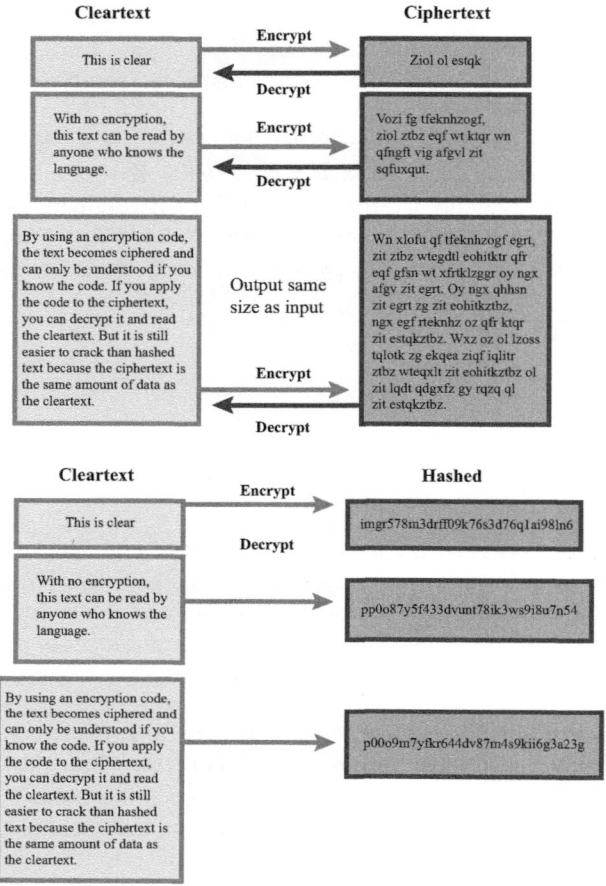

This diagram illustrates the output of encrypted content versus hashed content. When text is encrypted, the output will be the same size as the input, and it can be decrypted to show the original input. When text is hashed, input of any size will shrink to an output of a predetermined size, commonly 128 bits. The output cannot be decrypted, only authenticated by comparing it with a known hash value.

uses an algorithm to generate two keys, one to encrypt the data and another to decrypt it. The encryption key, also called the "public key," is made available to anyone who wishes to send the receiver a message. (For this reason, asymmetric-key encryption is also known as "public-key encryption.") The decryption key, or private key, remains known only to the receiver. It is also possible to encrypt data using the private key and decrypt it using the public key. However, the same key cannot be used to both encrypt and decrypt.

Asymmetric-key encryption works because the mathematical algorithms used to create the public and private keys are so complex that it is computationally impractical determine the private key based on the public key. This complexity also means that asymmetric encryption is slower and requires more processing power. First developed in the 1970s, asymmetric encryption is the standard form of encryption used to protect Internet data transmission.

Authentication and Security

Authentication is the process of verifying the identity of a sender or the authenticity of the data sent. A common method of authentication is a hashing algorithm, which translates a string of data into a fixed-length number sequence known as a "hash value." This value can be reverted to the original data using the same algorithm. The mathematical complexity of hashing algorithms makes it extremely difficult to decrypt hashed data without knowing the exact algorithm used. For example, a 128-bit hashing algorithm can generate 2^{128} different possible hash values.

In order to authenticate sent data, such as a message, the sender may first convert the data into a hash value. This value, also called a "message digest," may then be encrypted using a private key unique to the sender. This creates a digital signature that verifies the authenticity of the message and the identity of the sender. The original unhashed message is then encrypted using the public key that corresponds to the receiver's private key. Both the privately encrypted digest and the publicly encrypted message are sent to the receiver, who decrypts the original message using their private key and decrypts the message digest using the sender's public key. The receiver then hashes the original message using the same algorithm as the sender. If the message is authentic, the decrypted digest and the new digest should match.

Encryption Systems in Practice

One of the most commonly used encryption programs is Pretty Good Privacy (PGP). It was developed in 1991 and combines symmetric- and asymmetric-key encryption. The original message is encrypted using a unique one-time-only private key called a "session key." The session key is then encrypted using the receiver's public key, so that it can only be decrypted using the receiver's private key. This encrypted key is sent to the receiver along with the encrypted message. The receiver uses their private key to decrypt the session key, which can then be used to decrypt the message. For added security and authentication,

PGP also uses a digital signature system that compares the decrypted message against a message digest. The PGP system is one of the standards in personal and corporate security and is highly resistant to attack. The data security company Symantec acquired PGP in 2010 and has since incorporated the software into many of its encryption programs.

Encryption can be based on either hardware or software. Most modern encryption systems are based on software programs that can be installed on a system to protect data contained in or produced by a variety of other programs. Encryption based on hardware is less vulnerable to outside attack. Some hardware devices, such as self-encrypting drives (SEDs), come with built-in hardware encryption and are useful for high-security data. However, hardware encryption is less flexible and can be prohibitively costly to implement on a wide scale. Essentially, software encryption tends to be more flexible and widely usable, while hardware encryption is more secure and may be more efficient for high-security systems.

—*Micah L. Issitt*

BIBLIOGRAPHY

Bright, Peter. "Locking the Bad Guys Out with Asymmetric Encryption." *Ars Technica*. Condé Nast, 12 Feb. 2013. Web. 23 Feb. 2016.

Delfs, Hans, and Helmut Knebl. *Introduction to Cryptography: Principles and Applications*. 3rd ed. Berlin: Springer, 2015. Print.

History of Cryptography: An Easy to Understand History of Cryptography. N.p.: Thawte, 2013. *Thawte*. Web. 4 Feb. 2016.

"An Introduction to Public Key Cryptography and PGP." *Surveillance Self-Defense*. Electronic Frontier Foundation, 7 Nov. 2014. Web. 4 Feb. 2016.

Lackey, Ella Deon, et al. "Introduction to Public-Key Cryptography." *Mozilla Developer Network*. Mozilla, 21 Mar. 2015. Web. 4 Feb. 2016.

McDonald, Nicholas G. "Past, Present, and Future Methods of Cryptography and Data Encryption." *SpaceStation*. U of Utah, 2009. Web. 4 Feb. 2016.

ERROR HANDLING

FIELDS OF STUDY

Software Development; Coding Techniques; Software Engineering

ABSTRACT

Error handling is the process of responding appropriately to errors that occur during program execution. Errors may be addressed using unstructured or structured programming techniques. Errors can be handled by automatically running alternative code to accomplish the same task or by informing the user or developer of the error, ideally allowing them to correct the problem causing the error.

PRINCIPAL TERMS

- **debugging:** the process of locating and correcting flaws, or bugs, in a computer program.
- **error handler:** computer code that responds to errors that occur during the execution of a computer program.
- **exception:** in computer programming, an unpredictable or anomalous situation that occurs during the execution of a program.
- **logic error:** an error in which a program fails to operate in the way the programmer intended.
- **syntax error:** an error caused by a violation of one of the rules that govern how the elements of a programming language should be arranged.

HANDLING ERRORS IN A COMPUTER PROGRAM

Computer programmers and users are human, and mistakes may be introduced either in a program's code itself or through incorrect input or system requirements at run time. Errors occur when a computer program fails to compile, becomes unresponsive, fails to operate as intended, or terminates

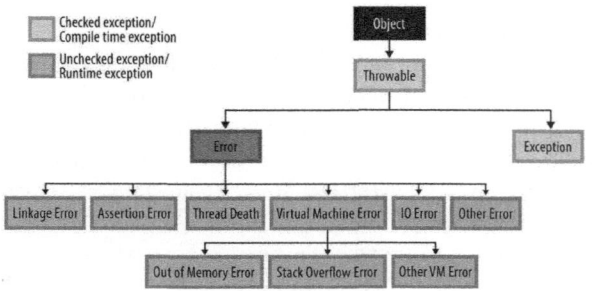

In the Java language, if an object belongs to the error superclass "throwable," it will either be an error or an exception. Error subclasses include Linkage Error, Assertion Error, Thread Death, Virtual Machine Error, IO Error, and others. Subclasses of Virtual Machine Error include Out of Memory Error, Stack Overflow Error, and others.

unexpectedly. Programs should be designed and written so that errors are minimized, but they must also deal effectively with the errors that do occur. Certain types of errors called exceptions, which occur during execution, may be anticipated in such a way that the program can respond appropriately. The process of finding and correcting errors is called debugging. Different techniques are employed to handle different types of errors when they are found.

Syntax errors are errors that occur due to violations of a programming language's syntax, or rules governing how elements such as words and symbols are arranged. For example, if a language specifies that all strings must be enclosed in double quotes, the following line of code would generate a syntax error due to the use of single quotes:

StringVariable1 = 'Pacific Ocean'

Syntax errors prevent a program from compiling. Compiling is the process of converting computer code into a form that can be executed and used.

Logic errors occur when code is syntactically correct, so the program compiles, but the resulting program does not function as intended. For example, if a program is designed to display the names of all employees hired during the month of January but it displays the names of employees hired in February instead, a logic error has occurred. Proofreading, testing, and feedback are all important steps in finding and handling syntax and logic errors. Programmers also use software debugging tools to find and correct these errors.

Run-time errors are errors that occur when a program is in use. They are typically caused by the program encountering issues outside of its source code, such as invalid parameters in the larger computer system or incorrect data input. Run-time errors are not detected when a program is compiled, and they may cause a program to become unresponsive or terminate unexpectedly. Typical run-time errors include divide-by-zero errors and attempting to access a required file that has been deleted. The negative effects of run-time errors can be minimized through the use of error handlers, or sections of code designed to correct errors. Correction may involve running alternative code to accomplish the requested task. The program may also handle errors by informing the user and allowing them to resolve the problem.

Structured and Unstructured Error Handling

Errors may be addressed through either structured or unstructured error handling. Unstructured error handling works by directing the program to a specific line of code or by instructing the program to proceed to the next line of code if an error occurs. This is a form of unstructured programming. Unstructured programs are difficult to debug, modify, and maintain. For this reason, programmers often use structured error handling instead.

Structured error handlers in many programming languages use a construct called a try/catch block. A try/catch block is structured as follows:

TRY
  Code that may cause run-time errors
CATCH
  Code that runs if a run-time error has occurred
END TRY

Try/catch blocks may contain multiple catch blocks that catch different types of errors.

While try/catch is an important technique, other methodologies can also be used to handle errors. Functions can use return codes to allow error conditions to be detected and dealt with. Guard clauses and data validation can be used to prevent errors from occurring by correcting error conditions automatically or by offering users the chance to correct errors before they occur.

While error handling offers many benefits, including making programs more robust and responsive, designing and implementing error handling can be time consuming. Schedules for software development projects must set aside time for error handling. Some bugs will escape detection during development, even with rigorous testing. User feedback is helpful in identifying error-causing bugs, which can be fixed or handled through later releases.

Error Handling in Practice

A retailer may use a computer program to calculate the average price of all items sold in one day. The program would prompt the user to input the number of items sold and the total dollar amount of sales, and then it would divide the dollar amount by the number of items. If the user enters 0 for the total number of items sold, a divide-by-zero error will occur. The following error handler could be used to handle this error:

```
TRY
  Prompt the user for number of items sold
  Prompt the user for total amount of sales in
  dollars
  Calculate the average item price by dividing by
  the total amount of sales in dollars by the total
  number of items sold
  Display the average item price
CATCH divide-by-zero error
  Inform the user that they must enter a nonzero
  value for the number of items sold
CATCH any other error
  Code that executes for any error other than
  divide by zero
END TRY
```

The try block includes the code that may cause the error, and the first catch block contains the code that will run if a divide-by-zero error occurs. The second catch block is included because it is good practice to add code to handle any potential errors that do not have a dedicated error handler.

Improving Software through Error Handling

For a computer program to accomplish the purpose for which it was designed, errors must be detected, minimized, and handled appropriately. Computer programs that do not handle errors well suffer from unexpected failures and are frustrating to use. Programmers must anticipate the types of errors that may occur in their programs and develop code with error handling in mind. Failure to do so will limit the stability and usefulness of the software they create. As long as errors are possible, error handling will remain a critical part of computer programming.

—*Maura Valentino, MSLIS*

Bibliography

Friedman, Daniel P., and Mitchell Wand. *Essentials of Programming Languages*. 3rd ed., MIT P, 2008.

Haverbeke, Marijn. *Eloquent JavaScript: A Modern Introduction to Programming*. 2nd ed., No Starch Press, 2015.

MacLennan, Bruce J. *Principles of Programming Languages: Design, Evaluation, and Implementation*. 3rd ed., Oxford UP, 1999.

Schneider, David I. *An Introduction to Programming Using Visual Basic*. 10th ed., Pearson, 2017.

Scott, Michael L. *Programming Language Pragmatics*. 4th ed., Elsevier, 2016.

Van Roy, Peter, and Seif Haridi. *Concepts, Techniques, and Models of Computer Programming*. MIT P, 2004.

EVENT-DRIVEN MARKETING (EDM)

Event-driven marketing (EDM), also called "event-based marketing" or "trigger-based marketing," is a marketing practice in which sellers cater to the specific circumstances of their customers. An event in this context is a change in an individual's life that can be detected or accurately predicted by the seller. In EDM, the best way to reach a customer is to send personalized messages about an event occurring in that customer's life at that particular moment. In order for the message to be relevant, the event must be of some significance and the customer must be contacted immediately, before the event is no longer current.

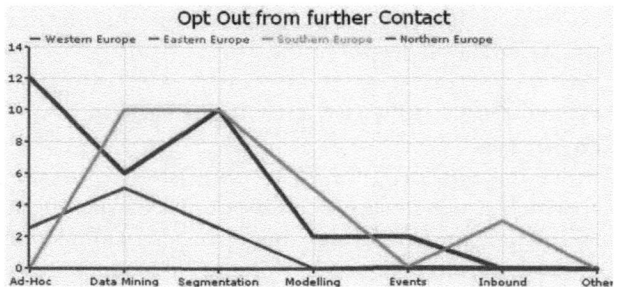
Sales results from using different CRM techniques

Overview

In event-driven marketing, sellers examine the transactional histories of their customers to determine upcoming events. Such events could be life changing, such as having a baby. Campaigns are then targeted toward those specific customers at those specific times. This marketing practice may be used for a smaller event as well. For example, in anticipation of winter, a seller might promote a sale on gloves by alerting the customer through e-mail.

There are several types of events. Significant events, such as a large bank deposit from a new job, indicate that the customer is likely to be receptive to communication at that time. Lifecycle events include marriage, the birth of a child, the purchase of a house, and other life-changing moments. In contrast, EDM can target simple, predicted events based on previous behavior, such as an individual's birthday. There are also trigger events, which happen to a customer today but are likely not significant; these involve binary actions, such as offering a magazine renewal before the current subscription runs out. Lastly, behavioral or super events, in which a series of events occur within a certain time or in a certain order (e.g., marrying, moving to a new house, and then having a baby), reveal a lot of information about the customer.

Sellers have always kept track of individual customer preferences to predict what potential customers will want to buy. However, this could not be achieved on a mass scale until the right technology became available. Direct mail—a related form of targeted marketing that involves letters, brochures, catalogs, and other promotional materials—began in the late nineteenth century. The development of computer databases and customer relationship management (CRM) software in the 1990s made it possible to send more specific direct mail, e-mail, and other communications.

Event-driven marketing first emerged as a specific marketing practice in 1995 at National Australia Bank (NAB), under the leadership of Fernando Ricardo of NAB and Ray O'Brien of Teradata. Before EDM, up-to-date information about customers had been limited, and NAB wanted a way to better anticipate customer needs. Ricardo and O'Brien collaborated on a solution, and their combined talents—marketing expertise and Teradata's ability to quickly load and process large volumes of data—was successful. In 2000, after NAB went public with their new marketing strategy, several other banks around the world implemented EDM practices. Since then, database, server, and CRM technology have continued to advance, making EDM possible via more machines, social media, cloud computing, and more organizations. The marketing practice has achieved significant success in retail, as the industry has recognized that critical events in the lives of customers can change their brand loyalties.

—*Julia Gilstein*

Bibliography

Estopace, Eden. "Event-Driven Marketing: Retail's Next Big Step?" *Enterprise Innovation*. Questex Media, 25 Mar. 2013. Web. 7 Oct. 2013.

"An Explanation of Event Driven Marketing." *Eventricity*. Eventricity Ltd, n.d. Web. 7 Oct. 2013.

Kalyanam, Kirthi, and Monte Zweben. "The Perfect Message at the Perfect Moment." *Harvard Business Review* 83.11 (2005): 112–20. Print.

Luckham, David C. *Event Processing for Business*. Hoboken: Wiley, 2012. Print.

Scott, David Meerman. *The New Rules of Marketing and PR*. 4th ed. Hoboken: Wiley, 2013. Print.

Stuntebeck, Vince, and Trevor Jones. "B2B Event-Driven Marketing: Triggers Your Analytics Shouldn't Miss." *MarketingProfs*. MarketingProfs, 22 Apr. 2013. Web. 7 Oct. 2013.

Van Bel, Egbert J., Ed Sander, and Alan Weber. *Follow That Customer! The Event-Driven Marketing Handbook.* Chicago: Racom, 2011. Print.

"What Is an Event?" *Eventricity.* Eventricity Ltd, n.d. Web. 7 Oct. 2013.

EXPECTANCY THEORY

Expectancy theory is a theory of motivation focusing on how and why people become motivated. Business expert Victor Vroom identified the attributes of expectancy theory in 1964, although some researchers believe the original theory dates back to Edward Tolman's work in the 1930s. Tolman, an American psychologist, suggested that human behavior is directly affected more by expectation than by any direct stimuli. Vroom's work, however, greatly impacted the understanding of how cognitive processes affect an employee's motivation to perform.

OVERVIEW

Expectancy theory consists of three basic components: (1) the employee's expectancy that working hard will lead to his or her desired level of performance; (2) the employee's expectancy that working hard will thus ensure that rewards will follow; and (3) whether or not the employee's perception that the outcome of working hard is worth the effort or value associated with hard work. Vroom's VIE expectancy theory is often formulated using the equation MF (motivational forces) = V (valence) x I (instrumentality) x E (expectancy). The effort exerted by an employee is directly influenced by the expectation (E) that by trying hard, the task can be successfully performed. If, however, the employee believes that trying hard will not generate successful performance (I), the effort exerted by the employee will decline due to a lack of anticipated rewards (V).

In the application of this formula, motivational forces (MF) directly relate to those internal and external variables affecting an individual's performance and effort. Employees consistently leverage the results of their efforts with the expectation that hard work will lead to intrinsic or extrinsic rewards. The perception of acquiring potential rewards based on their individual effort must then be measured in terms of value: is my effort worth the reward expected? Expectancy (E), therefore, is directly associated with the innate belief that successful performance will lead to rewards. In an organizational context, individuals maximize their instrumentality through the achievement of high performance goals with the understanding that successful performance will be followed by rewards. When employees perceive that rewards are not congruent with performance, the value (V) of their effort decreases due to the perception that the efforts gained as a result of hard work are not worth the effort expended. The influence of instrumentality and value subsequently affects the effort to which an employee will persist in hard work.

The theoretical implications of expectancy theory are complex and depend upon numerous factors that may possibly influence employee motivation. The degree to which individual employees assigns a value to his or her instrumentality will determine the level of effort devoted to a particular task. Numerous factors must be taken into consideration when measuring the perceived rewards and consequences of effort as postulated by Vroom's expectancy theory. Despite the complexities of expectancy theory, there are numerous applications in education, business, social sciences and government.

—*Patricia Hoffman-Miller, MPA, PhD*

What will you make of your life?

You have all the tools and resources you need
To become the person you want to become

All you have to do
Is to decide what you're going to do
With all these tools and resources?

THE CHOICE IS YOURS

Emphasizing choice in decision-making as a tool for achievement and empowerment.

BIBLIOGRAPHY

Baker-Eveleth, L., and R. Stone. "Expectancy Theory and Behavioral Intentions to Use Computer Applications." *Interdisciplinary Journal of Information* 3 (2008): 135–46. Print.

Cassar, G., and H. Friedman. "Does Self-Efficacy Affect Entrepreneurial Investment?" *Strategic Entrepreneurial Journal* 3.3 (2009): 241–60. Print.

Estes, Brent. "Predicting Productivity in a Complex Labor Market: A Sabermetric Assessment of Free Agency on Major League Baseball Performance." *Business Studies Journal* 3.1 (2011): 23–58. Print.

Estes, Brent, and Barbara Polnick. "Examining Motivation Theory in Higher Education: An Expectancy Theory Analysis of Tenured Faculty Productivity." *International Journal of Management, Business and Administration* 15.1 (2012): n.p. Print.

Jex, Steve M., and Thomas W. Britt. *Organizational Psychology: A Scientist-Practitioner Approach.* 3rd ed. Malden: Wiley, 2014. Print.

Lunenberg, Fred C. "Expectancy Theory of Motivation: Motivating by Altering Expectations." *International Journal of Management, Business and Administration* 15.1 (2011): n.p. Print.

Miles, Jeffrey A., ed. *New Directions in Management and Organizations Theory.* Newcastle upon Tyne: Cambridge Scholars, 2014. Print.

Nasri, Wadie, and Lanouar Charfeddine. "Motivating Salespeople to Contribute to Marketing Intelligence Activities: An Expectancy Theory Approach." *International Journal of Marketing Studies* 4.1 (2012): 168. Print.

Renko, Maija, K. Galen Kroeck, and Amanda Bullough. "Expectancy Theory and Nascent Entrepreneurship." *Small Business Economics* 39.3 (2012): 667–84. Print.

Shweiki, Ehyal, et al. "Applying Expectancy Theory to Residency Training: Proposing Opportunities to Understand Resident Motivation and Enhance Residency Training." *Advances in Medical Education and Practice* 6 (2015): 339–46. Print.

Ugah, Akobundu, and Uche Arua. "Expectancy Theory, Maslow's Hierarchy of Needs, and Cataloguing Departments." *Library Philosophy and Practice* 1 (2011): 51. Print.

EXPERIMENTER'S BIAS

The results of experiments can be flawed or skewed because of bias. Those designing, conducting, or analyzing an experiment often hold expectations regarding the experiment's outcome, such as hoping for an outcome that supports the initial hypothesis. Such expectations can shape how the experiment is structured, conducted, and/or interpreted, thereby affecting the outcome. This typically unconscious and unintentional phenomenon is known as experimenter's bias.

The main types of experimenter's bias include self-fulfilling prophecy, observer bias, and interpreter bias. Most modern social science and clinical experiments are designed with one or more safeguards in place to minimize the possibility of such biases distorting results.

OVERVIEW

In the mid- to late 1960s, psychologist Robert Rosenthal began uncovering and reporting on experimenter's bias in social science research. His most famous and controversial work was a 1968 study on teacher expectations. In it, Rosenthal and his colleagues gave students a standardized intelligence test, then randomly assigned some to a group designated "intellectual bloomers" and told teachers that these students were expected to perform very well academically. When tested eight months later, the "intellectual bloomers" had indeed done better than their peers, suggesting that teacher expectancy had affected the educational outcomes. This phenomenon, in which a person's behavior is shaped by and conforms to the expectations of others, came to be known as the Pygmalion effect, named for the play by George Bernard Shaw. Rosenthal's work shed light on issues of internal validity and launched a new area of research into methodologies.

The most widely recognized form of experimenter's bias, the self-fulfilling prophecy, occurs when an experimenter's expectancy informs his or her own behavior toward a study subject, eliciting predicted response and thereby confirming the original expectations. Among the subtle factors that can sway outcomes among human study participants are the experimenter's word choice, tone, body language, gestures, and expressions. Similarly, animal subjects may respond to experimenters' cues, differential handling, and, in the case of primates, nonverbal body language. The Pygmalion effect is one type of self-fulfilling prophecy. Another is the experimenter expectancy effect, in which the experimenter's expectations influence his or her own behavior toward the subjects in such a way as to increase the likelihood of those expectations being met.

Other biases arise not from the experimenter's interaction with the subjects but rather from his or her observations and interpretations of their responses. Observer bias occurs when the experimenter's assumptions, preconceptions, or prior knowledge affects what he or she observes and records about the results of the experiment. Interpreter bias is an error in data interpretation, such as a focus on just one possible interpretation of the data to the exclusion of all others.

To attempt to prevent bias, most social science and clinical studies are either single-blind studies, in which subjects are unaware of whether they are participating in a control or a study group, or double-blind studies, in which both experimenters and subjects are unaware of which subjects are in which groups. Other ways of avoiding experimenter's bias include standardizing methods and procedures to minimize differences in experimenter-subject interactions; using blinded observers or confederates as assistants, further distancing the experimenter from the subjects; and separating the roles of investigator and experimenter.

Experimenter's bias and the prevention thereof have implications for areas of research as diverse as social psychology, education, medicine, and politics.

—*Céleste Codington-Lacerte*

Bibliography

Becker, Lee A. "VIII. The Internal Validity of Research." *Effect Size Calculators*. U of Colorado: Colorado Springs, 16 Mar. 1998. Web. 28 July 2015.

Colman, Andrew M. *A Dictionary of Psychology*. 4th ed. New York: Oxford UP, 2015. Print.

Finn, Patrick. "Primer on Research: Bias and Blinding; Self-Fulfilling Prophecies and Intentional Ignorance." *ASHA Leader* June 2006: 16–22. Web. 3 Oct. 2013.

Gould, Jay E. *Concise Handbook of Experimental Methods for the Behavioral and Biological Sciences*. Boca Raton: CRC, 2002. Print.

Greenberg, Jerald, and Robert Folger. *Controversial Issues in Social Research Methods*. New York: Springer, 1988. Print.

Halperin, Sandra, and Oliver Heath. *Political Research: Methods and Practical Skills*. New York: Oxford UP, 2012. Print.

Jussim, Lee. *Social Perception and Social Reality: Why Accuracy Dominates Bias and Self-Fulfilling Prophecy*. New York: Oxford UP, 2012. Print.

Rosenthal, Robert, and Ralph L. Rosnow. *Artifacts in Behavioral Research*. New York: Oxford UP, 2009. Print.

Schultz, Kenneth F., and David A. Grimes. "Blinding in Randomised Trials: Hiding Who Got What." *Lancet* 359.9307 (2002): 696–700. *RHL: The WHO Reproductive Health Library*. Web. 11 June 2015.

Supino, Phyllis G. "Fundamental Issues in Evaluating the Impact of Interventions: Sources and Control of Bias." *Principles of Research Methodology: A Guide for Clinical Investigators*. Ed. Supino and Jeffrey S. Borer. New York: Springer, 2012. 79–110. Print.

EXTREME PROGRAMMING

FIELDS OF STUDY
Software Development; Programming Methodologies; Software Engineering

ABSTRACT
Extreme programming (XP) is a variation of the agile software development methodology based on five values: simplicity, communication, feedback, respect, and courage. Extreme programming features small teams of developers, managers, and customers who work closely together during short development cycles. It promotes the use of simple designs and practices and frequent small releases.

PRINCIPAL TERMS
- **iteration:** in software development, a single, self-contained phase of a product's overall development cycle, typically lasting one to three weeks, in which one aspect of the product is addressed from beginning to end before moving onto the next phase of development.
- **pair programming:** an agile software development practice in which two programmers work together at the same time and share a single workstation.
- **release:** the version of a software system or other product made available to consumers, or the act of making that version available.
- **story:** in software development, a description of requirements created by project stakeholders and used by developers to determine project time frames and schedules.
- **unit test:** the process of testing an individual component of a larger software system.

UNDERSTANDING EXTREME PROGRAMMING
Extreme programming (XP) is a version of the agile software development methodology. The name "extreme programming" derives from the fact that XP takes successful elements from proven software development methodologies to an extreme level. For example, it is common practice to use periodic code reviews to increase code quality. XP takes this to an extreme level by encouraging the use of pair programming, in which two programmers continually review each other's code in real time. XP is based on five values: simplicity, communication, feedback, courage, and respect.

Simplicity is promoted using simple designs and numerous iterations, concentrating on what can be done in the short term to meet immediate requirements. Future releases are used to respond to new requirements.

Unlike traditional software development methodologies, XP favors communication over documentation. Small teams work in close physical proximity and cooperate on all aspects of the development process. They share information verbally and use documentation only when necessary.

A system of continuous feedback is maintained by XP developers using unit tests, customer acceptance tests, and oversight of other team members' code. All team members are encouraged to respond positively to changes and to adapt their processes as needed.

XP requires team members to work courageously. They must be willing to take risks, adapt to changing situations, and be honest about progress and deadlines. To promote courageous behavior, team members support each other and do not waste time assigning or avoiding blame for failure.

The final value is respect for self and respect for others. This includes respect for the work being done and respect for the customer's requirements and stories. The contributions of all team members are valued.

TAKING THINGS TO EXTREMES
XP focuses on four activities: coding, testing, listening, and designing. Coding is the most important activity. The ultimate purpose of software development is to produce code. Testing is a core activity because code must be tested to ensure it meets the needs for which it was written. XP makes extensive use of various forms of testing, including unit testing, acceptance testing, and continuous code reviews. The listening activity directly supports the core value of communication. Team members should employ effective listening skills at all times. The final activity is designing. XP seeks to break the design-and-develop

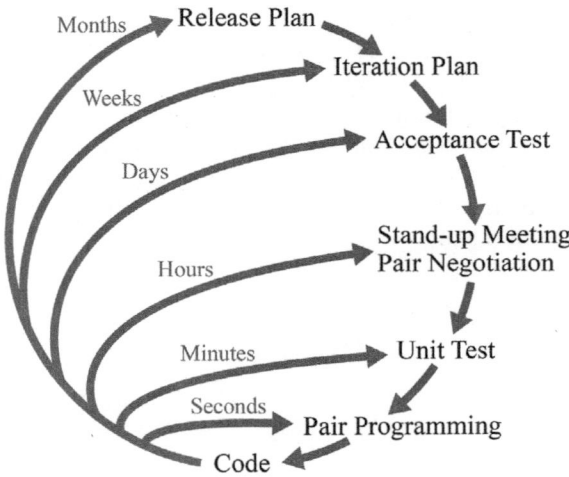

Planning/Feedback Loops

Under the process of extreme programming, feedback and plan updates occur quickly at each phase of the process so that a release can happen in months, iterations of that release happen in weeks, acceptance testing happens in days, and unit testing happens in minutes. The visual representation of this process is adapted from Don Wells, via Wikimedia Commons.

paradigm used by traditional development methodologies in favor of an iterative paradigm that embraces flexibility while stressing good and minimal design.

XP offers several advantages over other methodologies. Frequent small releases and short iterations enable code to be thoroughly tested. Unit testing ensures that defects in code are detected, and customer acceptance testing ensures that the system actually meets the customer's real needs. Unlike traditional development methodologies that work best when project requirements are fixed, XP adapts easily to changing requirements. It can lead to cost savings as unnecessary processes and documentation are avoided.

XP does have drawbacks. It requires extensive customer involvement in the development process, which may not be possible in all situations. In addition, a reliance on communication over documentation makes projects vulnerable to disruption if team members leave the organization and must be replaced. Finally, XP promotes the use of uncommon practices that may be uncomfortable for some developers, such as pair programming.

EXTREME PROGRAMMING IN THE REAL WORLD

XP can be used in a wide variety of real-world programming scenarios. For example, imagine a corporation that produces consumer electronic products decides to launch a subsidiary to produce electronic systems for the aviation industry. The subsidiary is located in another city and will be opened in four months. The information technology (IT) department has been tasked with developing a customer relationship management application to be used by the sales department at the new subsidiary. The application must interoperate between both locations. XP is a good choice for this type of project for the following reasons.

As the application will be used in a new line of business, it is likely that its requirements will change frequently as sales staff gain experience. XP's short iterations and frequent small releases are a good match for projects with changing requirements. In addition, the project has a high level of risk, assuming the IT staff has little experience in developing a distributed application that must interoperate between two remote locations, and the customer must have a functional system that meets their needs by a specific date. XP's values and practices reduce risk and help ensure project success. Finally, the sales staff that will be posted to the new office is available to devote their time to the project. This is important, because XP requires end users to be dedicated team members.

A FLEXIBLE DEVELOPMENT STRATEGY

XP is an effective methodology when used to develop solutions in high-risk environments. It is ideally suited for projects with changing requirements, and it works best when customers or other end users are available to be active members of the project team. XP requires developers and management to be open to new ways of thinking and new processes. As such, it is an effective methodology for use in a wide range of dynamic businesses, including information technology, social media, and mobile application development. XP's focus on flexibility and adaptability ensures that it will remain an important methodology as customers' needs change over time.

—*Maura Valentino, MSLIS*

BIBLIOGRAPHY

Bell, Michael. *Incremental Software Architecture: A Method for Saving Failing IT Implementations.* John Wiley & Sons, 2016.

Friedman, Daniel P., and Mitchell Wand. *Essentials of Programming Languages.* 3rd ed., MIT P, 2008.

Jayaswal, Bijay K., and Peter C. Patton. *Design for Trustworthy Software: Tools, Techniques, and Methodology of Developing Robust Software.* Prentice Hall, 2007.

MacLennan, Bruce J. *Principles of Programming Languages: Design, Evaluation, and Implementation.* 3rd ed., Oxford UP, 1999.

Scott, Michael L. *Programming Language Pragmatics.* 4th ed., Elsevier, 2016.

Van Roy, Peter, and Seif Haridi. *Concepts, Techniques, and Models of Computer Programming.* MIT P, 2004.

Wysocki, Robert K. *Effective Project Management: Traditional, Agile, Extreme.* 7th ed., John Wiley & Sons, 2014.

F

FIREWALLS

FIELDS OF STUDY

Information Systems; Privacy; Security

ABSTRACT

A firewall is a program designed to monitor the traffic entering and leaving a computer network or single device and prevent malicious programs or users from entering the protected system. Firewalls may protect a single device, such as a server or personal computer (PC), or even an entire computer network. They also differ in how they filter data. Firewalls are used alongside other computer security measures to protect sensitive data.

PRINCIPAL TERMS

- **application-level firewalls:** firewalls that serve as proxy servers through which all traffic to and from applications must flow.
- **host-based firewalls:** firewalls that protect a specific device, such as a server or personal computer, rather than the network as a whole.
- **network firewalls:** firewalls that protect an entire network rather than a specific device.
- **packet filters:** filters that allow data packets to enter a network or block them on an individual basis.
- **proxy server:** a computer through which all traffic flows before reaching the user's computer.
- **stateful filters:** filters that assess the state of a connection and allow or disallow data transfers accordingly.

History of Firewalls

In the early twenty-first century, increasing cybercrime and cyberterrorism made computer security a serious concern for governments, businesses and organizations, and the public. Nearly any computer system connected to the Internet can be accessed by malicious users or infected by harmful programs such as viruses. Both large networks and single PCs face this risk. To prevent such security breaches, organizations and individuals use various security technologies, particularly firewalls. Firewalls are programs or sometimes dedicated devices that monitor the data entering a system and prevent unwanted data from doing so. This protects the computer from both malicious programs and unauthorized access.

The term "firewall" is borrowed from the field of building safety. In that field it refers to a wall specially built to stop the spread of fire within a structure. Computer firewalls fill a similar role, preventing harmful elements from entering the protected area. The idea of computer firewalls originated in the 1980s. At that time, network administrators used routers, devices that transfer data between networks, to separate one network from another. This stopped problems in one network from spreading into others. By the early 1990s, the proliferation of computer viruses and increased risk of hacking made the widespread need for firewalls clear. Some of the advances in that era, such as increased access to the Internet and developments in operating systems, also introduced new vulnerabilities. Early firewalls relied heavily on the use of proxy servers. Proxy servers are servers through which all traffic flows before entering a user's computer or network. In the twenty-first century, firewalls can filter data according to varied criteria and protect a network at multiple points.

Types of Firewalls

All firewalls work to prevent unwanted data from entering a computer or network. However, they do so in different ways. Commonly used firewalls can be in various positions relative to the rest of the system. An individual computer may have its own personal

129

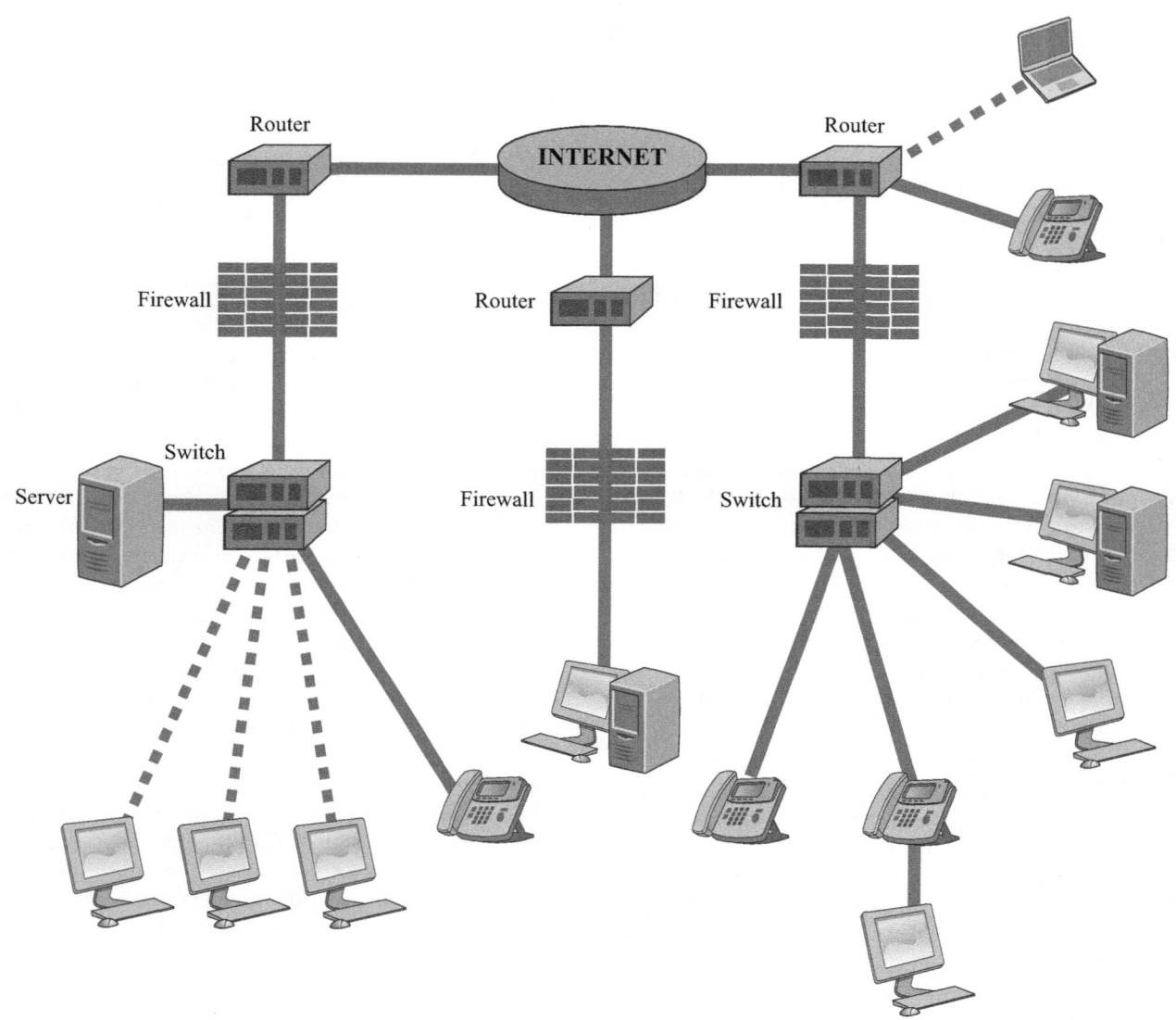

Firewalls are one of many protective measures used to prevent hackers from accessing computers or networks.

firewall, such as Windows Firewall or Macintosh OS X's built-in firewall. Other networked devices, such as servers, may also have personal firewalls. These are known as host-based firewalls because they protect a single host rather than the whole network. They protect computers and other devices not only from malicious programs or users on the Internet but also from viruses and other threats that have already infiltrated the internal network, such as a corporate intranet, to which they belong. Network firewalls, on the other hand, are positioned at the entrance to the internal network. All traffic into or out of that network must filter through them. A network firewall may be a single device, such as a router or dedicated computer, which serves as the entrance point for all data. It then blocks any data that is malicious or otherwise unwanted. Application-level firewalls, which monitor and allow or disallow data transfers from and to applications, may be host based or network based.

Firewalls also vary based on how they filter data. Packet filters examine incoming data packets individually and determine whether to block or allow each one to proceed. They decide this based on factors such as the origin and destination of the packets. Stateful filters determine whether to admit or block incoming data based on the state of the connection. Firewalls that use stateful filtering can identify whether data packets trying to enter the computer system are part of an ongoing, active connection and determine whether to let them in based on that context. This allows them to examine and filter incoming data more quickly than their stateless counterparts.

Firewalls and Computer Security

By preventing malicious programs or users from accessing systems, firewalls protect sensitive data stored in or transmitted via computers. They are used to protect personally identifying information, such as Social Security numbers, as well as proprietary trade or government information. Both the technology industry and the public have put increased emphasis on such protections in the early twenty-first century, as identity theft, fraud, and other cybercrimes have become major issues. In light of such threats, firewalls play an essential role in the field of computer security. However, experts caution that a firewall should not be the sole security measure used. Rather, firewalls should be used along with other computer security practices. These practices include using secure passwords, regularly updating software to install patches and eliminate vulnerabilities, and avoiding accessing compromised websites or downloading files from suspicious sources.

—*Joy Crelin*

Bibliography

"How Firewalls Work." *Boston University Information Services and Technology.* Boston U, n.d. Web. 28 Feb. 2016.

Ingham, Kenneth, and Stephanie Forrest. *A History and Survey of Network Firewalls.* Albuquerque: U of New Mexico, 2002. PDF file.

Morreale, Patricia, and Kornel Terplan, eds. *The CRC Handbook of Modern Telecommunications.* 2nd ed. Boca Raton: CRC, 2009. Print.

Northrup, Tony. "Firewalls." *TechNet.* Microsoft, n.d. Web. 28 Feb. 2016.

Stallings, William, and Lawrie Brown. *Computer Security: Principles and Practice.* 3rd ed. Boston: Pearson, 2015. Print.

Vacca, John, ed. *Network and System Security.* 2nd ed. Waltham: Elsevier, 2014. Print.

FIRMWARE

FIELDS OF STUDY

Embedded Systems; Software Engineering; System-Level Programming

ABSTRACT

Firmware occupies a position in between hardware, which is fixed and physically unchanging, and software, which has no physical form apart from the media it is stored on. Firmware is stored in nonvolatile memory in a computer or device so that it is always available when the device is powered on. An example can be seen in the firmware of a digital watch, which remains in place even when the battery is removed and later replaced.

PRINCIPAL TERMS

- **embedded systems:** computer systems that are incorporated into larger devices or systems to monitor performance or to regulate system functions.
- **flashing:** a process by which the flash memory on a motherboard or an embedded system is updated with a newer version of software.
- **free software:** software developed by programmers for their own use or for public use and distributed without charge; it usually has conditions attached that prevent others from acquiring it and then selling it for their own profit.
- **homebrew:** software that is developed for a device or platform by individuals not affiliated with the device manufacturer; it is an unofficial or

"homemade" version of the software that is developed to provide additional functionality not included or not permitted by the manufacturer.
- **nonvolatile memory:** computer storage that retains its contents after power to the system is cut off, rather than memory that is erased at system shutdown.

OVERVIEW OF FIRMWARE

Many consumer devices have become so complex that they need a basic computer to operate them. However, they do not need a fully featured computer with an operating system (OS) and specially designed software. The answer to this need is to use embedded systems. These systems are installed on microchips inside devices as simple as children's toys and as complex as medical devices such as digital thermometers. The term "embedded" is used because the chips containing firmware are ordinarily not directly accessible to consumers. They are installed within the device or system and expected to work throughout its lifespan.

Computers also use firmware, which is called the "basic input/output system," or BIOS. Even though the computer has its own OS installed and numerous programs to accomplish more specific tasks, there is still a need for firmware. This is because, when the computer is powered on, some part of it must be immediately able to tell the system what to do in order to set itself up. The computer must be told to check the part of the hard drive that contains the start-up sequence, then to load the OS, and so on. The firmware serves this purpose because, as soon as electric current flows into the system, the information stored in the computer's nonvolatile memory is loaded and its instructions are executed. Firmware is usually unaffected even when a different OS is installed. However, the user can also configure the BIOS to some extent and can boot the computer into the BIOS to make changes when necessary. For example, a computer that is configured to boot from the CD-ROM drive first could have this changed in the BIOS so that it would first attempt to read information from an attached USB drive.

MODIFYING AND REPLACING FIRMWARE

Sophisticated users of technology sometimes find that the firmware installed by a manufacturer does

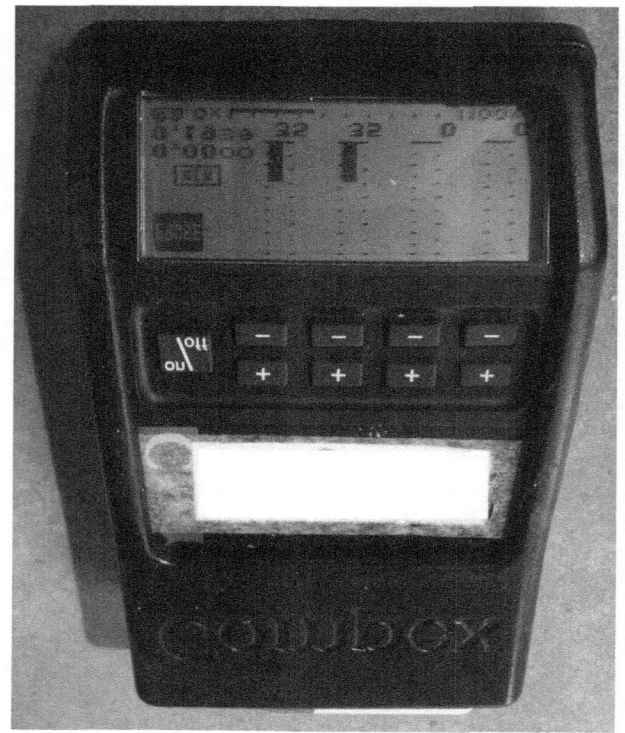

Picture of FES device, manufactured by Compe.

not meet all of their needs. When this occurs, it is possible to update the BIOS through a process known as flashing. When the firmware is flashed, it is replaced by a new version, usually with new capabilities. In some cases, the firmware is flashed because the device manufacturer has updated it with a new version. This is rarely done, as firmware functionality is so basic to the operation of the device that it is thoroughly tested prior to release. From time to time, however, security vulnerabilities or other software bugs are found in firmware. Manufacturers helping customers with troubleshooting often recommend using the latest firmware to rule out such defects.

Some devices, especially gaming consoles, have user communities that can create their own versions of firmware. These user-developed firmware versions are referred to as homebrew software, as they are produced by users rather than manufacturers. Homebrew firmware is usually distributed on the Internet as free software, or freeware, so that anyone can download it and flash their device. In the case of gaming consoles, this can open up new capabilities. Manufacturers tend to produce devices only for specialized functions. They exclude other functions because the functions would increase the cost or make it too easy to use the device for illegal or undesirable purposes. Flashing such devices with homebrew software can make these functions available.

Automobile Software

One of the market segments that has become increasingly reliant on firmware is automobile manufacturing. More and more functions in cars are now controlled by firmware. Not only are the speedometer and fuel gauge computer displays driven by firmware, but cars come with firmware applications for music players, real-time navigation and map display, and interfaces with passengers' cell phones.

Firmware as Vulnerability

Although firmware is not very visible to users, it has still been a topic of concern for computer security professionals. With homebrew firmware distributed over the Internet, the concern is that the firmware may contain "backdoors." A backdoor is a secret means of conveying the user's personal information to unauthorized parties. Even with firmware from official sources, some worry that it would be possible for the government or the device manufacturer to include security vulnerabilities, whether deliberate or inadvertent.

—*Scott Zimmer, JD*

Bibliography

Bembenik, Robert, Łukasz Skonieczny, Henryk Rybiński, Marzena Kryszkiewicz, and Marek Niezgódka, eds. *Intelligent Tools for Building a Scientific Information Platform: Advanced Architectures and Solutions.* New York: Springer, 2013. Print.

Dice, Pete. *Quick Boot: A Guide for Embedded Firmware Developers.* Hillsboro: Intel, 2012. Print.

Iniewski, Krzysztof. *Embedded Systems: Hardware, Design, and Implementation.* Hoboken: Wiley, 2012. Print.

Khan, Gul N., and Krzysztof Iniewski, eds. *Embedded and Networking Systems: Design, Software, and Implementation.* Boca Raton: CRC, 2014. Print.

Noergaard, Tammy. *Embedded Systems Architecture: A Comprehensive Guide for Engineers and Programmers.* 2nd ed. Boston: Elsevier, 2013. Print.

Sun, Jiming, Vincent Zimmer, Marc Jones, and Stefan Reinauer. *Embedded Firmware Solutions: Development Best Practices for the Internet of Things.* Berkeley: ApressOpen, 2015. Print.

FUNCTIONAL ELECTRICAL STIMULATION (FES)

Functional electrical stimulation (FES), also called functional neuromuscular stimulation (FNS) or neuromuscular electrical stimulation (NMES), is a medical technique that uses electricity to stimulate a patient's muscles or nerves for therapeutic purposes. It has proved effective in helping patients recover from or manage numerous different muscular or neurological ailments, including stroke, cerebral palsy, muscular dystrophy, spinal cord injury, and multiple sclerosis.

Overview

The use of electricity to stimulate nerves in order to restore physical functioning dates back to the early 1960s, when W. T. Liberson and his colleagues developed an electrotherapy technique to treat foot drop, a common result of neurological or muscular trauma. Damage to the brain, spinal cord, nerves, or muscles, as from a stroke or a degenerative disease such as muscular dystrophy, can result in the patient's inability to raise the front of his or her foot while walking. Liberson developed a device that sensed when the foot was raised from the ground, at which point it applied surface stimulation to the peroneal nerve, which connects to the muscles of the foot, causing the foot to flex and thus allowing the patient to walk normally. Liberson and colleagues originally called the technique "functional electrotherapy"; a year later, J. H. Moe and H. W. Post coined the term "functional electrical stimulation" in a 1962 paper.

Since then, FES has been used to treat numerous disorders affecting muscles, nerves, or both, through either surface stimulation or implanted devices. Neurological conditions that benefit from FES include strokes, multiple sclerosis, cerebral palsy, and paralysis resulting from spinal cord injuries. In patients who are struggling to regain physical functioning during stroke recovery, applying electrical stimulation to affected areas activates the nerves that cause the muscles to move, which both rebuilds muscle strength and helps the brain relearn how to initiate the movement on its own. In cases of multiple sclerosis, FES can improve the functioning of damaged nerves that control various muscle groups. For patients with cerebral palsy, FES can improve muscle control and reduce spasticity in the extremities. Multiple sclerosis and cerebral palsy patients, like those who have had strokes, are also prone to foot drop, for which FES remains a common treatment.

For patients experiencing paralysis due to spinal cord injury, FES can be used to exercise the muscles of paralyzed limbs, thus preventing atrophy and increasing overall fitness. In some cases, it may restore some level of mobility and control over bodily functions or even allow paralyzed extremities to function with assistance. One notable case was that of Nan Davis, a young woman who was in a car crash in 1978, on the night of her high school graduation, and was paralyzed below her rib cage. While attending Wright State University in Ohio, Davis volunteered for an experimental program headed by Jerrold Petrofsky, a biomedical engineering professor at the school. In 1983, with the aid of a device developed by Petrofsky that used FES to stimulate the muscles in her legs, Davis was able to walk across the stage at her college graduation to receive her diploma.

FES has also shown success as a treatment for disorders that affect the muscles themselves, such as the various types of muscular dystrophy, which are a group of diseases in which the muscles grow progressively weaker. As in cases of muscular atrophy due to neurological causes, FES can be used to exercise the affected muscles, halting deterioration and improving muscle strength.

—*Randa Tantawi, PhD*

Bibliography

Campbell, Joyce M. "General Considerations in the Clinical Application of Electrical Stimulation." *International Functional Electrical Stimulation Society.* IFESS, n.d. Web. 7 Oct. 2013.

Chisari, Carmelo, et al. "Chronic Muscle Stimulation Improves Muscle Function and Reverts the Abnormal Surface EMG Pattern in Myotonic Dystrophy: A Pilot Study." *Journal of NeuroEngineering and Rehabilitation* 10.1 (2013): 94. Print.

"Functional Electrical Stimulation (FES) Factsheet." *Multiple Sclerosis Trust.* Multiple Sclerosis Trust, Dec. 2012. Web. 7 Oct. 2013.

Liberson, W. T., et al. "Functional Electrotherapy: Stimulation of the Peroneal Nerve Synchronized with the Swing Phase of the Gait of Hemiplegic Patients." *Archives of Physical Medicine and Rehabilitation* 42 (1961): 101–5. Print.

Peng, Chih-Wei, et al. "Review: Clinical Benefits of Functional Electrical Stimulation Cycling Exercise for Subjects with Central Neurological Impairments." *Journal of Medical and Biological Engineering* 31.1 (2011): 1–11. Print.

Prescott, Bonnie. "Improved Method of Electrical Stimulation Could Help Treat Damaged Nerves." *Beth Israel Deaconess Medical Center.* Beth Israel Deaconess Med. Ctr., 21 Nov. 2011. Web. 7 Oct. 2013.

Ragnarsson, K. T. "Functional Electrical Stimulation after Spinal Cord Injury: Current Use, Therapeutic Effects and Future Directions." *Spinal Cord* 46.4 (2008): 255–74. Print.

Tansey, Keith, et al. "Restorative Neurology of Motor Control after Spinal Cord Injury." *Restorative Neurology of Spinal Cord Injury.* Ed. Milan R. Dimitrijevic et al. Oxford: Oxford UP, 2012. 43–64. Print.

Vrbová, Gerta, Olga Hudlicka, and Kristin Schaefer Centofanti. *Application of Muscle/Nerve Stimulation in Health and Disease.* Dordrecht: Springer, 2008. Print. Advances in Muscle Research 4.

Wagner, Sean. "'Functional Electrical Stimulation' Treatment Improves Walking Ability of Parkinson's Patients." *Medical News Today.* MediLexicon Intl., 3 June 2008. Web. 7 Oct. 2013.

G

GAME PROGRAMMING

FIELDS OF STUDY
Software Engineering; Programming Language

ABSTRACT
Game programming is a type of software engineering used to develop computer and video games. Depending on the type of game under development, programmers may be required to specialize in areas of software development not normally required for computer programmers. These specialties include artificial intelligence, physics, audio, graphics, input, database management, and network management.

PRINCIPAL TERMS

- **game loop:** the main part of a game program that allows the game's physics, artificial intelligence, and graphics to continue to run with or without user input.
- **homebrew:** a slang term for software made by programmers who create games in their spare time, rather than those who are employed by software companies.
- **object-oriented programming:** a type of programming in which the source code is organized into objects, which are elements with a unique identity that have a defined set of attributes and behaviors.
- **prototype:** an early version of software that is still under development, used to demonstrate what the finished product will look like and what features it will include.
- **pseudocode:** a combination of a programming language and a spoken language, such as English, that is used to outline a program's code.
- **source code:** the set of instructions written in a programming language to create a program.

HOW GAME PROGRAMMING WORKS

While many video games are developed as homebrew projects by individuals working in their spare time, most major games are developed by employees of large companies that specialize in video games. In such a company, the process of creating a video game begins with a basic idea of the game's design. If enough people in the company feel the idea has merit, the company will create a prototype to develop the concept further. The prototype gives form to the game's basic story line, graphics, and programming. It helps developers decide whether the game is likely to do well in the marketplace and what resources are needed for its development.

Games that move past the prototyping stage proceed along many of the same pathways as traditional software development. However, video games differ from most other software applications in that elements of the game must continue to run in the absence of user input. Programmers must therefore create a game loop, which continues to run the program's graphics, artificial intelligence, and audio when the user is not active.

Game programmers often use object-oriented programming (OOP). This is an approach to software development that defines the essential objects of a program and how they can interact with each other. Each object shares attributes and behaviors with other objects in the same class. Because many video games run hundreds of thousands of lines of code, an object-oriented approach to game development can help make the code more manageable. OOP makes programs easier to modify, maintain, and extend. This approach can make it easier to plan out and repurpose elements of the game.

Large numbers of people are involved in writing the source code of a game intended for wide

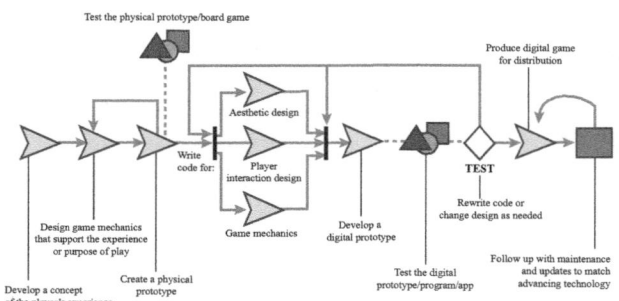

The development of a quality game begins with defining the player's experience and the game mechanics. Aesthetics and the digital prototypes of game pieces and environments are actually determined much later in the process.

distribution. They are typically organized in teams, with each team specializing in certain aspects of the game's mechanics or certain types of programming. To facilitate collaboration, detailed comments and pseudocode are used to document features of the program and explain choices about coding strategies. Pseudocode is a combination of a programming language and a natural language such as English. It is not executable, but it is much easier to read and understand than source code.

Game Programming Tools

Game programmers use the same basic tools as most software developers. These tools include text editors, debuggers, and compilers. The tools used are often determined by the type of device the game is intended to run on. Games can be developed for a particular computer platform, such as Windows, Apple, or Linux. They can also be developed for a gaming console, such as the Wii, Xbox, or PlayStation. Finally, games may be developed for mobile platforms such as Apple's iOS or Google's Android operating system.

A wide variety of programming languages are used to create video games. Some of the most commonly used programming languages in game development are C++, C#, and Java. Many video games are programmed with a combination of several different languages. Assembly code or C may be used to write lower-level modules. The graphics of many games are programmed using a shading language such as Cg or HLSL.

Game developers often use application programming interfaces (APIs) such as OpenGL and DirectX. They also rely heavily on the use of libraries and APIs to repurpose frequently used functions. Many of these resources are associated with animation, graphics, and the manipulation of 3D elements. One driving force in game development has been the use of increasingly sophisticated visual elements. The video game industry has come to rival motion pictures in their use of complex plotlines, dazzling special effects, and fully developed characters.

A Popular but Demanding Career

Game programming has been a popular career choice among young people. Due to the popularity of video games, it is not surprising that large numbers of people are interested in a career that will allow them to get paid to work on games. Becoming a game programmer requires the ability to code and a familiarity with several different programming languages. Game programmers can specialize in developing game engines, artificial intelligence, audio, graphics, user interfaces, inputs, or gameplay.

Game programmers regularly report high levels of stress as they are pressured to produce increasingly impressive games every year. There is intense competition to produce better special effects, more exciting stories, and more realistic action with each new title. Some game companies have been criticized for the high demands and heavy workloads placed upon game developers.

—*Scott Zimmer, JD*

Bibliography

Harbour, Jonathan S. *Beginning Game Programming*. 4th ed. Boston: Cengage, 2015. Print.

Kim, Chang-Hun, et al. *Real-Time Visual Effects for Game Programming*. Singapore: Springer, 2015. Print.

Madhav, Sanjay. *Game Programming Algorithms and Techniques: A Platform-Agnostic Approach*. Upper Saddle River: Addison, 2014. Print.

Marchant, Ben. "Game Programming in C and C++." *Cprogramming.com*. Cprogramming.com, 2011. Web. 16 Mar. 2016.

Nystrom, Robert. *Game Programming Patterns*. N.p.: Author, 2009–14. Web. 16 Mar. 2016.

Yamamoto, Jazon. *The Black Art of Multiplatform Game Programming*. Boston: Cengage, 2015. Print.

GAMIFICATION

Gamification, a term coined in the early 2000s, refers to the use of video-game logic and psychology in real-world environments, most prominently in marketing, education, and the corporate world. The theory of gamification holds that people—whether consumers, coworkers, or students—respond naturally and efficiently to competition, reward, and simulated risk of the type that have made video games such a cultural phenomenon since the 1980s. The concept applies especially to the generation of Americans born after 1975, many of whom were raised playing video games, who began assuming positions of prominence in businesses and organizations in the early twenty-first century. These video-game aficionados brought with them many of the assumptions and strategies of gaming—incentivized decision making, rapid problem solving, the self-evident logic of specific tasks and short-term rewards, an adrenaline response to simulated risk, and the perception of achievement as a measure of self-expression—all guided by the assumption that operating in such a matrix is both fun and profitable.

Overview

Gamification assumes that productivity and efficiency can be enhanced by creating artificial narratives and a game feel around otherwise routine endeavors, thus increasing engagement, raising skill levels, and positively influencing those who are participating. Though the paradigm of gamification emerged in a very short time through weekend seminars, online training sessions, and social media, its application soon became widespread. Examples of gamification include fast-food restaurants that offer a free meal after ten visits, community organizations that attract volunteers by creating an artificial system of participation levels, and banks that create ATM games that award points for savvy deposit decisions. Further examples of gamification include entrepreneurs who offer a system of points to potential investors as a way to measure their short-term returns and to map otherwise entirely artificial levels of success; fitness clubs that sponsor competitions that pit members against each other to meet their fitness goals; and companies that maximize productivity by engaging coworkers in competition against one another, not to earn traditional benefits such as better office space or salary bumps, but rather in the simple spirit of competition. In all these situations, gaming principles are brought into an otherwise non-gaming context.

Psychologists point out that the essential strategy of gamification is just a newer version of old-school operant conditioning that leverages desired behavior by reducing people to pawns and supervisors to crass manipulators. Some companies and educational institutions that must consider long-term developments fear that gamification narrows people's focus to short-term gains and to necessarily limited, even vague, objectives. A number of educators in particular have expressed concerns that the emphasis on external rewards may decrease students' intrinsic motivation for learning, although the extent to which this is a genuine concern remains in dispute; many argue that proper gamification of learning involves using extrinsic rewards to support, rather than supplant, intrinsic motivation.

Some more traditional individuals, particularly those born before the advent of video games, find the premise of gamification insulting and do not believe that the serious work of real life should mimic the simulated thrills of video games. Critics of

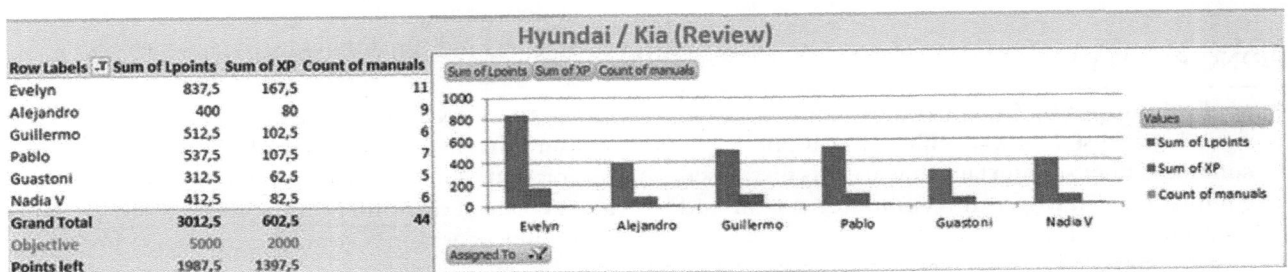

This is a chart that shows the progress of gamification.

gamification believe that simply motivating people to act in a certain way—whether customers, students, or coworkers—is a potentially catastrophic strategy because motion can be mistaken for progress, activity for achievement, and competition for teamwork.

—*Joseph Dewey, MA, PhD*

BIBLIOGRAPHY

Burke, Brian. "The Gamification of Business." *Forbes.* Forbes.com, 21 Jan. 2013. Web. 20 Aug. 2013.

Burke, Brian. *Gamify: How Gamification Motivates People to Do Extraordinary Things.* Brookline: Bibliomotion, 2014. Print.

Kapp, Karl M. *The Gamification of Learning and Instruction: Game-Based Methods and Strategies for Training and Education.* San Francisco: Pfeiffer, 2012. Print.

McCormick, Ty. "Gamification: A Short History." *Foreign Policy.* FP Group, 24 June 2013. Web. 20 Aug. 2013.

McGonigal, Jane. *Reality Is Broken: Why Games Make Us Better and How They Can Change the World.* New York: Penguin, 2011. Print.

Walker, Tim. "Gamification in the Classroom: The Right or Wrong Way to Motivate Students?" *NEA Today.* Natl. Educ. Assn., 23 June 2014. Web. 28 July 2015.

Werbach, Kevin, and Dan Hunter. *For the Win: How Game Thinking Can Revolutionize Your Business.* Philadelphia: Wharton, 2012. Print.

Zichermann, Gabe, and Christopher Cunningham. *Gamification by Design: Implementing Game Mechanics in Web and Mobile Apps.* Sebastopol: O'Reilly, 2011. Print.

Zichermann, Gabe, and Joselin Linder. *The Gamification Revolution: How Leaders Leverage Game Mechanics to Crush the Competition.* New York: McGraw, 2013. Print.

GRAPHICAL USER INTERFACE (GUI)

FIELDS OF STUDY

Computer Science; Applications; System-Level Programming

ABSTRACT

Graphical user interfaces (GUIs) are human-computer interaction systems. In these systems, users interact with the computer by manipulating visual representations of objects or commands. GUIs are part of common operating systems like Windows and Mac OS. They are also used in other applications.

PRINCIPAL TERMS

- **application-specific GUI:** a graphical interface designed to be used for a specific application.
- **command line:** an interface that accepts text-based commands to navigate the computer system and access files and folders.
- **direct manipulation interfaces:** computer interaction format that allows users to directly manipulate graphical objects or physical shapes that are automatically translated into coding.
- **interface metaphors:** linking computer commands, actions, and processes with real-world actions, processes, or objects that have functional similarities.
- **object-oriented user interface:** an interface that allows users to interact with onscreen objects as they would in real-world situations, rather than selecting objects that are changed through a separate control panel interface.
- **user-centered design:** design based on a perceived understanding of user preferences, needs, tendencies, and capabilities.

GRAPHICS AND INTERFACE BASICS

A user interface is a system for human-computer interaction. The interface determines the way that a user can access and work with data stored on a computer or within a computer network. Interfaces can be either text-based or graphics-based. Text-based systems allow users to input commands. These commands may be text strings or specific words that activate functions. By contrast, graphical user interfaces (GUIs) are designed so that computer functions are tied to graphic icons (like folders, files, and drives). Manipulating an icon causes the computer to perform certain functions.

History of Interface Design

The earliest computers used a text-based interface. Users entered text instructions into a command line. For instance, typing "run" in the command line would tell the computer to activate a program or process. One of the earliest text-based interfaces for consumer computer technology was known as a "disk operating system" (DOS). Using DOS-based systems required users to learn specific text commands, such as "del" for deleting or erasing files or "dir" for listing the contents of a directory. The first GUIs were created in the 1970s as a visual "shell" built over DOS system.

GUIs transform the computer screen into a physical map on which graphics represent functions, programs, files, and directories. In GUIs, users control an onscreen pointer, usually an arrow or hand symbol, to navigate the computer screen. Users activate computing functions by directing the pointer over an icon and "clicking" on it. For instance, GUI users can cause the computer to display the contents of a directory (the "dir" command in DOS) by clicking on a folder or directory icon on the screen. Modern GUIs combine text-based icons, such as those found in menu bars and movable windows, with linked text icons that can be used to access programs and directories.

Elements of GUIs and Other Object Interfaces

Computer programs are built using coded instructions that tell the computer how to behave when given inputs from a user. Many different programming languages can be used to create GUIs. These include C++, C#, JavaFX, XAML, XUL, among others. Each language offers different advantages and disadvantages when used to create and modify GUIs.

User-centered design focuses on understanding and addressing user preferences, needs, capabilities, and tendencies. According to these design principles, interface metaphors help make GUIs user

Graphical user interfaces (GUIs) became popular in the early 1980s. Early uses of a GUI included analog clocks, simple icons, charts, and menus.

friendly. Interface metaphors are models that represent real-world objects or concepts to enhance user understanding of computer functions. For example, the desktop structure of a GUI is designed using the metaphor of a desk. Computer desktops, like actual desktops, might have stacks of documents (windows) and objects or tools for performing various functions. Computer folders, trash cans, and recycle bins are icons whose functions mirror those of their real-world counterparts.

Object-oriented user interfaces (OOUIs) allow a user to manipulate objects onscreen in intuitive ways based on the function that the user hopes to achieve. Most modern GUIs have some object-oriented functionality. Icons that can be dragged, dropped, slid, toggled, pushed, and clicked are "objects." Objects include folders, program shortcuts, drive icons, and trash or recycle bins. Interfaces that use icons can also be direct manipulation interfaces (DMI). These interfaces allow the user to adjust onscreen objects as though they were physical objects to get certain results. Resizing a window by dragging its corner is one example of direct manipulation used in many GUIs.

Current and Future of Interface Design

GUIs have long been based on a model known as WIMP. WIMP stands for "windows, icons, menus, and pointer objects," which describes the ways that users can interact with the interface. Modern GUIs are a blend of graphics-based and text-based functions, but this system is more difficult to implement on modern handheld computers, which have less space to hold icons and menus. Touch-screen interfaces represent the post-WIMP age of interface design. With touch screens, users more often interact directly with objects on the screen, rather than using menus and text-based instructions. Touch-screen design is important in a many application-specific GUIs. These interfaces are designed to handle a single process or application, such as self-checkout kiosks in grocery stores and point-of-sale retail software.

Current computer interfaces typically require users to navigate through files, folders, and menus to locate functions, data, or programs. However, voice activation of programs or functions is now available on many computing devices. As this technology becomes more common and effective, verbal commands may replace many functions that have been accessed by point-and-click or menu navigation.

—*Micah L. Issitt*

Bibliography

"Graphical User Interface (GUI)." *Techopedia.* Techopedia, n.d. Web. 5 Feb. 2016.

Johnson, Jeff. *Designing with the Mind in Mind.* 2nd ed. Waltham: Morgan, 2014. Print.

Lohr, Steve. "Humanizing Technology: A History of Human-Computer Interaction." *New York Times: Bits.* New York Times, 7 Sept. 2015. Web. 31 Jan. 2016.

Reimer, Jeremy. "A History of the GUI." *Ars Technica.* Condé Nast, 5 May 2005. Web. 31 Jan. 2016.

"User Interface Design Basics." *Usability.* US Dept. of Health and Human Services, 2 Feb. 2016. Web. 2 Feb. 2016.

Wood, David. *Interface Design: An Introduction to Visual Communication in UI Design.* New York: Fairchild, 2014. Print.

GRAPHICS FORMATS

Fields of Study

Information Systems; Digital Media; Graphic Design

Abstract

Graphics formats are standardized forms of computer files used to transfer, display, store, or print reproductions of digital images. Digital image files are divided into two major families, vector and raster files. They can be compressed or uncompressed for storage. Each type of digital file has advantages and disadvantages when used for various applications.

Principal Terms

- **compressed data:** data that has been encoded such that storing or transferring the data requires fewer bits of information.
- **lossless compression:** data compression that allows the original data to be compressed and reconstructed without any loss of accuracy.
- **lossy compression:** data compression that uses approximation to represent content and therefore reduces the accuracy of reconstructed data.
- **LZW compression:** a type of lossless compression that uses a table-based algorithm.
- **RGB:** a color model that uses red, green, and blue to form other colors through various combinations.

Digital Imaging

A digital image is a mathematical representation of an image that can be displayed, manipulated, and modified with a computer or other digital device. It can also be compressed. Compression uses algorithms to reduce the size of the image file to facilitate sharing, displaying, or storing images. Digital images may be stored and manipulated as raster or vector images. A third type of graphic file family, called "metafiles," uses both raster and vector elements.

The quality and resolution (clarity) of an image depend on the digital file's size and complexity. In

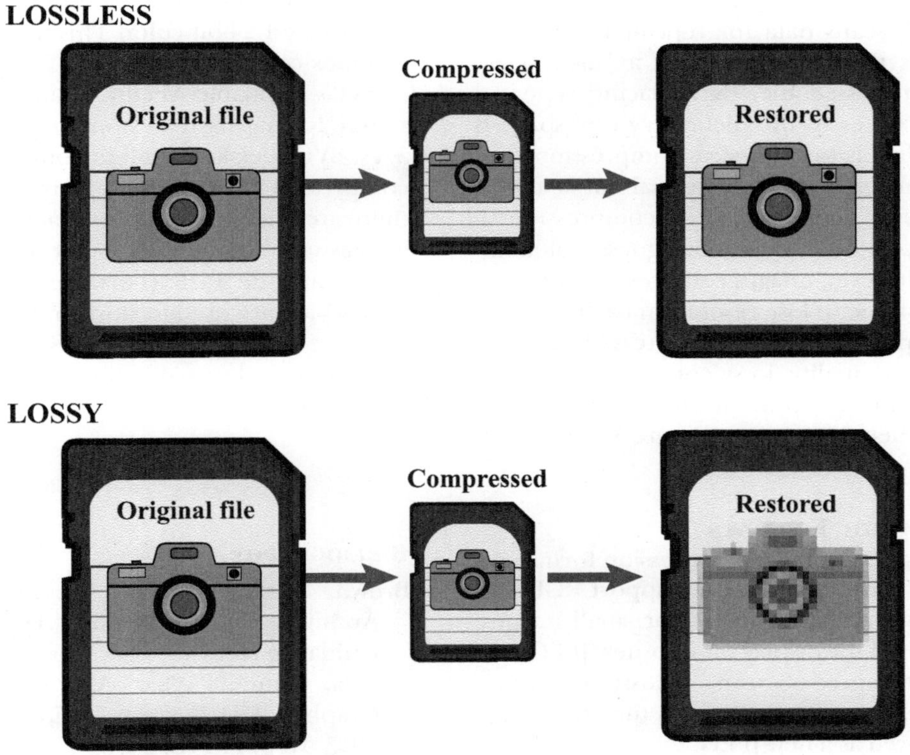

LOSSLESS

LOSSY

Depending on the image format, data may be lost after compression and restoration. Loss of image data reduces the quality of the image.

raster graphics, images are stored as a set of squares, called "pixels." Each pixel has a color value and a color depth. This is defined by the number of "bits" allocated to each pixel. Pixels can range from 1 bit per pixel, which has a monochrome (two-color) depth, to 32-bit, or "true color." 32-bit color allows for more than four billion colors through various combinations. Raster graphics have the highest level of color detail because each pixel in the image can have its own color depth. For this reason, raster formats are used for photographs and in image programs like Adobe Photoshop. However, the resolution of a raster image depends on size because the image has the same number of pixels at any magnification. For this reason, raster images cannot be magnified past a certain point without losing resolution. Vector graphics store images as sets of polygons that are not size-dependent and look the same at any magnification. For relatively simple graphics, like logos, vector files are smaller and more precise than raster images.

However, vector files do not support complex colors or advanced effects, like blurring or drop shadows.

Two basic color models are used to digitally display various colors. The RGB color model, also called "additive color," combines red, green, and blue to create colors. The CMYK model, also called "subtractive color," combines the subtractive primary colors cyan, magenta, yellow, and black to absorb certain wavelengths of light while reflecting others.

Image Compression

Image compression reduces the size of an image to enable easier storage and processing. Lossless compression uses a modeling algorithm that identifies repeated or redundant information contained within an image. It stores this information as a set of instructions can be used to reconstruct the image without any loss of data or resolution. One form of lossless compression commonly used is the LZW compression algorithm developed in the 1980s. The LZW algorithm uses a "code table" or "dictionary" for compression. It scans data for repeated sequences and then adds these sequences to a "dictionary" within the compressed file. By replacing repeated data with references to the dictionary file, space is saved but no data is lost. Lossless compression is of benefit when image quality is essential but is less efficient at reducing image size. Lossy compression algorithms reduce file size by removing less "valuable" information. However, images compressed with lossy algorithms continue to lose resolution each time the image is compressed and decompressed. Despite the loss of image quality, lossy compression creates smaller files and is useful when image quality is less important or when computing resources are in high demand.

COMMON GRAPHIC FORMATS

JPEG is a type of lossy image compression format developed in the early 1990s. JPEGs support RGB and CMYK color and are most useful for small images, such as those used for display on websites. JPEGs are automatically compressed using a lossy algorithm. Thus, some image quality is lost each time the image is edited and saved as a new JPEG.

GIF (Graphics Interchange Format) files have a limited color palette and use LZW compression so that they can be compressed without losing quality. Unlike JPEG, GIF supports "transparency" within an image by ignoring certain colors when displaying or printing. GIF files are open source and can be used in a wide variety of programs and applications. However, most GIF formats support only limited color because the embedded LZW compression is most effective when an image contains a limited color palette. PNGs (Portable Network Graphics) are open-source alternatives to GIFs that support transparency and 24-bit color. This makes them better at complex colors than GIFs.

SVGs (Scalable Vector Graphics) are an open-source format used to store and transfer vector images. SVG files lack built-in compression but can be compressed using external programs. In addition, there are "metafile" formats that can be used to share images combining both vector and raster elements. These include PDF (Portable Document Format) files, which are used to store and display documents, and the Encapsulated PostScript (EPS) format, which is typically used to transfer image files between programs.

—*Micah L. Issitt*

BIBLIOGRAPHY

Brown, Adrian. *Graphics File Formats.* Kew: Natl. Archives, 2008. PDF file. Digital Preservation Guidance Note 4.
Celada, Laura. "What Are the Most Common Graphics File Formats." *FESPA.* FESPA, 27 Mar. 2015. Web. 11 Feb. 2016.
Costello, Vic, Susan Youngblood, and Norman E. Youngblood. *Multimedia Foundations: Core Concepts for Digital Design.* New York: Focal, 2012. Print.
Dale, Nell, and John Lewis. *Computer Science Illuminated.* 6th ed. Burlington: Jones, 2016. Print.
"Introduction to Image Files Tutorial." *Boston University Information Services and Technology.* Boston U, n.d. Web. 11 Feb. 2016.
Stuart, Allison. "File Formats Explained: PDF, PNG and More." *99Designs.* 99Designs, 21 May 2015. Web. 11 Feb. 2016.

GUARD CLAUSE

FIELDS OF STUDY

Coding Techniques; Software Development; Computer Science

ABSTRACT

A guard clause is a computer construct that allows or disallows sections of code to run based on preconditions. This is achieved with conditional expressions, Boolean operators, check functions, and other elements of computer programming language. Guard clauses can be used for a wide variety of purposes, including preventing errors from occurring, responding to errors that do occur, and refactoring code to replace nested conditional statements.

Guard clause

PRINCIPAL TERMS

- **Boolean operator:** one of a predefined set of words—typically AND, OR, NOT, and variations—that is used to combine expressions that resolve to either-or values.
- **check functions:** computer functions that check for a specific data type.
- **conditional expression:** a statement executed based on whether a condition is found to be true or false.
- **preconditions:** in computer programming, conditions that must be true for a block of code to execute.

What Are Guard Clauses?

A guard clause is a computer construct that uses conditional expressions to determine whether to allow blocks of code to execute. They allow simple, local processing of exceptions and exit from subroutines. Some common forms of guard clauses are combined Boolean expressions, Boolean operators, and check functions. All essentially work by determining whether preconditions are true or false. A clause "guards" against invalid preconditions at the beginning of a block of code. It allows subsequent code to run if preconditions are met.

Guard clauses serve many purposes. For example, they are often used to check that the data passed to procedures such as functions and methods is of the correct data type, or to validate the data in some other way. This prevents the execution of code for no reason or the execution of code that will generate errors or exceptions. Since guard clauses are often used to report on errors or throw exceptions, they can also serve as a type of error handling code. Many programmers also favor using guard clauses in place of nested conditional statements. This can be done when writing code or when refactoring, or redesigning, existing code. It often makes code easier to read, understand, and maintain.

Replacing Nested Conditional Statements

The same functionality can often be provided using a series of nested conditional statements or using a guard clause. However, multiple nested conditional statements grow increasingly difficult to read as more are added. In most programming languages and development environments, each if-statement is usually indented from the surrounding statement in the following way:

If(
  If(
    If(
      If(
        If(

In complex code, this leads to developers having to scroll back and forth while reading, which can be inefficient. This also makes it difficult to immediately see where any needed modifications to the code must be inserted. Using guard statements to replace nested if-statements can make the code flatter, shorter, and simpler.

However, this approach is not universally accepted. Many traditional approaches to programming operate under the principle that a procedure should do whatever it was designed to accomplish and then provide a single point of return to the calling code. For this reason, nested conditional statements are favored by those who follow a strict structured programming paradigm. Advocates of pure structured programming do not believe that the readability benefits of guard clauses justify abandoning the single point of return approach.

Supporters of the guard clause approach counter that use of the single point of return has diminished due to the increased use of shorter procedures. They also challenge the idea that the single point of return approach is superior to the guard clause approach on a fundamental level. To many programmers, the guard clause approach makes the functionality of a procedure easier to understand.

Creating an Effective Guard Clause

A common source of errors in a computer program occurs when procedures are passed null values when called. Guard clauses can be used to prevent these types of errors from occurring. For example, take a software developer for a movie theater chain creating a function to determine the total daily attendance based on a given number of days. This takes the following form:

```
Function calcAverageAttendance (x As Integer, y As Integer) As Integer
   calcAverageAttendance = x / y
   Exit Function
End Function
```

143

where x is the total attendance and y is the given number of days.

A guard clause could be used to check that neither of the input values is null before processing the calculation:

```
Function calcAverageAttendance (x As Integer, y
As Integer) As Integer
  If (IsNull(x)) Then
  Exit Function
  If (IsNull(y)) Then
  Exit Function
  calcAverageAttendance = x / y
  Exit Function
End Function
```

The IsNull function returns true if the value of the parameter passed to the function is null. The guard clause uses the IsNull function to check if either x or y is null. If so, it immediately exits the function, preventing an error from occurring. If neither is null, the average is calculated and returned.

Using Guard Clauses to Create More Efficient Code

Guard clauses are a useful part of a developer's toolkit. As they can help programs avoid and respond to errors, they are used for a wide variety of purposes. They can validate data to prevent errors from occurring when updating database tables or when storing values in data structures such as trees structures and arrays. They can be used to ensure that a long section of code in a large program does not waste system resources by executing unnecessarily. They can improve the development process by refactoring code to make it more efficient. With all of these benefits to offer, guard clauses are sure to remain in widespread use in many development environments.

—*Maura Valentino, MSLIS*

Bibliography

Friedman, Daniel P., and Mitchell Wand. *Essentials of Programming Languages*. 3rd ed., MIT P, 2008.

Haverbeke, Marijn. *Eloquent JavaScript: A Modern Introduction to Programming*. 2nd ed., No Starch Press, 2014.

MacLennan, Bruce J. *Principles of Programming Languages: Design, Evaluation, and Implementation*. 3rd ed., Oxford UP, 1999.

Schneider, David I. *Introduction to Programming Using Visual Basic*. 10th ed., Pearson, 2016.

Scott, Michael L. *Programming Language Pragmatics*. 4th ed., Morgan Kaufmann Publishers, 2016.

Van Roy, Peter, and Seif Haridi. *Concepts, Techniques, and Models of Computer Programming*. Massachusetts Institute of Technology, 2004.

HTTP COOKIE

An HTTP (Hypertext Transfer Protocol) cookie is a piece of data that is stored on a user's computer when it accesses a website and is then later retrieved by that site. Also known as an "Internet cookie," "web cookie," or "browser cookie," this data may record and track a user's browsing history, account details, or form entries, among other information. While some types of cookies are necessary for web browsers to function, their use has raised concerns about violation of privacy.

Overview

HTTP cookies were introduced by Netscape Communications in the first edition of its Netscape Navigator browser, released in 1994. The name comes from "magic cookie," a computing term used since the 1970s to describe a piece of data exchanged between programs, often for identification purposes.

In basic HTTP functioning, every time a browser interacts with the server hosting a particular website, the server treats the connection as a brand new request, not recognizing it from previous interactions. As such, HTTP is considered a "stateless" protocol, meaning it stores no information on its own. In order for a website or other application to remember such things as the identity of a logged-in user or the items placed in a virtual shopping cart, when a browser connects with a host server for the first time, the server stores a cookie on the browser's computer. The next time the browser connects to the server, the cookie reminds the application of the stored information—that is, its state. If the user does something to change the application's state, such as adding an additional item to his or her virtual shopping cart, the server updates the information in the cookie.

Different types of cookies are used for different purposes. All cookies fall into one of two categories: session cookies, which are stored only for the length of a user's browsing session and are deleted when the browser is closed, and persistent cookies, which remain stored on the user's computer until they either reach a predetermined expiration date or are manually deleted. First-party cookies are those created and stored by a site the user chooses to visit, while third-party cookies are installed by some entity other than the site the user is visiting, often by companies advertising on that site. First-party cookies include authentication cookies, which are created when a user logs into an account on a particular website and identify that user until he or she logs out, and may be either session cookies or persistent cookies. Third-party cookies are usually persistent. One common type is the third-party tracking cookie; these cookies maintain a record of a user's browsing history, which companies may then use to gather consumer data or to more precisely target advertisements. Other types of cookies include HTTP-only cookies, which are only used when HTTP requests are being transmitted and thus are more secure; flash cookies or local shared objects (LSOs), which are stored by websites that use Adobe Flash and are more difficult to delete; and opt-out cookies, which prevent advertising companies from showing users targeted ads.

The use of third-party tracking cookies has raised concerns among users who do not want companies to be able to monitor their online habits. Responses to these concerns include the European Union's Directive on Privacy and Electronic Communications, introduced in 2002 and updated in 2009, which requires companies to obtain consent before installing unnecessary cookies on a user's computer. In addition, most browsers have the ability to block third-party cookies, though some companies have developed methods of circumventing that block. In 2012,

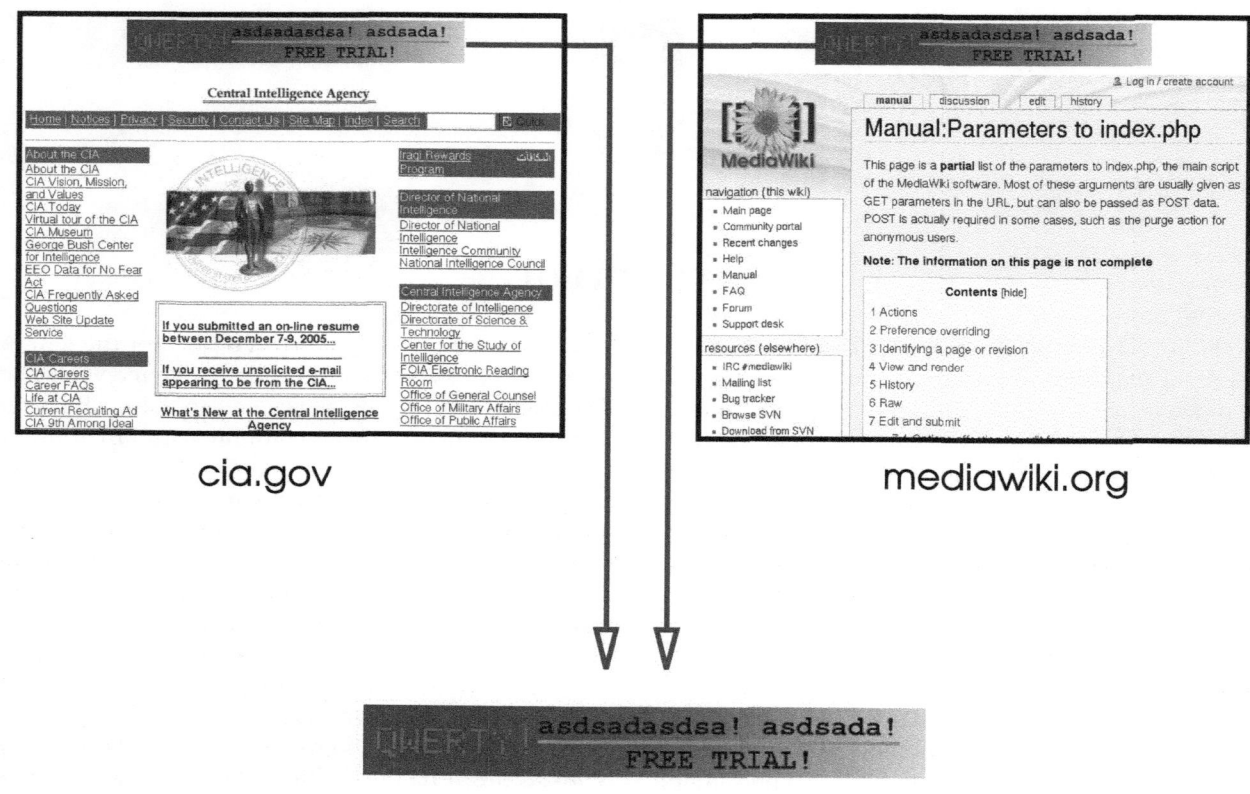

Third party HTTP cookies.

for example, Google was discovered to have been deliberately defying the Safari browser's default privacy setting, which bans the installation of third-party cookies.

—*Randa Tantawi, PhD*

BIBLIOGRAPHY

"Cookies: Leaving a Trail on the Web." *OnGuard Online*. US Federal Trade Commission, Nov. 2011. Web. 25 Sept. 2013.

Gourley, David, et al. "Client Identification and Cookies." *HTTP: The Definitive Guide*. Sebastopol, CA: O'Reilly Media, 2002. 257–76. Print.

Hofmann, Markus, and Leland R. Beaumont. "Content Transfer." *Content Networking: Architecture, Protocols, and Practice*. San Francisco: Elsevier, 2005. 25–52. Print.

Kristol, David M. "HTTP Cookies: Standards, Privacy, and Politics." *ArXiv.org*. Cornell U Lib., 9 May 2001. Web. 25 Sept. 2013.

Singel, Ryan. "Google Busted with Hand in Safari-Browser Cookie Jar." *Wired*. Condé Nast, 17 Feb. 2012. Web. 25 Sept. 2013.

Singel, Ryan. "You Deleted Your Cookies? Think Again." *Wired*. Condé Nast, 10 Aug. 2009. Web. 25 Sept. 2013.

Zakas, Nicholas C. "Cookies and Security." *NCZOnline*. Zakas, 12 May 2009. Web. 25 Sept. 2013.

Zakas, Nicholas C. "HTTP Cookies Explained." *NCZOnline*. Zakas, 5 May 2009. Web. 25 Sept. 2013.

I

IMAGINED COMMUNITIES

Imagined communities are populations of individuals that—though they may never come into contact with the vast majority of the group's other members—all self-identify under a shared community identity. The concept was first presented as a frame for understanding nationalism in Benedict Anderson's 1983 book *Imagined Communities: Reflections on the Origin and Spread of Nationalism*. The term is now applied more broadly, and can be used to refer to communities defined by a variety of common causes, interests, and characteristics.

OVERVIEW

In forming the imagined communities concept, Anderson sought to explain nationalism through a modernist lens, describing it as a socially constructed value rooted in political and economic interests. This viewpoint stands in opposition to the primordialist take on nationalism that was popular throughout the early twentieth century, but faced heavy criticism in the years following World War II. Primordialism holds that nations are inevitable formations that evolve organically due to shared cultural traits—language, for example—that create a sense of community knowledge and cohesion. Anderson's view of nationalism also holds language in high regard, but he is more concerned with language media and the use of language as a political tool.

Anderson asserts that the increasingly widespread use of print media in the eighteenth and early

Liberty Leading the People (Eugène Delacroix, 1830) is a famous example of nationalist art.

Richard March Hoe's 1864 printing press. Benedict Anderson's imagined communities grew out of nationalism fueled by the Industrial Revolution and printing press capitalism.

nineteenth centuries—the creation of the first truly mass medium—also led to the development of mass literacy, making the masses of Europe more aware of the context of their existence. Through this awareness, a communal identity formed, and the historic concepts of dynastic monarchy and rule by divine right lost some of their power. In this way, the concept of the nation-state—an imagined entity with finite boundaries, sovereign and with the right to self-determination by its people—came into existence. Confirming this idea were the revolutions in France and the British colonies in America, which established sovereign republics with governments ostensibly answerable to their people—populations with by then a sense of national identity, or imagined community.

Over time, the term "imagined community" became more general in meaning. It is now used in academic circles to apply to communities of common interest, online communities, and other groups lacking regular face-to-face interaction formed around any number of shared identities, from sexual orientation to political affiliation.

—*Steven Miller*

Bibliography

Adamczyk, Maria. "Forum for the Ugly People—Study of an Imagined Community." *Sociological Review* 58 (Dec. 2010): 97–113. Print.

Beck, Ulrich. "Cosmopolitanism as Imagined Communities of Global Risk." *American Behavioral Scientist* 55.10 (Oct. 2011): 1346–61. Print.

Gruzd, Anatoliy, et al. "Imagining Twitter as an Imagined Community." *American Behavioral Scientist* 55.10 (Oct. 2011): 1294–318. Print.

Pentecost, Kathryn. "Imagined Communities in Cyberspace." *Social Alternatives* 30.2 (2011): 44–47. Print.

Shahzad, Farhat. "Forging the Nation as an Imagined Community." *Nations and Nationalism* 18.1 (Jan. 2012): 21–38. Print.

INCREMENTAL DEVELOPMENT

FIELDS OF STUDY
Software Development; Programming Methodologies; Software Engineering

ABSTRACT
Incremental development is a software development methodology in which the functionality of the system is divided into multiple, separate modules. The module with the system's core functionality is typically the first deployed. Later modules that add functionality are then deployed and integrated with the system until all required functionality is achieved.

PRINCIPAL TERMS
- **integration:** in computer science, the process of combining individual system components, whether hardware or software, into a single, unified system.
- **software development life cycle (SDLC):** the phases of a software development model and the order in which those phases are executed.
- **waterfall development:** a traditional, linear software development methodology based on a series of ordered phases.

Understanding Incremental Development
Incremental development is a software development methodology in which the software development life cycle (SDLC) combines aspects of the iterative development and traditional waterfall development models. In incremental development, first a detailed set of specifications and requirements is created to address the functionality required in the completed system. The required functionality is then divided into multiple parts called "modules" or "builds." Each build is then developed separately.

Usually, the first module developed and placed in production contains the system's core functionality. Once the first module is designed, developed, tested, and deployed, additional modules are built and combined with all earlier builds. This integration process continues until all required functionality has been achieved.

Modules are developed using a multiphase, linear design methodology similar to the waterfall development methodology. Each module typically passes through the phases of design, development, and testing before being deployed and integrated with earlier builds. As this waterfall pattern of ordered phases is repeated multiple times, the SDLC of the incremental method can be thought of as an incremental methodology within an iterative framework.

Advantages and Disadvantages of Incremental Development
Incremental development offers many of the benefits of waterfall development while addressing some of its shortcomings. Like the waterfall model, incremental development is easy to understand and manage. Each phase within a build has clear, agreed-upon deliverables and a formal review process, and is completed before the next phase begins. Like projects developed using the waterfall methodology, projects that use the incremental methodology are usually well documented. There is minimal customer and end-user involvement after the initial specifications, requirements, and needed functionality have been determined. Because the design is finalized early in the development cycle and changes little in development, the incremental model works well for projects where systems are integrated with other systems.

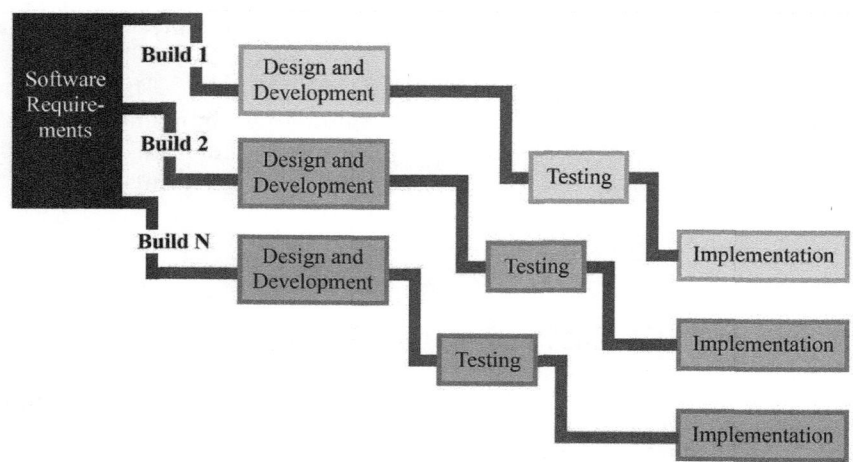

Software is built through multiple cycles of design and development, testing, and implementation. Each build completes a portion or increment of the requirements for the full software design.

Incremental development also offers advantages that address many of the drawbacks to waterfall development. Unlike the waterfall model, the incremental model delivers a functional system early in the SDLC. This can reduce initial delivery time and costs. Later builds offer the chance to reevaluate the project's scope and requirements based on stakeholder feedback and incorporate needed changes. This offers greater flexibility to meet changing circumstances and newly discovered needs. Furthermore, the multiple, smaller modules make builds easier to manage, develop, test, and debug.

However, incremental development does share some of the same disadvantages of the waterfall model. Both models require detailed planning and analysis at the outset to ensure that all project specifications, requirements, and needed functionality have been addressed in the overall system design. Failure to do so can lead to delays and other problems as requirements are redefined and the system is redesigned. For projects with loosely defined requirements that will likely change in development and testing, iterative development models such as rapid application development (RAD) and spiral development may be a more effective choice. On the other hand, if the project is very stable and requirements are fully understood, the waterfall model may be preferable as it has fewer steps and lower overall cost than incremental development.

Incremental Development in Practice

Incremental development is most effective for projects where the system's specifications and requirements are mostly known at the outset and where a subset of functionality can provide the core experience required by the user. Competitive or financial pressures can also make the incremental model an effective strategy, particularly when the product must be released quickly. A functional system can be delivered early in the project's lifetime and refined later in the SDLC.

For example, a mobile application start-up might have an idea for a game for use on smartphones and tablets. The mobile application (app) industry is highly competitive, with new and established companies releasing thousands of apps a year. Production of a marketable app in the least amount of time provides a key business advantage, as it reduces the time in which a competitor might release a similar app that could take market share. Releasing a functional product quickly therefore helps prevent capital and other resources from being wasted. As incremental development endeavors to provide a system with core functionality in the least amount of time, it is an excellent methodology for use in such cases.

The Best of Both Worlds

Incremental development combines some of the best features of waterfall and iterative development methodologies, balancing between flexibility and clear structure. By offering greater speed and flexibility than waterfall development, the incremental model suits fast-moving, highly competitive environments like those found in Web and mobile app development. While less flexible than other iterative development methodologies, incremental development provides a simple structure that ensures systems are relatively easy to manage, require less user and stakeholder involvement, and are usually well documented. These qualities make incremental development a powerful tool in a developer's repertoire. It remains a popular methodology that can be used and studied to provide insight into the broader world of software development.

—*Maura Valentino, MSLIS*

Bibliography

Bell, Michael. *Incremental Software Architecture: A Method for Saving Failing IT Implementations*. John Wiley & Sons, 2016.

Friedman, Daniel P., and Mitchell Wand. *Essentials of Programming Languages*. 3rd ed., MIT P, 2008.

Jayaswal, Bijay K., and Peter C. Patton. *Design for Trustworthy Software: Tools, Techniques, and Methodology of Developing Robust Software*. Prentice Hall, 2007.

MacLennan, Bruce J. *Principles of Programming Languages: Design, Evaluation, and Implementation*. 3rd ed., Oxford UP, 1999.

Scott, Michael L. *Programming Language Pragmatics*. 4th ed., Morgan Kaufmann Publishers, 2016.

Van Roy, Peter, and Seif Haridi. *Concepts, Techniques, and Models of Computer Programming*. MIT P, 2004.

Wysocki, Robert K. *Effective Project Management: Traditional, Agile, Extreme*. 7th ed., John Wiley & Sons, 2014.

INFORMATION TECHNOLOGY (IT)

FIELDS OF STUDY

Computer Science; Network Design

ABSTRACT

Information technology (IT) is devoted to the creation and manipulation of information using computers and other types of devices. It also involves the installation and use of software, the sets of instructions computers follow in order to function. At one time, most computers were designed with a single purpose in mind, but over time, IT has transformed into a discipline that can be used in almost any context.

PRINCIPAL TERMS

- **device:** equipment designed to perform a specific function when attached to a computer, such as a scanner, printer, or projector.
- **hardware:** the physical parts that make up a computer. These include the motherboard and processor, as well as input and output devices such as monitors, keyboards, and mice.
- **network:** two or more computers being linked in a way that allows them to transmit information back and forth.
- **software:** the sets of instructions that a computer follows in order to carry out tasks. Software may be stored on physical media, but the media is not the software.
- **system:** either a single computer or, more generally, a collection of interconnected elements of technology that operate in conjunction with one another.
- **telecom equipment:** hardware that is intended for use in telecommunications, such as cables, switches, and routers.

History of Information Technology

Information technology (IT) encompasses a wide range of activities, from pure theory to hands-on jobs. At one end of the spectrum are IT professionals who design software and create system-level network designs. These help organizations to maximize their efficiency in handling data and processing information. At the other end are positions in which physical hardware, from telecom equipment to devices such as routers and switches, are connected to one another to form networks. They are then tested to make sure that they are working correctly. This range includes many different types of employees. For example, there are computer technicians, system administrators, programmers at the system and application levels, chief technology officers, and chief information officers.

One way of studying the history of IT is to focus on the ways that information has been stored. IT's history can be divided into different eras based on what type of information storage was available. These eras include prehistoric, before information was written down, and early historical, when information started to be recorded on stone tablets. In the middle historical period, information was recorded on paper and stored in libraries and other archives. In the modern era, information has moved from physical storage to electronic storage. Over time, information storage has become less physical and more abstract. IT now usually refers to the configuration of computer hardware in business networks. These allow for the manipulation and transfer of electronically stored information.

Dot-Com Bubble

IT gained prominence in the 1990s, as the Internet began to grow rapidly and become more user-friendly than it had been in the past. Many companies arose to try to take advantage of the new business models that it made possible. Computer programmers and network technology experts found themselves in high demand. Startup companies tried to build online services quickly and effectively. Investment in technology companies put hundreds of millions of dollars into IT research. Even established companies realized that they needed to invest in their IT infrastructure and personnel if they wanted to stay competitive. As the IT sector of the economy grew rapidly, financial experts began to worry that it was forming an economic bubble. An economic bubble occurs when a market grows rapidly and then that growth declines abruptly. The bubble eventually "pops," and investors pull their money out. This did happen, and many Internet startups shut down.

While the dot-com bubble, as it came to be known, passed quickly, IT remained a central part of life. Simple tasks that used to be done without sophisticated technology, such as banking, shopping, and even reading a book, now involve computers, mobile phones, tablets, or e-readers. This means that the average person must be more familiar with IT in the twenty-first century than in any previous era. Because of this, IT has become a topic of general interest. For example, an average person needs to know a bit about network configuration in order to set up a home system.

Data Production Growth

With IT, new information is constantly being created. Once it was possible for a single person to master all of society's knowledge. In the modern world, more data is produced every year than a person could assimilate in lifetime. It is estimated that by 2020, there will be more than five thousand gigabytes (GB) of data for each and every person on earth.

The availability of IT is the factor most responsible for the explosion in data production. Most cell phone plans measure customer data in how many GB per month may be used, for example. This is because of the many photos, videos, and social media status updates people create and share on the Internet every day. It is estimated that every two days, human beings create as much information as existed worldwide before 2003. The pace of this data explosion increases as time goes on.

—*Scott Zimmer, JD*

Bibliography

Black, Jeremy. *The Power of Knowledge: How Information and Technology Made the Modern World.* New Haven: Yale UP, 2014. Print.

Bwalya, Kelvin J., Nathan M. Mnjama, and Peter M. I. I. M. Sebina. *Concepts and Advances in Information Knowledge Management: Studies from Developing and Emerging Economies.* Boston: Elsevier, 2014. Print.

Campbell-Kelly, Martin, William Aspray, Nathan Ensmenger, and Jeffrey R. Yost. *Computer: A History of the Information Machine.* Boulder: Westview, 2014. Print.

Fox, Richard. *Information Technology: An Introduction for Today's Digital World.* Boca Raton: CRC, 2013. Print.

Lee, Roger Y., ed. *Applied Computing and Information Technology.* New York: Springer, 2014. Print.

Marchewka, Jack T. *Information Technology Project Management.* 5th ed. Hoboken: Wiley, 2015. Print.

INFORMATION VISUALIZATION

Information visualization helps the human eye perceive huge amounts of data in an understandable way. It enables researchers to represent, draft, and display data visually in a way that allows humans to more efficiently identify, visualize, and remember it. Techniques of information visualization take advantage of the broad bandwidth pathway of the human eye. Information visualization is an indispensable part of applied research and problem solving, helping people graphically display, spatially organize, and locate huge amounts of data and information to facilitate fast and efficient categorization, classification, organization, and systematization of information.

OVERVIEW

Both quantitative and qualitative methods of research use information visualization techniques extensively. Graphs, charts, scatterplots, box plots, tables, diagrams, maps, three-dimensional figures, pictograms, schemes, tree diagrams, and coordinate plots are just few examples of information visualization techniques that help illustrate data. Information visualization reinforces human cognition by intuitively representing

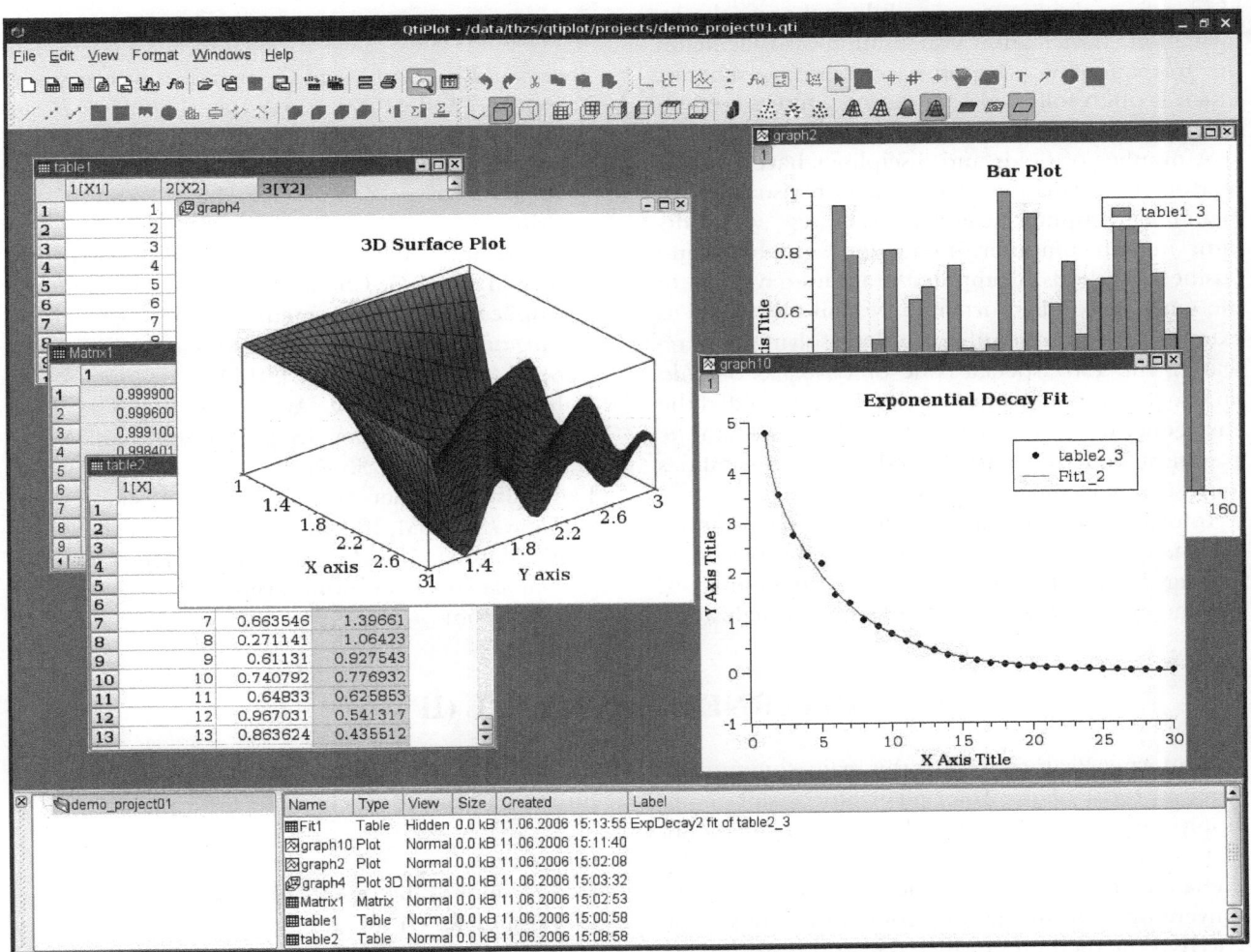

A screenshot from QtiPlot 0.8.5 running on GNU/Linux.

abstract information, including both numerical and nonnumerical data, textual information, and geographical dislocation.

Statistical research uses information visualization most vividly. The results of hypothesis testing, regression analysis, and instrumental variables analysis are all displayed via information visualization techniques to make the numbers and figures meaningful for data interpretation, analysis, and relevant decision making. Data grouping, clustering, classifying, and organizing are other examples of data mining and visual presentation.

Information visualization helps a human eye creatively structure and order seemingly unstructurable information. It yields unexpected insights into a vast array of data and opens up the orders and systems hidden from the human eye while gathering the information. Information visualization also facilitates the formation of a hypothesis-generating scheme, which is often followed by a more complex method of hypothesis testing.

A number of fields and disciplines have contributed to the formation of information visualization, including computer science, psychology, visual design, human-computer interaction, graphics, and business methods. Computer graphics gave rise to the emergence of information visualization and its extensive use in scientific problem solving and research; the 1987 special issue of *Computer Graphics* on visualization in scientific computing heralded the emergence of information visualization as a scientific graphical technique and opened up vast possibilities for researchers and explorers.

Information visualization is used in scientific research, data representation, digital libraries, data mining, financial data analysis, social data analysis, and many other fields. Cloud-based visualization services, digital media, computer-mediated communication, dynamic interactive representations, cognitive ethnography, multiscale software, and video data analysis are all fast-growing and developing fields that use information visualization extensively. In addition, effective system design is based on dynamic interactive representations, and scientists refer to information visualization in order to understand their cognitive and computational characteristics. Information visualization is also increasingly used in modern artworks, and its principles are applied to exhibit displays and configurations in modern art museums throughout the world.

—*Mariam Orkodashvili, PhD*

BIBLIOGRAPHY

Freeman, Eric, and David Gelernter. "Lifestreams: A Storage Model for Personal Data." *SIGMOD Record* 25.1 (1996): 80–86. Print.

Gong, Yihong. *Intelligent Image Databases: Towards Advanced Image Retrieval.* Hingham: Kluwer, 1998. Print.

Hauptmann, Alexander G., Michael J. Witbrock, and Michael G. Christel. "News-on-Demand: An Application of Informedia Technology." *D-Lib Magazine.* Corp. for Natl. Research Initiatives, Sept. 1995. Web. 24 Feb. 2014.

Koblin, Aaron. "Q&A: Aaron Koblin: The Data Visualizer." Interview by Jascha Hoffman. *Nature* 486.7401 (2012): 33. Print.

Lau, Albert. "Data Sets: Narrative Visualization." *Many Eyes.* IBM, 18 May 2011. Web. 24 Feb. 2014.

Lima, Manuel. *Visual Complexity: Mapping Patterns of Information.* Princeton: Princeton Architectural, 2013. Print.

INTERNET PROTOCOL (IP)

The Internet Protocol (IP) is the main communications protocol of the Internet protocol suit (commonly called the TCP/IP), the networking model that establishes the Internet. A network enables the exchange of information, while a protocol is a set of conventions for formatting communications data. The IP is the method by which data is transferred between computers via the Internet. Thus, its routing system, or system for directing, is what makes the Internet a place for exchanging information.

OVERVIEW

Two versions of the Internet Protocol are in use: four (IPv4) and six (IPv6). IPv4 is the most common, as nations have been slow to adopt IPv6. Each computer connected to the Internet has an IP address, a set of numbers that uniquely identify the computer. IPv4 addresses consist of 32 bits of digital space, while IPv6 uses 128-bit addresses. There are two parts to the address identification: the specific computer network and the specific device within that network. On the Internet itself, between routers that move packets of information, only the network address needs to be accessed. Both pieces of information are necessary for sending the message from a network point directly to a computer.

When data such as a website page or an e-mail is sent via the Internet, the message is divided into packets called "datagrams." Each datagram is composed of a header and a payload. The header contains the source (sender) and destination (receiver) IP addresses, each of which include the computer's unique network and device addresses, and other metadata needed for routing and delivering the packet. The payload is the message data itself.

The datagram is first sent to a gateway computer—a network point that functions as the entranceway to another network—which reads the destination address and forwards the message to the next gateway. This gateway reads and forwards the address to yet another gateway, and so on, until the message reaches the closest gateway to its destination. The last gateway recognizes the datagram's address as belonging to its domain, or the set of network addresses under its control, and forwards the message directly to the final device destination.

IP routing service is considered unreliable because of the dynamic nature of the Internet and the possibility that any network element may also be unreliable. Therefore, the Internet Protocol has no continuing connection between end points. Each packet of data is treated independently from all other packets of data. Individual datagrams may not necessarily travel the same route across the Internet, as they are trying to get past any errors along the way, and they may arrive in a different order than they were sent. The IP is just a delivery service, so another Internet protocol, the Transmission Control Protocol (TCP), reorganizes the datagrams into their original order.

—*Julia Gilstein*

BIBLIOGRAPHY

Abraham, Prabhakaran, Mustafa Almahdi Algaet, and Ali Ahmad Milad. "Performance and Efficient Allocation of Virtual Internet Protocol Addressing in Next Generation Network Environment." *Australian Journal of Basic & Applied Sciences* 7.7 (2013): 827–32. Print.

Blank, Andrew G. *TCP/IP Jumpstart: Internet Protocol Basics*. 2nd ed. San Francisco: Sybex, 2002. Print.

Cirani, Simone, Gianluigi Ferrari, and Luca Veltri. "Enforcing Security Mechanism in the IP-Based Internet of Things: An Algorithmic Overview." *Algorithms* 6.2 (2013): 197–226. Print.

Clark, Martin P. *Data Networks, IP, and the Internet*. Hoboken: Wiley, 2003. Print.

Coleman, Liv. "'We Reject: Kings, Presidents, and Voting': Internet Community Autonomy in Managing the Growth of the Internet." *Journal of Information Technology & Politics* 10.2 (2013): 171–89. Print.

Gaffin, Julie C. *Internet Protocol 6*. New York: Novinka, 2007. Print.

Loshin, Peter. *IPv6: Theory, Protocol, and Practice*. 2nd ed. San Francisco: Morgan Kaufmann, 2004. Print.

Oki, Eiji, et al. *Advanced Internet Protocols, Services, and Applications*. Hoboken: Wiley, 2012. Print.

Yoo, Christopher S. "Protocol Layering and Internet Policy." *University of Pennsylvania Law Review* 161.6 (2013): 1707–71. Print.

INVERSION OF CONTROL (HOLLYWOOD PRINCIPLE)

FIELDS OF STUDY

Software Development; Coding Techniques; Computer Science

ABSTRACT

Inversion of control is a software development methodology that reverses the control flow used by traditional procedural programming. In inversion of control, the program responds to calls from a framework rather than controlling program flow itself. This is often used in object-oriented programming.

PRINCIPAL TERMS

- **call:** the process of invoking a subroutine, such as a function or method.

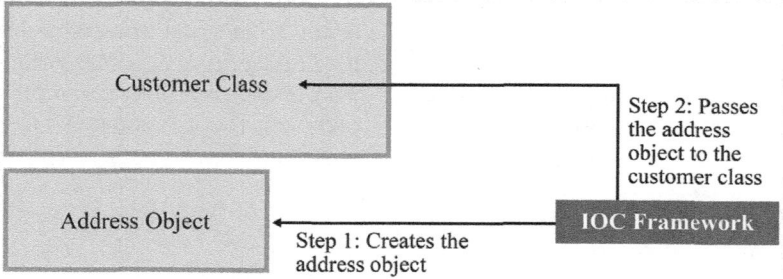

Using an IoC framework, an address object is not created by a class; rather, it is created by a framework, which then passes that object to the customer class.

- **coupling:** the extent to which software components rely on other components in order to function.
- **dependency:** a relationship between software components that defines the ways in which one component relies on others to function correctly.
- **software architecture:** the overall structure of a software system, including the components that make up the system, the relationships among them, and the properties of both those components and their relationships.

What Is Inversion of Control?

Inversion of control is an approach to software design that is used to create a software architecture with looser coupling and fewer dependencies. Inversion of control reverses the typical control flow (the order in which a program executes commands) found in applications developed using more traditional methods. In procedural programming, for example, a program will call procedures during execution. Upon being called, those procedures, also called "functions" or "subroutines," will respond by performing specific predefined tasks, then returning one or more values to the larger program. The overall program is designed to be project specific, while the procedures are typically called from a library of reusable code.

When using inversion of control, this process is reversed: rather than calling procedures from a library, the program responds to calls issued by a software framework that provides generic functionality. In this scenario, the entity issuing the calls (the framework), not the library of procedures that respond to the calls, can be used by other applications. Because the program waits for the framework to initiate calls, inversion of control is sometimes called the "Hollywood principle" in reference to the old movie-industry saying "Don't call us, we'll call you."

A key concept underlying inversion of control is that software components are bound to one another at run time. This differs from traditional programming, where components are statically bound together when the program is compiled. The run-time binding found in inversion of control is possible because interactions between components are defined through abstractions.

Another core principle of inversion of control is that the execution of a task should be decoupled from its implementation. In other words, an application can invoke and use the functionality of another software component without knowing how that component accomplishes its functionality. In addition, individual components should contain only the code that provides the specific functionality they were designed to implement, allowing other components to provide other functionality as needed. Inversion of control is ideally suited for use in object-oriented programming, where looser coupling and reduced dependencies make it easier to alter or reuse objects and classes.

Benefits and Drawbacks

Inversion of control has several advantages over the traditional approach to control flow. It supports decoupling of the various layers and components of a system. This allows developers to focus on developing a component without having to consider how the frameworks implement the functionality they provide. Decoupling reduces the chance of errors as components are changed over time, because any changes in an individual component's code will neither require nor cause further changes in programs that use that component, as long as the inputs that the component takes and the functionality it provides remain unchanged. In addition, the run-time binding of a program to the components it uses frees the program from having to maintain its dependencies. Code is also easier to test because sections of code can be tested in isolation from other components. Lastly, components can be replaced without requiring the entire program to be recompiled.

However, inversion of control also presents some drawbacks. Frameworks and other components can be large, complex, and difficult to understand. New frameworks and components are constantly being created, and changes and updates to existing frameworks and components happen periodically. Thus, a developer who implements inversion of control may have to spend extra time keeping up with the changes. Many frameworks are proprietary, so the internal code used in the framework cannot be studied or modified. Lastly, because inversion of control differs greatly from traditional development models, developers who have used traditional approaches for a long time may be reluctant to embrace this paradigm.

Programming, Hollywood Style

Inversion of control is commonly used by programs that are designed to run on operating systems such as Microsoft Windows. Such programs must respond to events initiated by the user, such as clicking a mouse button or pressing a key on the keyboard. For example, a program might need to display a message box when the user presses the mouse button. Using the Visual Basic programming language, this can be accomplished as follows.

First, a procedure is created that detects when the user clicks the mouse and then sets the variable *mouseClicked* to "true" to indicate to the program that the mouse click has occurred:

```
Private Sub Butttonl_Click (sender As System.
Object, e As System.EventArgs) Handles Button1.
Click
    mouseClicked = True
End Sub
```

Next, a subroutine is created to respond to the mouse click. This routine uses a loop that executes until the mouse is clicked and the variable *mouseClicked* is set to "true" by the above function. Inside loop, Application.DoEvents() is called, which keeps the program from freezing while the code waits for the mouse to be clicked. When the mouse button is clicked and *mouseClicked* is set to "true," the loop exits and the MsgBox function is used to display the message to the user:

```
Sub respondToClick()
  Do Until mouseClicked = True
    Application.DoEvents()
  Loop
  MsgBox "Please wait while the file is updated."
End Sub
```

This is a classic example of inversion of control, as the specific functionality (display a message to the user) is invoked when the mouse is clicked and the framework notifies the program that a mouse click has occurred. In this scenario, the program responds to events generated by the framework (mouse clicks) rather than controlling the flow of actions itself.

It Can Be a Good Thing to Lose Control

Inversion of control brings the power of frameworks and other components to programming, allowing programs to greatly expand the functionality they provide. Due to the widespread use of inversion of control, numerous frameworks and other components are available for use. As such, programs being developed for a wide variety of business sectors can benefit from the use of inversion of control. As new frameworks and components are developed regularly, inversion of control should remain a relevant methodology for implementing software systems for some time to come.

—*Maura Valentino, MSLIS*

Bibliography

Bell, Michael. *Incremental Software Architecture: A Method for Saving Failing IT Implementations.* John Wiley & Sons, 2016.

Friedman, Daniel P., and Mitchell Wand. *Essentials of Programming Languages.* 3rd ed., MIT P, 2008.

Jayaswal, Bijay K., and Peter C. Patton. *Design for Trustworthy Software: Tools, Techniques, and Methodology of Developing Robust Software.* Prentice Hall, 2007.

MacLennan, Bruce J. *Principles of Programming Languages: Design, Evaluation, and Implementation*. 3rd ed., Oxford UP, 1999.

Scott, Michael L. *Programming Language Pragmatics*. 4th ed., Elsevier, 2016.

Van Roy, Peter, and Seif Haridi. *Concepts, Techniques, and Models of Computer Programming*. MIT P, 2004.

Wysocki, Robert K. *Effective Project Management: Traditional, Agile, Extreme*. 7th ed., John Wiley & Sons, 2014.

IOS

FIELDS OF STUDY

Operating Systems; Software Engineering; Mobile Platforms

ABSTRACT

Apple's iOS is an operating system designed for mobile computing. It is used on the company's iPhone and iPad products. The system, which debuted in 2007, is based on the company's OS X. iOS was the first mobile operating system to incorporate advanced touch-screen controls.

PRINCIPAL TERMS

- **jailbreaking:** the process of removing software restrictions within iOS that prevent a device from running certain kinds of software.
- **multitasking:** in computing, the process of completing multiple operations concurrently.
- **multitouch gestures:** combinations of finger movements used to interact with touch-screen or other touch-sensitive displays in order to accomplish various tasks. Examples include double-tapping and swiping the finger along the screen.
- **platform:** the underlying computer system on which an application is designed to run.
- **3D Touch:** a feature that senses the pressure with which users exert upon Apple touch screens.
- **widgets:** small, self-contained applications that run continuously without being activated like a typical application.

A NEW GENERATION OF MOBILE OPERATING SYSTEMS

Apple's iOS is an operating system (OS) designed for use on Apple's mobile devices, including the iPhone, iPad, Apple TV, and iPod Touch. In 2016, iOS was the world's second most popular mobile OS after the Android OS. Introduced in 2007, iOS was one of the first mobile OSs to incorporate a capacitive touch-screen system. The touch screen allows users to activate functions by touching the screen with their fingers. The Apple iOS was also among the first mobile OSs to give users the ability to download applications (apps) to their mobile devices. The iOS is therefore a platform for hundreds of thousands third-party apps.

The first iOS system and iPhone were unveiled at the 2007 Macworld Conference. The original iOS had a number of limitations. For example, it was unable to run third-party apps, had no copy and paste functions, and could not send e-mail attachments. It was also not designed for multitasking, forcing users to wait for each process to finish before beginning another. However, iOS introduced a sophisticated capacitive touch screen. The iOS touch features allowed users to activate most functions with their fingers rather than needing a stylus or buttons on the device. The original iPhone had only five physical buttons. All other functions, including the keyboard, were integrated into the device's touch screen. In addition, the iOS system supports multitouch gestures. This allows a user to use two or more fingers (pressure points) to activate additional functions. Examples include "pinching" and "stretching" to shrink or expand an image.

JAILBREAKING

Computer hobbyists soon learned to modify the underlying software restrictions built into iOS, a process called jailbreaking. Modified devices allow users greater freedom to download and install apps. It also allows users to install iOS on devices other than Apple devices. Apple has not pursued legal action against those who jailbreak iPhones or other devices. In 2010, the US Copyright Office authorized an

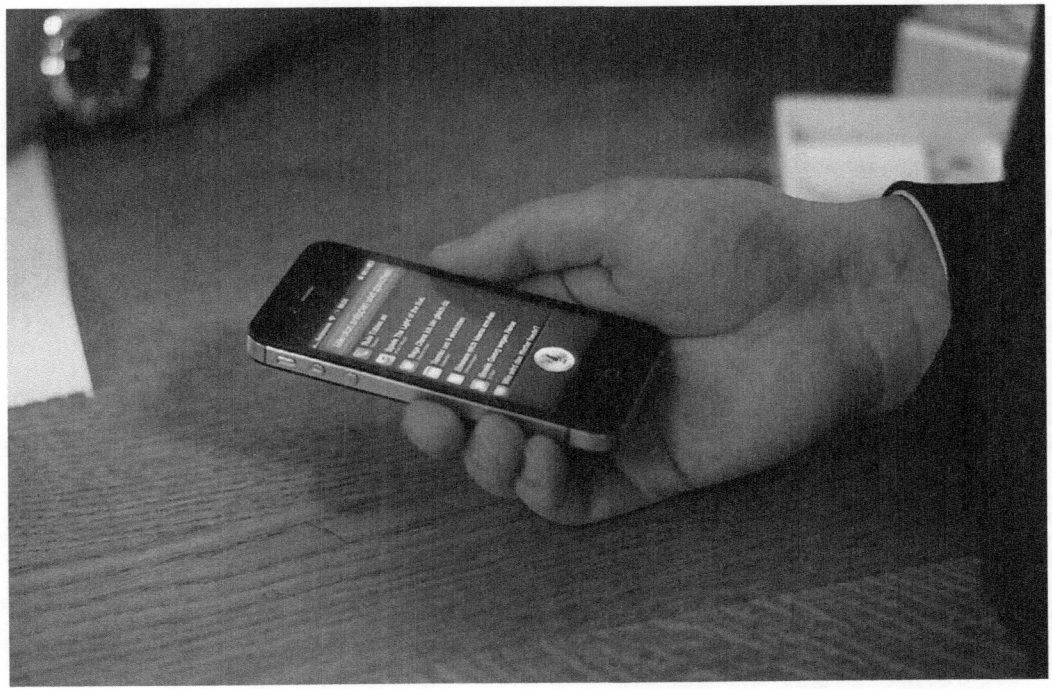

iOS integrated Siri, a voice-activated search engine, for the iPhone. Through multiple updates, its capabilities have expanded to work extremely well with a number of popular websites with all Apple apps, as well as with many popular third-party apps.

exception permitting users to jailbreak their legally owned copies of iOS. However, jailbreaking iOS voids Apple warranties.

Version Updates

The second version of iOS was launched in July 2008. With iOS 2, Apple introduced the App Store, where users could download third-party apps and games. In 2009, iOS 3 provided support for copy and paste functions and multimedia messaging. A major advancement came with the release of iOS 4 in 2010. This update introduced the ability to multitask, allowing iOS to begin multiple tasks concurrently without waiting for one task to finish before initiating the next task in the queue. The iOS 4 release was also the first to feature a folder system in which similar apps could be grouped together on the device's home screen (called the "springboard"). FaceTime video calls also became available with iOS 4.

The release of iOS 5 in 2011 integrated the voice-activated virtual assistant Siri as a default app. Other iOS 5 updates include the introduction of iMessage, Reminders, and Newsstand. In 2012, iOS 6 replaced Google Maps with Apple Maps and redesigned the App Store, among other updates. Released in 2013, iOS 7 featured a new aesthetic and introduced the Control Center, AirDrop, and iTunes Radio.

New Innovations

With the release of iOS 8, Apple included third-party widget support for the first time in the company's history. Widgets are small programs that do not need to be opened and continuously run on a device. Examples including stock tickers and weather widgets that display current conditions based on data from the web. Widgets had been a feature of Android and Windows mobile OSs for years. However, iOS 8 was the first iOS version to support widgets for Apple. Since their release, Apple has expanded the availability of widgets for users.

The release of iOS 9 in 2015 marked a visual departure for Apple. This update debuted a new typeface for iOS called San Francisco. This specially tailored font replaced the former Helvetica Neue.

The release of iOS 9 also improved the battery life of Apple devices. This update introduced a low-power mode that deactivates high-energy programs until the phone is fully charged. Low-power mode can extend battery life by as much as an hour on average.

Coinciding with the release of iOS 9, Apple also debuted the iPhone 6S and iPhone 6S Plus, which introduced 3D Touch. This new feature is built into the hardware of newer Apple devices and can sense how deeply a user is pressing on the touch screen. 3D Touch is incorporated into iOS 9 and enables previews of various functions within apps without needing to fully activate or switch to a new app. For instance, within the camera app, lightly holding a finger over a photo icon will bring up an enlarged preview without needing to open the iPhoto app.

—*Micah L. Issitt*

BIBLIOGRAPHY

Heisler, Yoni. "The History and Evolution of iOS, from the Original iPhone to iOS 9." *BGR*. BGR Media, 12 Feb. 2016. Web. 26 Feb. 2016.

"iOS: A Visual History." *Verge*. Vox Media, 16 Sept. 2013. Web. 24 Feb. 2016.

Kelly, Gordon, "Apple iOS 9: 11 Important New Features." *Forbes*. Forbes.com, 16 Sept. 2015. Web. 28 Feb. 2016.

Parker, Jason, "The Continuing Evolution of iOS." *CNET*. CBS Interactive, 7 May 2014. Web. 26 Feb. 2016.

Williams, Rhiannon. "Apple iOS: A Brief History." *Telegraph*. Telegraph Media Group, 17 Sept. 2015. Web. 25 Feb. 2016.

Williams, Rhiannon, "iOS 9: Should You Upgrade?" *Telegraph*. Telegraph Media Group, 16 Sept. 2015. Web. 25 Feb. 2016.

ITERATIVE CONSTRUCTS

FIELDS OF STUDY

Algorithms; Coding Techniques

ABSTRACT

Iterative constructs are structures used in computer programming to repeat the same computer code multiple times. Loops are the most commonly used iterative constructs. They provide mechanisms for controlling the number of times the code in the loop executes before the loop ends.

PRINCIPAL TERMS

- **event loop:** a segment of computer code that executes in response to external events.
- **for loop:** a segment of computer code that executes repeatedly a set number of times.
- **while loop:** a segment of computer code that executes repeatedly for as long as a given condition is true.

DEFINING LOOPS

A loop is an iterative construct that allows a computer to execute a section of code zero, one, or multiple times based on whether a condition is true or false. Loops can also be designed to execute a section of code a specified number of times or to respond to external events. A counter variable may be used to track the repetitions. Without loops, programmers would need to include the same section of code multiple times within the program. Duplicating code within a program creates unnecessary work for the programmer, increases the chance of errors, and makes debugging and maintenance more difficult.

Three types of loops are commonly used in programming. (Note that different languages may refer to these by slightly different names, but they behave the same.) The first type of loop is the while loop. The statements in a while loop execute repeatedly as long as a condition is true. When the condition becomes false, the loop ends and program execution continues with the next statement. There are two variations of the while loop. In the first, the condition is checked before the code in the loop is executed. The other type, a do while loop, checks whether the condition is true after the code in the loop has executed once. Unlike the standard while loop, a do while loop always executes at least once.

The second type of loop is the for loop. The for loop is used when the requisite number of iterations is a known value. The for loop states concisely the starting value, number of iterations, and the change that occurs over that period. For loops can also be nested in order to create complex repetitions with multiple variables.

The third type of loop is the event loop. The code in an event loop waits for an external event to occur, such as the user pressing a key on the keyboard. When the loop detects that an event has occurred, it then executes code based on the event it has detected. For example, when the user presses the B key on the keyboard, an event loop can detect this action and direct the program to execute code that displays the letter *b* on the monitor. Multiple events, or tasks, may be registered. In that case, the event loop executes code based on the order in which the tasks were received. An event loop processes tasks until it receives a command to stop.

Event loops also allow different programs to communicate and interact. For example, one program might generate an event that triggers an event loop in another program. This form of interaction between software components is commonly used in software development.

Considerations with Loops

Loops help programmers accomplish many common programming tasks. They allow programs to access databases efficiently, respond to system events, interact with the user, and operate with one another. However, programmers must choose the correct type of loop with care. If the code contained within the loop must execute at least once, then the programmer should use a do while loop. If, however, the code in a loop must only be executed if the loop's condition is true, a while loop must be used. Use of the wrong loop can result in incorrect program execution.

Loops must also be written such that when the tasks a loop is designed to accomplish are complete, the loop transfers control to the next statement. Failure to do so can result in an infinite loop. To prevent infinite loops, the programmer must ensure that the condition for all while loops will eventually be set to false, that the number of iterations specified in for loops be reached, or that the loop end for another reason, such as the presence of a break or exit

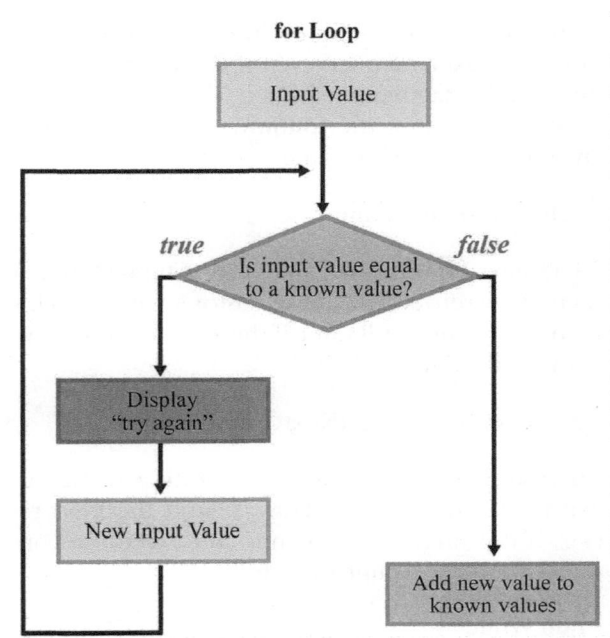

A for loop is an iterative construct that allows for a program loop to continue until a specific criteria is met. In this case, a new input value is compared to known values. If it matches a known value, the program loops through a request to try again until a new value does not match known values. At that point, the program continues by adding the new value to the known values.

command. Otherwise, infinite loops can result in the program becoming unresponsive.

Loops in Practice

Loops are used to accomplish a wide variety of real-world programming tasks. For example, a loop might be used to display a list of customer names using JavaScript. Before the loop begins, the first name in the customer database is retrieved by calling the getFirstCustomerName() function.

customer_name = getFirstCustomerName()

If there are no customers in the database, the function returns a null value. If there is at least one customer in database, control should pass to the code within the loop so the list of customer names can be displayed. However, if there are no customers in the database, the code in the loop should not execute. To ensure this, a while loop is used to check the condition at the beginning of the loop.

While customer_name Not Equal to Null

Note that if a do loop were used, the code in the loop would execute once even if there were no customers in the database.

Within the loop, the following line of code displays the first customer name.

Echo customer_name

The next line uses the getNextCustomerName() function to retrieve the next customer name. This function returns a null value if there are no more customers in the database.

customer_name = getNextCustomerName()

If customer_name is not null, another name has been retrieved from the database and the loop repeats. If it is null, the loop ends and execution continues with the next line of code.

End While

The programmer must ensure that at some point, customer_name is set to null to avoid an infinite loop error.

Loops in Software Development

Loops are used to implement many of the algorithms used in computer programming. Without loops, code would need to be repeated many times within the same program, leading to reduced performance and greater difficulty in debugging and maintenance. While loops and for loops allow programmers to implement repetitive actions flexibly and efficiently. Event loops make user interaction possible. They are thus critical to the graphical user interfaces in modern operating systems and to responsive web browsers.

—*Maura Valentino, MSLIS*

Bibliography

Friedman, Daniel P., and Mitchell Wand. *Essentials of Programming Languages.* 3rd ed., MIT P, 2008.

Haverbeke, Marijn. *Eloquent JavaScript: A Modern Introduction to Programming.* 2nd ed., No Starch Press, 2014.

MacLennan, Bruce J. *Principles of Programming Languages: Design, Evaluation, and Implementation.* 3rd ed., Oxford UP, 1999.

Schneider, David I. *An Introduction to Programming using Visual Basic.* 10th ed., Pearson, 2016.

Scott, Michael L. *Programming Language Pragmatics.* 4th ed., Morgan Kaufmann Publishers, 2016.

Van Roy, Peter, and Seif Haridi. *Concepts, Techniques, and Models of Computer Programming.* MIT P, 2004.

J

JAVA PROGRAMMING LANGUAGE

In anticipation of the growing need for a modern programming language designed to work with the Internet and the underlying processes of devices with embedded microcomputer systems (such as DVD players and cellular phones), the Java cross-platform software was developed by Sun Microsystems in the 1990s and released to programmers in 1996. As of 2013, Java is the underlying technology driving many embedded devices as well as many programs for desktop computers. Java is an object-oriented programming language, which functions by "byte-compiling" code into a format that can run on many different types of computers.

Overview

Although the underlying language of computers is binary code, it is difficult and time consuming to write programs this way. For this reason, programming languages allow programmers to write, or code, in a format that is easier to read and comprehend. This code is then converted by another computer program, known as a compiler, into binary code so the computer can execute instructions. Java is an example of a particular language, although it is compiled quite differently from more traditional languages.

Sun Microsystems and a team of engineers known as the Green Team proposed Java during the 1990s. The language was developed based on borrowing its syntax—that is, its linguistic structure and rules—from the dominant programming language on the market at the time of its development, C++. Like C++, Java is an object-oriented programming language, meaning that it relies upon complex data structures to increase the level of organization in code. The team developing Java noticed that with the growth of the Internet, there was a growing need to be able to generate code that could run on more than one type of computer, for instance, computers running Microsoft Windows with Intel processors, and computers running Apple's Macintosh OS with PowerPC processors.

Although Java code looks very similar to code written in C++, it functions differently from C++ and

The cover illustration for the Java Programming book on Wikibooks.

other previously existing programming languages because it is neither interpreted nor compiled. Interpreted code is converted to binary instructions as the user uses the program, which can often make a program slow. Compiled code is converted to binary before it is actually run. In a stark departure from both these methods, Java was developed to compile code first to an intermediate format called bytecode, which is then executed not by a computer itself, but by the Java Virtual Machine (JVM). Using this method, code can be written without knowing what hardware it will be run on. As long as a version of the JVM exists for that hardware, it can execute this bytecode. Versions of the JVM were quickly distributed to many Web-connected devices and platforms, and Java technology was added to the influential Netscape Navigator web browser, hastening the adoption of Java-related technology.

In the twenty-first century, Java is used in numerous formats; it is the primary language used to write programs for the Android mobile operating system, and fulfills its originally designed role as a language for embedded devices like disc players. Since 2010, JVM has been developed by Oracle, which owns the rights to the original technologies that comprise Java, although other versions of the JVM exist, produced both by competing companies and independent open-source developers.

—*Luke E. A. Lockhart*

BIBLIOGRAPHY

Antoy, Sergio, and Michael Hanus. "Functional Logic Programming." *Communications of the ACM* 53 (2010): 74–85. Print.

Cazzola, Walter, and Edoardo Vacchi. "@Java: Bringing a Richer Annotation Model to Java." *Computer Languages, Systems & Structures* 40.1 (2014): 2–18. Print.

Chrząszcz, Jacek, Patryk Czarnik, and Aleksy Schubert. "A Dozen Instructions Make Java Bytecode." *Electronic Notes in Theoretical Computer Science* 264.4 (2011): 19–34. Print.

Emurian, Henry H., and Peng Zheng. "Programmed Instruction and Interteaching Applications to Teaching Java: A Systematic Replication." *Computers in Human Behavior* 26.5 (2010): 1166–75. Print.

"History of Java Technology." *Oracle.com*. Oracle, n.d. Web. 22 Aug. 2013.

Javary, Michèle. "Evolving Technologies and Market Structures: Schumpeterian Gales of Creative Destruction and the United Kingdom Internet Service Providers' Market." *Journal of Economic Issues* 38.3 (2004): 629–57. Print.

Reynolds, Mark C. "Modeling the Java Bytecode Verifier." *Science of Computer Programming* 78.3 (2013): 327–42. Print.

Reza, Juan Rolando. "Java Supervenience." *Computer Languages, Systems & Structures* 38.1 (2012): 73–97. Print.

Schildt, Herbert. *Java: The Complete Reference*. 8th ed. New York: McGraw, 2011. Print.

JAVASCRIPT

JavaScript is one of the most popular programming languages for applications on the web and has been a staple of web technology since the late 1990s. JavaScript files can be identified by their .js file extension. The language enables users to perform many of the tasks we now think of as being synonymous with the Internet. From asynchronous communication and productivity applications to photo sharing and social games, JavaScript enables users to interact directly with web content. Features such as recognizing keystrokes and expediting responsiveness by virtue of local hosting go beyond the capabilities of HTML (hypertext mark-up language). Web browsers are the most common hosting environments for JavaScript, but JavaScript is also used outside of webpages in things like small, locally run desktop applications and PDF documents.

OVERVIEW

JavaScript was developed by Brendan Eich in the mid-1990s for the Netscape Navigator web browser with the hope that it could function as a sort of distributed operating system that could compete with Microsoft. It was called Mocha, and later Livescript

Brendan Eich, Mozilla Corporation.

during its initial development. Netscape agreed on a license with Sun Microsystems, the creator of the programming language Java, to finally call it JavaScript. According to Eich, as quoted in an interview for *InfoWorld*, "the idea was to make it a complementary scripting to go with Java." Individuals not familiar with web programming sometimes confuse JavaScript with Java. While JavaScript takes some of its naming conventions from Java, the languages have little else to do with one another. Unlike Java, which is intended for professional computer programmers, JavaScript was aimed at web designers and other non-programmers.

JavaScript is not the only technology that freed web users from the constraints of HTML. Flash, originally built in the 1990s by Macromedia (and later managed by Adobe), was a popular solution for web designers and developers seeking to add interactivity to their web pages. In certain respects, Flash exceeded many of the early capabilities of JavaScript, and JavaScript had fallen out of style among many designers and developers. Nonetheless, communities of JavaScript developers worked to create many popular frameworks and technologies that are ubiquitous on the web in the 2010s, such as Ajax, jQuery, Dojo, and Prototype.

Flash had its own constraints that kept web applications from meeting the needs of users. For instance, Flash requires a special extension to be added to each browser, and the extension requires frequent updates. Flash applications were also self-contained are also generally divorced from all the other content on the web page and thus their contents are not indexed by search engines.

When the Flash plug-in was excluded from the Apple iPhone's Safari browser, the web development community further renewed its interest in JavaScript. In 2010 JavaScript became a trademark of the Oracle Corporation and was licensed by Netscape Communications and the Mozilla Foundation.

—*Marjee Chmiel, MA, PhD*

BIBLIOGRAPHY

Goodman, Danny, Michael Morrison, and Brendan Eich. *Javascript Bible.* New York: Wiley, 2007. Print.

Krill, Paul. "JavaScript Creator Ponders Past, Future." *InfoWorld.* Infoworld, 23 June 2008. Web. 5 Aug. 2013.

Lane, Jonathan, et al. "JavaScript Primer." *Foundation Website Creation with HTML5, CSS3, and JavaScript.* New York: Apress, 2012. Print.

Perfetti, Christine, and Jared M. Spool. "Macromedia Flash: A New Hope for Web Applications." *UIE. com.* User Interface Engineering, 2002. Web. 5 Aug. 2013.

Severance, Charles. "JavaScript: Designing a Language in 10 Days." *Computer* 45.2 (2012): 7–8. Print.

Simpson, Kyle. *JavaScript and HTML5 Now.* Sebastopol: O'Reilly Media, 2012. Kindle file.

Wright, Tim. *Learning JavaScript: A Hands-On Guide to the Fundamentals of Modern JavaScript.* Boston: Addison, 2012. Print.

Yule, Daniel, and Jamie Blustein. "Of Hoverboards and Hypertext." *Design, User Experience, and Usability. Design Philosophy, Methods, and Tools.* Berlin: Springer, 2013. Print.

KNOWLEDGE WORKER

A knowledge worker is a person who works in the knowledge-based economy, or knowledge economy, which is believed to be the next step of economic development in the progression from agriculture to manufacturing to services and then to information or knowledge. The existence of the knowledge economy is predicated on the fact that in most consumer goods markets, supply of goods and services now exceeds demand and so, for sales to take place, some form of high-value knowledge should be embedded in the product involved or in its creation. This allows the product to be differentiated from its competitors and hence to appeal to consumers more, because it provides a value proposition.

Overview

The types of activity included in the knowledge economy vary from country to country, reflecting different historical and cultural factors that have contributed to the production of unique items. While in Western countries the knowledge economy is expected to concentrate on media or creative productions, information provision in advanced banking products, and so forth, in other countries the knowledge might be related to jewelry creation or rug-making. The concept of knowledge work is the same in both cases, since they feature the use of knowledge that is asymmetrically available—whether from the heritage of a long tradition of artisanship or through sophisticated dredging of databases—in order to produce items that are unique in a way that appeals to consumers. Put another way, this indicates that the knowledge economy is not quite such a modern concept as it might sometimes be portrayed.

Irrespective of the type of knowledge used, there are some characteristics of the nature of knowledge work that unite all those people involved in it. First, there must be some degree of freedom in the process of working that permits more information gathering, experimentation, and innovation and the recognition therefore that this can lead to failure from time to time. No one can predict accurately what innovations will succeed in the marketplace, which is evident from the number of new product failures associated with even the best-resourced companies in consumer goods markets. There should therefore be some freedom for the knowledge worker and different criteria for determining personal success.

Since people learn in different ways, there should also be some latitude in hours and workplace behavior. This is well known in high-tech firms in the United States, for example, which encourage many kinds of nonstandard behavior at work as a means of stimulating creativity. There is, nevertheless, something of a contradiction inherent in companies requiring designated individuals to be creative to order, especially when the company expects to retain all the benefits of that creativity. Individual compensation deals and performance assessments are common in this kind of arrangement and can lead to some lack of solidarity with colleagues.

There is an argument that the benefits of the products of knowledge workers are in fact useless, because they stimulate otherwise nonexistent demand by creating unique items that have no other appeal other than their unique nature. In such cases, the knowledge workers involved are wasteful of resources and the process unsustainable.

—*John Walsh, PhD*

Bibliography

Drucker, Peter F. "Knowledge-Worker Productivity: The Biggest Challenge." *California Management Review* 41.2 (1999): 79–94. Print.

Elliott, Larry, and Dan Atkinson. *Fantasy Island: Waking Up to the Incredible Economic, Political and Social Illusions of the Blair Legacy*. London: Constable, 2007. Print.

Horibe, Frances. *Managing Knowledge Workers: New Skills and Attitudes to Unlock the Intellectual Capital in your Organization*. New York: Wiley, 1999. Print.

Jemielniak, Dariusz. *The New Knowledge Workers*. Cheltenham: Elgar, 2012. Print.

Liu, Alan. *The Laws of Cool: Knowledge Work and the Culture of Information*. Chicago: U of Chicago P, 2004. Print.

O'Brien, James, and George Marakas. *Management Information Systems*. 10th ed. New York: McGraw-Hill, 2010. Pint.

Reinhardt, W., et al. "Knowledge Worker Roles and Actions—Results of Two Empirical Studies." *Knowledge and Process Management* 18.3 (2011): 150–174. Print.

LEVELS OF PROCESSING THEORY

The levels of processing theory is a model used to describe the development of memory, contrasting with the two-process or "multi-level" theory and the "working memory" models. The levels of processing model holds that the "level of processing" that an individual uses to process incoming data determines how deeply the information is encoded into memory. In comparison to the "multi-level" theory, the levels of processing model holds that there is only a single store of memory, without the process of transferring information between short and long-term memory, but that information may be encoded in a more detailed manner depending how the information is received and processed.

Overview

The levels of processing model was developed by cognitive psychologist Fergus Craik and his colleague Robert Lockhart, and first explored in a paper published in 1972 in the *Journal of Verbal Learning and Verbal Behavior*. Craik and Lockhart were motivated by problems with the "two-system" memory model, which describes memory as an interplay between separate systems for short- and long-term memory storage. Craik and Lockhart theorized that the length of time a person stores memory might be better explained by examining the type of processing used on incoming data.

Craik and Lockhart proposed a difference between shallow processing, which does not involve any exploration of meaning, and deep processing, which involves processing meaning and therefore leads to longer-term comprehension. Craik and Tuving used linguistic tools to measure and demonstrate the levels of processing by asking subjects to remember lists of words after being asked questions about the words. Questions about simple physical properties, such as "does the word contain the letter 'a,'" produced shallow processing and lower recall, whereas questions like "does the word fit into this sentence" involved semantic processing and resulted in better recall.

As far as cognition is concerned, the levels of processing theory seems logical. People have better recall when the facts they are asked to remember fit with things that they already know about, thereby causing them to consider the further implications of the new knowledge they've acquired. Another key concept in the levels of processing approach is "elaboration," which is defined as continually processing information on a deep level, thereby further aiding in committing the information to long-term recall.

Elaborative rehearsal is the process of relating new material to information that is already stored through deep processing. Elaborative rehearsal has been repeatedly demonstrated to be an effective method of enhancing recall. Simple repetition of information is called "maintenance rehearsal," and results in low-levels of recall, such as repeating a phone number after learning it. Elaborative rehearsal, by contrast, involves creating associations with stored memory, such as associating the numbers in a phone number with important dates or other numbers that are familiar and already committed to memory.

The levels of processing theory was never intended to be a comprehensive theory of memory, but rather was designed to provide a framework for viewing the relationship between processing and the formation of memory. One of the strengths of the levels of processing model is that it is dynamic, focusing on methods of handling incoming information rather than on interaction between theoretical storage mechanisms that passively hold information.

—*Micah Issitt*

BIBLIOGRAPHY

Anzulewicz, Anna, et al. "Does Level of Processing Affect the Transition from Unconscious to Conscious Perception." *Consciousness and Cognition* 36 (2015): 1–11. Print.

Craik, Fergus I. M., and Endel Tulving, "Depth of Processing and the Retention of Words in Episodic Memory," *Journal of Experimental Psychology*, 1975, 104(3), 268–94. Print.

Craik, Fergus I. M., and Robert. S. Lockhart. "Levels of Processing: A Framework for Memory Research." *Journal of Verbal Learning and Verbal Behavior*, 1972, 11, 671–84. Print.

Dehn, Milton J. *Working Memory and Academic Learning: Assessment and Intervention.* Hoboken: Wiley, 2011. Print.

Rose, Nathan S., and Craik, Fergus I. M. "A Processing Approach to the Working Memory/Long-Term Memory Distinction: Evidence from the Levels-of-Processing Span Task." *Journal of Experimental Psychology: Learning, Memory, and Cognition* 38.4 (2012): 1019–029. Print.

Schwartz, Bennett L. *Memory: Foundations and Applications.* Thousand Oaks: Sage, 2011. Print.

Surprenant, Aimee M., and Ian Neath. *Principles of Memory.* New York: Psychology P, 2009. Print.

Weisberg, Robert W. and Lauretta M. Reeves. *Cognition: From Memory to Creativity.* Hoboken: Wiley. Print.

LOGIC SYNTHESIS

Logic synthesis is a branch of computer programming in which initial concepts of what functions computers might be expected to perform are restructured and translated into the vocabulary and language of computers. Computers must be told specifically what their programmers want them to do; abstract ideas or concepts that encompass a range of computer functions cannot, in practical terms, be conveyed as is in the very limited language of computers themselves. In the field of software engineering, logic synthesis refers to the process by which software engineers convert the abstract concept of the function a computer circuit will be expected to execute into specific instructions that computer circuits will accept.

In the ever-increasingly competitive field of software engineering, many programmers focus on minimizing the design cycle, which is the time between idea (the conception of what a computer program will execute) and actualization (a workable program that can be introduced into the highly competitive software market). By systematically working the painstaking step-by-step process of algorithmic thinking that is essential to logic synthesis, software engineers convert otherwise abstract instructions into desired circuit behavior.

OVERVIEW

The basis of logic synthesis is rooted in computer-programming precepts that date back to the mid-nineteenth century—specifically to the algebraic functions first described by British mathematician George Boole (1815–64), who conceived of the complex architecture of abstract mathematical thought in either-or terms. Computer circuits, the workhorses that actualize ideas into real-time computer functions, operate according to specific binary instructions—mapped sequences of logical steps, also called logic gates, in which each step is posed as a YES/ NO or an EITHER/OR statement. This enables the circuit to complete the intended instruction, optimizing function by encoding the intended process map into minimization steps that will speak the specific language of the computer.

For example, say an appliance manufacturer wants to develop an integrated circuit capable of continuously monitoring the temperature of a microwave cooking chamber. The software engineer assigned to the project must translate that abstract concept into a language a computer circuit will understand, carefully applying sequential Boolean logic to move the idea from desired behavior to a specific conceptual structure and then to a workable physical layout or application. Using logic synthesis, a software engineer converts that high-level abstract concept into an optimized gate representation, which is a design implementation that breaks down the concept into binary steps that a circuit will accept and execute effectively and efficiently. Translating processes into

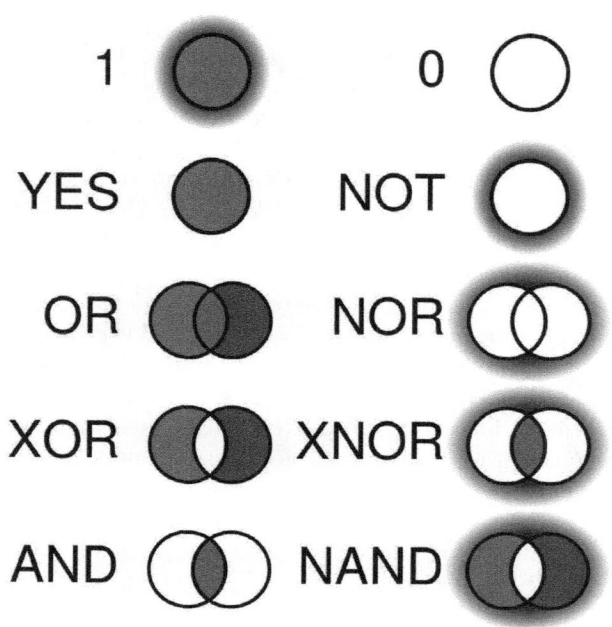

Visual diagram of various logic gates. The parts of the diagram correspond to the four permutations possible given two Boolean variables.

George Boole.

specific hardware and software description is at the heart of logic synthesis. While this is often done by software engineers, it can also be delegated to sophisticated computer programs that in turn can create new and efficient program software.

—*Joseph Dewey, PhD*

BIBLIOGRAPHY

Ciletti, Michael D. *Advanced Digital Design with the Verilog HDL*. 2nd ed. Upper Saddle River: Prentice, 2010. Print.

De Micheli, Giovanni. *Synthesis and Optimization of Digital Circuits*. New York: McGraw, 1994. Print.

Devadas, Srinivas, Abhijit Ghosh, and Kurt Keutzer. *Logic Synthesis*. New York: McGraw, 1994. Print.

Hachtel, Gary D., and Fabio Somenzi. *Logic Synthesis and Verification Algorithms*. New York: Springer, 2006. Print.

Hassoun, Soha, and Tsutomu Sasao, eds. *Logic Synthesis and Verification*. Norwell: Kluwer, 2002. Print.

Rushton, Andrew. *VHDL for Logic Synthesis*. 3rd ed. Hoboken: Wiley, 2011. Print.

Sasao, Tsutomu. *Switching Theory for Logic Synthesis*. New York: Springer, 1999. Print.

LOGISTICS

Logistics involves managing the movement of resources between a point of origin and a final destination. These resources can include tangible goods, such as equipment, products, and personnel, or intangible commodities, such as information, energy, and time. Logistics management processes are typically designed to optimize efficiencies through the integrated planning, implementation, control, and monitoring of key aspects required to move resources across a supply chain. This includes production,

Logistics

ARABIAN GULF (Oct. 20, 2011) Logistics Specialist Seaman Krystal K. Weed inventories supplies in a storeroom aboard the aircraft carrier USS George H.W. Bush (CVN 77).

packaging, inventory, transportation, staffing, and communication. Essentially, logistics is used to get the correct product to its intended destination at the right time for the best price.

Overview

Logistics has its roots in military supply. Early armies used the principles of logistics to ensure that weapons, food, and other supplies were available when and where required by traveling troops. Logistics management in the modern-day military uses multiple variables to predict demand, cost, consumption, and replacement requirements of goods and equipment to create a robust and easily supportable supply-chain system.

The application of logistics in the business sector is tied to globalization. As supply chains became increasingly complex, with resources and supplies located in countries across the world, businesses started looking to logistics as a way to more effectively manage their flow of resources.

Logistics in business encompasses multiple specialties. For example, *inbound logistics* focuses on purchasing and coordinating the movement of materials coming into a company, factory, or store. *Outbound logistics* pertains to processes involved in the storage and movement of resources from the business to the end user. Within a business, logistics also may be applied to a specific project. Regardless of the scope, most logistics efforts incorporate to some degree the management of both tangible and intangible assets.

Procurement logistics, *distribution logistics*, and *production logistics* target specific activities. Procurement logistics addresses market research, requirements planning, manufacture-versus-purchase decisions, supplier management, and order control. Distribution logistics involves order fulfillment, warehousing, and transportation tasks associated with delivering the final product or resource to the end user. Production logistics is the management of processes occurring after procurement and prior to distribution. The areas covered by production logistics include planning of layouts and production and the organization and control of processes.

Other common types of logistics applied in business operations include *disposal logistics, reverse logistics, green logistics,* and *emergency logistics.* Disposal logistics seeks to enhance service while reducing expenses associated with the disposal of waste products resulting from operations. Reverse logistics encompasses practices tied to channeling the business's surplus resources for reuse or disposal as necessary. Green logistics is aimed at minimizing the environmental impact of all of a company's logistical practices. Emergency logistics is used less frequently but is activated when circumstances such as major weather events or significant production delays warrant a change in standard logistics processes to continue to accommodate supply-chain needs with minimal disruption.

Businesses sometimes outsource some or all of their logistics activities to external providers. A typical example is when a company hires a third-party, or nonaffiliated, transportation operator to deliver goods that are otherwise produced and managed by the business itself.

—*Shari Parsons Miller, MA*

Bibliography

Coyle, John J., et al. *Supply Chain Management: A Logistics Perspective.* 9th ed. Mason: South-Western, 2013. Print.

de Jong, Gerard, and Moshe Ben-Akiva. "Transportation and Logistics in Supply Chains." *Supply Chain Management, Marketing and Advertising, and Global Management.* Ed. Hossein Bidgoli. Hoboken: Wiley, 2010. 146–58. Print. Vol. 2 of *The Handbook of Technology Management.* 3 vols.

Farahani, Reza Zanjirani, Shabnam Rezapour, and Laleh Kardar, eds. *Logistics Operations and Management: Concepts and Models.* Waltham: Elsevier, 2011. Print.

McKinnon, Alan, et al., eds. *Green Logistics: Improving the Environmental Sustainability of Logistics.* 3rd ed. Philadelphia: Kogan, 2015. Print.

McMillan, Charles. "Global Logistics and International Supply Chain Management." *Supply Chain Management, Marketing and Advertising, and Global Management.* Ed. Hossein Bidgoli. Hoboken: Wiley, 2010. 68–88. Print. Vol. 2 of *The Handbook of Technology Management.* 3 vols.

Sadler, Ian. *Logistics and Supply Chain Integration.* Thousand Oaks: Sage, 2007. Print.

Schönsleben, Paul. *Integral Logistics Management: Operations and Supply Chain Management in Comprehensive Value-Added Networks.* 3rd ed. Boca Raton: Auerbach, 2007. Print.

M

MACHINE LEARNING

Machine learning is a branch of computer science algorithms that allow the computer to display behavior learned from past experience, rather than human instruction. Machine learning is essential to the development of artificial intelligence, but it is also applicable to many everyday computing tasks. Common examples of programs that employ machine learning include e-mail spam filters, optical character recognition, and news feeds on social networking sites that alter their displays based on previous user activity and preferences.

Overview

One of the earliest attempts to enable machine learning was the perceptron algorithm, developed by Frank Rosenblatt in 1957. The algorithm was intended to teach pattern recognition and was based on the structure of a neural network, which is a computing model designed to imitate an animal's central nervous system. While the perceptron model showed early promise, Marvin Minsky and Seymour Papert demonstrated in their 1969 book *Perceptrons: An Introduction to Computational Geometry* that it had significant limitations, as there were certain classes of problems the model was unable to learn. Consequently, researchers did not pursue this area of study for some time.

In the 1980s, after other avenues for developing artificial intelligence had resulted in only limited success, scientists began to revisit the perceptron model. Multilayer perceptrons, or neural networks composed of multiple layered computational units, proved to have the processing power to express problems that Rosenblatt's single-layer, or linear, perceptrons could not. Around the same time, John Ross Quinlan introduced decision-tree algorithms, which use predictive models to determine a variable's value based on available data.

Since then, numerous machine-learning algorithms have been developed. Among those most commonly used are support vector machines (SVMs) and naive Bayes classifiers. SVMs, introduced by Vladimir N. Vapnik and Corinna Cortes in 1995 and based on an algorithm previously created by Vapnik, are used to recognize patterns in data and classify the various data points. Naive Bayes classifiers are applications of Bayes's theorem, named for the eighteenth-century mathematician and reverend Thomas Bayes, which deals with conditional probabilities. This algorithm was used in one of the earliest e-mail spam filters, iFile, released by Jason Rennie in 1996. Many e-mail clients still employ Bayesian spam filtering, which works by determining the probability that an e-mail containing certain keywords is spam.

Machine-learning algorithms can be divided into categories based on how they train the machine. These categories include supervised learning, in which the machine learns from inputs that are mapped to desired outputs; unsupervised learning, in which the machine analyzes input without knowledge of the desired output; semi-supervised learning, in which some of the input is paired with a desired output and some is not; transduction, in which the machine tries to predict new outputs based on training with previous inputs and outputs; reinforcement learning, in which the machine must form a policy on how to act based on observing how certain actions affect its environment; and learning to learn, which teaches inductive bias based on previous experience. SVMs, multilayer perceptrons, decision trees, and naive Bayes classifiers all fall into the category of supervised learning.

It is important to distinguish between machine learning and data mining. Although the two concepts are related and use similar, often overlapping, methods, data mining is focused on discovering

Thomas Bayes (1702-1761) Nonconformist minister and mathematician. Originator of the statistical theory of probability, the basis of most market research and opinion poll techniques.

information about given data, while machine learning focuses more on learning from the given data in order to make predictions about other data in the future. Many consider data mining to be a subset of machine learning.

—*Randa Tantawi, PhD*

Bibliography

Abu-Mostafa, Yaser S. "Machines That Think for Themselves." *Scientific American* July 2012: 78-81. Print.

Brodley, Carla E. "Challenges and Opportunities in Applied Machine Learning." *AI Magazine* 33.1 (2012): 11-24. Print.

Domingos, Pedro. "A Few Useful Things to Know about Machine Learning." *Communications of the ACM* 55.10 (2012): 78-87. Print.

Heaven, Douglas. "Higher State of Mind." *New Scientist* 10 Aug. 2013: 32-35. Print.

James, Mike. "The Triumph of Deep Learning." *I Programmer*. I-programmer.info, 14 Dec. 2012. Web. 27 Sept. 2013.

Marsland, Stephen. *Machine Learning: An Algorithmic Perspective*. Boca Raton: Taylor, 2009. Print.

Mitchell, Tom M. *The Discipline of Machine Learning*. Pittsburgh: Carnegie Mellon U, 2006. PDF file.

Mitchell, Tom M. *Machine Learning*. New York: McGraw, 1997. Print.

Piore, Adam. "Mind in the Machine." *Discover* June 2013: 52-59. Print.

MALWARE

FIELDS OF STUDY

Software Engineering; Security

ABSTRACT

Malware, or malicious software, is a form of software designed to disrupt a computer or to take advantage of computer users. Creating and distributing malware is a form of cybercrime. Criminals have frequently used malware to conduct digital extortion.

PRINCIPAL TERMS

- **adware:** software that generates advertisements to present to a computer user.

Malware

Malicious Programming

Malware, or malicious software, is a name given to any software program or computer code that is used for malicious, criminal, or unauthorized purposes. While there are many different types of malware, all malware acts against the interests of the computer user, either by damaging the user's computer or extorting payment from the user. Most malware is made and spread for the purposes of extortion. Other malware programs destroy or compromise a user's data. In some cases, government defense agencies have developed and used malware. One example is the 2010 STUXNET virus, which attacked digital systems and damaged physical equipment operated by enemy states or organizations.

History of Malware

The earliest forms of malware were viruses and worms. A virus is a self-replicating computer program that attaches itself to another program or file. It is transferred between computers when the infected file is sent to another computer. A worm is similar to a virus, but it can replicate itself and send itself to another networked computer without being attached to another file. The first viruses and worms were experimental programs created by computer hobbyists in the 1980s. As soon as they were created, computer engineers began working on the first antivirus programs to remove viruses and worms from infected computers.

Public knowledge about malware expanded rapidly in the late 1990s and early 2000s due to several well-publicized computer viruses. These included the Happy99 worm in 1999 and the ILOVEYOU worm in May 2000, the latter of which infected nearly 50 million computers within ten days. According to research from the antivirus company Symantec in 2015, more than 317 million new malware programs were created in 2014. Yet despite public awareness of malware, many large organizations are less careful than they should be. In a 2015 study of seventy major companies worldwide, Verizon reported that almost 90

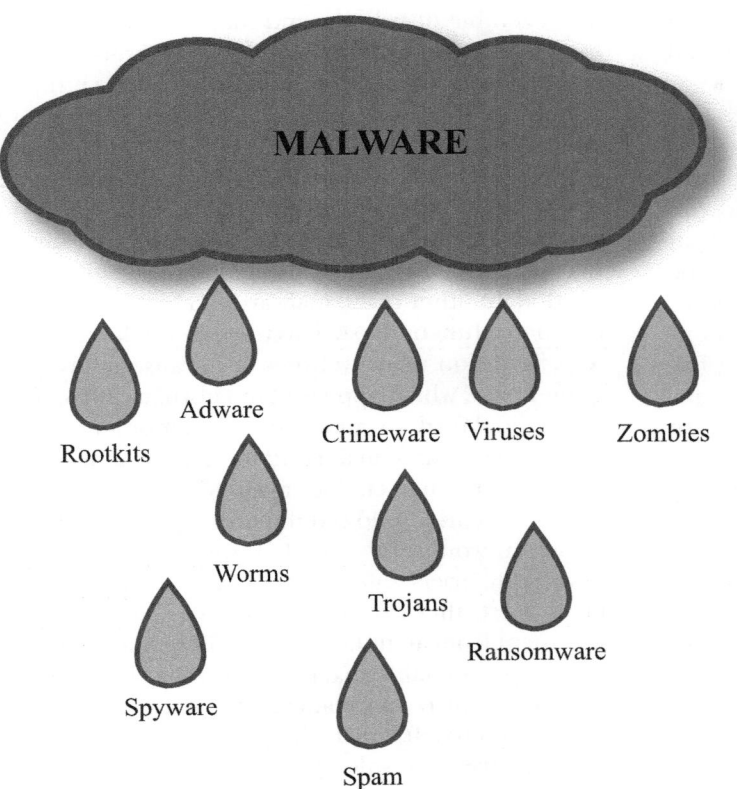

Malware consists of any software designed to cause harm to a device, steal information, corrupt data, confiscate or overwhelm the processor, or delete files. Examples of malware include adware, viruses, worms, Trojans, spam, and zombies.

- **ransomware:** malware that encrypts or blocks access to certain files or programs and then asks users to pay to have the encryption or other restrictions removed.
- **scareware:** malware that attempts to trick users into downloading or purchasing software or applications to address a computer problem.
- **spyware:** software installed on a computer that allows a third party to gain information about the computer user's activity or the contents of the user's hard drive.
- **worm:** a type of malware that can replicate itself and spread to other computers independently; unlike a computer virus, it does not have to be attached to a specific program.
- **zombie computer:** a computer that is connected to the Internet or a local network and has been compromised such that it can be used to launch malware or virus attacks against other computers on the same network.

percent of data breaches in 2014 exploited known vulnerabilities that were reported in 2002 but had not yet been patched.

Types of Malware

One of the most familiar types of malware is adware. This refers to programs that create and display unwanted advertisements to users, often in pop-ups or unclosable windows. Adware may be legal or illegal, depending on how the programs are used. Some Internet browsers use adware programs that analyze a user's shopping or web browsing history in order to present targeted advertisements. A 2014 survey by Google and the University of California, Berkeley, showed that more than five million computers in the United States were infected by adware.

Another type of malware is known as spyware. This is a program that is installed on a user's computer to track the user's activity or provide a third party with access to the computer system. Spyware programs can also be legal. Many can be unwittingly downloaded by users who visit certain sites or attempt to download other files.

One of the more common types of malware is scareware. Scareware tries to convince users that their computer has been infected by a virus or has experienced another technical issue. Users are then prompted to purchase "antivirus" or "computer cleaning" software to fix the problem.

Although ransomware dates back as far as 1989, it gained new popularity in the 2010s. Ransomware is a type of malware that encrypts or blocks access to certain features of a computer or programs. Users with infected computers are then asked to pay a ransom to have the encryption removed.

Addressing the Threat

Combating malware is difficult for various reasons. Launching malware attacks internationally makes it difficult for police or national security agencies to target those responsible. Cybercriminals may also use zombie computers to distribute malware. Zombie computers are computers that have been infected with a virus without the owner's knowledge. Cybercriminals may use hundreds of zombie computers simultaneously. Investigators may therefore trace malware to a computer only to find that it is a zombie distributor and that there are no links to the program's originator. While malware is most common on personal computers, there are a number of malware programs that can be distributed through tablets and smartphones.

Often creators of malware try to trick users into downloading their programs. Adware may appear in the form of a message from a user's computer saying that a "driver" or other downloadable "update" is needed. In other cases, malware can be hidden in social media functions, such as the Facebook "like" buttons found on many websites. The ransomware program Locky, which appeared in February 2016, used Microsoft Word to attack users' computers. Users would receive an e-mail containing a document that prompted them to enable "macros" to read the document. If the user followed the instructions, the Locky program would be installed on their computer. Essentially, users infected by Locky made two mistakes. First, they downloaded a Word document attachment from an unknown user. Then they followed a prompt to enable macros within the document—a feature that is automatically turned off in all versions of Microsoft Word. Many malware programs depend on users downloading or installing programs. Therefore, computer security experts warn that the best way to avoid contamination is to avoid opening e-mails, messages, and attachments from unknown or untrusted sources.

—*Micah L. Issitt*

Bibliography

Bradley, Tony. "Experts Pick the Top 5 Security Threats for 2015." *PCWorld*. IDG Consumer & SMB, 14 Jan. 2015. Web. 12 Mar. 2016.

Brandom, Russell. "Google Survey Finds More than Five Million Users Infected with Adware." *The Verge*. Vox Media, 6 May 2015. Web. 12 Mar. 2016.

Franceschi-Bicchierai, Lorenzo. "Love Bug: The Virus That Hit 50 Million People Turns 15." *Motherboard*. Vice Media, 4 May 2015. Web. 16 Mar. 2016.

Gallagher, Sean. "'Locky' Crypto-Ransomware Rides In on Malicious Word Document Macro." *Ars Technica*. Condé Nast, 17 Feb. 2016. Web. 16 Mar. 2016.

MASSIVE OPEN ONLINE COURSE (MOOC)

A massive open online courses (MOOC) is an evolved version of distance and online education, both of which have been around in various forms for decades. Correspondence education dates back as far as the late nineteenth century, while computers and the Internet have been used in education since their earliest stages. As early as 1972, Athabasca University in Alberta, Canada was delivering distance education using print, television, and eventually online content to teach courses to national and international students. MOOCs promote independent learning by exploring a variety of learning possibilities in the digitally connected world. They enable access to a collective space where knowledge is constructed in a participatory, interactive, contextualized, and creative framework, made possible by free-of-charge, web-based, open-access online courses.

Overview

The term *MOOC* was coined by Dave Cormier, a professor at the University of Prince Edward Island, for a course designed and taught in 2008 by George Siemens and Stephen Downes. The course, Connectivism and Connective Knowledge, was offered for credit to a few registered fee-paying students, and free of charge with no credit to about 2,300 online students from the general public. Subsequently, MOOCs gained momentum as an increasingly popular form of education. Coursera, Udacity, and edX arose as three of the leading MOOC providers, collaborating and partnering with universities to offer course content to a broader array of students.

Coursera and Udacity, among others, used a commercial model, while edX was one of several nonprofit MOOC providers. Both models demonstrated the ability to draw in students, with Coursera attracting five million students and edX over 1.3 million by 2013. The market for MOOCs grew rapidly in 2012—called "the year of the MOOC" by the *New York Times*—and the phenomenon received prominent media coverage due to the potential disruptive impact on traditional education. By 2013, Coursera offered more than 300 courses in several languages from different disciplines (e.g., math, science, humanities, technology) instructed by professors at sixty-two universities and colleges around the world, including Stanford, Duke, and the Hong Kong University of Science and Technology.

MOOCs maintain some traditional schooling conventions, but they depart from these standards in significant ways. Characteristics such as course duration time and specified topics of interest are the same. Unlike regular courses offered by most universities, however, a MOOC is free, it provides access to open and shared content, and it requires no prerequisites other than an interest in learning, Internet access, and some digital skills. Most MOOCs do not offer formal accreditation, although some offer a certificate of course completion, generally for a fee. The most significant difference from traditional institutionalized education is the theoretical foundation for knowledge and learning. In terms of knowledge, for example, to a certain extent, MOOCs break away from the traditional practice of purchasing prepackaged knowledge from educational institutions and field experts. Instead, knowledge is facilitated by an instructor and freely constructed and negotiated among participants. Knowledge is "emergent" from the active engagement between participants and facilitators through dialogue and discussion. Learning occurs as a function of participation, engagement, and exploration in an online network that is distributed across participants, texts, blogs, articles, and websites. Some speculate that networked contributions may

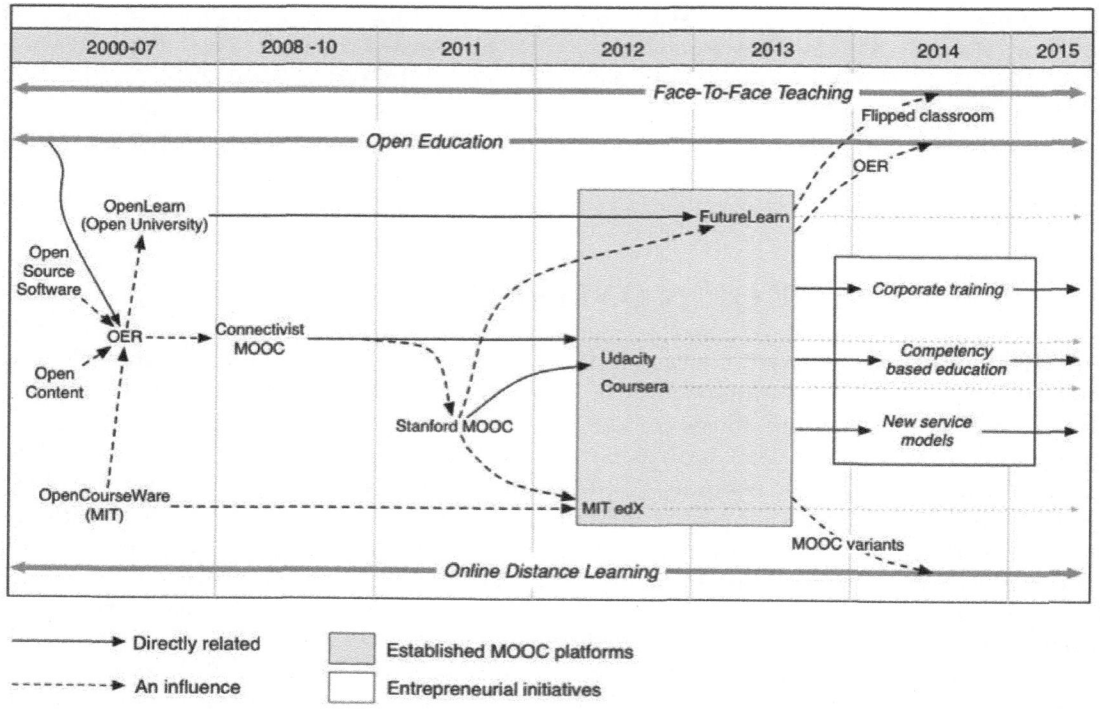

Displays a timeline of the development of moocs and open education with respect to various organisational efforts in the areas.

offer interdisciplinary information exchange that will expand to different fields, thus benefiting both learners and disciplines.

The MOOC concept can augment existing teaching and learning practices, and it has the potential to make education more attainable on a global level. It also models and reflects a commitment to lifelong learning by including learners in the knowledge-building process. However, not all learners benefit equally. Due to the massive numbers of enrolled students, the environment does not support individualized instructor attention and an individual learner's specific needs and abilities. Students with no prerequisite information or basic digital skills and those who are not intrinsically motivated may struggle in a MOOC, while others may be excluded completely given a lack of access to technology. Critics have claimed that while MOOCs aim to bring equal education to all populations, particularly internationally, studies have shown that most participants are from developed, Western countries and already have a college degree. Some observers also claim that by spreading the course content of elite universities, MOOCs many actually reinforce the homogeneity of higher education and limit cross-cultural transmission of knowledge.

MOOCs face additional challenges as well. Educators must have the time, effort, and flexibility to alter courses constructed for the traditional classroom to fit the digital MOOC format. Designers are still grappling with serious issues ranging from high dropout rates, accreditation, accessibility, quality standards, assessments, valuation, plagiarism, and copyright and legal issues. University libraries are tasked with locating alternative material that is sharable in an open public space. Critics warn that the open access ideals of MOOCs may be compromised by textbook publishers and other business interests that view the millions of students participating in MOOCs as a source of revenue for downloadable reading material, flashcards, and other learning supplements. Although MOOCs have been widely studied and debated throughout their existence, empirical research on the phenomenon is relatively scarce, and the future, while promising, is unpredictable.

—*Hitaf R. Kady, MA,
and Jennifer A. Vadeboncoeur, MA, PhD*

BIBLIOGRAPHY

"The Attack of the MOOCs." *The Economist.* Economist, 20 July 2013. Web. 29 July 2013.

Butler, Brandon. "Massive Open Online Courses: Legal and Policy Issues for Research Libraries." *Association of Research Libraries.* Assoc. of Research Libraries, 22 Oct. 2012. Web. 5 Aug. 2013.

Cooper, Steve, and Mehran Salami. "Reflections on Stanford's MOOCs." Communications of the ACM 56.2 (2013): 28-30. *Association of Computing Machinery.* Web. 5 Aug. 2013.

Daniel, John. "Making Sense of MOOCs: Musings in a Maze of Myth, Paradox and Possibility." *JIME Journal of Interactive Media in Education.* JIME, Dec. 2012. Web. 5 Aug. 2013.

Fowler, Geoffrey A. "An Early Report Card on Massive Open Online Courses." *Wall Street Journal.* Dow Jones, 8 Oct. 2013. Web. 18 Jun. 2015.

Gais, Hannah. "Is the Developing World 'MOOC'd Out'?" *Al Jazeera America.* Al Jazeera America, 17 July 2014. Web. 18 Jun. 2015.

Howard, Jennifer. "Publishers See Online Mega-Courses as Opportunity to Sell Textbooks." *Chronicle of Higher Education.* Chronicle of Higher Ed., Sept. 2010. Web. 5 Aug. 2013.

Hu, Helen. "MOOCs Changing the Way We Think about Higher Education." Diverse Issues in Higher Education. Diverse Issues in Higher Ed., April 2013. Web. 29 July 2013.

Sharples, Mike, et al., "Innovating Pedagogy 2012: Exploring New Forms of Teaching, Learning, and Assessment, to Guide Educators and Policy Makers." *The Open University.* Open U, 2012. Web. 5 Aug. 2013.

Zemsky, Robert. "With a MOOC MOOC Here and a MOOC MOOC There, Here a MOOC, There a MOOC, Everywhere a MOOC MOOC." *Jour. of General Ed.* 63.4 (2014): 237–243. Print.

META-ANALYSIS

Meta-analysis is a research method that systematically combines information from a number of research studies to draw one or more conclusions from the body of available evidence. Meta-analysis is most useful in addressing questions for which a substantial body of high-quality research already exists, because the quality of the conclusions drawn by a meta-analysis depends in large part on the quality of the studies included. Meta-analysis can be particularly useful when studies have produced inconsistent or conflicting results. As the number of published research studies on a particular topic increases, meta-analysis is an important tool to help make sense of the available evidence.

OVERVIEW

Meta-analysis is sometimes referred to as an "analysis of analyses" because it draw conclusions by analyzing the results from a body of previous analyses. The concept of using mathematical methods to systematically examine the results of an existing body of studies, and to draw conclusions from those studies, dates back at least to the work of the British statistician Ronald A. Fisher in the first half of the twentieth century. However, meta-analysis did not become a common research technique until the late 1970s, following the 1977 publication by Mary Lee Smith and Gene V. Glass of a meta-analysis of hundreds of studies examining the outcomes of counseling and psychotherapy. Meta-analysis begins with a statement of the research question. For instance, does psychological counseling or psychotherapy have beneficial effects for those who receive it? The next step is to conduct research to find studies addressing the question. Ideally, a meta-analysis will include both published and unpublished studies in order to reduce the effects of publication bias, which refers to the tendency for studies that produced significant results to

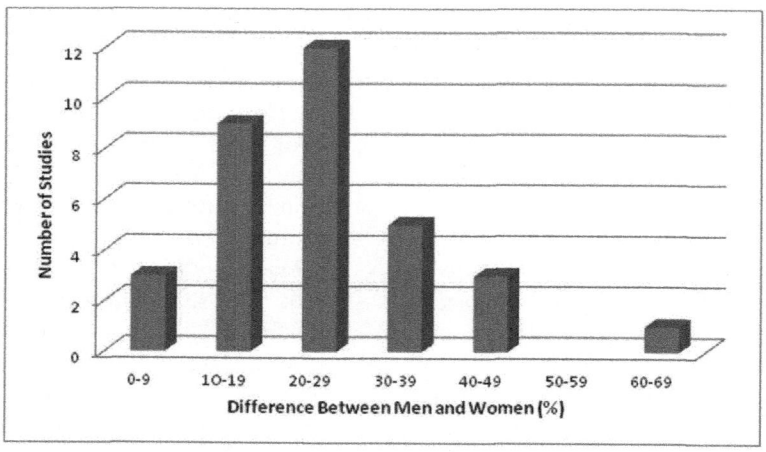

Results from a meta-analysis of sex differences in jealousy showing moderate gender differences.

be published more frequently than studies that did not produce significant results.

Once a body of studies has been selected, they must be evaluated for quality, following a systematic set of procedures. For instance, is a particular research design or sample size required for a study to be included in the meta-analysis? This step is crucial, because the results of the meta-analysis can be quite different depending on the inclusion criteria applied. For this reason, a published meta-analysis generally includes a table describing all the studies located during the initial search, along with the inclusion criteria, the number of studies excluded, and the reasons for their exclusion. Sometimes two or more meta-analyses are performed addressing the same research question, using differing inclusion criteria in order to see how much the conclusions are affected by selection criteria. Finally, the results of the studies are synthesized quantitatively. Often, this is done by expressing the results of each study in terms of an effect size, a single number that includes the outcome of the study, the sample size, and the sample variability. The effect sizes from all the studies can then be analyzed, for instance to see how many found a positive effect and the average size of that effect.

—Sarah E. Boslaugh, MPH, PhD

Bibliography

Card, Noel A. *Applied Meta-analysis for Social Science Research.* New York: Guilford, 2012. Print.

Chen, Ding-Geng, and Karl E. Peace. *Applied Meta-analysis with R.* Boca Raton: CRC, 2013. Print.

Cheung, Mike W. L. *Meta-Analysis: A Structural Equation Modeling Approach.* Malden: Wiley, 2015. Print.

Higgins, Julian P. T., and Sally Green. *Cochrane Handbook for Systematic Reviews of Interventions.* Hoboken: Wiley, 2008. Print.

Hunt, Morton. *How Science Takes Stock: The Story of Meta-analysis.* New York: Russell Sage Foundation, 1997.

Koricheva, Julia, Jessica Gurevitch, and Kerrie Mengersen, eds. *Handbook of Meta-analysis in Ecology and Evolution.* Princeton: Princeton UP, 2013. Print.

Littell, Julia H., Jacqueline Corcoran, and Vijayan Pillai. *Systematic Reviews and Meta-analysis.* New York: Oxford UP, 2008. Print.

Pigott, Terri D. *Advances in Meta-analysis.* New York: Springer, 2012. Print.

Ringquist, Evan J. *Meta-analysis for Public Management and Policy.* San Francisco: Jossey, 2013. Print.

Schmidt, Frank L., and John E. Hunter. *Methods of Meta-Analysis.* 3rd ed. Los Angeles: Sage, 2015. Print.

Smith, Mary Lee, and Gene V. Glass. "Meta-analysis of Psychotherapy Outcome Studies." *American Psychologist* 32 (1977): 752–60. Print.

Zoccai, Giuseppe Biondi. *Network Meta-Analysis: Evidence Synthesis with Mixed Treatment Comparison.* Hauppage: Nova, 2014. Print.

METACOMPUTING

FIELDS OF STUDY
Computer Science; Information Systems

ABSTRACT
Metacomputing is the use of computing to study and design solutions to complex problems. These problems can range from how best to design large-scale computer networking systems to how to determine the most efficient method for performing mathematical operations involving very large numbers. In essence, metacomputing is computing about computing. Metacomputing makes it possible to perform operations that individual computers, and even some supercomputers, could not handle alone.

PRINCIPAL TERMS

- **domain-dependent complexity:** a complexity that results from factors specific to the context in which the computational problem is set.
- **meta-complexity:** a complexity that arises when the computational analysis of a problem is compounded by the complex nature of the problem itself.
- **middle computing:** computing that occurs at the application tier and involves intensive processing of data that will subsequently be presented to the user or another, intervening application.
- **networking:** the use of physical or wireless connections to link together different computers and computer networks so that they can communicate with one another and collaborate on computationally intensive tasks.
- **supercomputer:** an extremely powerful computer that far outpaces conventional desktop computers.
- **ubiquitous computing:** an approach to computing in which computing activity is not isolated in a desktop, laptop, or server, but can occur everywhere and at any time through the use of microprocessors embedded in everyday devices.

REASONS FOR METACOMPUTING
The field of metacomputing arose during the 1980s. Researchers began to realize that the rapid growth in networked computer systems would soon make it difficult to take advantage of all interconnected computing resources. This could lead to wasted resources unless an additional layer of computing power were developed. This layer would not work on a computational problem itself; instead, it would determine the most efficient method of addressing the problem. In other words, researchers saw the potential for using computers in a manner so complicated that only a computer could manage it. This new metacomputing layer would rest atop the middle computing layer that works on research tasks. It would ensure that the middle layer makes the best use of its resources and that it approaches calculations as efficiently as possible.

One reason an application may need a metacomputing layer is the presence of complexities. Complexities are elements of a computational problem that make it more difficult to solve. Domain-dependent complexities arise due to the context of the computation. For example, when calculating the force and direction necessary for an arrow to strike its target, the effects of wind speed and direction would be a domain-dependent complexity. Meta-complexities are those that arise due to the nature of the computing problem rather than its context. An example of a meta-complexity is a function that has more than one possible solution.

METACOMPUTING AND THE INTERNET
Metacomputing is frequently used to make it possible to solve complex calculations by networking between many different computers. The networked computers can combine their resources so that each one works on part of the problem. In this way, they become a virtual supercomputer with greater capabilities than any individual machine. One successful example of this is a project carried out by biochemists studying the way proteins fold and attach to one another. This subject is usually studied using computer programs that model the proteins' behavior. However, these programs consume a lot of time and computing power. Metacomputing allowed the scientists to create a game that users all over the world can play that generates data about protein folding at the same time. Users try to fit shapes together in different ways, contributing their time and computing power to the project in a fun and easy way.

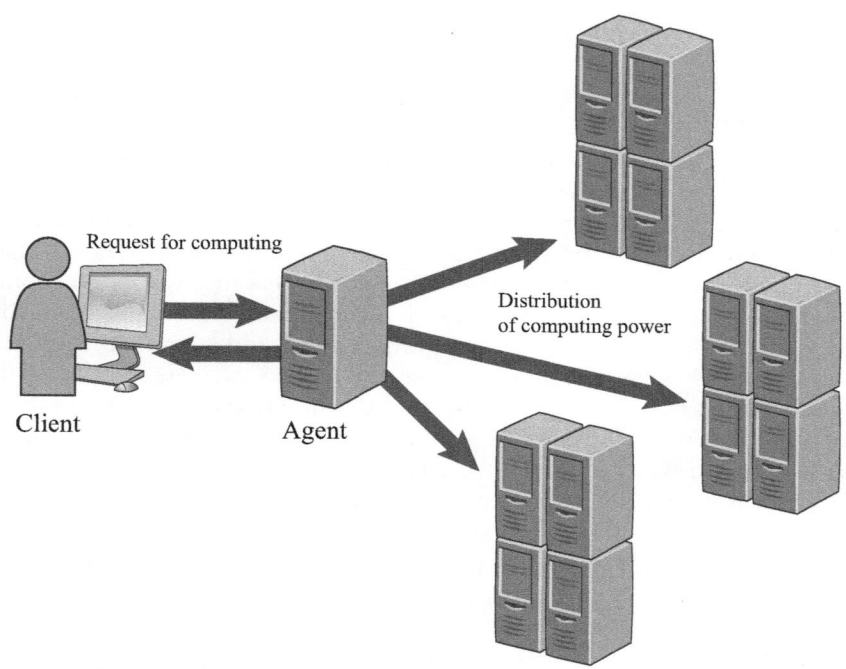

Metacomputing is a system design concept that allows multiple computers to share the responsibility of computation to complete a request efficiently.

Ubiquitous Metacomputing

Another trend with great potential for metacomputing is ubiquitous computing, meaning computing that is everywhere and in everything. As more and more mundane devices are equipped with Internet-connected microprocessors, from coffee makers to cars to clothing, there is the potential to harness this computing power and data. For example, a person might use multiple devices to monitor different aspects of their health, such as activity level, water intake, and calories burned. Metacomputing could correlate all of this independently collected information and analyze it. This data could then be used to diagnose potential diseases, recommend lifestyle changes, and so forth.

One form of ubiquitous metacomputing that already exists is the way that various smartphone applications use location data to describe and predict traffic patterns. One person's data cannot reveal much about traffic. However, when many people's location, speed, and direction are reported simultaneously, the data can be used to predict how long one person's commute will be on a given morning.

Mimicking the Brain

Metacomputing is often used when there is a need for a computer system that can "learn." A system that can learn is one that can analyze its own performance to make adjustments to its processes and even its architecture. These systems bear a strong resemblance to the operation of the human brain. In some cases they are intentionally designed to imitate the way the brain approaches problems and learns from its past performance. Metacomputing, in this sense, is not too dissimilar from metacognition.

Metacomputing sometimes conjures up fears of the dangers posed by artificial intelligence in science fiction. In reality, metacomputing is just another category of computer problem to be solved, not the beginning of world domination by machines. Humans can conceive of computer problems so complex that it is nearly impossible to solve them without the aid of another computer. Metacomputing is simply the solution to this dilemma.

—*Scott Zimmer, JD*

Bibliography

Loo, Alfred Waising, ed. *Distributed Computing Innovations for Business, Engineering, and Science.* Hershey: Information Science Reference, 2013. Print.

Mallick, Pradeep Kumar, ed. *Research Advances in the Integration of Big Data and Smart Computing.* Hershey: Information Science Reference, 2016. Print.

Mason, Paul. *Understanding Computer Search and Research.* Chicago: Heinemann, 2015. Print.

Nayeem, Sk. Md. Abu, Jyotirmoy Mukhopadhyay, and S. B. Rao, eds. *Mathematics and Computing: Current Research and Developments.* New Delhi: Narosa, 2013. Print.

Segall, Richard S., Jeffrey S. Cook, and Qingyu Zhang, eds. *Research and Applications in Global Supercomputing.* Hershey: Information Science Reference, 2015. Print.

Tripathy, B. K., and D. P. Acharjya, eds. *Global Trends in Intelligent Computing Research and Development.* Hershey: Information Science Reference, 2014. Print.

METADATA

Metadata is descriptive, structural, or administrative information included as part of a digital file, electronic record, or other resource. It often includes information such as the file or resource's title, creator, and structure. Metadata is used to catalog and preserve information. By including keywords and descriptions within the metadata, creators or archivists enable users to search for specific records as well as discover similar ones. Metadata is related to the fields of both computer science and library science, as it combines traditional cataloging methods with new electronic file types and distribution methods.

OVERVIEW

Often described as data about data, metadata consists of information either embedded in a digital file or stored as a separate record in a database or similar directory. The concept is closely related to the cataloging systems used for centuries in libraries and archives, which typically consist of collections of listings—either physical or, following the advent of computers, digital—that include the title, the creator, and a description of each work or artifact, among other key facts. Metadata is intended to assist people in searching for and retrieving the files or information they seek. Similarly, this data enables archivists and catalogers to ensure that information will be well organized and preserved for future reference.

Metadata is an evolving concept, and the term has consequently been used in various ways by different organizations and in different fields. Considered broadly, metadata can be divided into three areas based on the type of information being preserved. Information such as the title and author of an e-book falls under the classification of descriptive metadata, while data regarding the structure of the e-book—the order of the pages and chapters—is considered structural metadata. Administrative metadata is somewhat more loosely defined but generally includes information regarding when the file was created. This category may also include the subcategories of technical metadata, which concerns file formats and sizes; preservation metadata, which can include information about the file's relationships to other files; and intellectual property rights metadata, which is used to capture copyright information. In other cases, technical, preservation, and intellectual property rights metadata are considered distinct categories.

When an item exists in only physical form, as in the case of physical artifacts and texts that have not been digitized, the metadata for that item is typically stored in a searchable database. When dealing with electronic files and other resources, however, creators and catalogers often embed metadata within the files. In hypertext markup language (HTML) files such websites, metadata is typically embedded in the files through the use of specialized HTML tags. Image, audio, and video files can likewise include embedded metadata, usually added by the creator or, in the case of commercially distributed films or music, the studio or distributor. Some cameras, including many smartphone cameras, automatically embed metadata such as the date, time, and place a photograph was taken in each file they produce. This capability has led to concerns that sensitive personal information could be distributed in the form of metadata without an individual's knowledge or permission, and the use of metadata thus remains a topic of debate among individuals and institutions concerned about privacy.

In 2013 National Security Agency (NSA) contractor Edward Snowden revealed that the agency was collecting, analyzing, and storing bulk telephone and online metadata from US citizens and others under the auspices of the USA PATRIOT Act (2001). Though the agency argued that the practice was necessary to detect and fight terrorism, the US

183

Court of Appeals ruled that such bulk data collection and retention was illegal in May 2015. In June 2015 President Barack Obama signed the USA Freedom Act, which put new restrictions on the NSA surveillance of phone metadata.

—*Joy Crelin*

BIBLIOGRAPHY

Baca, Murtha, ed. *Introduction to Metadata*. 2nd ed. Los Angeles: Getty Research Inst., 2008. Print.

Cox, Mike, Ellen Mulder, and Linda Tadic. *Descriptive Metadata for Television*. Burlington: Focal, 2006. Print.

García-Barriocanal, Elena, et al., eds. *Metadata and Semantic Research*. New York: Springer, 2011. Print.

Hider, Philip. *Information Resource Description: Creating and Managing Metadata*. Chicago: Amer. Lib. Assoc., 2012. Print.

Hillmann, Diane I., and Elaine L. Westbrooks. *Metadata in Practice*. Chicago: Amer. Lib. Assn., 2004. Print.

National Information Standards Organization. *Understanding Metadata*. Bethesda: NISO P, 2004. Digital file.

Park, Jung-ran. *Metadata Best Practices and Guidelines*. London: Routledge, 2011. Print.

Roberts, Dan, and Spencer Ackerman. "NSA Mass Phone Surveillance Revealed by Edward Snowden Ruled Illegal." *Guardian*. Guardian News and Media, 7 May 2015. Web. 29 June 2015.

Smiraglia, Richard P., ed. *Metadata: A Cataloger's Primer*. London: Routledge, 2012. Print.

Steinhauer, Jennifer, and Jonathan Weisman. "U.S. Surveillance in Place since 9/11 Is Sharply Limited." *New York Times*. New York Times, 2 June 2015. Web. 29 June 2015.

MICROPROCESSORS

FIELDS OF STUDY

Computer Engineering; System-Level Programming

ABSTRACT

Microprocessors are part of the hardware of a computer. They consist of electronic circuitry that stores instructions for the basic operation of the computer and processes data from applications and programs. Microprocessor technology debuted in the 1970s and has advanced rapidly into the 2010s. Most modern microprocessors use multiple processing "cores." These cores divide processing tasks between them, which allows the computer to handle multiple tasks at a time.

PRINCIPAL TERMS

- **central processing unit (CPU):** electronic circuitry that provides instructions for how a computer handles processes and manages data from applications and programs.
- **clock speed:** the speed at which a microprocessor can execute instructions; also called "clock rate."
- **data width:** a measure of the amount of data that can be transmitted at one time through the computer bus, the specific circuits and wires that carry data from one part of a computer to another.
- **micron:** a unit of measurement equaling one millionth of a meter; typically used to measure the width of a core in an optical figure or the line width on a microchip.
- **million instructions per second (MIPS):** a unit of measurement used to evaluate computer performance or the cost of computing resources.
- **transistor:** a computing component generally made of silicon that can amplify electronic signals or work as a switch to direct electronic signals within a computer system.

BASICS OF MICROPROCESSING

Microprocessors are computer chips that contain instructions and circuitry needed to power all the basic functions of a computer. Most modern microprocessors consist of a single integrated circuit, which is a set of conducting materials (usually silicon) arranged on a plate. Microprocessors are designed to

Microprocessors contain all the components of a CPU on a single chip; this allows new devices to have higher computing power in a smaller unit.

receive electronic signals and to perform processes on incoming data according to instructions programmed into the central processing unit (CPU) and contained in computer memory. They then to produce output that can direct other computing functions. In the 2010s, microprocessors are the standard for all computing, from handheld devices to supercomputers. Among the modern advancements has been development of integrated circuits with more than one "core." A core is the circuitry responsible for calculations and moving data. As of 2016, microprocessors may have as many as eighteen cores. The technology for adding cores and for integrating data shared by cores is a key area of development in microprocessor engineering.

Microprocessor History and Capacity

Before the 1970s, and the invention of the first microprocessor, computer processing was handled by a set of individual computer chips and transistors. Transistors are electronic components that either amplify or help to direct electronic signals. The first microprocessor for the home computer market was the Intel 8080, an 8-bit microprocessor introduced in 1974. The number of bits refers to the storage size of each unit of the computer's memory. From the 1970s to the 2010s, microprocessors have followed the same basic design and concept but have increased processing speed and capability. The standard for computing in the 1990s and 2000s was the 32-bit microprocessor. The first 64-bit processors were introduced in the 1990s. They have been slow to spread, however, because most basic computing functions do not require 64-bit processing.

Computer performance can be measured in million instructions per second (MIPS). The MIPS measurement has been largely replaced by measurements using floating-point operations per second (FLOPS) or millions of FLOPS (MFLOPS). Floating-point operations are specific operations, such as performing a complete basic calculation. A processor with a 1 gigaFLOP (GFLOP) performance rating can perform one billion FLOPS each second. Most modern microprocessors can perform 10 GFLOPS per second. Specialized computers can perform in the quadrillions of operations per second (petaFLOPS) scale.

Characteristics of Microprocessors

The small components within modern microprocessors are often measured in microns or micrometers, a unit equaling one-millionth of a meter. Microprocessors are usually measured in line width, which measures the width of individual circuits. The earliest microprocessor, the 1971 Intel 4004, had a minimum line width of 10 microns. Modern microprocessors can have line width measurements as low as 0.022 microns.

All microprocessors are created with a basic instruction set. This defines the various instructions that can be processed within the unit. The Intel 4004 chip, which was installed in a basic calculator, provided instructions for basic addition and subtraction. Modern microprocessors can handle a wide variety of calculations.

Different brands and models of microprocessors differ in bandwidth, which measures how many bits of data a processor can handle per second. Microprocessors also differ in data width, which measures the amount of data that can be transferred between two or more components per second. A computer's bus describes the parts that link the processor to other parts and to the computer's main memory. The size of a computer's bus is known as the width. It determines how much data can be transferred each second. A computer's bus has a clock speed, measured in megahertz (MHz) or gigahertz (GHz). All other factors being equal, computers with larger data width and a faster clock speed can transfer data faster and thus run faster when completing basic processes.

MICROPROCESSOR DEVELOPMENT

Intel cofounder Gordon Moore noted that the capacity of computing hardware has doubled every two years since the 1970s, an observation now known as Moore's law. Microprocessor advancement is complicated by several factors, however. These factors include the rising cost of producing microprocessors and the fact that the ability to reduce power needs has not grown at the same pace as processor capacity. Therefore, unless engineers can reduce power usage, there is a limit to the size and processing speed of microprocessor technology. Data centers in the United States, for instance, used about 91 billion kilowatt-hours of electricity in 2013. This is equal to the amount of electricity generated each year by thirty-four large coal-burning power plants. Computer engineers are exploring ways to address these issues, including alternatives for silicon in the form of carbon nanotubes, bioinformatics, and quantum computing processors.

—*Micah L. Issitt*

BIBLIOGRAPHY

Ambinder, Marc. "What's Really Limiting Advances in Computer Tech." *Week.* The Week, 2 Sept. 2014. Web. 4 Mar. 2016.

Borkar, Shekhar, and Andrew A. Chien. "The Future of Microprocessors." *Communications of the ACM.* ACM, May 2011. Web. 3 Mar. 2016.

Delforge, Pierre. "America's Data Centers Consuming and Wasting Growing Amounts of Energy." *NRDC.* Natural Resources Defense Council, 6 Feb. 2015. Web. 17 Mar. 2016.

McMillan, Robert. "IBM Bets $3B That the Silicon Microchip Is Becoming Obsolete." *Wired.* Condé Nast, 9 July 2014. Web. 10 Mar. 2016.

"Microprocessors: Explore the Curriculum." *Intel.* Intel Corp., 2015. Web. 11 Mar. 2016.

"Microprocessors." *MIT Technology Review.* MIT Technology Review, 2016. Web. 11 Mar. 2016.

Wood, Lamont. "The 8080 Chip at 40: What's Next for the Mighty Microprocessor?" *Computerworld.* Computerworld, 8 Jan. 2015. Web. 12 Mar. 2016.

MIXED METHODS RESEARCH (MMR)

Mixed methods research (MMR), sometimes called multimethodology is an approach to a research problem that leverages the advantages of both qualitative and quantitative research methods to better understand the subject than any individual approach could offer on its own. It recognizes that qualitative and quantitative data can each contribute to different aspects of understanding of a problem, and it integrates both approaches within a holistic methodological approach.

If a problem is viewed from only one perspective, only a single facet of the situation will be seen, which may be misleading. MMR allows research problems to be viewed from multiple perspectives, giving a clearer picture of the issue.

OVERVIEW

There is often a dividing line between quantitative research and qualitative research. It is often assumed that quantitative research draws on positivistic ontologies of research, whereas by comparison qualitative research draws more upon interpretative, critical, and constructivist research traditions. Within the field of research, there is a broad spectrum of opinion

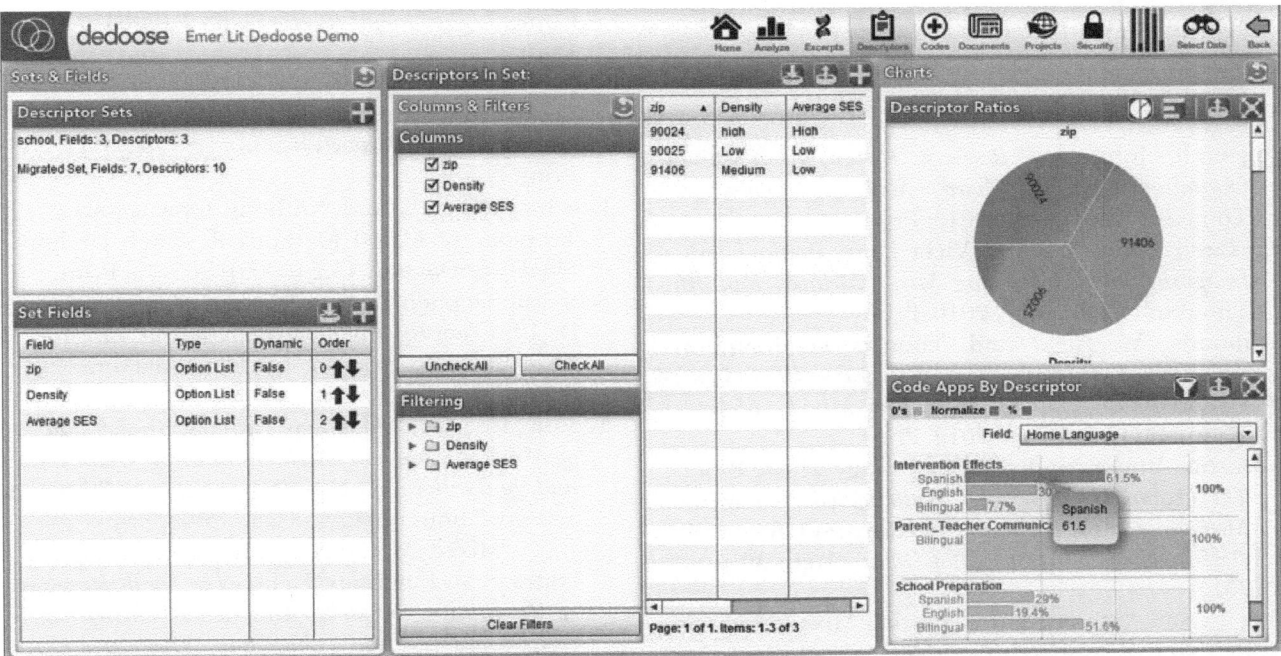

Integration of quantitative data for qualitative analysis and mixed methods research in a web-based Computer-aided Qualitative Data Analysis Software (CAQDAS).

and debate as far as these approaches to research are concerned. Some see them as diametrically opposed, whereas others see that with a pragmatic approach, different techniques can work together. The division between quantitative and qualitative research led Charles Teddlie and Abbas Tashakkori to characterize MMR as "the third methodological movement."

Using data from multiple methodological approaches is referred to as triangulation. Triangulation involves using more than two methods in order to cross check results; the idea is that if different research methodologies produce broadly similar results, then the strength of the research is reinforced. If the results from different research methods produce conflicting results, then the researcher has the opportunity to revise how the methods used may have impacted the results or determine whether or not the problem needs to be reframed. Triangulation can also involve using multiple researchers, multiple theoretical approaches to a problem, and different empirical data for the same problem. While triangulation focuses on the convergence of results, equally interesting are cases where different methods produce divergent or dissimilar results. This in itself can yield interesting insights into the issue being examined.

While MMR aims to compensate for the weaknesses of using a single method, the technique poses some challenges. Often quantitative and qualitative researchers have advanced challenging, opposed, and contrary epistemologies allied to their research approaches. Proponents of MMR have countered this with notions of methodological eclecticism framed within a research paradigm of pragmatism. Rather than adopting fixed ontological viewpoints, proponents of MMR argue for paradigm pluralism; that is, choosing research paradigms that are appropriate for different aspects of a problem.

The practical challenges of MMR include requiring the researcher to have a broad skill set in order to be able to apply a range of different methods competently or to work as part of an organized research team that can deploy a range of methods in search of common answers.

—*Gavin D. J. Harper, MSc, MIET*

Bibliography

Bergman, Manfred Max. *Advances in Mixed Methods Research: Theories and Applications.* Thousand Oaks: Sage, 2008, Print.

Cresswell, John, W. *Research Design: Qualitative, Quantitative, and Mixed Methods Approaches.* Thousand Oaks: Sage, 2013. Print

Cresswell, John, W., and Vicki Lynn Plano Clark. *Designing and Conducting Mixed Methods Research* Thousand Oaks: Sage, 2010. Print

Edmonds, W. Alex, and Thomas D. Kennedy. *An Applied Reference Guide to Research Designs: Quantitative, Qualitative, and Mixed Methods* Thousand Oaks: Sage, 2012. Print.

Johnson, Burke, and Sharlene Nagy Hesse-Biber. *The Oxford Handbook Of Multimethod And Mixed Methods Research Inquiry.* Oxford: Oxford UP, 2015. *eBook Collection (EBSCOhost).* Web. 30 June 2015.

———, and Anthony Onwuegbuzie. "Mixed Methods Research: A Research Paradigm Whose Time Has Come." *Educational Researcher* 33.7 (2004): 14–26. Print.

———, Anthony Onwuegbuzie, and Lisa A. Turner. "Toward a Definition of Mixed Methods Research." *Journal of Mixed Methods Research* 1.2 (2007): 112–33. Print.

Leech, Nancy L., and Anthony Onwuegbuzie. "A Typology of Mixed Methods Research Designs." *Quality & Quantity* 43.2 (2009): 265–75. Print.

Schensul, Stephen L., Jean J. Schensul, and Margaret D. Le Compte. *Initiating Ethnographic Research: A Mixed Methods Approach.* Lanham: AltaMira, 2012. Print.

Teddlie, Charles, and Abbas Tashakkori. *Foundations of Mixed Methods Research: Integrating Quantitative and Qualitative Approaches in the Social and Behavioral Sciences.* Thousand Oaks: Sage, 2008, Print.

———. "Mixed Methods Research: Contemporary Issues in an Emerging Field." *The SAGE Handbook of Qualitative Research.* 4th ed. Eds. Norman K. Denzin and Yvonna S. Norman. Thousand Oaks: Sage, 2011. 285–300. Print.

MOBILE APPS

FIELDS OF STUDY

Applications; Mobile Platforms

ABSTRACT

Mobile apps are programs designed to run on smartphones, tablets, and other mobile devices. These apps usually perform a specific task, such as reporting the weather forecast or displaying maps for navigation. Mobile devices have special requirements because of their small screens and limited input options. Furthermore, a touch screen is typically the only way to enter information into a mobile device. Programmers need special knowledge to understand the mobile platform for which they wish to create apps.

PRINCIPAL TERMS

- **applications:** programs that perform specific functions that are not essential to the operation of the computer or mobile device; often called "apps."
- **emulators:** programs that mimic the functionality of a mobile device and are used to test apps under development.
- **mobile website:** a website that has been optimized for use on mobile devices, typically featuring a simplified interface designed for touch screens.
- **platform:** the hardware and system software of a mobile device on which apps run.
- **system software:** the operating system that allows programs on a mobile device to function.
- **utility programs:** apps that perform basic functions on a computer or mobile device such as displaying the time or checking for available network connections.

TYPES OF MOBILE SOFTWARE

Mobile applications, or apps, are computer programs designed specifically to run on smartphones, tablets, and other mobile devices. Apps must be designed for a specific platform. A mobile platform is the hardware and system software on which a mobile device

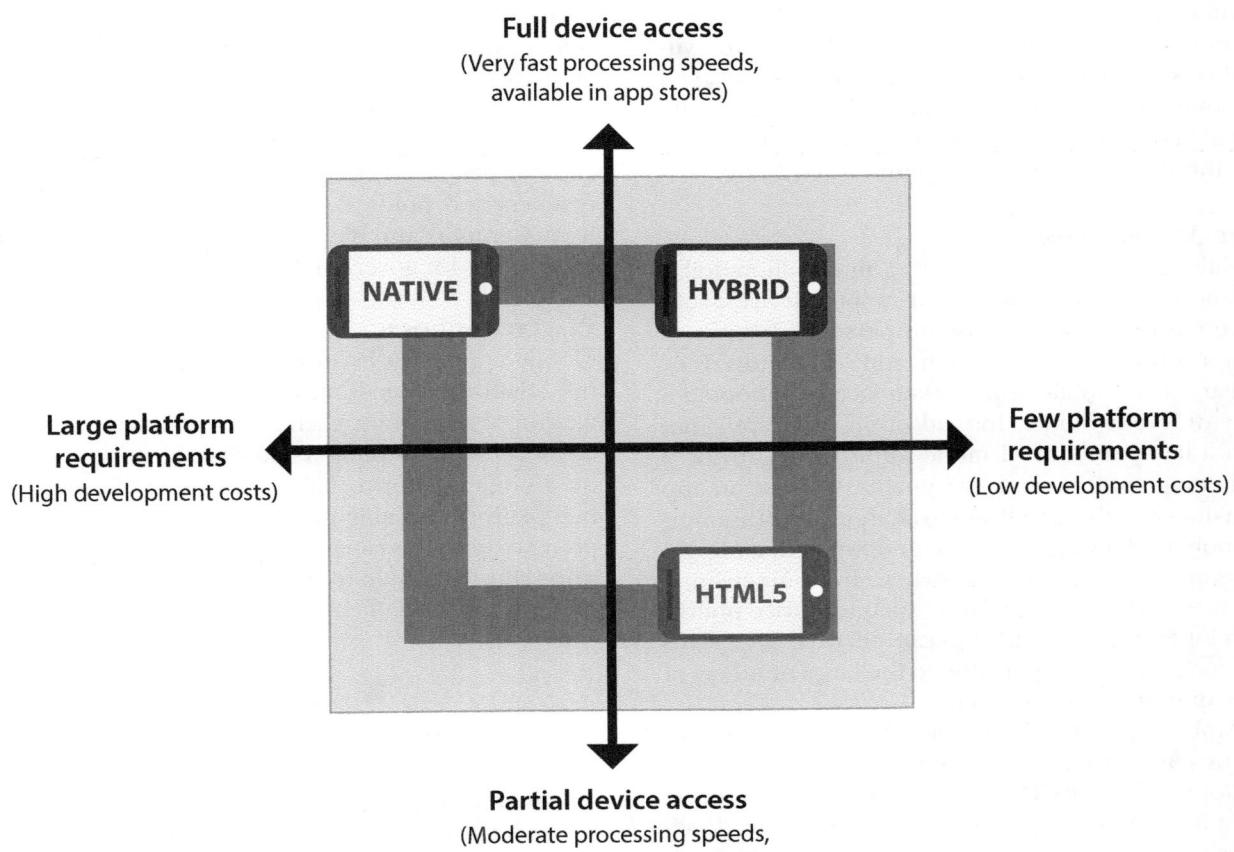

Native mobile apps and hybrid apps offer some features and capabilities unavailable with standard web programming, and they offer faster connectivity than is available with traditional computer software.

operates. Some of the most widely used mobile platforms include Google's Android, Apple's iOS, and Microsoft's Windows. Mobile devices support a variety of software. At the most basic level, a platform's system software includes its operating system (OS). The OS supports all other programs that run on that device, including apps. In addition to the OS, smartphones and tablets come with a variety of preinstalled utility programs, or utilities, that manage basic functions. Examples of utilities include a clock and a calendar, photo storage, security programs, and clipboard managers. While utilities are not essential to the functionality of the OS, they perform key tasks that support other programs.

However, the real power of mobile devices has come from the huge assortment of mobile apps they can run. The app stores for various mobile devices contain hundreds of thousands of different apps to download. Each app has been designed with different user needs in mind. Many are games of one sort or another. There are also vast numbers of apps for every pursuit imaginable, including video chat, navigation, social networking, file sharing and storage, and banking.

Developers of mobile apps use various approaches to design their software. In some cases, an app is little more than a mobile website that has been optimized for use with small touch screens. For example, the Facebook app provides essentially the same functionality as its website, although it can be integrated with the device's photo storage for easier uploading. Other mobile apps are developed specifically for the mobile devices they run on. Programmers must program their apps for a specific mobile platform. App developers usually use a special software development kit and an emulator for testing the app on a virtual version of a mobile device. Emulators provide a way of easily testing mobile apps. Emulators generate detailed output as the app runs, so the developer can use this data to diagnose problems with the app.

The App Economy

Mobile apps have evolved into a multibillion-dollar business. Before the advent of mobile devices, software was developed for use on personal computers and a software package often sold for hundreds of dollars. The mobile app marketplace has adopted a very different model. Instead of creating apps that cost a large amount of money and try to provide a wide range of functions, the goal is to create an app that does one thing well and to charge a small amount of money. Many apps are free to download, and paid apps are typically priced anywhere from ninety-nine cents to a few dollars. Despite such low price points, developers of successful apps can still earn large sums of money. This is in part due to the large numbers of smartphone and tablet users.

Mobile apps also have a low cost of distribution. In the past, software was sold on physical media such as floppy disks or CD-ROMs. These had to be packaged and shipped to retailers at the software developer's expense. With mobile apps, there are no such overhead costs because apps are stored online by the platform's app store and then downloaded by users. Aside from the time and effort of developing an app, the only other financial cost to the developer is an annual registration fee to the platform's app store and a percentage of revenue claimed by the platform. App stores typically vet apps to make sure they do not contain malware, violate intellectual property, or make false advertising claims. App developers can earn revenue from the download fee, in-app advertisements, and in-app purchases. There have been several developers who have produced an unexpectedly popular app, catapulting them to fame and wealth.

Mobile Apps and Social Change

Mobile apps can do much more than entertain and distract. For example, Twitter is a microblogging app that allows users to post short messages and read updates from others. Twitter has played a significant role in social movements such as the Arab Spring of 2011. Because it relies on telecommunications technology that continues to function even when there are disruptions in other media, the Twitter app has allowed people to communicate even during major disasters and political upheavals. Other apps allow users to report and pinpoint environmental damage or potholes for government agencies to fix.

Platform Lock

Mobile apps must be designed for a particular mobile platform. Sometimes, a developer will make a version of an app for each major platform. In other cases, however, developers only create an app to run on a single platform. This leaves users of other mobile platforms unable to use the app. In the case of popular apps, this can cause frustration among those who want to be able to use whatever apps they want on their platform of choice.

—*Scott Zimmer, JD*

Bibliography

Banga, Cameron, and Josh Weinhold. *Essential Mobile Interaction Design: Perfecting Interface Design in Mobile Apps.* Upper Saddle River: Addison-Wesley, 2014. Print.

Glaser, J. D. *Secure Development for Mobile Apps: How to Design and Code Secure Mobile Applications with PHP and JavaScript.* Boca Raton: CRC, 2015. Print.

Iversen, Jakob, and Michael Eierman. *Learning Mobile App Development: A Hands-On Guide to Building Apps with iOS and Android.* Upper Saddle River: Addison-Wesley, 2014. Print.

Miller, Charles, and Aaron Doering. *The New Landscape of Mobile Learning: Redesigning Education in an App-Based World.* New York: Routledge, 2014. Print.

Salz, Peggy Anne, and Jennifer Moranz. *The Everything Guide to Mobile Apps: A Practical Guide to Affordable Mobile App Development for Your Business.* Avon: Adams Media, 2013. Print.

Takahashi, Dean. "The App Economy Could Double to $101 Billion by 2020." *VB.* Venture Beat, 10 Feb. 2016. Web. 11 Mar. 2016.

MOBILE TECHNOLOGY

Mobile technology is the technology that powers mobile devices such as smartphones and tablets. The category of mobile technology encompasses both the hardware elements of mobile devices, such as processors and memory cards, and the operating systems and other software that allow these devices to function. The various networks that enable mobile devices to place calls, send text messages, and connect to the Internet can also be considered part of this technology. Mobile technology has evolved significantly over the years and is expected to continue to develop as numerous companies and individuals seek new ways of using it.

Overview

Mobile technology is a constantly growing field that is responsible for putting numerous technological innovations in the hands of the public. While making a phone call, browsing the Internet, or playing a video game once required an individual to remain in one place indoors, mobile technology has enabled legions of consumers to carry out such activities while out of the house. In addition to promoting communication and media consumption, mobile technology has allowed consumers to perform everyday tasks such as shopping, banking, searching for jobs, or booking transportation or accommodations from anywhere wireless or mobile Internet service is available.

The category of mobile technology is quite broad and includes many different types of devices, as well as their hardware components and the software and networks they require to function. Cellular phones are the most common form of mobile device and can

A smartphone is a tool of mobile computing, capable of web browsing, e-mail access, video playback, document editing, file transfer, image editing, and many other tasks.

A Kyoto taxi cab equipped with GPS navigation system.

be divided into several categories. Traditional cell phones function as telephones and may also have text-messaging capabilities, while feature phones offer a wider range of functions, often allowing users to play games, listen to music, and send e-mail. Smartphones allow for extensive Internet use and also give users the ability to download applications, or apps, devoted to specific functions. Tablets are often considered a form of mobile technology as well, as they are portable and allow for a degree of mobile computing beyond that of cell phones. E-book readers, which allow users to download and read digital texts via mobile technology, are a subset of tablet computers.

Numerous components form the internal and external workings of mobile devices, including high-resolution displays, small but powerful cameras, miniature microphones and speakers, and protective outer casings. Also inside mobile devices are processors, memory cards, wireless connectivity devices, and other components specifically optimized for mobile use.

The operating systems used by mobile devices are a key aspect of mobile technology. These systems include Android, Windows Phone, and iOS, the latter of which is only available on devices produced by Apple. Many mobile devices allow users to download apps for a wide range of purposes, including gaming, social networking, shopping, and entertainment.

The networks that mobile devices use to make telephone calls, send text messages, download apps, check e-mail, and browse the Internet are a crucial element of mobile technology. Although most mobile devices can connect to standard wireless Internet networks, such networks are not always available, particularly when users are on the go. However, high-speed mobile broadband communication networks are available in many locations, allowing people whose mobile devices are equipped with the latest telecommunications standards to use their full range of functions regardless of their location.

—*Joy Crelin*

BIBLIOGRAPHY

Gleason, Ann Whitney. *Mobile Technologies for Every Library*. Lanham: Rowman, 2015. *eBook Collection (EBSCOhost)*. Web. 1 July 2015.

Goggin, Gerard. *Cell Phone Culture: Mobile Technology in Everyday Life*. New York: Routledge, 2006. Print.

Grant, August E., and Jennifer H. Meadows, eds. *Communication Technology Update and Fundamentals*. 13th ed. Waltham: Focal, 2012. Print.

Pérez, André. *Mobile Networks Architecture*. Hoboken: Wiley, 2012. Print.

Peters, Thomas A., and Lori Bell, eds. *The Handheld Library: Mobile Technology and the Librarian.* Santa Barbara: ABC-CLIO, 2013. Print.

Santos, Raul Aquino, and Arthur Edwards Block, eds. *Embedded Systems and Wireless Technology.* Boca Raton: CRC, 2012. Print.

Sauter, Martin. *From GSM to LTE: An Introduction to Mobile Networks and Mobile Broadbands.* Hoboken: Wiley, 2011. Print.

Wilken, Rowan, and Gerard Goggin, eds. *Mobile Technology and Place.* New York: Routledge, 2012. Print.

Zelkowitz, Marvin V., ed. *The Internet and Mobile Technology.* Burlington: Academic, 2011. Print.

MOTHERBOARDS

FIELDS OF STUDY

Computer Engineering; Information Technology

ABSTRACT

The motherboard is the main printed circuit board inside a computer. It has two main functions: to support other computer components, such as random access memory (RAM), video cards, sound cards, and other devices; and to allow these devices to communicate with other parts of the computer by using the circuits etched into the motherboard, which are linked to the slots holding the various components.

PRINCIPAL TERMS

- **core voltage:** the amount of power delivered to the processing unit of a computer from the power supply.
- **crosstalk:** interference of the signals on one circuit with the signals on another, caused by the two circuits being too close together.
- **printed circuit board:** a flat copper sheet shielded by fiberglass insulation in which numerous lines have been etched and holes have been punched, allowing various electronic components to be connected and to communicate with one another and with external components via the exposed copper traces.
- **trace impedance:** a measure of the inherent resistance to electrical signals passing through the traces etched on a circuit board.
- **tuning:** the process of making minute adjustments to a computer's settings in order to improve its performance.

EVOLUTION OF MOTHERBOARDS

The motherboard of a computer is a multilayered printed circuit board (PCB) that supports all of the computer's other components, which are secondary to its functions. In other words, it is like the "mother" of other, lesser circuit boards. It is connected, either directly or indirectly, to every other part of the computer.

In the early days of computers, motherboards consisted of several PCBs connected either by wires or by being plugged into a backplane (a set of interconnected sockets mounted on a frame). Each necessary computer component, such as the central processing unit (CPU) and the system memory, required one or more PCBs to house its various parts. With the advent and refinement of microprocessors, computer components rapidly shrank in size. While a CPU in the late 1960s consisted of numerous integrated circuit (IC) chips attached to PCBs, by 1971 Intel had produced a CPU that fit on a single chip. Other essential and peripheral components could also be housed in a single chip each. As a result, the motherboard could support a greater number of components, even as it too was reduced in size. Sockets were added to support more peripheral functions, such as mouse, keyboard, and audio support.

In addition to being more cost effective, this consolidation of functions helped make computers run faster. Sending information from point to point on a computer takes time. It is much faster to send information directly from the motherboard to a peripheral device than it is to send it from the CPU PCB across the backplane to the memory PCB, and then from there to the device.

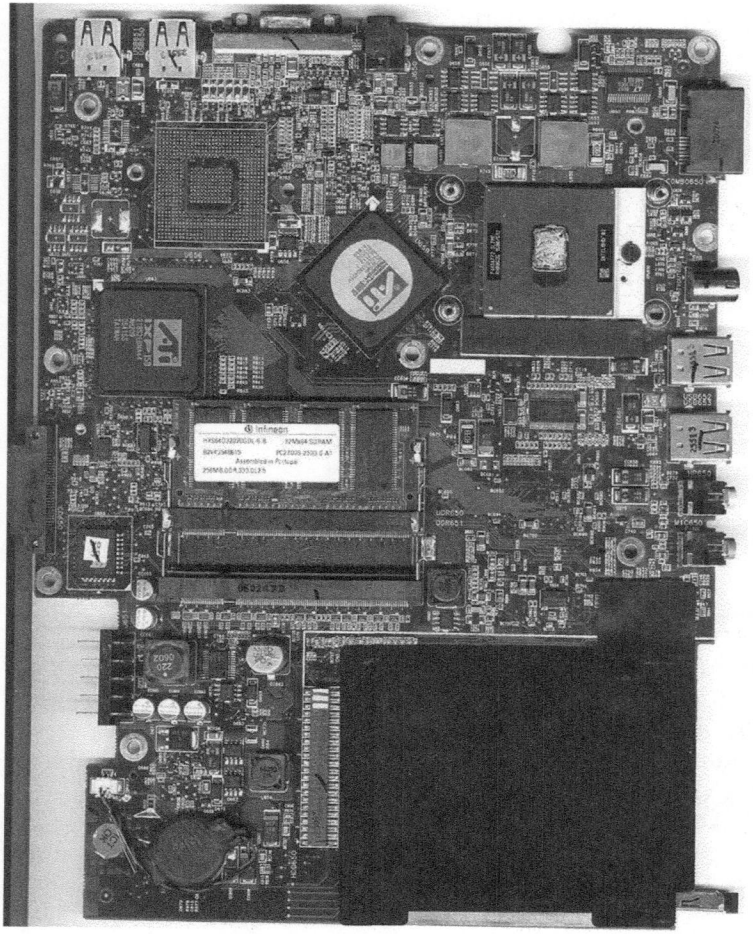

A motherboard is the main printed circuit board (PCB) of a computer. Also known as a logic board or mainboard, it connects the CPU to memory and to peripherals. Often it includes hard drives, sound cards, network cards, video cards, and other components.

Motherboard Design

Designing a motherboard is quite challenging. The main issues arise from the presence of a large number of very small circuits in a relatively small area. One of the first considerations is how best to arrange components on the motherboard's various layers. A typical PCB consists of sheets of copper separated by sheets of fiberglass. The fiberglass insulates the copper layers from each other. Most motherboards consist of six to twelve layers, though more or fewer layers are also possible. Certain layers typically have specific functions. The outer layers are signal layers, while other layers carry voltage, ground returns, or carry memory, processor, and input/output data.

Lines etched in the fiberglass insulating each layer allow the familiar copper lines, or traces, to show through. These traces conduct the electrical signals. Most motherboard designers use computer simulations to determine the optimal length, width, and route of the individual traces. For example, a motherboard will have a target trace impedance value, often fifty or sixty ohms, which must be kept constant. Widening a trace will decrease impedance, while narrowing it will make it greater. Another issue is crosstalk resulting from the high level of circuit density. If this happens, traces must be either better insulated or moved farther apart so that interference will diminish.

Some sophisticated computer users may try tuning their systems by adding or removing motherboard components, adjusting power settings, or "overclocking" the CPU to make it run faster. Overclocking can be risky, as it typically requires increasing the core voltage. Most motherboards have a built-in voltage regulator to ensure that the core voltage does not exceed the recommended voltage for the CPU and other processors. However, some regulators allow users to adjust their settings. While there is usually a buffer zone between the recommended voltage and the maximum safe voltage, setting the voltage too high can still cause processors to overheat or even burn out.

Form Factor

Though somewhat standardized, motherboards still come in different sizes and shapes. The main distinction is between motherboards for laptops and those for desktops. Motherboards designed for one of these categories generally will not fit into the other category. Most large desktop computer cases have enough room inside for just about any model of desktop motherboard, though smaller motherboards leave more space for peripherals to be added later.

BIOS

A motherboard will have some basic software embedded in a read-only memory (ROM) chip. This software is called the BIOS, which stands for "basic input/output system." When the power button is pressed, the BIOS tells the computer what devices to activate in order to locate the operating system and begin running it. If a computer malfunctions, it may be necessary to use the BIOS to change how the motherboard behaves while the system starts up.

—Scott Zimmer, JD

BIBLIOGRAPHY

Andrews, Jean. *A+ Guide to Hardware: Managing, Maintaining, and Troubleshooting.* 6th ed. Boston: Course Tech., 2014. Print.

Andrews, Jean. *A+ Guide to Managing and Maintaining Your PC.* 8th ed. Boston: Course Tech., 2014. Print.

Cooper, Stephen. "Motherboard Design Process." *MBReview.com.* Author, 4 Sept. 2009. Web. 14 Mar. 2016.

Englander, Irv. *The Architecture of Computer Hardware, Systems Software, & Networking: An Information Technology Approach.* 5th ed. Hoboken: Wiley, 2014. Print.

Mueller, Scott. *Upgrading and Repairing PCs.* 22nd ed. Indianapolis: Que, 2015. Print.

Roberts, Richard M. *Computer Service and Repair.* 4th ed. Tinley Park: Goodheart, 2015. Print.

White, Ron. *How Computers Work: The Evolution of Technology.* Illus. Tim Downs. 10th ed. Indianapolis: Que, 2015. Print.

MULTIPROCESSING OPERATING SYSTEMS (OS)

FIELDS OF STUDY

Computer Engineering; Operating Systems

ABSTRACT

A multiprocessing operating system (OS) is one in which two or more central processing units (CPUs) control the functions of the computer. Each CPU contains a copy of the OS, and these copies communicate with one another to coordinate operations. The use of multiple processors allows the computer to perform calculations faster, since tasks can be divided up between processors.

PRINCIPAL TERMS

- **central processing unit (CPU):** sometimes described as the "brain" of a computer, the collection of circuits responsible for performing the main operations and calculations of a computer.
- **communication architecture:** the design of computer components and circuitry that facilitates the rapid and efficient transmission of signals between different parts of the computer.
- **parallel processing:** the division of a task among several processors working simultaneously, so that the task is completed more quickly.
- **processor coupling:** the linking of multiple processors within a computer so that they can work together to perform calculations more rapidly. This can be characterized as loose or tight, depending on the degree to which processors rely on one another.
- **processor symmetry:** multiple processors sharing access to input and output devices on an equal basis and being controlled by a single operating system.

MULTIPROCESSING VERSUS SINGLE-PROCESSOR OPERATING SYSTEMS

Multiprocessing operating systems (OSs) perform the same functions as single-processor OSs. They schedule and monitor operations and calculations in order to complete user-initiated tasks. The difference is that multiprocessing OSs divide the work up into various subtasks and then assign these subtasks to different central processing units (CPUs). Multiprocessing uses a distinct communication architecture to accomplish this. A multiprocessing OS needs a mechanism for the processors to interact with one another as they schedule tasks and coordinate their completion. Because multiprocessing OSs rely on parallel processing, each processor involved in a task must be able to inform the others about

Multiprocessing operating systems can handle tasks more quickly, as each CPU that becomes available can access the shared memory to complete the task at hand so all tasks can be completed the most efficiently.

how its task is progressing. This allows the work of the processors to be integrated when the calculations are done such that delays and other inefficiencies are minimized.

For example, if a single-processor OS were running an application requiring three tasks to be performed, one taking five milliseconds, another taking eight milliseconds, and the last taking seven milliseconds, the processor would perform each task in order. The entire application would thus require twenty milliseconds. If a multiprocessing OS were running the same application, the three tasks would be assigned to separate processors. The first would complete the first task in five milliseconds, the second would do the second task in eight milliseconds, and the third would finish its task in seven milliseconds. Thus, the multiprocessing OS would complete the entire task in eight milliseconds. From this example, it is clear that multiprocessing OSs offer distinct advantages.

Coupling

Multiprocessing OSs can be designed in a number of different ways. One main difference between designs is the degree to which the processors communicate and coordinate with one another. This is known as processor coupling. Coupling is classified as either "tight" or "loose." Loosely coupled multiprocessors mostly communicate with one another through shared devices rather than direct channels. For the most part, loosely coupled CPUs operate independently. Instead of coordinating their use of devices by directly communicating with other processors, they share access to resources by queueing for them. In tightly coupled systems, each CPU is more closely bound to the others in the system. They coordinate operations and share a single queue for resources.

One type of tightly coupled multiprocessing system has processors share memory with each other. This is known as symmetric multiprocessing (SMP). Processor symmetry is present when the multiprocessing OS treats all processors equally, rather than prioritizing a particular one for certain operations. Multiprocessing OSs are designed with special features that support SMP, because the OS must be able to take advantage of the presence of more than one processor. The OS has to "know" that it can divide up certain types of tasks among different processors. It must also be able to track the progress of each task so that the results of each operation can be combined once they conclude. In contrast, asymmetric multiprocessing occurs when a computer assigns system maintenance tasks to some types of processors and application tasks to others. Because the type of task assigned to each processor is not the same, they are out of symmetry. SMP has become more commonplace because it is usually more efficient.

Multitasking

The advent of multiprocessing OSs has had a major influence on how people perform their work. Multiprocessing OSs can execute more than one program at a time. This enables computers to use more user-friendly interfaces based on graphical representations of input and output. It allows users with relatively little training to perform computing tasks that once were highly complex. They can even perform many such tasks at once.

Multiprocessing OSs, though once a major innovation, have become the norm rather than the exception. As each generation of computers must run more and more complex applications, the processing workload becomes greater and greater. Without the advantages offered by multiple processors and OSs tailored to take advantage of them, computers would not be able to keep up.

—*Scott Zimmer, JD*

Bibliography

Garrido, José M., Richard Schlesinger, and Kenneth E. Hoganson. *Principles of Modern Operating Systems*. 2nd ed. Burlington: Jones, 2013. Print.

Gonzalez, Teofilo, and Jorge Díaz-Herrera, eds. *Computing Handbook: Computer Science and Software Engineering*. 3rd ed. Boca Raton: CRC, 2014. Print.

Sandberg, Bobbi. *Networking: The Complete Reference*. 3rd ed. New York: McGraw, 2015. Print.

Silberschatz, Abraham, Peter B. Galvin, and Greg Gagne. *Operating Systems Concepts*. 9th ed. Hoboken: Wiley, 2012. Print.

Stallings, William. *Operating Systems: Internals and Design Principles*. 8th ed. Boston: Pearson, 2014. Print.

Tanenbaum, Andrew S., and Herbert Bos. *Modern Operating Systems*. 4th ed. Boston: Pearson, 2014. Print.

MULTI-USER OPERATING SYSTEM (OS)

FIELDS OF STUDY

Computer Science; Operating Systems

ABSTRACT

A multi-user operating system (OS) is one that can be used by more than one person at a time while running on a single machine. Different users access the machine running the OS through networked terminals. The OS can handle requests from users by taking turns among connected users. This capability is not available in a single-user OS, where one user interacts directly with a machine with a single-user operating system installed on it.

PRINCIPAL TERMS

- **multiterminal configuration:** a computer configuration in which several terminals are connected to a single computer, allowing more than one person to use the computer.
- **networking:** connecting two or more computers to one another using physical wires or wireless connections.

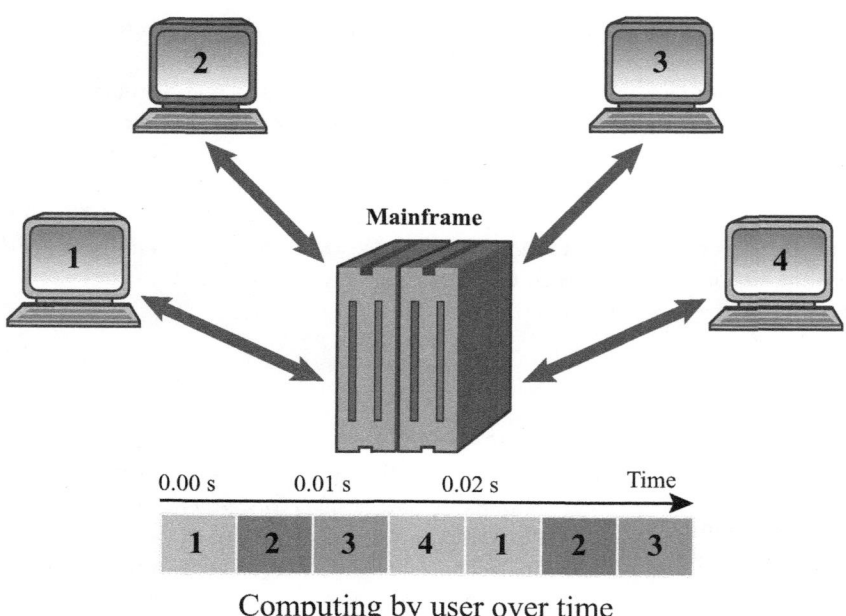

Computing by user over time

Multi-user operating systems are designed to have multiple terminals (monitor, keyboard, mouse, etc.) all connected to a single mainframe (a powerful CPU with many microprocessors) that allocates time for each user's processing demands so that it appears to the users that they are all working simultaneously.

- **resource allocation:** a system for dividing computing resources among multiple, competing requests so that each request is eventually fulfilled.
- **system:** a computer's combination of hardware and software resources that must be managed by the operating system.
- **terminals:** a set of basic input devices, such as a keyboard, mouse, and monitor, that are used to connect to a computer running a multi-user operating system.
- **time-sharing:** a strategy used by multi-user operating systems to work on multiple user requests by switching between tasks in very small intervals of time.

Multi-User Operating Systems

A computer's operating system (OS) is its most fundamental type of software. It manages the computer system (its hardware and other installed software). An OS is often described as the computer's "traffic cop." It regulates how the different parts of the computer can be used and which users or devices may use them. Many OS functions are invisible to the user of the computer. This is either because they occur automatically or because they happen at such a low level, as with memory management and disk formatting.

A multi-user OS performs the same types of operations as a single-user OS. However, it responds to requests from more than one user at a time. When computers were first developed, they were huge, complex machines that took up a great deal of physical space. Some of the first computers took up whole rooms and required several people to spend hours programming them to solve even simple calculations. These origins shaped the way that people thought about how a computer should work. Computers became more powerful and able to handle more and more complex calculations in shorter time periods. However, computer scientists continued to think of a computer as a centralized machine usable by more than one person at a time through multiple terminals connected by networking. This is why some of the earliest OSs developed, such as UNIX, were designed with a multiterminal configuration in mind. Given the nature of early computers, it made more sense to share access to a single computer. Only years later, when technology advanced and PCs became widely available and affordable, would the focus switch to single-user OSs.

Shared Computing

In order for an OS to be able to serve multiple users at once, the system needs to have either multiple processors that can be devoted to different users or a mechanism for dividing its time between multiple users. Some multi-user OSs use both strategies, since there comes a point at which it becomes impractical to continue adding processors. A multi-user OS may appear to be responding to many different requests at once. However, what is actually happening inside the machine is that the computer is spending a very small amount of time on one task and then switching to another task. This is called time-sharing. This task switching continues at a speed that is too fast for the user to detect. It therefore appears that separate tasks are being performed at once. Multi-user OSs function this way because if they handled one task at a time, users would have to wait in line for their requests to be filled. This would be inefficient because users would be idle until their requests had been processed. By alternating between users, a multi-user OS reduces the amount of time spent waiting. Time-sharing is one aspect of resource allocation. This is how a system's resources are divided between the different tasks it must perform.

One example of a multi-user OS is the software used to run the servers that support most webmail applications. These systems have millions or even billions of users who continually log on to check their messages, so they require OSs that can handle large numbers of users at once. A typical webmail application might require hundreds of computers, each running a multi-user OS capable of supporting thousands of users at once.

Economy of Scale

Many companies are moving back toward the use of multi-user OSs in an effort to contain costs. For a large organization to purchase and maintain full-featured computers for each employee, there must be a sizable investment in personnel and computer hardware. Companies seeking to avoid such expense find that it can be much more cost effective to deploy minimally equipped terminals for most users. This allows them to connect to servers running multi-user OSs. Backing up user data is also simpler with a multi-user OS because all of the data to be backed up is on the same machine. Therefore, it is less likely that users will lose their data and it saves time and money for the organization.

—*Scott Zimmer, JD*

Bibliography

Anderson, Thomas, and Michael Dahlin. *Operating Systems: Principles and Practice*. West Lake Hills: Recursive, 2014. Print.

Garrido, José M, Richard Schlesinger, and Kenneth E. Hoganson. *Principles of Modern Operating Systems*. Burlington: Jones, 2013. Print.

Holcombe, Jane, and Charles Holcombe. *Survey of Operating Systems*. New York: McGraw, 2015. Print.

Silberschatz, Abraham, Peter B. Galvin, and Greg Gagne. *Operating System Concepts Essentials*. 2nd ed. Wiley, 2014. Print.

Stallings, William. *Operating Systems: Internals and Design Principles*. Boston: Pearson, 2015. Print.

Tanenbaum, Andrew S. *Modern Operating Systems*. Boston: Pearson, 2015. Print.

NAMING CONVENTIONS

FIELDS OF STUDY

Coding Techniques; Software Development; Computer Science

ABSTRACT

Naming conventions are systems of rules to use when naming variables, constants, methods, and other components of computer code. When applied consistently, naming conventions make code easier to read, understand, and search. They take little time to apply while coding and save valuable development time when code is reviewed.

PRINCIPAL TERMS

- **class:** in object-oriented programming, a category of related objects that share common variables and methods.
- **constant:** in computer programming, a value that does not change, or an identifier assigned to represent such a value.
- **method:** in object-oriented programming, a procedure or function specific to objects of a particular class.
- **package:** in object-oriented programming, a namespace, or set of symbols, that is used to organize related classes and interfaces in a logical manner.
- **table:** a data structure that organizes information according to predefined rules, most often in rows and columns.
- **variable:** in computer programming, a symbolic name that refers to data stored in a specific location in a computer's memory, the value of which can be changed.

What Are Naming Conventions?

Naming conventions are the rules that programmers are expected to follow when choosing identifiers to represent different data types and structures, such as variables and constants. Different programming languages follow different naming conventions, and companies often have their own in-house conventions as well. Failure to adhere to naming conventions will not prevent a program from compiling or executing, assuming that the name does not violate the syntax of the language used. It will, however, make the program code more difficult for people to read and understand. Following established conventions is especially important when multiple programmers are working on the same project or when the code is likely to be reviewed, changed, or reused in the future.

When designing a naming system, several factors must be considered. Will names be restricted to a fixed length or a minimum or maximum length? What information will each name contain? Names that are very short, such as one-letter identifiers, do not provide enough information. Very long names can be onerous to type repeatedly, which may lead to errors. Modern code-editing software provides autocomplete functions that reduce these types of errors. If these tools are used, some of the drawbacks of using longer names are mitigated. Names should be easy to understand and remember. The naming system must be logical to make it easier to learn, remember, and use.

Naming conventions require clarity. Names should convey enough information to avoid possible confusion with similar objects or methods. Because of this, multiword names are often preferred. If a programmer is writing a procedure that will add up all purchases made in a week, rather than naming the procedure sum(), the programmer might use the name sumWeeklyTotal(), which is clearer and describes the procedure's purpose. Beginning each word in a name with a capital letter, a style known

SAMPLE PROBLEM

A retailer needs a method created that will apply a 10 percent discount to purchases over $200. The source code is as follows:

```
Method()
  disc = .1
  IF amount > 200 THEN
    New amount = amount – (disc * amount)
  ELSE
    New amount = amount
  END IF

determineDiscount()
  PERCENT_DISCOUNT = .1
  IF customerTotal > 200 THEN
    finalCustomerTotal = customerTotal – (PERCENT_DISCOUNT * customerTotal)
  ELSE
    finalCustomerTotal = customerTotal
  END IF
```

Rename the method, the variables, and the constant using the following rules:

> Do not use spaces in names.
> Methods are represented by a verb-noun pair, such as getDate(), and are set in lower camel case, with all words capitalized except for the first word.
> Variables should be easy to read and remember and should be longer than one character. They should also be set in lower camel case.
> Constants should be set in all capital letters, and words should be separated by underscores.

Answer:

One possible answer is below.

```
determineDiscount()
  PERCENT_DISCOUNT = .1
  IF customerTotal &gt; 200
  THEN finalCustomerTotal = customerTotal –
    (PERCENT_DISCOUNT *
  customerTotal)
  ELSE
  finalCustomerTotal = customerTotal
  END IF
```

as "camel case," makes such names more readable. Other methods for distinguishing words in a multiword name include separating the words with hyphens or underscores.

While meaningful identifiers are preferred, redundancy within hierarchical systems, such as classes or tables, should be avoided. In a class called FootballPlayer, for example, a programmer should avoid naming a method footballPlayerRun(), as this name provides redundant information. In this case, the name run() would be preferable.

Naming schemes that include abbreviations, numbers, or codes can be confusing to read and difficult to remember, wasting development time and possibly leading to errors. When abbreviations are used to create names, it is easy to make mistakes because of inconsistent usage. This can result in searches for a particular abbreviation returning incorrect results, making code more difficult to write, debug, and maintain. Naming conventions that require a key to determine codes to use when creating names should be avoided, as they are time consuming and onerous to use and implement. Avoid slang, inside jokes, or humorous names when naming objects.

Why Use Naming Conventions?

Naming conventions make code clearer and easier to read and understand. They also add descriptive information, or metadata, that helps explain each component's purpose. For example, a method can be named using a verb-noun combination that expresses the purpose of the method. A consistent naming convention also allows for more efficient searches and avoids confusion between items in the code.

For naming conventions to be effective, they must be used by all members of a development team in a consistent fashion. If one naming structure is expected and another is found, confusion can arise. In order to encourage consistent use of a naming convention, rules should be clear, and names should be easy to create.

A good naming convention can also help programmers more clearly define the purpose of their own code. If a developer finds it difficult to create a name for a component, they might need to clarify the reason the component is being used. In this way, an effective naming convention can aid in the development of robust, error-free code.

Naming Conventions in Action

Naming conventions are important to the ensure clear and consistent code. A naming convention should be simple and easy to remember and use. Properly designed naming conventions make code easier to read without imposing undue burdens on developers.

While naming conventions differ among languages and organizations, and sometimes even among different groups in the same organization, there are certain common elements among them. One such element is the aversion to single-letter identifiers, except sometimes in the case of temporary variables. Another element common to naming conventions that favor delimiting words by camel case is that different types of identifiers follow different capitalization rules. For example, it is a widely used convention in Java that class names should be set in upper camel case, where the first word is capitalized (UpperCamelCase); method and variable names in lower camel case, where the first word is lowercase (lowerCamelCase); and names of constants in all capital letters, with words separated by underscores (ALL_CAPS). Package names should be set all in lowercase, typically in a hierarchical order, with levels separated by periods and multiword level names run together (domain.product.feature or companyname.productname.feature).

Why Are Naming Conventions Important?

Naming conventions may seem secondary to the logic of the code itself, but if the code created cannot be easily understood, it is more difficult to create, debug, and maintain. Naming conventions make it easier for programmers to work with each other's code. They can also help developers understand their own code while writing complex programs over long periods of time. Effective naming conventions reduce the time wasted reinterpreting the same information multiple times. With good naming conventions, code is easily understood and can be explained to less technical collaborators.

Some older programming languages put strict limits on names, making them less easily understood. However, as technology has advanced and the value of descriptive names has been established, these restrictions have been largely eliminated. While conventions vary among institutions and even projects, using a naming convention is an accepted aspect of most development methodologies.

—Maura Valentino, MSLIS

BIBLIOGRAPHY

Friedman, Daniel P., and Mitchell Wand. *Essentials of Programming Languages.* 3rd ed., MIT P, 2008.

Haverbeke, Marijn. *Eloquent JavaScript: A Modern Introduction to Programming.* 2nd ed., No Starch Press, 2015.

"Java Programming Style Guidelines." *GeoSoft,* Geotechnical Software Services, Apr. 2015, geosoft.no/development/javastyle.html. Accessed 14 Feb. 2017.

MacLennan, Bruce J. *Principles of Programming Languages: Design, Evaluation, and Implementation.* 3rd ed., Oxford UP, 1999.

Schneider, David I. *An Introduction to Programming Using Visual Basic.* 10th ed., Pearson, 2017.

Scott, Michael L. *Programming Language Pragmatics.* 4th ed., Elsevier, 2016.

Van Roy, Peter, and Seif Haridi. *Concepts, Techniques, and Models of Computer Programming.* MIT P, 2004.

Way, Jeffrey. "9 Confusing Naming Conventions for Beginners." *Envato Tuts+,* Envato, 22 Oct. 2010, code.tutsplus.com/articles/9-confusing-naming-conventions-for-beginners–net-15584. Accessed 14 Feb. 2017.

NET NEUTRALITY

Net neutrality is the idea that Internet service providers and governments should treat all online data the same, regardless of the website, platform, user, or application it comes from. Proponents of net neutrality believe it should be assured by law so that service providers cannot fragment access to users in order to make them pay more for different services, such as using File Transfer Protocol (FTP), playing online games, or establishing high-quality peer-to-peer (P2P) communications. To do so would mean establishing "tiered service," which many advocates of net neutrality say curtails the ideals of freedom of information that the Internet was founded on. Tiered service would require a service provider's customers to purchase different packages in order to obtain the level of access they want to information and services usually freely available to everyone on the Internet. Service providers would be able to block certain content and slow down or prohibit certain types of services to those who do not pay for such package upgrades.

OVERVIEW

The most marked characteristic of the Internet is its decentralized structure. Theoretically, all nodes in the network are potentially equal in power and influence, although in practice this rarely happens because of pre-existing social and economic structures. There are three dimensions to the Internet: hardware, software, and infrastructure. Hardware consists of the devices people use to access the Internet, whether it is a smartphone, a desktop computer, or the massive server farms where multitudes of data are stored. Software consists of the programs and applications people use to access or interact online. Infrastructure refers to the systems that connect hardware to other hardware. There are two main types of infrastructure: wired systems and wireless systems. Wired systems use phone lines or fiber optic cable networks, while wireless systems use allotted frequencies on the radio spectrum to transmit Internet access.

The explosion of the Internet as a tool of mass communication was accompanied by socioeconomic trends such as vertical integration. This means that the companies that provide access to the Internet for a fee have also taken over other companies that supply content to the Internet. This has led to concerns over the concentration of power and the control of information flowing through the Internet. Some argue that concentration of power limits the democratic exchange of data and the quality of civic participation on the Internet. Net neutrality refers to the struggle to keep the Internet a public good,

Symbol of Network Neutrality as poster for promotional purposes.

rather than a resource to be used for the personal gain of a relatively small group of people.

One of the emergent issues in the net neutrality debate is the new challenges created by data-rich media forms transmitted via the World Wide Web. For instance, video transmission can congest Internet infrastructures at peak use periods, and vertically integrated ISPs (Internet service providers) have the ability to favor their own content during these times. This gives them an unfair advantage in the competition for consumers' attention. ISPs that have developed infrastructure can limit the performance of competing ISPs that use their infrastructure, as well as restricting bandwidth of peer-to-peer file sharing programs.

Opponents of net neutrality cite various reasons for their position. One of the most common arguments is that regulation of the Internet will lead to decreased competition, which in turn will reduce the type of rapid innovation that made the Internet successful and revolutionary in the first place. A related notion is that regulation will cause a drop in investments that support not only Internet service providers but many peripheral industries as well. Many claim that the fear of privileged Internet access for the wealthy and poor service for others is a false scenario. Other argue that regulation may cause an increase in taxation or that neutrality will cause an overuse of Internet bandwidth.

In the debate about net neutrality for wireless communications, one of the crucial issues is the equitable and safe distribution of the frequencies on the radio spectrum. If the radio spectrum is considered a natural resource, how does one make sure one company does not monopolize it and divert Internet traffic to sites the company controls? Antitrust laws that ensure fairness in competition among businesses are an integral part of the net neutrality debate at this point.

The debate over net neutrality is affected by the intense and swift technological innovation in the Internet and data services industries. For example, in 2013, the company that developed a standardized system for data exchange between computers worked on technology to allow data exchange between radio frequencies, phone lines, and cell phones. This emerging technology is called IPICS, or the Internet Protocol Interoperability and Collaboration System. This changes information and communications technology dramatically and will likely impact policy in order to ensure a fair and a robust online public sphere.

Because of this rapid pace of technological development, relevant laws and policy often lag behind. For example, though issues regarding net neutrality have been debated since 2000, it was not until 2010 that the US Federal Communications Commission (FCC) proposed the Open Internet Order to preserve net neutrality. In November 2014, President Barack Obama called on the FCC to regulate Internet service providers like public utilities, imploring the agency to create strict rules to enforce net neutrality. In response the FCC investigated the possibility of implementing Title II of the Communications Act of 1934, regarding common carriers, and the Telecommunications Act of 1996 to the Internet in order to support neutrality. While a vote held on February 26, 2015, went in favor of net neutrality, in April, 2018, those rules were rolled back, resulting in the "death of net neutrality."

—*Trevor Cunnington, MA, PhD*

BIBLIOGRAPHY

Ammori, Marvin. "The Case for Net Neutrality." *Foreign Affairs* 93.4 (2014): 62–73. Print.

Anderson, Steve. "Net Neutrality: The View from Canada." *Media Development* 56.1 (2009): 8–11. Print.

Cook, Vickie S. "Net Neutrality: What Is It and Why Should Educators Care?" *Delta Kappa Gamma Bulletin* 80.4 (2014): 46–49. Print.

Downes, Larry. "Unscrambling the FCC's Net Neutrality Order: Preserving the Open Internet—But Which One?" *CommLaw Conspectus* 20.1 (2011): 83–128. Print.

Feder, Samuel L., and Luke C. Platzer. "FCC Open Internet Order: Is Net Neutrality Itself Problematic for Free Speech?" *Communications Lawyer* 28.1 (2011): 1–26. Print.

Jackson, Charles L. "Wireless Efficiency Versus Net Neutrality." *Federal Communications Law Journal* 63.2 (2011): 445–80. Print.

Lessig, Lawrence. *Remix: Making Art and Commerce Thrive in the Hybrid Economy.* New York: Penguin, 2008. Print.

Renda, Andrea. "Competition-Regulation Interface in Telecommunications: What's Left of the Essential Facility Doctrine." *Telecommunications Policy* 34.1 (2010): 23–35. Print.

Ruiz, Rebecca R. "F.C.C. Sets Net Neutrality Rules." New York Times. New York Times, 12 Mar. 2015. Web. 2 Apr. 2015.

Shin, Dong-Hee. "A Comparative Analysis of Net Neutrality: Insights Gained by Juxtaposing the US and Korea." *Telecommunications Policy* 38.11 (2014): 1117–133. Print.

Silicon Valley Historical Association. "Cisco Systems." *Silicon Valley Historical Association.* Silicon Valley Historical Assoc., 2013. Web. 29 July 2013.

Sommer, Jeff. "What the Net Neutrality Rules Say." *New York Times.* New York Times, 12 Mar. 2015. Web. 2 Apr. 2015.

United States. Federal Communications Commission. *Preserving the Open Internet: Report and Order.* Washington: GPO, 2010. Print.

NETWORK SECURITY

Computer networks are an essential aspect of modern telecommunications, linking together many computers to facilitate rapid communication and the convenient storage of large quantities of information. However, computer networks are also prone to attack from people who seek to gain unauthorized access to the network, sometimes for malicious or fraudulent purposes, other times as a sort of intellectual exercise or prank, and the sophistication of these attacks continues to increase. In order to prevent or limit the damages caused by such attacks, most networks use a multilayered approach to security, with multiple components working independently to prevent different types of attacks.

Overview

The first level of network security is allowing only authorized individuals who have authorized usernames to access the system and requiring them to use a password to confirm their identity. A password is a sequence of characters presumably known only by the user, and a number of rules may be used to increase password security, including requiring longer passwords that use a blend of letters, numbers, and symbols, to make it more difficult for unauthorized users to guess an authorized user's password. Higher levels of security may require information that specifically identifies a person (e.g., a retinal scan) or a personal possession (e.g., a mobile phone).

Most computer security systems also include a firewall or barrier between an internal network and other networks to prevent unauthorized data packets from entering the internal system. Intrusion detection and prevention systems may also be used to monitor activity within a network and identify suspicious patterns of activity. In addition, the use of security software to detect computer viruses and malware (i.e., "malicious software" intended to damage the network or gather information without the user's knowledge) are in common use to protect network security.

The Computer Security Institute (CSI), a private professional organization, and the San Francisco Federal Bureau of Investigation together conducted an annual survey of people working in computer security to determine trends in network security. Results from the survey in 2010–11, found that most respondents (60.4 percent) worked for an organization that

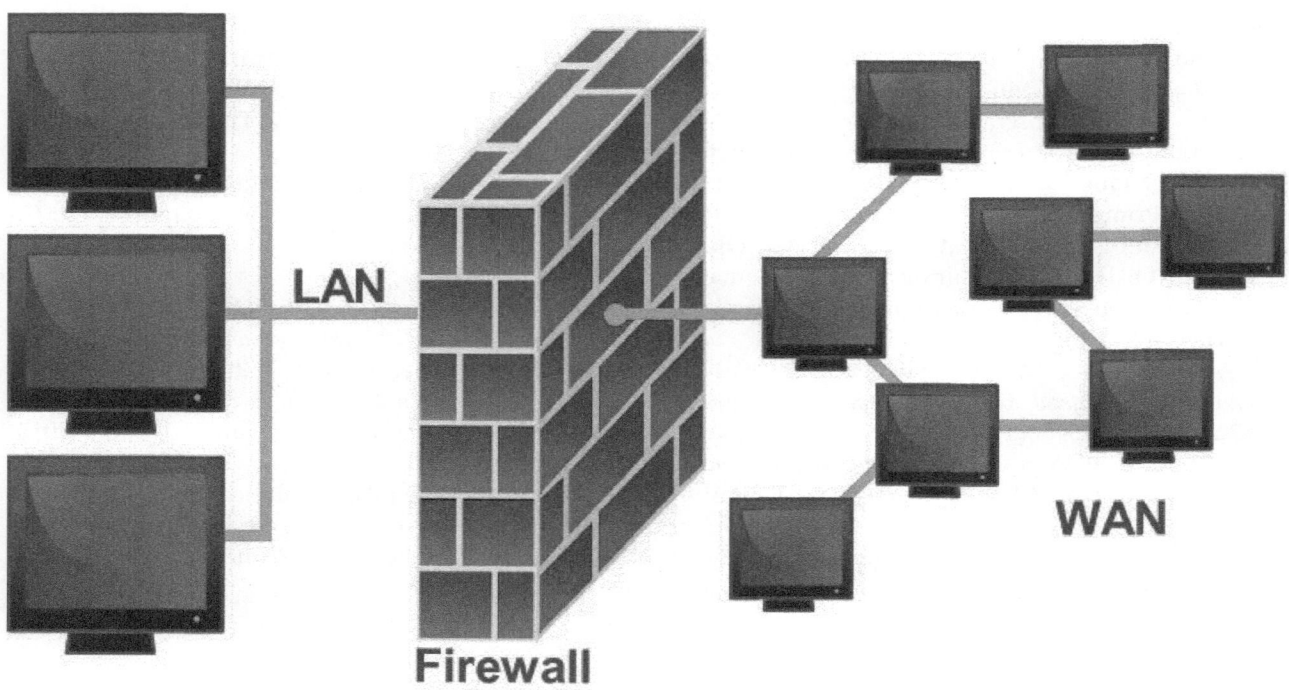

Simulação da participação de um Firewall entre uma LAN e uma WAN.

had a formal security policy, while most of the rest said they had either an informal policy (14.5 percent) or were in the process of developing a formal policy (17.2 percent). The most common type of security breach reported was malware infection, with over two-thirds (67.1 percent) reporting at least one such incident. Other common types of security breaches reported included being fraudulently represented as a sender of phishing messages (an attempt to get personal information such as credit card information through fraudulent emails; 39 percent), internal attack by bots or zombies (software programs that perform automated tasks, which may include taking over a user's computer for purposes such as sending spam; 29 percent), and denial of service attacks (17 percent). Almost half (45.6 percent) of those reporting a security incident said they had been the victim of at least one targeted attack, but reports of financial fraud incidents were relatively rare (8.7 percent).

—*Sarah E. Boslaugh, PhD, MPH*

BIBLIOGRAPHY

Brenner, Susan W. *Cybercrime and the Law: Challengers, Issues, and Outcomes.* Boston: Northeastern UP, 2012. Print.

Computer Security Institute. *15th Annual 2010/2011 Computer Crime and Security Survey.* New York: Computer Security Inst., n.d. PDF file.

Davidoff, Sherri, and Jonathan Ham. *Network Forensics: Tracking Hackers through Cyberspace.* Upper Saddle River: Prentice, 2012. Print.

Donahue, Gary A. *Network Warrior.* Sebastopol: O'Reilly, 2011. Print.

Federal Bureau of Investigation. "Cyber Crime." *FBI. gov.* Department of Justice, n.d. Web 30 July 2013.

Maiwald, Eric. *Network Security: A Beginner's Guide.* 3rd ed. New York: McGraw, 2013. Print.

McClure, Stuart, Joel Scambray, and George Kurtz. *Hacking Exposed: Network Security Secrets & Solutions.* 7th ed. New York: McGraw, 2012. Print.

Stalling, William. *Network Security Essentials: Applications and Standards.* 5th ed. Upper Saddle River: Prentice, 2014. Print.

NEURO-LINGUISTIC PROGRAMMING (NLP)

Neuro-linguistic programming (NLP) is a philosophy and approach to mental training developed in the 1970s as an alternative method of therapy. NLP attempts to understand and duplicate the methods used by "effective" individuals who have greater success in communication and influencing others. NLP concerns the way that information is received through the senses, translated into linguistic patterns, and ultimately developed into systematized patterns of interaction directed towards the achievement of goals. The majority of psychological researchers consider NLP to be largely a pseudoscientific philosophy, with little support for the basic tenets of the system and little in the way of measurable outcomes from NLP application in therapeutic techniques.

Overview

Neuro-linguistic programming was first developed by linguist and author John Grinder and psychologist Richard Bandler in the 1970s. Grinder and Bandler combined concepts from gestalt therapy and ideas from family therapists Virginia Satir and Milton Erickson. Before developing NLP Bandler and Grinder wrote materials examining Erickson's hypnosis techniques. The basic idea of NLP is to examine human behavior by breaking it down into simpler components, which can then be analyzed and evaluated for effectiveness.

The first step in NLP as a therapeutic technique is to develop a rapport with the client, which is completed through a variety of processes. One of the basic rapport-building processes is "mirroring," in which the therapist attempts to mirror the patient's breathing, facial expressions and other body language. There is also "cross-over" mirroring, in which the therapist uses another movement to mimic that of the patient, such as matching breathing to the rise and fall of a patient's vocal pitch. Mirroring is thought to bring about a sort of sympathetic relationship between the patient and the therapist, leaving the patient in a suggestive state.

Reframing is another technique in which a certain behavior, like a phobic response, is analyzed for its effectiveness in one of two ways. First, the behavior can be isolated from its context to determine what benefit (or perceived benefit) the behavior brings. Or, working within the same context, the therapist can guide the patient to ascribe a different meaning to the same behavior. The idea is to shift perspectives to better understand others' and one's own behavior. Eventually, the patient is taught to explore alternative behaviors that can be connected to similar benefits.

NLP had a rapid rise in popularity during the 1980s, which can be considered part of a nationwide explosion of interest in pop psychology and alternative therapy. There has been no clinical data supporting the effectiveness of NLP therapy and scientific evaluations of NLP ideas regarding brain function have revealed that many of the central tenets of the system are flawed. Some of the basic techniques for behavior modification are similar to those in gestalt therapy, and which may explain some trends of effectiveness, but NLP as a unique system has failed to demonstrate significant merit. However, a variety of therapists and self-help authors continue to promote NLP as a therapeutic and motivational tool.

—*Micah Issitt*

Richard Bandler, co-creator of neuro-linguistic programming, at a NLP seminar, 2007.

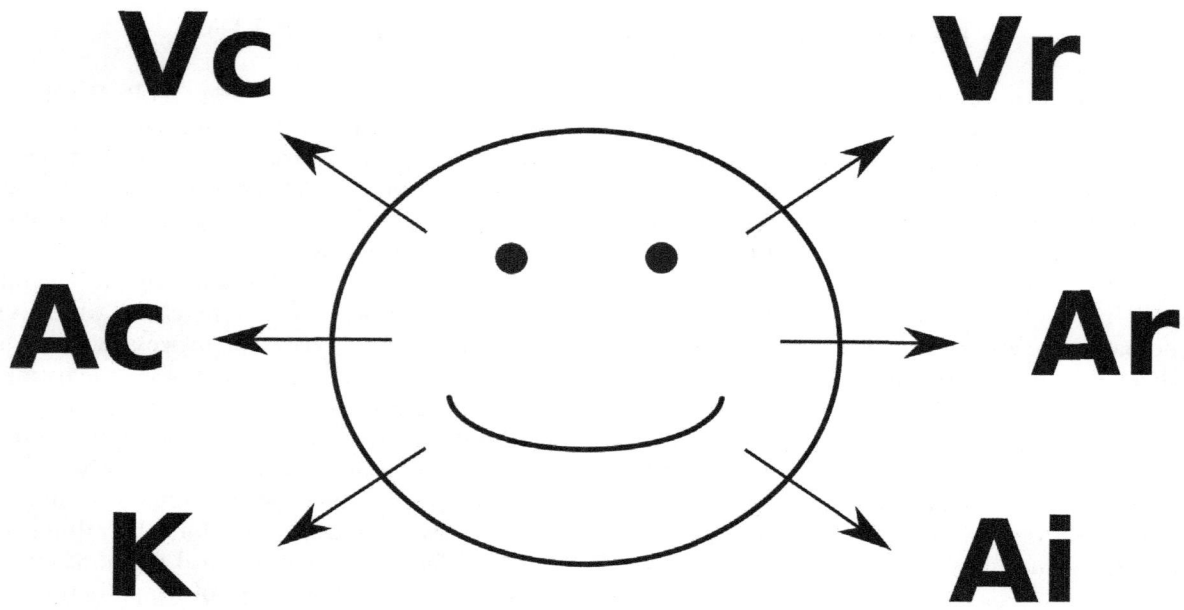

An NLP "eye accessing cue chart." The six directions represent "visual construct," "visual recall," "auditory construct," "auditory recall," "kinesthetic," and "auditory internal dialogue."

Bibliography

Anderson, Mike, and Sergio Della Salla. *Neuroscience in Education: The Good, the Bad, and the Ugly.* New York: Oxford UP, 2012. Print.

Bandler, Richard, and John T. Grinder, *Frogs into Princes: Neuro Linguistic Programming,* Moab: Real People P, 1979. Print.

Eisner, Donald A. *The Death of Psychotherapy: From Freud to Alien Abductions.* Westport: Greenwood, 2000. Print.

Gibson, Barbara P. *The Complete Guide to Understanding and Using NLP.* Print. Ocala: Atlantic, 2011. Print.

Linder-Pelz, Susie. *NLP Coaching: An Evidence-Based Approach for Coaches, Leaders and Individuals.* Philadelphia: Kogan, 2010. Print.

O'Connor, Joseph, and John Seymour. *Introducing NLP: Psychological Skills for Understanding and Influencing People.* San Francisco: Conary P, 2011. Print.

Richards, Jack C., and Theodore S. Rodgers. *Approaches and Methods in Language Teaching.* New York: Cambridge UP, 2001. Print.

Wake, Lisa, *The Role of Brief Therapy in Attachment Disorders,* London: Karnac, 2010. Print.

NEUROMARKETING

Neuromarketing is an emerging field of marketing research that uses neuroimaging technologies to record consumers' emotional, physiological, and neurological responses to marketing stimuli. Researchers use medical technologies such as functional magnetic resonance imaging (fMRI), electroencephalography (EEG), steady-state topography (SST), galvanic skin response, eye tracking, and other techniques to measure sensorimotor, cognitive, and affective reactions to advertisements, product packaging, and other marketing materials. Marketers then interpret the data collected by these neurotechnologies in order to design marketing campaigns that better influence and predict consumer behavior.

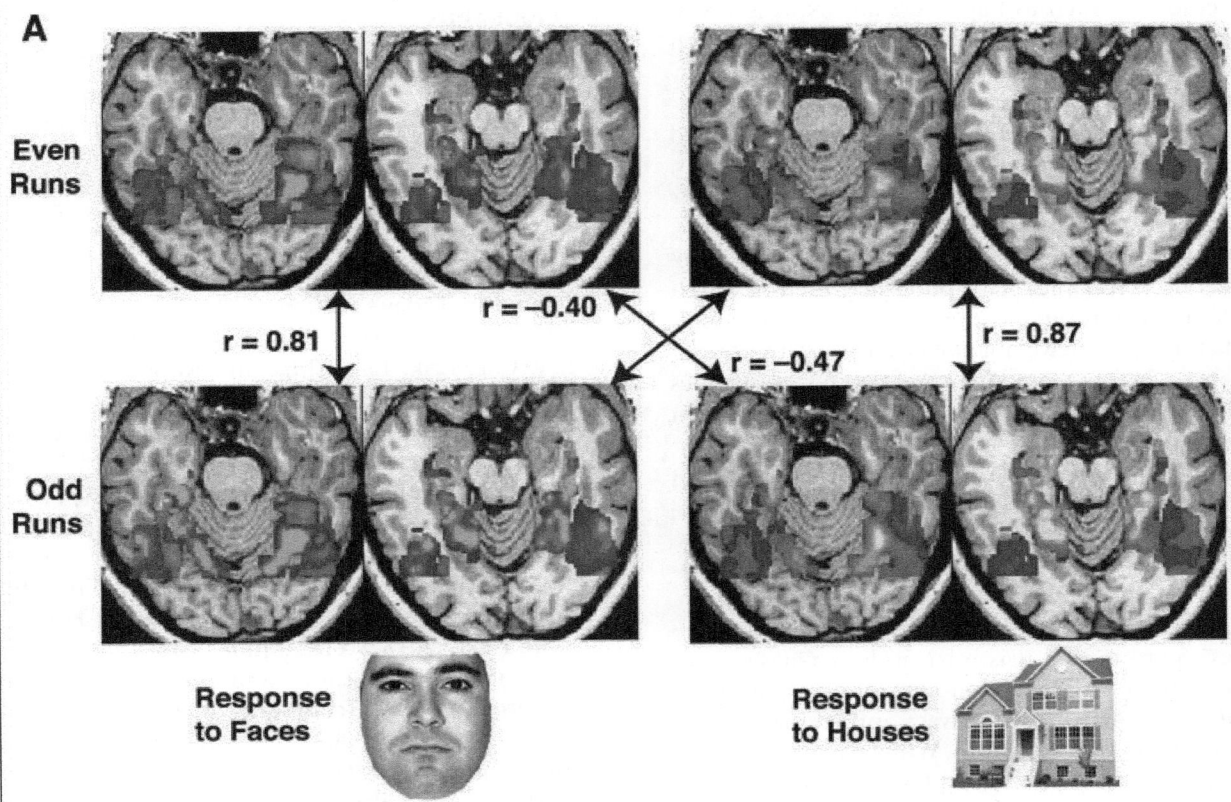

Eye tracking device

OVERVIEW

The term "neuromarketing," first coined in 2002, can be applied both to the neuroimaging techniques used to measure consumer responses to marketing materials and to the advertisements that are designed on the basis of this research. The basic framework underpinning the field is the theory that consumers' decision-making processes are more strongly influenced by subconscious, emotional cognitive systems than by rational, conscious cognitive systems. These subconscious responses to advertisements are thought to be unquantifiable in traditional focus groups or customer surveys, thereby necessitating the use of neuroimaging techniques to measure consumers' brain activity and other physiological responses to advertisements. Neuroimaging techniques are also thought to elicit more honest responses from consumers who might otherwise have compelling reasons to posture or give the responses that they believe are desired.

Eye-tracking devices can precisely quantify which part of an advertisement first catches the consumer's attention and which parts are uninteresting or distracting. For example, using eye-tracking techniques, marketers can determine whether consumers spent more time looking at a model's face in an advertisement than the product or brand logo. Functional magnetic resonance imaging (fMRI), electroencephalography (EEG), and steady-state topography (SST) are also used for neuromarketing research; these technologies measure brain activity, from which neuromarketers can extrapolate customers' attention level, memory processing, and emotional response. However, because these neuroimaging technologies measure surrogate signals for brain activity, such as the blood-oxygen levels of certain brain regions, it takes a significant amount of statistical analysis and image processing to interpret the results, which enables neuromarketers to over- or underemphasize certain responses. Further, cultural differences in how emotions such as happiness are expressed can affect a marketing campaign's effectiveness, even with neurological data supporting the same brain activity in people of different backgrounds.

While neuromarketing techniques have offered advertisers useful consumer and market insights, the field of neuromarketing has drawn criticism for overstating its scientific and academic credentials. Some critics have pointed to the fact that few neuromarketing companies have published their research results in peer-reviewed journals, and many neuromarketing claims have not been backed by supporting evidence. In response, neuromarketing companies contend that they have not published their results in order to protect their proprietary research.

Much as the prospect of subliminal messaging did in the mid-twentieth century, the field of neuromarketing has also raised a number of ethical concerns. Neuromarketing insights have been applied to political campaigns and advertisements for prescription medications and junk food, prompting some critics to argue that it is intrusively persuasive and a threat to consumer autonomy. Other critics have highlighted the issues of privacy and confidentiality in neuromarketing research. While some neuromarketing claims have been overblown and the available technology is far from being able to read minds or predict future behavior, the field has offered advertisers important tools for interpreting consumer behavior.

—*Mary Woodbury Hooper*

BIBLIOGRAPHY

Ariely, Dan, and Gregory S. Berns. "Neuromarketing: The Hope and Hype of Neuroimaging in Business." *Nature Reviews Neuroscience* 11.4 (2010): 284–92. Print.

Barkin, Eric. "The Prospects and Limitations of Neuromarketing." *CRM* July 2013: 46–50. Print.

Crain, Rance. "Neuromarketing Threat Seems Quaint in Today's Ad Landscape." *Advertising Age* 8 July 2013: 22. Print.

Fisher, Carl Erik, Lisa Chin, and Robert Klitzman. "Defining Neuromarketing: Practices and Professional Challenges." *Harvard Review of Psychiatry* 18.4 (2010): 230–37. Print.

Morin, Christophe. "Neuromarketing: The New Science of Consumer Behavior." *Society* 48.2 (2011): 131–35. Print.

Nobel, Carmen. "Neuromarketing: Tapping into the 'Pleasure Center' of Consumers." *Forbes*. Forbes.com, 1 Feb. 2013. Web. 29 June 2015.

Orzan, G., I. A. Zara, and V. L. Purcarea. "Neuromarketing Techniques in Pharmaceutical Drugs Advertising: A Discussion and Agenda for Future Research." *Journal of Medicine and Life* 5.4 (2012): 428–32. Print.

Shaw, Hollie. "Advertisers Are Looking inside Your Brain: Neuromarketing Is Here and It Knows What You Want." *Financial Post*. Natl. Post, 17 Apr. 2015. Web. 29 June 2015.

Singer, Natasha. "Making Ads That Whisper to the Brain." *New York Times*. New York Times, 13 Nov. 2010. Web. 7 Oct. 2013.

NEUROMORPHIC CHIPS

FIELDS OF STUDY

Computer Engineering; Information Systems

ABSTRACT

Neuromorphic chips are a new generation of computer processors being designed to emulate the way that the brain works. Instead of being locked into a single architecture of binary signals, neuromorphic chips can form and dissolve connections based on their environment, in effect "learning" from their surroundings. These chips are needed for complex tasks such as image recognition, navigation, and problem solving.

PRINCIPAL TERMS

- **autonomous:** able to operate independently, without external or conscious control.
- **Human Brain Project:** a project launched in 2013 in an effort at modeling a functioning brain by 2023; also known as HBP.

Neuromorphic chips are designed to detect and predict patterns in data and processing pathways to improve future computing. They simulate the brain's neuroplasticity, allowing for efficient abstraction and analysis of visual and auditory patterns. Each of the chips on this board has hundreds of millions of connections mimicking the synapses that connect neurons.

- **memristor:** a memory resistor, a circuit that can change its own electrical resistance based on the resistance it has used in the past and can respond to familiar phenomena in a consistent way.
- **nervous (neural) system:** the system of nerve pathways by which an organism senses changes in itself and its environment and transmits electrochemical signals describing these changes to the brain so that the brain can respond.
- **neuroplasticity:** the capacity of the brain to change as it acquires new information and forms new neural connections.

BRAIN-BASED DESIGN

Neuromorphic chips have much in common with traditional microprocessor chips. Both kinds of chip control how a computer receives an input, processes that information, and then produces output either in the form of information, action, or both. The difference is that traditional chips consist of millions of tiny, integrated circuits. These circuits store and process information by alternating between binary "on" and "off" states. Neuromorphic chips are designed to mimic the way that the human body's nervous (neural) system handles information. They are designed not only to process information but to learn along the way. A system that can learn may be more powerful than a system that must be programmed to respond in every possible situation. Systems that can learn can function autonomously, that is, on their own without guidance. A good example of this is car navigation systems. These systems have to store or access detailed maps as well as satellite data about their current position. With this data, a navigation system can then plot a course. The neuromorphic chips may not need all of that background data. They may be better able to understand information about their immediate surroundings and use it to predict outcomes based on past events. They would function much the way someone dropped off a few blocks from home could figure out a way to get there without consulting a map.

In order to design neuromorphic chips, engineers draw upon scientific research about the brain and how it functions. One group doing such research is the Human Brain Project (HBP). HBP is trying to build working models of a rodent brain. Eventually HBP will try to build a fully working model of a human brain. Having models like these will allow scientists to test hypotheses about how the brain works.

Their research will aid in the development of computer chips that can mimic such operations.

The Limits of Silicon

One reason researchers have begun to develop neuromorphic chips is that the designs of traditional chips are approaching the limits of their computational power. For many years, efforts were focused on developing computers capable of the type of learning and insights that the human brain can accomplish. These efforts were made on both the hardware and the software side. Programmers designed applications and operating systems to use data storage and access algorithms like those found in the neural networks of the brain. Chip designers found ways to make circuits smaller and smaller so they could be ever more densely packed onto conventional chips. Unfortunately, both approaches have failed to produce machines that have either the brain's information processing power or its neuroplasticity. Neuroplasticity is the brain's ability to continuously change itself and improve at tasks through repetition.

Neuromorphic chips try to mimic the way the brain works. In the brain, instead of circuits, there are about 100 billion neurons. Each neuron is connected to other neurons by synapses, which carry electrical impulses. The brain is such a powerful computing device because its huge amount of neurons and synapses allow it to take advantage of parallel processing. Parallel processing is when different parts of the brain work on a problem at the same time. Parallel processing allows the brain to form new connections between neurons when certain pathways have proven especially useful. This forming of new neural pathways is what happens when a person learns something new or how to do a task more efficiently or effectively. The goal of neuromorphic chip designers is to develop chips that approach the brain's level of computational density and neuroplasticity. For example, researchers have proposed the use of a memristor, a kind of learning circuit. To bring such ideas about, a chip architecture completely different from the traditional binary chips is needed.

Complex Tasks

Neuromorphic chips are especially suitable for computing tasks that have proven too intense for traditional chips to handle. These tasks include speech-to-text translation, facial recognition, and so-called smart navigation. All of these applications require a computer with a large amount of processing power and the ability to make guesses about current and future decisions based on past decisions. Because neuromorphic chips are still in the design and experimentation phase, many more uses for them have yet to emerge.

The Way of the Future

Neuromorphic chips represent an answer to the computing questions that the future poses. Most of the computing applications being developed require more than the ability to process large amounts of data. This capability already exists. Instead, they require a device that can use data in many different forms to draw conclusions about the environment. The computers of the future will be expected to act more like human brains, so they will need to be designed and built more like brains.

—*Scott Zimmer, JD*

Bibliography

Burger, John R. *Brain Theory from a Circuits and Systems Perspective: How Electrical Science Explains Neuro-Circuits, Neuro-Systems, and Qubits.* New York: Springer, 2013. Print.

Human Brain Project. Human Brain Project, 2013. Web. 16 Feb. 2016.

Lakhtakia, A., and R. J. Marti?n-Palma. *Engineered Biomimicry.* Amsterdam: Elsevier, 2013. Print.

Liu, Shih-Chii, Tobi Delbruck, Giacomo Indiveri, Adrian Whatley, and Rodney Douglas. *Event-Based Neuromorphic Systems.* Chichester: Wiley, 2015. Print.

Prokopenko, Mikhail. *Advances in Applied Self-Organizing Systems.* London: Springer, 2013. Print.

Quian, Quiroga R., and Stefano Panzeri. *Principles of Neural Coding.* Boca Raton: CRC, 2013. Print.

Rice, Daniel M. *Calculus of Thought: Neuromorphic Logistic Regression in Cognitive Machines.* Waltham: Academic, 2014. Print.

OBJECTIVITY

FIELDS OF STUDY
Research Theory; Research Design

ABSTRACT
In science, objectivity, or the quality of being uninfluenced by individual perspectives or biases, is considered a major goal. Yet researchers are unavoidably individuals and therefore always subjective to some degree. For this reason, both natural and social sciences devote considerable effort to ensuring steps are taken to make research as objective as possible. Part of this is a conscious decision by researchers, who generally strive to be objective in following the scientific method. Yet biases remain, often unconsciously. It is important to recognize inherent biases in order to minimize them and their impacts on research findings.

PRINCIPAL TERMS

- **cognitive bias:** an ingrained irrational or illogical judgment based on subjective perspective.
- **cultural bias:** the interpretation of a phenomenon from the perspective of the interpreter's own culture.
- **double-blind:** the condition of a study in which neither the researchers nor the subjects know which experimental unit is a control group or a test group.
- **qualitative data:** information that describes but does not numerically measure the attributes of a phenomenon.
- **quantitative data:** information that expresses attributes of a phenomenon in terms of numerical measurements, such as amount or quantity.
- **subjectivity:** the quality of being true only according to the perspective of an individual subject, rather than according to any external criteria.

UNDERSTANDING OBJECTIVITY

Regardless of their factual knowledge, every researcher approaches a project as an individual with unique perspectives that have evolved throughout their life experiences. Some of this subjectivity is trivial, such as a preference for coffee over tea. Such trivial subjectivity either does not impact the validity of research or can be consciously dismissed if relevant to a study. Education, training, and experience all push researchers to be as detached as possible from their individual preferences and perspectives. This quality, known as objectivity, is generally seen as critical to scientific research. The scientific method, with its insistence on testing and reproduction of results, is often considered a way to strive for objectivity.

However, maintaining objectivity is far from simple. In addition to basic subjective preferences, all researchers and all human subjects of research are subject to subtler biases, including sampling bias, cognitive bias, and cultural bias. These influences, which are typically unconscious and are thought to be deeply ingrained in human psychology, can detract from objectivity in powerful ways. For example, a white researcher living in a predominantly white town may recruit only white people for a study. If the study is focused on a local phenomenon, this may not be problematic. However, if the researcher attempts to generalize the results of the study to the whole US population, both sampling and cultural bias will likely compromise objectivity and therefore the validity of the findings. Systematic bias can affect the development and structure of experiments, the accuracy of measurements, and the interpretation of results.

Objectivity is particularly at issue when working with qualitative data, which is by nature more subjective than quantitative data. Some have argued that qualitative research can never be truly objective, as it necessarily involves interpretation of subjective data. Other philosophers of science claim that even quantitative research cannot be fully free from bias. Some research ethicists suggest that arguments over objectivity in research are simply the result of scholars attempting to define the various paradigms of their disciplines. Regardless, scientists agree that objectivity should be maintained whenever possible.

MAINTAINING OBJECTIVITY
Scientists strive to maintain objectivity at every step in a study, from initial conception to evaluation of results. Experts on research integrity agree that researchers should never begin with the idea that they know exactly what they will find. The purpose of research is to arrive at answers that reflect actual findings. Thus, findings may support initial hypotheses, but they may also turn a researcher in a new direction. In the initial stages of research, when hypotheses are being formed and research questions are being stated, it is essential that language be as unbiased as possible so as not to distort results. The size and makeup of samples in a study are also important. Randomization and other methods are used to make research subjects as representative of the general population as possible.

One of the most effective methods for maintaining objectivity in data collection is to set up blind research projects. Single-blind studies eliminate bias among research subjects by keeping them unaware of whether they are members of a control group or the test group. A double-blind study attempts to remove bias among researchers as well, and is considered the gold standard for much research. However, blind trials are not always possible and require significant time and investment.

Once data is collected, researchers face additional objectivity challenges in analyzing and reporting their findings. Possible issues include failure to record data correctly, errors in statistical analysis, and cultural biases in interpretation. Objectivity may also be lost if a researcher opts for a method of analysis that does not reflect what was actually learned. Replication of research and peer review of material submitted for publication are two common methods used to encourage the highest possible degree of objectivity. Even so, studies that take steps to maintain objectivity may still later be found to be biased.

REAL-WORLD RESEARCH
An ongoing concern is whether or not it is possible for research conducted or sponsored by private, for-profit companies to be as objective as that undertaken by government or nonprofit organizations. Scholars have identified a "funding effect" bias, noting that results of industry-funded studies consistently present the industry's products more favorably than do results of research funded by outside parties. Examples include research funded by the tobacco industry that downplays the health effects of smoking and research funded by fossil-fuel companies that questions climate change. Critics claim that this apparent lack of objectivity threatens the integrity of scientific research.

Because of major problems with conflicts of interest in research, significant interest has been directed toward the promotion of objectivity. Various academic, national, and international organizations have formed to uphold good research practices. Much research itself remains devoted to studying the causes and effects of objectivity, subjectivity, and bias. By better understanding the issues and their implications, scientists can better pursue truth.

RESEARCH ETHICS
Objectivity remains vital to science even as some researchers question its limits. The belief that natural science lends itself to objectivity while social science is more likely to be subjective remains common, even to the point of contention between the two broad fields. Yet it is apparent that objectivity is an important goal, if perhaps one impossible to fully achieve, across the sciences and beyond. It has important applications to other disciplines, from philosophy to international relations, and even to everyday life. As a concept, objectivity stands as a critical idea that illustrates the great complexity of human consciousness.

—*Elizabeth Rholetter Purdy, PhD*

BIBLIOGRAPHY
Agazzi, Evandro. *Scientific Objectivity and Its Contexts.* Springer, 2014.
Grinnell, Frederick. "Research Integrity and Everyday Practice of Science." *Science and Engineering Ethics*, vol. 19, no. 3, 2013, pp. 685–701. *Academic Search Complete*, search.ebscohost.com/login.aspx?direct=

true&db=a9h&AN=89583462&site=ehost-live. Accessed 19 Apr. 2017.

Harding, Sandra. *Objectivity and Diversity: Another Logic of Scientific Research.* U of Chicago P, 2015.

Krimsky, Sheldon. "Do Financial Conflicts of Interest Bias Research? An Inquiry into the 'Funding Effect' Hypothesis." *Science, Technology, & Human Values*, vol. 38, no. 4, 2013, pp. 566–87.

Letherby, Gayle, et al. *Objectivity and Subjectivity in Social Research.* Sage Publications, 2013.

Steck, Andreas J., and Barbara Steck. *Brain and Mind: Subjective Experience and Scientific Objectivity.* Springer, 2016.

Yetiv, Steve A. *National Security through a Cockeyed Lens: How Cognitive Bias Impacts US Foreign Policy.* Johns Hopkins UP, 2013.

OBJECT-ORIENTED DESIGN (OOD)

FIELDS OF STUDY

Software Engineering; Programming Language; Computer Science

ABSTRACT

Object-oriented design (OOD) is an approach to software design that uses a process of defining objects and their interactions when planning code to develop a computer program. Programmers use conceptual tools to transform a model into the specifications required to create the system. Programs created through OOD are typically more flexible and easier to write.

PRINCIPAL TERMS

- **attributes:** the specific features that define an object's properties or characteristics.
- **class-based inheritance:** a form of code reuse in which attributes are drawn from a preexisting class to create a new class with additional attributes.
- **class:** a collection of independent objects that share similar properties and behaviors.
- **method:** a procedure that describes the behavior of an object and its interactions with other objects.
- **object:** an element with a unique identity and a defined set of attributes and behaviors.
- **prototypal inheritance:** a form of code reuse in which existing objects are cloned to serve as prototypes.

Advancing Software Development

Object-oriented design (OOD) was developed to improve the accuracy of code while reducing software development time. Object-oriented (OO) systems are made up of objects that work together by sending messages to each other to tell a program how to behave. Objects in software design represent real-life objects and concerns. For example, in the code for a company's human resources website, an object might represent an individual employee. An object is independent of all other objects and has its own status. However, all objects share attributes or features with other objects in the same class. A class describes the shared attributes of a group of related objects.

OOD typically follows object-oriented analysis (OOA), which is the first step in the software development process when building an OO system. OOA involves planning out the features of a new program. OOD involves defining the specific objects and classes that will make up that program. The first OO language, Simula, was developed in the 1960s, followed by Smalltalk in 1972. Examples of well-known OO programming languages include Ruby, C++, Java, PHP, and Smalltalk.

OOD consists of two main processes, system design and object design. First, the desired system's architecture is mapped out. This involves defining the system's classes. Using the example of a human resources website, a class might consist of employees, while an object in that class would be a specific employee. Programmers plan out the essential attributes of a class that are shared across all objects within the class. For example, all employees would have a name, a position, and a salary. The class defines these attributes but does not specify their values. Thus, the class would have a field for the employee's name, and the individual object for a specific employee might have the name Joe Smith. All the objects within a class also

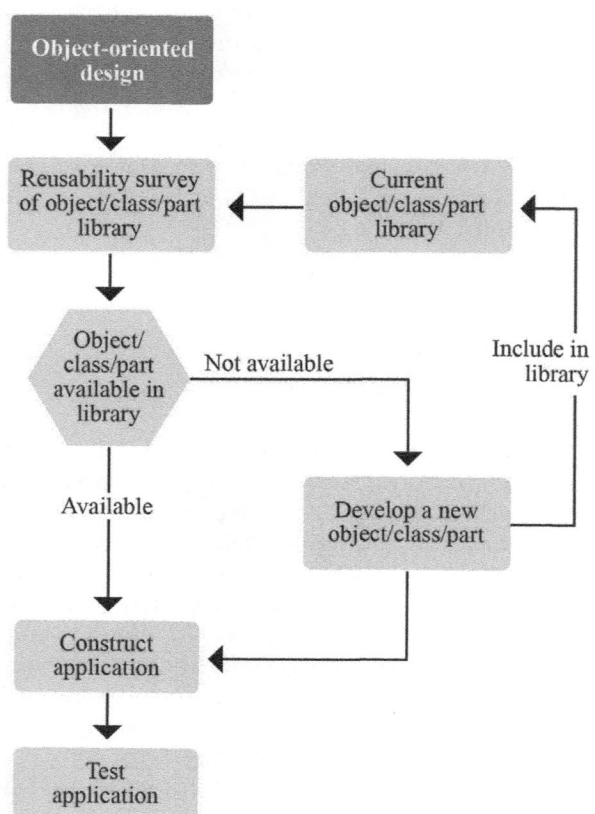

The benefit of object-oriented design lies in the reusability of the object database. Because each object has only one job, the code for an object remains applicable for any instance where the object is needed.

existing objects, called prototypes, instead of drawing from existing classes.

OOD Principles and Applications

Software engineers often follow a set of OOD principles identified by the acronym SOLID. These principles were first compiled by software engineer and author Robert Cecil Martin. The SOLID principles provide guidance for better software development, maintenance, and expansion. These principles are:

Single responsibility principle: Each class should have only one job.

Open-closed principle: A class should be open for extension but closed to modification (that is, it can be easily extended with new code without modifying the class itself).

Liskov substitution principle: Each subclass can be substituted for its parent class.

Interface segregation principle: Do not force clients to implement an interface (a list of methods) they do not use. Interfaces should be as small as possible. Having multiple smaller interfaces is better than having one large one.

Dependency inversion principle: High-level modules must not be dependent on low-level modules. Attempt to minimize dependencies between objects.

OOD is used to create software that solves real-world problems, such as addressing a user's needs when interfacing with an automatic teller machine (ATM). In this example, the software engineer first defines the classes, such as the transaction, screen, keypad, cash dispenser, and bank database. Classes or objects can be thought of as nouns. Some classes, such as transactions, might be further categorized into more specific subclasses, such as inquiry, withdrawal, deposit, or transfer.

Relationships between classes are then defined. For example, a withdrawal is related to the card reader, where the user inserts an ATM card, and the keypad, where the user enters their personal identification number (PIN). Next, classes are assigned attributes. For a withdrawal, these might be the account number, the balance, and the amount of cash to withdraw.

Finally, methods or operations are assigned to each class, based on user needs identified in the OOA phase. Methods can be thought of as verbs, such as execute, display, or withdraw. Detailed diagrams describe changes that happen within the objects, the

share methods, or specific behaviors. For example, employee objects could be terminated, given a raise, or promoted to a different position.

Object design is based on the classes that are mapped out in the system design phase. Once software engineers have identified the required classes, they are able to write code to create the attributes and methods that will define the necessary objects. As development progresses, they may decide that new classes or subclasses need to be created. New classes can inherit features from existing classes through class-based inheritance. Inheritance is a form of code reuse. Class-based inheritance draws attributes from an existing class to create a new class. New attributes can then be added to the new class. This is distinct from prototypal inheritance, which involves cloning activities they carry out, and how they interact with each other within the system.

PRECAUTIONS AND DRAWBACKS

Successful development with OOD requires realistic expectations. Possible problems with this approach include insufficient training, which can lead to the belief that OOD will clear up every development delay, resulting in missed deadlines. Further issues arise when timelines are shortened in expectation of OOD's promise of speed. Thorough training is a programmer's best course to understanding the limitations and subtleties of OOD.

OOD has been a mainstay in software programming and will be used for years to come, but advances in technology will require programming languages that can deliver faster search results, more accurate calculations, and better use of limited bandwidth for the applications users increasingly rely upon. In addition, OOD is not the best choice for every requirement. It is appropriate for systems with a large amount of graphics, interfaces, and databases, but for simpler systems, task-oriented design (TOD) might be a better choice.

—*Teresa E. Schmidt*

BIBLIOGRAPHY

Booch, Grady, et al. *Object-Oriented Analysis and Design with Applications.* 3rd ed. Upper Saddle River: Addison, 2007. Print.

Dennis, Alan, Barbara Haley Wixom, and David Tegarden. *Systems Analysis and Design: An Object-Oriented Approach with UML.* 5th ed. Hoboken: Wiley, 2015. Print.

Garza, George. "Working with the Cons of Object Oriented Programming." Ed. Linda Richter. *Bright Hub.* Bright Hub, 19 May 2011. Web. 6 Feb. 2016.

Harel, Jacob. "SynthOS and Task-Oriented Programming." *Embedded Computing Design.* Embedded Computing Design, 2 Feb. 2016. Web. 7 Feb. 2016.

Metz, Sandi. *Practical Object-Oriented Design in Ruby: An Agile Primer.* Upper Saddle River: Addison, 2012. Print.

Oloruntoba, Samuel. "SOLID: The First 5 Principles of Object Oriented Design." *Scotch.* Scotch.io, 18 Mar. 2015. Web. 1 Feb. 2016.

Puryear, Martin. "Programming Trends to Look for This Year." *TechCrunch.* AOL, 13 Jan. 2016. Web. 7 Feb. 2016.

Weisfeld, Matt. *The Object-Oriented Thought Process.* 4th ed. Upper Saddle River: Addison, 2013. Print.

OBJECT-ORIENTED PROGRAMMING (OOP)

FIELDS OF STUDY

Software Development; Programming Methodologies; Computer Science

ABSTRACT

Object-oriented programming (OOP) is a paradigm based on objects that act as containers for data and functionality that interoperate to create the final application. Object-oriented programming languages (OOPLs) are based on the OOP paradigm. They were designed to improve upon traditional procedural programming.

PRINCIPAL TERMS

- **class:** in object-oriented programming, a category of related objects that share common variables and methods.
- **interface:** in object-oriented programming, a structure that specifies the behaviors, or methods, and properties required of the objects in a class; also called a "contract."
- **object:** in object-oriented programming, a self-contained module of data belonging to a particular class that shares the variables and methods assigned to that class.
- **package:** in object-oriented programming, a namespace, or set of symbols, that is used to organize related classes and interfaces in a logical manner.

UNDERSTANDING OBJECT-ORIENTED PROGRAMMING

Object-oriented programming (OOP) is a paradigm based on objects that are containers for the data and functionality. These objects work together to form

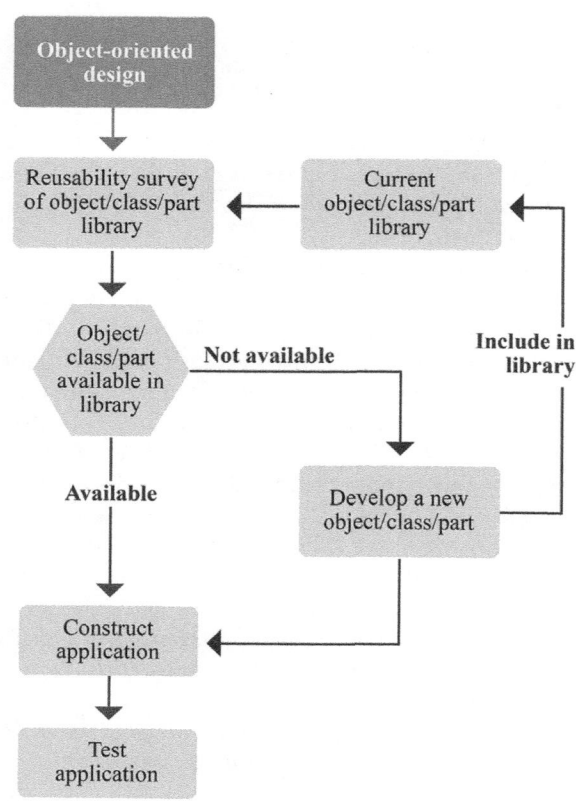

The benefit of object-oriented design lies in the reusability of the object database. Because each object has only one job, the code for an object remains applicable for any instance where the object is needed.

the completed program. A class defines what properties an object has and what functionality it provides. The properties represent the object's state, which is stored in a field or variable. The functions, or methods, define the object's behavior. For example, an airplane object would have properties (states) such as wingspan, speed, weight, and color and have methods (behaviors) such as taxi, take off, climb, descend, turn, and land. The methods and properties that an object must make available for use by other components in the program are known as its interface. Related classes and interfaces are organized into groups called packages.

In the real world, many objects share the same properties and functionality. For example, all aircraft can take off and land and share common properties, such as speed. However, specific types of aircraft offer added functionality and properties that distinguish them from other types. For example, helicopters can hover as well as take off and land. In this way, a helicopter can be thought of as a specialist class of aircraft. OOP models this concept using a technique called "inheritance." Using inheritance, an aircraft class could be created that includes a takeoff and land method, and then a helicopter child class could inherit that functionality from the aircraft parent class and add the hover functionality. Similarly, an interface for aircraft would require that any object within the aircraft class have the methods for take off and land.

Object-oriented programming languages (OOPLs) are based on the OOP paradigm. There are different types of OOPLs. Pure OOPLs such as Java and Ruby are based solely on OOP. Hybrid languages, such as Python, are based primarily on the OOP approach but support elements of the procedural programming paradigm. The final type are procedural languages that have been extended to include OOP concepts, such as COBOL and Fortran.

IMPROVING SOFTWARE DEVELOPMENT
OOPLs were designed to improve upon procedural languages. They allow greater modularity of code than do procedural languages. This makes it easier to prevent and correct problems within the program. OOPLs offer strong data integrity and data security because an object's methods and properties can be made private, or hidden. OOPLs work well with all types of data, unlike string-oriented languages, which are designed to work with strings of characters. OOP programs are also easy to extend as new features are needed. OOPLs also promote the reuse of code through inheritance. They are easy to scale and are a good fit for applications that need to adapt to future changes. OOPLs are good for projects where developers who work independently on very different systems need to collaborate with one another, as the details do not need to be shared or accessed by every developer.

The different types of OOPLs have different strengths and weaknesses. Pure languages offer the benefits of OOP while preventing the errors and inefficiencies that occur in procedural languages. However, they lack the ability to use procedural approaches when they would be the best choice. Hybrid languages offer the benefits of OOP while affording the programmer the flexibility to use certain procedural techniques when appropriate.

SAMPLE PROBLEM

A developer is creating code that will be used to perform basic calculations like addition and multiplication. The following is an early version that only offers the ability to perform the addition of two numbers.

```
Public Class Calculator
    Private number1 As Integer
    Private number2 As Integer

    Property input1() As Integer
    Get
        Return number1
    End Get
    Set (value As Integer)
        number1 = value
    End Set
    End Property

    Property input2() As Integer
    Get
        Return number2
    End Get
    Set (value As Integer)
        number2 = value
    End Set
    End Property

    Public Function Add() As Integer
        Dim result As Integer
        result = number1 + number 2
        Return result
    End Function
End Class

Function addNumbers(x As Integer, y As Integer) As Integer
    Dim myCalculator As Calculator = New Calculator()
    Dim myResult As Integer
    myCalculator.input1 = x
    myCalculator.input2 = y
    myResult = myCalculator.Add()
    Return myResult
End Function
```

Identify the classes, instance variables, and methods within the code.

Answer:

The first section of code defines a class, Calculator, that has two properties (input1 and input2) and one method (Add) that adds two numbers together and returns the result. Two variables are defined in the class, number1 and number2.

Two variables are defined in the instance of the calculator class created in the addNumbers() method. The first is the myCalculator variable. This variable is used to reference the instance of the Calculator object created in the method. The second variable is the myResult variable, which is used to store the value returned when the function calls the myCalculator.Add method.

SUCCESSFUL SOFTWARE

OOP offered a new programming paradigm that addressed many of the shortcomings of procedural and string-oriented programming. It also allowed the development of hybrid languages that gave programmers access to the best features of both approaches. OOPLs, both pure and hybrid, have become the most widely used languages and grown to dominate software development. The ability of OOPLs to incorporate other paradigms will likely allow them to remain relevant as new programming paradigms are developed.

—Maura Valentino, MSLIS

BIBLIOGRAPHY

Belton, Padraig. "Coding the Future: What Will the Future of Computing Look Like?" *BBC News*, BBC, 15 May 2015, www.bbc.com/news/business-32743770. Accessed 24 Feb. 2016.

Friedman, Daniel P., and Mitchell Wand. *Essentials of Programming Languages*. 3rd ed., MIT P, 2008.

Haverbeke, Marijn. *Eloquent JavaScript: A Modern Introduction to Programming*. 2nd ed., No Starch Press, 2014.

MacLennan, Bruce J. *Principles of Programming Languages: Design, Evaluation, and Implementation*. 3rd ed., Oxford UP, 1999.

Schneider, David I. *An Introduction to Programming Using Visual Basic*. 10th ed., Pearson, 2016.

Scott, Michael L. *Programming Language Pragmatics*. 4th ed., Morgan Kaufmann Publishers, 2016.

Van Roy, Peter, and Seif Haridi. *Concepts, Techniques, and Models of Computer Programming*. MIT P, 2004.

Weisfeld, Matt. *The Object-Oriented Thought Process*. 4th ed., Addison-Wesley, 2013.

Zakas, Nicholas C. *The Principles of Object-Oriented JavaScript*. No Starch Press, 2014.

P

PRIVACY RIGHTS

Before the 1960s, the right to privacy was not articulated in US law. It was generally understood that US citizens had a right to privacy that protected their persons and property, but it was unclear what was involved in that right. In 1965, in part as the result of widespread cultural changes, the Supreme Court acknowledged the right to privacy for the first time in *Griswold v. Connecticut*, which gave married couples the right to make their own decisions about obtaining birth control. Eight years later in *Roe v. Wade*, the Supreme Court based its support for the right to obtain an abortion on the right to privacy. Since that time, the concept of the right of privacy has been expanded to encompass everything from protection from illegal searches to the sexual rights of homosexuals. Following the 9/11 terrorist attacks on the United States by members of Al Qaeda, the war on terror stood the right to privacy on its head and initiated a lengthy battle between civil libertarians and those who believed privacy should take a back seat to national security.

OVERVIEW

Following acceptance of the right to privacy by state and federal courts, privacy debates erupted over such diverse issues as the right of gay men to engage in consensual sex within the privacy of their own homes, which was in violation of many state sodomy laws (*Bowers v. Hardwick*) and the right of high school students to be free from drug testing without cause for suspicion (*Board of Education v. Earls*). The Supreme Court did not uphold the privacy right in either

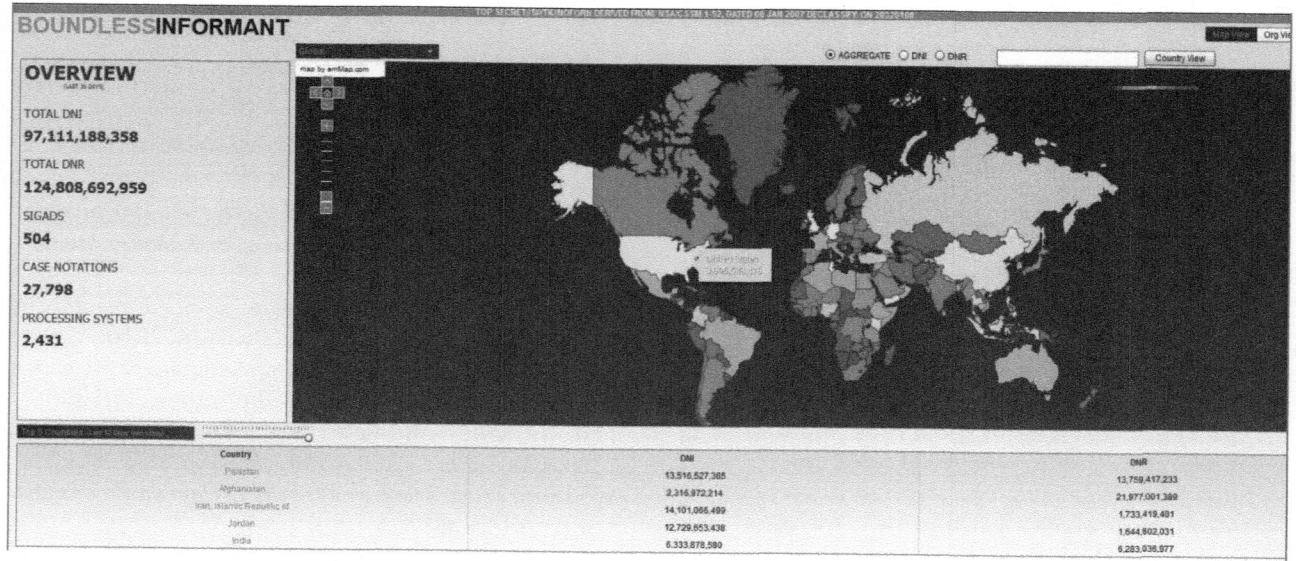

Privacy International 2007 privacy ranking.

case. However, in the 2003 decision *Lawrence v. Texas* the Supreme Court overturned *Bowers v. Hardwick*, holding that state sodomy laws were unconstitutional.

The privacy right of individuals versus the need of law enforcement to build tight cases against suspected offenders creates an ongoing dilemma. New technologies have made it easier to gather information, but privacy advocates contend that they have also made it easier to infringe on the rights of innocent individuals. In the early twenty-first century, one ongoing battle concerns the use of Automatic License Plate Readers (LPRs), which randomly read and analyze data on all vehicles, not just those of suspected criminals. Another battle involves routine collection of DNA of individuals arrested but not convicted of any crime.

Privacy rights of celebrities versus the freedom of the press guaranteed in the First Amendment has also created heated debate in the United States and abroad. When paparazzi were suspected of causing the death of Diana, Princess of Wales, in France in August 1997, there was a global backlash against intrusive press tactics. In 2009, California passed the Paparazzi Reform Initiative, establishing fines for paparazzi that crossed the line into offensiveness. Similar laws were already in effect in Europe.

Two years later, another international scandal over privacy rights erupted when it was reported that British media mogul Rupert Murdoch, who owns Fox News, the *New York Post* and the *Wall Street Journal*, was encouraging his reporters for the *Sun* and *News of the World* to tap the telephones of celebrities such as Prince William, actor Hugh Grant, Harry Potter author J. K. Rowling, and non-celebrity individuals and families involved in news stories. The resulting scandal brought down *News of the World* and toppled careers of top executives in the Murdoch empire while strengthening support for privacy rights in the United Kingdom and elsewhere.

In the United States, passage of the Privacy Act of 1974 restricted the right of government agencies to collect and disseminate information. However, the terrorist attacks of 9/11 ripped away privacy rights as well as the sense of security that many Americans took for granted. The post-9/11 era was defined by George W. Bush's war on terror, in which Congress complied with the president's request for unprecedented authority by passing the Uniting and Strengthening America by Providing Appropriate Tools Required to Intercept and Obstruct Terrorism, better known as the USA PATRIOT Act. Amid accusations of bypassing established rules of due process, the act contained more than a thousand provisions that gave the government the right to tap telephones, carry out surveillance of electronic media, track library activity, and collect DNA samples indiscriminately.

The ongoing battle over privacy took on added heat in the summer of 2013 when Edward Snowden, a former contractor with the National Security Agency (NSA) fled the country after leaking classified information about the government's gathering information on private telephone calls and emails. Snowden's actions led to national outrage concerning the secret Foreign Intelligence Surveillance Court, which was established in 1978 and was being used to allow the NSA to collect information on a variety of activities. Privacy groups ranging from church groups to human rights activists responded by filing suits designed to halt government spying without just cause.

The FBI and the technology company Apple got into a dispute in 2016 over privacy rights on Apple products. On December 2, 2015, Syed Rizwan Farook and Tashfeen Malik killed fourteen people and injured twenty-two at an office holiday party in San Bernardino, California. Both shooters were later killed by police. The FBI was unable to access the data on Farook's iPhone because of the security features on the phone. The federal agency requested that Apple help them hack into the phone, which the company refused on the grounds of maintaining customer privacy and concerns over setting precedents. The FBI eventually dropped the request to Apple and bypassed the company completely by hiring outside coders to write a program to hack the phone. As of April 2016, the FBI is deciding whether or not to share the code with Apple so they can fix the vulnerability in their products. The case continues to explore the gray area of citizens' rights to privacy with new technology in the face of potential threats to national security.

—*Elizabeth Rholetter Purdy, MA, PhD*

Bibliography

Alderman, Ellen and Caroline Kennedy. *The Right to Privacy*. New York: Knopf, 1995. Print.

Allen, Anita L. *Unpopular Privacy: What Must We Do?* New York: Oxford UP, 2011. Print.

Bernal, Paul. *Internet Privacy Rights: Right to Protect Autonomy.* New York: Cambridge UP, 2014. Print.

Brill, Steven. *After: How America Confronted the September 12 Era.* New York: Simon, 2003. Print.

Epstein, Lee, and Thomas G. Walker. *Constitutional Law for a Changing America: A Short Course.* 6th ed. Los Angeles: Sage, 2015. Print.

Keizer, Garret. *Privacy.* New York: Picador, 2012. Print.

Kemper, Bitsy. *The Right to Privacy: Interpreting the Constitution.* New York: Rosen, 2015. Print.

Krimsky, Sheldon, and Tania Simoncelli. *Genetic Justice: DNA Data Banks, Criminal Investigations, and Civil Liberties.* New York: Columbia UP, 2011. Print.

Lichtblau, Eric. "In Secret, Court Vastly Broadens Powers of N.S.A." *New York Times* 7 July 2013, New York ed.: A1. Print.

Strahilevitz, Lior Jacob. "Toward a Positive Theory of Privacy Law." *Harvard Law Review* 126.7 (2013): 2010–42. Print.

Wheeler, Leigh Ann. *How Sex Became a Civil Liberty.* New York: Oxford UP, 2013. Print.

Wright, Oliver. "Hacking Scandal: Is This Britain's Watergate?" *Independent* 9 July 2011, Independent.co.uk. Web. 1 Aug. 2013.

Zetter, Kim, and Brian Barrett. "Apple to FBI: You Can't Force us to Hack the San Bernardino iPhone." *Wired.* Condé Nast, 25 Feb. 2016. Web. 15 Apr. 2016.

PROGRAMMING LANGUAGES

FIELDS OF STUDY

Computer Engineering; Software Engineering; Applications

ABSTRACT

A programming language is a code used to control the operation of a computer and to create computer programs. Computer programs are created by sets of instructions that tell a computer how to do small but specific tasks, such as performing calculations or processing data.

PRINCIPAL TERMS

- **abstraction:** a technique used to reduce the structural complexity of programs, making them easier to create, understand, maintain, and use.
- **declarative language:** language that specifies the result desired but not the sequence of operations needed to achieve the desired result.
- **imperative language:** language that instructs a computer to perform a particular sequence of operations.
- **semantics:** rules that provide meaning to a language.
- **syntax:** rules that describe how to correctly structure the symbols that comprise a language.
- **Turing complete:** a programming language that can perform all possible computations.

WHAT ARE PROGRAMMING LANGUAGES?

Programming languages are constructed languages that are used to create computer programs. They relay sets of instructions that control the operation of a computer. A computer's central processing unit operates through machine code. Machine code is based on numerical instructions that are incredibly difficult to read, write, or edit. Therefore, higher-level programming languages were developed to simplify the creation of computer programs. Programming languages are used to write code that is then converted to machine code. The machine code is then executed by a computer, smartphone, or other machine.

There are many different types of programming languages. First-generation, or machine code, languages are processed by computers directly. Such languages are fast and efficient to execute. However, they are difficult for humans to read and require advanced knowledge of hardware to use. Second-generation languages, or assembly languages, are more easily read by humans. However, they must be converted into machine code before being executed by a computer. Second-generation languages are used more often than first-generation ones because

TIOBE Programming Community Index
Source: www.tiobe.com

On the TIOBE index for various programming languages, C and Java are the highest-rated languages.

they are easier for humans to use while still interacting quickly and efficiently with hardware.

Both first- and second-generation languages are low-level languages. Third-generation languages are the most widely used programming languages. Third-generation, or high-level programming, languages are easier to use than low-level languages. However, they are not as fast or efficient. Early examples of such languages include Fortran, COBOL, and ALGOL. Some of the most widely used programming languages in the twenty-first century are third-generation. These include C++, C#, Java, JavaScript, and BASIC. Programming languages with higher levels of abstraction are sometimes called fourth-generation languages. Higher levels of abstraction increases platform independence. Examples include Ruby, Python, and Perl.

Programming languages can be based on different programming paradigms or styles. Imperative languages, such as COBOL, use statements to instruct the computer to perform a specific sequence of operations to achieve the desired outcome. Declarative languages specify the desired outcome but not the specific sequence of operations that will be used to achieve it. Structured Query Language (SQL) is one example.

Programming languages can also be classified by the number of different computations they can perform. Turing complete programming languages can perform all possible computations and algorithms. Most programming languages are Turing complete. However, some programming languages, such as Charity and Epigram, can only perform a limited number of computations and are therefore not Turing complete.

How Programming Languages Are Structured

The basic structural rules of a programming language are defined in its syntax and semantics. A programming language's syntax is the grammar that defines the rules for how its symbols, such as words, numbers, and punctuation marks, are used. A programming language's semantics provide the rules used to interpret the meaning of statements constructed using its syntax. For example, the statement 1 + pizza might comply with a programming language's syntax. However, adding a number and a word together (as opposed to adding two numbers) might be semantically meaningless. Programs created with a programming language must comply with the structural rules established by the language's syntax and semantics if they are to execute correctly.

Abstraction reduces the structural complexity of programs. More abstract languages are easier to understand and use. Abstraction is based on the principle that any piece of functionality that a program uses should be implemented only once and never duplicated. Abstraction focuses only on the essential requirements of a program. Abstraction can be implemented using subroutines. A subroutine is a sequence of statements that perform a specific task, such as checking to see if a customer's name exists in a text file. If abstraction is not used, the sequence of statements needed to check if a customer's name exists in the file would need to be repeated every place in the program where such a check is needed. With subroutines, the sequence of statements exists in only one place, within the subroutine. Thus, it does not need to be duplicated.

Using Pseudocode

The programming language statements that comprise a program are called "code." Code must comply with all of the programming language's syntax and semantic rules. Pseudocode uses a combination of a programming language and a natural language such as English to simply describe a program or algorithm. Pseudocode is easier to understand than code and is often used in textbooks and scientific publications.

The Future of Programming Languages

Programming languages are growing in importance with the continued development of the Internet and with the introduction of new programmable machines including household appliances, driverless cars, and remotely controlled drones. Such systems will increase the demand for new programming languages and paradigms designed for larger, more complex, and highly interconnected programs.

—*Maura Valentino, MSLIS*

SAMPLE PROBLEM

The pseudocode below describes an algorithm designed to count the number of words in a text file and to delete the file if no words are found:

Open the file
For each word in file
counter = counter + 1;
If counter = 0 Then
Delete file
Close the file

Use the pseudocode to describe an algorithm that counts the number of words in a file and then prints the number of words if the number of words is greater than 300.

Answer:

Open the file
For each word in file
wordcount = wordcount + 1;
If wordcount > 300 Then
print wordcount
Close the file

The pseudocode above uses natural language for statements such as Open the file and programming language for statements such as wordcount = wordcount + 1. Pseudocode cannot be executed by a computer. However, it is helpful for showing the outline of a program or algorithm's operating principles.

Bibliography

Belton, Padraig. "Coding the Future: What Will the Future of Computing Look Like?" *BBC News*. BBC, 15 May 2015. Web. 24 Feb. 2016.

Friedman, Daniel P., and Mitchell Wand. *Essentials of Programming Languages*. Cambridge: MIT P, 2006. Print.

Harper, Robert. *Practical Foundations for Programming Languages*. Cambridge: Cambridge UP, 2013. Print.

MacLennan, Bruce J. *Principles of Programming Languages: Design, Evaluation, and Implementation*. Oxford: Oxford UP, 1999. Print.

Scott, Michael L. *Programming Language Pragmatics*. Burlington: Kaufmann, 2009. Print.

Van Roy, Peter. *Concepts, Techniques, and Models of Computer Programming*. Cambridge: MIT P, 2004. Print.

Watt, David A. *Programming Language Design Concepts*. West Sussex: Wiley, 2004. Print.

Woods, Dan. "Why Adopting the Declarative Programming Practices Will Improve Your Return from Technology." *Forbes*. Forbes.com, 17 Apr. 2013. Web. 2 Mar. 2016.

PROTOTYPING

FIELDS OF STUDY

Software Development; Programming Methodologies; Software Engineering

ABSTRACT

Prototyping is a software development methodology that focuses on the use of working models that are repeatedly refined based on feedback from the end user. Prototyping is most often used to develop systems that include significant end-user interaction and complex user interfaces.

PRINCIPAL TERMS

- **horizontal prototype:** a working software model that focuses on the user interface and provides a broad view of the overall system.
- **integration:** in computer science, the process of combining individual system components, whether hardware or software, into a single, unified system.
- **requirement:** a necessary characteristic of a software system or other product, such as a service it must provide or a constraint under which it must operate.
- **vertical prototype:** a working software model that focuses on a specific component of the overall system and details how that component will work.

UNDERSTANDING SOFTWARE PROTOTYPING

Prototyping is a software development methodology that involves making working models of the application as it is being developed. These working models, or prototypes, are repeatedly refined and improved based on feedback from users and other stakeholders. Prototyping is an iterative development method in which prototypes are designed, built, evaluated, and refined repeatedly until the system functions as required.

Prototypes are working models, but they are not complete. They include only partial functionality and are intended to be revised, expanded, and improved during each cycle in the development process. There are two categories of prototypes. Horizontal prototypes model the entire software system under development. They often focus on the user interface, offering a broad view of the overall system with limited internal functioning. Vertical prototypes are used to model individual parts of the system in an in-depth manner. They offer a more detailed view of how individual components, such as a specific system function or subsystem, will work.

Prototyping typically consists of four main steps: requirements identification, prototype development, user evaluation, and prototype revision. In the first step, the basic requirements that the software must fulfill are determined. The focus at this stage is on the users' needs and the design of the user interface, not on developing comprehensive and detailed requirements, as would be the case in traditional methodologies such as waterfall development. In the second step, the prototype is created, based on the project's initial requirements and on feedback received during any previous prototyping cycles. In the user evaluation phase, the customer and other stakeholders test the prototype and provide feedback to the development team. The fourth step involves revision or enhancement of the prototype based on that feedback. Steps two through four are then repeated until the application meets the needs of the users and other stakeholders.

There are different types of prototyping. In throwaway prototyping, also known as rapid or close-ended

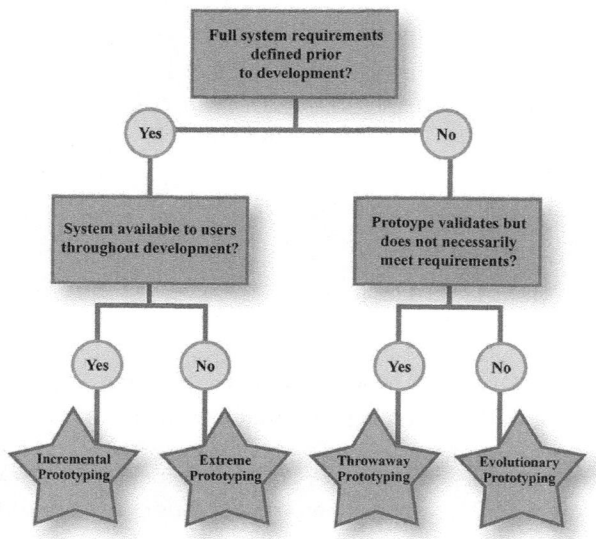

Choosing the correct approach to prototyping will depend on how much time is invested up front to define system requirements, whether or not the system will be available for use throughout development, and whether or not the prototype will be used to validate or fulfill a requirement during the development process.

prototyping, the first prototype is created very early in the process, after only basic requirements have been determined. The creation of this preliminary prototype allows developers and stakeholders to further understand and refine the requirements. The prototype is then discarded, and work begins on the actual system. This method is often used to quickly test a small portion of a larger software system. Once the throwaway prototype has been used to verify that this small portion functions correctly, its functionality is incorporated into the final system. This process of incorporating functionality is called integration.

Another type is called evolutionary prototyping. In evolutionary prototyping, the first prototype is based only on those requirements that are well defined and clearly understood at the outset. The initial prototype is not discarded. Instead, new functionality is added to it as more requirements are clarified and understood. This process is repeated until the initial, minimally functional prototype has been expanded into the final, fully functional system.

In incremental prototyping, multiple prototypes are built for each of the various system components. These components are developed separately and then integrated to form the completed application.

Extreme prototyping is a three-step method commonly used for web development. First, a static, typically HTML-based prototype of the entire website or application is developed. Next, a second prototype that simulates the complete user interface is developed. This prototype does not contain the underlying functionality required for the application to be operational. Actual functionality is added in the final step by integrating the prototypes from the first two steps.

Emphasis on User Interaction

Prototyping offers several advantages over other methodologies. As users are involved in all stages of the development process, user enthusiasm for, understanding of, and commitment to the project is maximized. User involvement also allows design flaws and missing functionality to be detected early in the development process. This increases the chance that the final application will meet the real-world needs of the end user.

While prototyping offers considerable benefits, there are also downsides to this methodology. Failure to determine detailed specifications at the beginning of the design process can result in constantly changing specifications and the inclusion of additional features that have a negative impact on project costs and scheduling. The system being developed may grow in size and complexity, causing stakeholders to lose sight of the project's original objectives. In addition, projects developed using prototyping must be managed to ensure that resources are used efficiently when building multiple prototypes.

Prototyping differs greatly from linear development methodologies such as the waterfall model. These methodologies define system specifications in detail at the beginning of the design process and minimize changes during the development process. Prototyping is more closely related to other iterative development models. Prototypes are often used in conjunction with iterative methodologies such as rapid application development (RAD) or the spiral model.

Software Prototyping in Practice

Prototyping is commonly used to develop web applications that feature extensive user interaction with the system. For example, a retailer may need to develop a virtual storefront that includes all of the products featured in their stores. Such an application will be focused on a diverse set of users of varying backgrounds and technical abilities, and the success of the application will rest on the ease with which end users can successfully navigate and use the system. The prototyping development model is ideal for such an application, as it allow users to be involved with and influence its development from project inception to final completion. This greatly increases the likelihood that the final system will meet the users' needs.

The Power of User Involvement

Prototyping facilitates the development of software systems that are easy to use and focused on meeting the end user's needs. As such, it is an ideal methodology for developing systems that require extensive user interaction and easy-to-use interfaces, such as web applications and mobile consumer applications. Prototyping allows for early identification of weaknesses in a system's design, and it encourages increased buy-in from project stakeholders. As software systems grow increasingly important in the lives of everyday users, prototyping, with its focus on user involvement and satisfaction, will remain an important software design methodology.

—*Maura Valentino, MSLIS*

Bibliography

Bell, Michael. *Incremental Software Architecture: A Method for Saving Failing IT Implementations.* John Wiley & Sons, 2016.

Friedman, Daniel P., and Mitchell Wand. *Essentials of Programming Languages.* 3rd ed., MIT P, 2008.

Jayaswal, Bijay K., and Peter C. Patton. *Design for Trustworthy Software: Tools, Techniques, and Methodology of Developing Robust Software.* Prentice Hall, 2007.

MacLennan, Bruce J. *Principles of Programming Languages: Design, Evaluation, and Implementation.* 3rd ed., Oxford UP, 1999.

Scott, Michael L. *Programming Language Pragmatics.* 4th ed., Elsevier, 2016.

Van Roy, Peter, and Seif Haridi. *Concepts, Techniques, and Models of Computer Programming.* MIT P, 2004.

Wysocki, Robert K. *Effective Project Management: Traditional, Agile, Extreme.* 7th ed., John Wiley & Sons, 2014.

QUANTUM COMPUTING

FIELDS OF STUDY
Computer Science; System-Level Programming; Computer Engineering

ABSTRACT
Quantum computing is an emerging field of computer engineering that uses charged particles rather than silicon electrical circuitry to process signals. Engineers believe that quantum computing has the potential to advance far beyond the limitations of traditional computing technology.

PRINCIPAL TERMS
- **entanglement:** the phenomenon in which two or more particles' quantum states remain linked even if the particles are later separated and become part of distinct systems.
- **quantum logic gate:** a device that alters the behavior or state of a small number of qubits.
- **quantum bit (qubit):** a basic unit of quantum computation that can exist in multiple states at the same time, and can therefore have multiple values simultaneously.
- **state:** a complete description of a physical system at a specific point in time, including such factors as energy, momentum, position, and spin.
- **superposition:** the principle that two or more waves, including waves describing quantum states, can be combined to give rise to a new wave state with unique properties. This allows a qubit to potentially be in two states at once.

SUBATOMIC COMPUTATION THEORIES
Quantum computing is an emerging field of computing that uses subatomic particles rather than silicon circuitry to transmit signals and perform calculations. Quantum physics studies particle behavior at the subatomic scale. At extremely small scales, subatomic particles such as photons (the basic unit of light) exhibit properties of both particles and waves. This phenomenon, called wave-particle duality, gives subatomic particles unique properties. Traditional computer algorithms are constrained by the physical properties of digital electrical signals. Engineers working on quantum computing hope that quantum algorithms, based on the unique properties of quantum mechanics, will be able to complete computations faster and more efficiently.

THE BASICS OF QUANTUM DATA
Digital computing uses electrical signals to create binary data. Binary digits, called bits, have two possible values: 0 or 1. Digital computers also use logic gates. These are electronic circuits that process bits of data by amplifying or changing signals. Logic gates in digital computers accept one or more inputs and produce only one output. In quantum computing, digital bits are replaced by quantum bits (qubits). Qubits are created by manipulating subatomic particles.

The value of a qubit represents its current quantum state. A quantum state is simply all known data about a particle, including its momentum, physical location, and energetic properties. To be used as a qubit, a particle should have two distinct states, representing the binary 0 and 1. For example, if the qubit is a photon, the two states would be horizontal polarization (0) and vertical polarization (1). However, a particle in a quantum system can exist in two or more states at the same time. This principle is called superposition. Thus, a qubit is not limited to binary values of either 0 or 1. Instead, it can have a

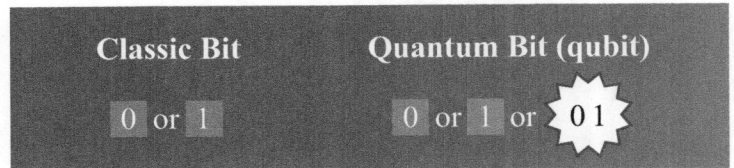

Quantum computing uses quantum bits (qubits). Classic bits can be in one of two states, 0 or 1, but qubits can be in state 0, state 1, or superstate 01.

value of 0, 1, or any superposition of both 0 and 1 at the same time.

Quantum particles also display a property known as entanglement. This is when two or more particles are linked in such a way that changing the state of one particle changes the state of the other(s), even after they are physically separated. Entanglement could potentially allow for the development of quantum computers that can instantly transmit information across great distances and physical barriers.

Practical Design of Quantum Computers

Current designs for quantum computers use energetic particles such as electrons or photons as qubits. The states of these particles are altered using quantum logic gates, much like digital logic gates alter electrical signals. A quantum gate may operate using energy from lasers, electromagnetic fields, or several other methods. These state changes can then be used to calculate data.

One avenue of research is the potential derivation of qubits from ion traps. Ions are atoms that have lost or gained one or more electrons. Ion traps use electric and magnetic fields to catch, keep, and arrange ions.

The Potential of Quantum Computing

As of 2016, the practical value of quantum computing had only been demonstrated for a small set of potential applications. One such application is Shor's algorithm, created by mathematician Peter Shor, which involves the mathematical process of factorization. Factorization is used to find two unknown prime numbers that, when multiplied together, give a third known number. Shor's algorithm uses the properties of quantum physics to speed up factorization. It can perform the calculation twice as fast as a standard algorithm. Researchers have also demonstrated that quantum algorithms might improve the speed and accuracy of search engines. However, research in this area is incomplete, and the potential benefits remain unclear.

There are significant challenges to overcome before quantum computing could become mainstream. Existing methods for controlling quantum states and manipulating particles require highly sensitive materials and equipment. Scientists working on quantum computers argue that they may make the biggest impact in technical sciences, where certain math and physics problems require calculations so extensive that solutions could not be found even with all the computer resources on the planet. Special quantum properties, such as entanglement and superposition, mean that qubits may be able to perform parallel computing processes that would be impractical or improbable with traditional computer technology.

—*Micah L. Issitt*

Bibliography

Ambainis, Andris. "What Can We Do with a Quantum Computer?" *Institute Letter* Spring 2014: 6–7. *Institute for Advanced Study*. Web. 24 Mar. 2016.

Bone, Simon, and Matias Castro. "A Brief History of Quantum Computing." *SURPRISE* May–June 1997: n. pag. *Department of Computing, Imperial College London*. Web. 24 Mar. 2016.

Crothers, Brooke. "Microsoft Explains Quantum Computing So Even You Can Understand." *CNET*. CBS Interactive, 25 July 2014. Web. 24 Mar. 2016.

Gaudin, Sharon. "Quantum Computing May Be Moving out of Science Fiction." *Computerworld*. Computerworld, 15 Dec. 2015. Web. 24 Mar. 2016.

"The Mind-Blowing Possibilities of Quantum Computing." *TechRadar*. Future, 17 Jan. 2010. Web. 26 Mar. 2016.

"A Quantum Leap in Computing." *NOVA*. WGBH/PBS Online, 21 July 2011. Web. 24 Mar. 2016.

R

RANDOM ACCESS MEMORY (RAM)

FIELDS OF STUDY
Computer Engineering; Information Technology

ABSTRACT
Random-access memory (RAM) is a form of memory that allows the computer to retain and quickly access program and operating system data. RAM hardware consists of an integrated circuit chip containing numerous transistors. Most RAM is dynamic, meaning it needs to be refreshed regularly, and volatile, meaning that data is not retained if the RAM loses power. However, some RAM is static or nonvolatile.

PRINCIPAL TERMS

- **direct-access storage:** a type of data storage in which the data has a dedicated address and location on the storage device, allowing it to be accessed directly rather than sequentially.
- **dynamic random-access memory (DRAM):** a form of RAM in which the device's memory must be refreshed on a regular basis, or else the data it contains will disappear.
- **nonvolatile random-access memory (NVRAM):** a form of RAM in which data is retained even when the device loses access to power.
- **read-only memory (ROM):** a type of nonvolatile data storage that can be read by the computer system but cannot be modified.
- **shadow RAM:** a form of RAM that copies code stored in read-only memory into RAM so that it can be accessed more quickly.
- **static random-access memory (SRAM):** a form of RAM in which the device's memory does not need to be regularly refreshed but data will still be lost if the device loses power.

HISTORY OF RAM
The speed and efficiency of computer processes are among the most areas of greatest concern for computer users. Computers that run slowly (lag) or stop working altogether (hang or freeze) when one or more programs are initiated are frustrating to use. Lagging or freezing is often due to insufficient computer memory, typically random-access memory (RAM). RAM is an essential computer component that takes the form of small chips. It enables computers to work faster by providing a temporary space in which to store and process data. Without RAM, this data would need to be retrieved from direct-access storage or read-only memory (ROM), which would take much longer.

Computer memory has taken different forms over the decades. Early memory technology was based on vacuum tubes and magnetic drums. Between the 1950s and the mid-1970s, a form of memory called "magnetic-core memory" was most common. Although RAM chips were first developed during the same period, they were initially unable to replace core memory because they did not yet have enough memory capacity.

A major step forward in RAM technology came in 1968, when IBM engineer Robert Dennard patented the first dynamic random-access memory (DRAM) chip. Dennard's original chip featured a memory cell consisting of a paired transistor and capacitor. The capacitor stored a single bit of binary data as an electrical charge, and the transistor read and refreshed the charge thousands of times per second. Over the following years, semiconductor companies such as Fairchild and Intel produced DRAM chips of varying capacities, with increasing numbers of memory cells per chip. Intel also introduced DRAM with three transistors per cell, but over time the need for smaller and smaller computer components made this design

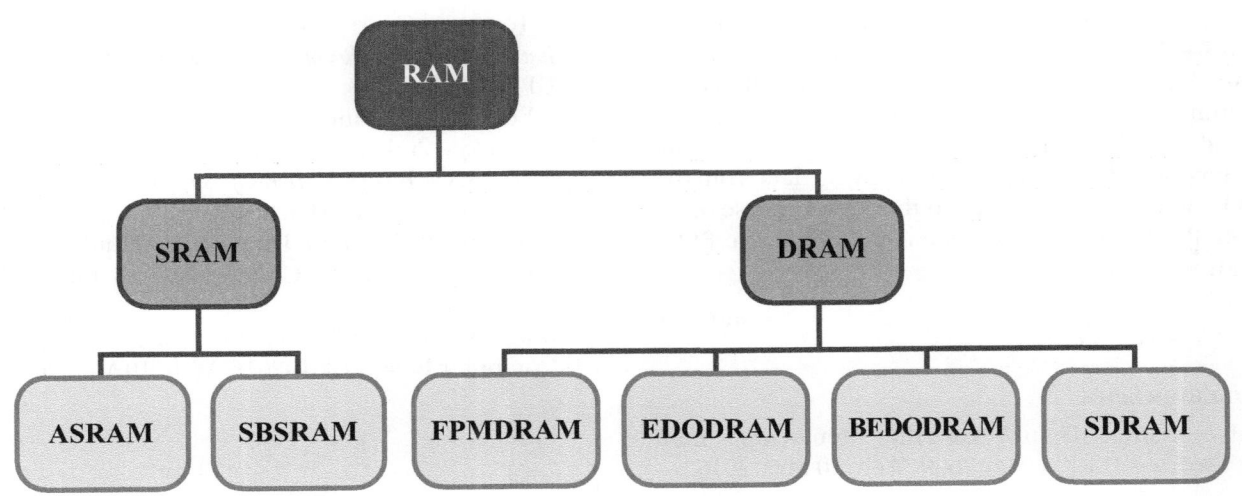

There are two major categories of random-access memory: static RAM (SRAM) and dynamic RAM (DRAM). Static RAM may be asynchronous SRAM (ASRAM) or synchronous SRAM with a burst feature (SBSRAM). Dynamic RAM may come in one of four types: fast page mode DRAM (FPMDRAM), extended data out DRAM (EDODRAM), extended data out DRAM with a burst feature (BEDODRAM), or synchronous DRAM (SDRAM).

less practical. In the 2010s, commonly used RAM chips incorporate billions of memory cells.

Types of RAM

Although all RAM serves the same basic purpose, there are a number of different varieties. Each type has its own unique characteristics. The RAM most often used in personal computers is a direct descendant of the DRAM invented by Dennard and popularized by companies such as Intel. DRAM is dynamic, meaning that the electrical charge in the memory cells, and thus the stored data, will fade if it is not refreshed often. A common variant of DRAM is speed-focused double data rate synchronous DRAM (DDR SDRAM), the fourth generation of which entered the market in 2014.

RAM that is not dynamic is known as static random-access memory (SRAM). SRAM chips contain many more transistors than their DRAM counterparts. They use six transistors per cell: two to control access to the cell and four to store a single bit of data. As such, they are much more costly to produce. A small amount of SRAM is often used in a computer's central processing unit (CPU), while DRAM performs the typical RAM functions.

Just as the majority of RAM is dynamic, most RAM is also volatile. Thus, the data stored in the RAM will disappear if it is no longer being supplied with electricity—for instance, if the computer in which it is installed has been turned off. Some RAM, however, can retain data even after losing power. Such RAM is known as nonvolatile random-access memory (NVRAM).

Using RAM

RAM works with a computer's other memory and storage components to enable the computer to run more quickly and efficiently, without lagging or freezing. Computer memory should not be confused with storage. Memory is where application data is processed and stored. Storage houses files and programs. It takes a computer longer to access program data stored in ROM or in long-term storage than to access data stored in RAM. Thus, using RAM enables a computer to retrieve data and perform requested functions faster. To improve a computer's performance, particularly when running resource-intensive programs, a user may replace its RAM with a higher-capacity chip so the computer can store more data in its temporary memory.

Shadow RAM

While RAM typically is used to manage data related to the applications in use, at times it can be used to assist in performing functions that do not usually involve RAM. Certain code, such as a computer's basic input/output system (BIOS), is typically stored within the computer's ROM. However, accessing data saved in ROM can be time consuming. Some computers can address this issue by copying data from the ROM and storing the copy in the RAM for ease of access. RAM that contains code copied from the ROM is known as shadow RAM.

—*Joy Crelin*

BIBLIOGRAPHY

Adee, Sally. "Thanks for the Memories." *IEEE Spectrum*. IEEE, 1 May 2009. Web. 10 Mar. 2016.

Hey, Tony, and Gyuri Pápay. *The Computing Universe: A Journey through a Revolution*. New York: Cambridge UP, 2015. Print.

ITL Education Solutions. *Introduction to Information Technology*. 2nd ed. Delhi: Pearson, 2012. Print.

"Shadow RAM Basics." *Microsoft Support*. Microsoft, 4 Dec. 2015. Web. 10 Mar. 2016.

Stokes, Jon. "RAM Guide Part I: DRAM and SDRAM Basics." *Ars Technica*. Condé Nast, 18 July 2000. Web. 10 Mar. 2016.

"Storage vs. Memory." *Computer Desktop Encyclopedia*. Computer Lang., 1981–2016. Web. 10 Mar. 2016.

RAPID APPLICATION DEVELOPMENT (RAD)

FIELDS OF STUDY

Software Development; Software Engineering

ABSTRACT

Rapid application development (RAD) is a software development methodology that relies on rapid prototyping instead of the extensive planning used in traditional software development. RAD is an incremental methodology that uses short iterations, reusable prototypes, and multiple software modules. These are combined in a process called integration to form the complete solution.

PRINCIPAL TERMS

- **integration:** in computer science, the process of combining individual system components, whether hardware or software, into a single, unified system.
- **iteration:** in software development, a single, self-contained phase of a product's overall development cycle, typically lasting one to three weeks, in which one aspect of the product is addressed from beginning to end before moving onto the next phase of development.
- **prototype:** a working model that serves as a basis for a final design.
- **software module:** an individual component of a software system that is integrated with other modules to form the final, completed design.

WHAT IS RAPID APPLICATION DEVELOPMENT (RAD)?

Rapid application development (RAD) is a software development methodology that relies on rapid prototyping instead of the extensive planning used in traditional software development methodologies. RAD is an incremental and iterative methodology. It uses short iterations, reusable prototypes, and multiple software modules developed in parallel. These elements are then combined in a process called integration to form the complete system or solution. Developers work closely with managers, customers, and other stakeholders in small teams.

RAD typically uses a five-phase development cycle. The first phase is known as business modeling. During this phase, requirements are identified and analyzed. The second phase is data modeling, during which the information gathered in the previous phase is used to define the required data objects. Next is the process modeling phase, in which the data models developed in the previous phase are converted to carry out the business requirements identified in the first phase. Processes for interacting with the data sets

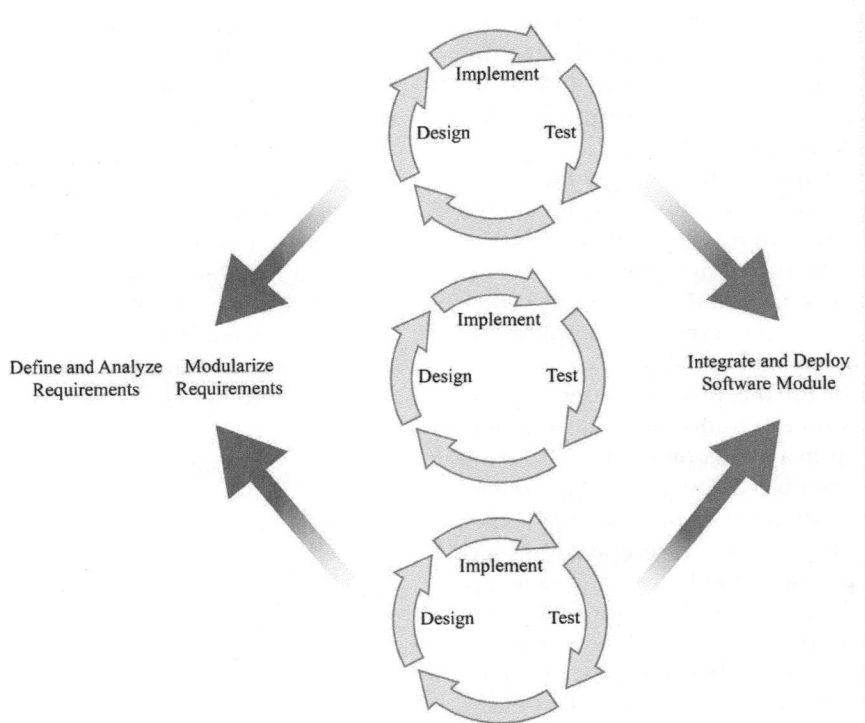

In the rapid application development model, after the initial requirements are defined and analyzed, they are grouped into modules for teams to work on separately. Each module will go through design, implement, and test cycles until the output meets the needs of the end user, at which point the modules are integrated into the system and deployed.

are defined. The fourth phase is the application generation phase, in which prototypes are created. The process and data models developed in the previous phases are converted to code using automation tools. The final phase is testing and turnover. Components are tested to ensure that the system functions correctly as new components are integrated into the system. RAD requires less final testing than other methodologies, as prototypes undergo independent testing. All phases are repeated during every iteration.

Quick, Adaptable, and Effective

RAD's iterative and incremental processes allow developers to quickly respond to changing requirements and deliver solutions. The methodology encourages the development of reusable components. Frequent integration of components makes it easier to minimize the impact of any bugs that are introduced to the system as new components are deployed and integrated. In addition, the customer and other stakeholders are involved throughout the design cycle, improving the chances that the solution will meet the customer's needs.

Successful RAD depends on strong development teams. Team members must correctly identify business processes and requirements, and a high level of expertise is needed, particularly with modeling. Systems that are not easily modularized are not good candidates for RAD, as modularization of functionality is a core concept of this methodology. RAD is most appropriate when developing medium to large solutions. The costs associated with managing and implementing RAD may be excessive for use on smaller, less complex designs.

Using RAD to Develop Solutions

RAD is an appropriate methodology in many situations. It is ideal when short development times are necessary, when proper expertise is available to analyze business processes and data usage, and when the costs associated with the use of automated development tools can be justified. RAD is also a good choice for projects with changing requirements.

For example, the information technology (IT) department for a large, multinational corporation might be charged with developing an application that will integrate the company's human resource (HR) management system with that of another company that it has recently acquired. The application must be delivered by the end of the year.

RAD is an ideal fit for this project for several reasons. The project is time sensitive, and the solution must be delivered within a matter of months. Based on an initial analysis, the required design will lend itself to modularization. The company employs a large, experienced IT staff that has completed several large-scale projects and has expertise in modeling and system architecture design. The project is a high priority for the HR and accounting departments, which are willing to make staff available to provide business expertise, customer testing, and feedback for the project. Given the corporation's size, the cost of RAD tools is not prohibitive. There is a strong likelihood that the investment can be leveraged on future projects. These factors all support the use of the RAD methodology.

The Future Is RAD

RAD helps deliver effective software solutions quickly and flexibly. Functionality can be developed and added as needed, resulting in adaptive solutions that are more likely to meet project requirements. For this reason, RAD is ideally suited for large projects with changing requirements. The extensive use of modeling, prototypes, and automation speeds the development process. Customer involvement in the design and testing process provides valuable feedback and leads to solutions that are more likely to meet the real needs of the customer. Companies in fast-moving, highly competitive business sectors, such as mobile application development and web commerce, need solutions that can be deployed quickly and changed easily as new customer needs are discovered and new business opportunities present themselves. With the rapidly evolving world of software becoming increasingly important to companies of all kinds, RAD will remain an important methodology for many developers.

—*Maura Valentino, MSLIS*

Bibliography

Bell, Michael. *Incremental Software Architecture: A Method for Saving Failing IT Implementations.* John Wiley & Sons, 2016.

Friedman, Daniel P., and Mitchell Wand. *Essentials of Programming Languages.* 3rd ed., MIT P, 2008.

Jayaswal, Bijay K., and Peter C. Patton. *Design for Trustworthy Software: Tools, Techniques, and Methodology of Developing Robust Software.* Prentice Hall, 2007.

MacLennan, Bruce J. *Principles of Programming Languages: Design, Evaluation, and Implementation.* 3rd ed., Oxford UP, 1999.

Scott, Michael L. *Programming Language Pragmatics.* 4th ed., Elsevier, 2016.

Van Roy, Peter, and Seif Haridi. *Concepts, Techniques, and Models of Computer Programming.* MIT P, 2004.

Wysocki, Robert K. *Effective Project Management: Traditional, Agile, Extreme.* 7th ed., John Wiley & Sons, 2014.

RATIONAL CHOICE THEORY

Rational choice theory is at the heart of microeconomics, and despite the fact that it has received little empirical support in any field of academic inquiry, it continues to dominate economics and political discourse in many countries. Rational choice theory holds that human beings are unique in the world because they use reason to guide their actions rather than relying wholly on instinct as animals do. Numerous philosophers throughout the centuries have argued that reason is drawn from God, human nature, or some other source and explains why people will act in ways that are not wholly consistent with self-gratification.

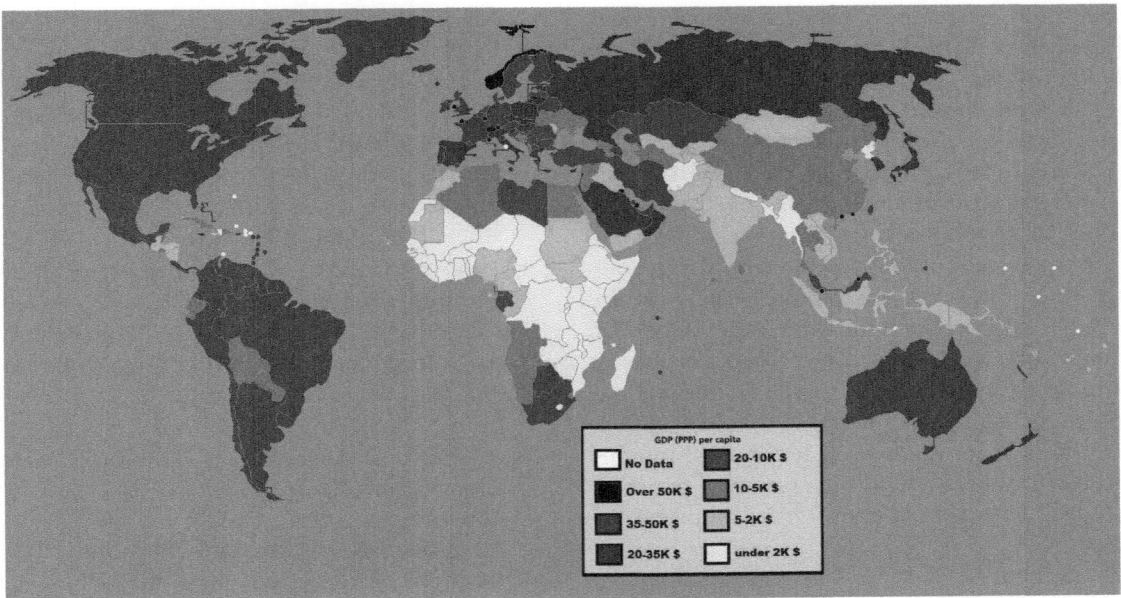

GDP per capita (data shown comes from Worldbank Data) income and resources have a direct impact on the kinds of choices an individual is able to make.

Overview

Rational choice theory explains why, for example, parents will sacrifice their time and efforts for the sake of their children, rulers will give up some ceremony in order to develop their holdings for the future, and children will forego play in favor of studying to obtain a better education and future prospects. At lower levels, of course, these actions are indistinguishable from those of animals. Where the rational choice theory differs is in supposing that people have agency over their actions and sovereignty over their decisions.

The agency aspect of rational choice theory involves both the ability to affect the future and the self-awareness that people have in their ability to change what will happen. It is possible to say that animals have the power to change future conditions, but they are not aware of this and do not use it to plan their behavior. Humans, on the other hand, are aware that they have different options available to them and will choose the option that they believe will maximize their returns, as measured by the concept of utility. Utility is a form of general purpose benefit that will be configured differently according to the inclinations or character of the individuals involved. Some people will have a preference for monetary gain, some for physical pleasure, and some for long-term security. The choices they make about how they behave will reflect these preferences.

The sovereignty aspect of rational choice theory indicates that people have the power to make these kinds of decisions. That is, they have the time and the resources to gather information about the different options, to assess reasonably accurately the likely outcomes of the various choices, and then to make their choices according to their own preferences (tastes and habits) and beliefs (perceptions about cause and effect). However, in real life, people most often have insufficient time and resources to accurately assess the possible results of their actions. Thus, they will routinize most decisions, repeatedly choosing an option that has provided an acceptable outcome in the past. Given the asymmetries of information that exist between providers and consumers, it is not surprising that people are rarely able to make accurate and rational choices. Even if they are able to calculate all their options in detail, numerous familial, social, cultural, and legal constraints exist that inhibit most people from behaving entirely according to their own desires and appetites. People for whom this does not apply are known as sociopaths.

Despite its shortcomings, rational choice theory remains a useful tool for predicting how most people will behave and is particularly relevant to market situations in which consumers can choose from among multiple options that can satisfy their demand. Along with lifestyle and routine activities theories, which deal with the targets of and motivations for victimization, rational choice theory has informed criminology, wherein punishments are believed to deter rational people from choosing to engage in criminal activity because the cost of the activity would outweigh the gain. However, the contradictions within the theory become problematic when rational choice is extended for application in non-market-based decision making to attempt to explain relations between partners, family members, and other important social interactions. This has led many humanities scholars outside the field of economics to assert that there is little empirical basis for the theory.

—*John Walsh, PhD*

BIBLIOGRAPHY

Ahmad, Janice, and Traqina Q. Emeka. "Rational Choice Theory." *Encyclopedia of Criminology and Criminal Justice* (Jan. 2014): 1–5. Web. 5 July 2015.

Friedman, Debra, and Michael Hechter. "The Contribution of Rational Choice Theory to Macrosociological Research." *Sociological Theory* 6.2 (1988): 201–18. Print.

Grassl, Wolfgang. "Aquinas on Management and Its Development." *Journal of Management Development* 29.7–8 (2010): 706–15. Print.

Hernstein, Richard J. "Rational Choice Theory: Necessary but Not Sufficient." *American Psychologist* 45.3 (1990): 356–67. Print.

Portmore, Douglas W. "Imperfect Reasons and Rational Options." *Noûs* 46.1 (2012): 24–60. Print.

Redd, Steven P., and Alex Mintz. "Policy Perspectives on National Security and Foreign Policy Decision Making." *Policy Studies Journal* 41.5 (2013): 511–37. Print.

Scott, John. "Rational Choice Theory." *Understanding Contemporary Society: Theories of the Present*. Ed. Gary Browning, Abigail Halcli, and Frank Webster. London: Sage, 2000. 126–37. Print.

Thompson. A. "Applying Rational Choice Theory to International Law: The Promise and Pitfalls." *Journal of Legal Studies* 31.1, part 2 (2002): n. pag. Print.

Williams, Christine B., and Jane Fedorowicz. "Rational Choice and Institutional Factors Underpinning State-Level Interagency Collaboration Initiatives." *Transforming Government: People, Process and Policy* 6.1 (2012): 13–26. Print.

S

SEARCH ENGINE OPTIMIZATION (SEO)

Search Engine Optimization (SEO) is an Internet marketing technique that tailors a website—through the inclusion of keywords and indexing terms and the manipulation of HTLM or other coding—to position it to receive a high, organic (unpaid) ranking within search engines such as Google or Bing. Competition for user attention on the Internet is fierce; therefore, website designers employ search engine optimization to make sure their websites are listed within the first three or four screens, because research has shown that few users will search much beyond that point. The goal of optimizing a website's position in a search engine's results, then, is to attract a higher volume of web traffic that, in turn, translates into higher visibility and/or greater profits. Although the technology behind SEO reflects some of the most sophisticated computer science programming, its logic reflects the conventional wisdom of more than a century of advertising by creating an audience or market through effective product placement.

Overview

Many websites simply pay for advertising, either through the search engines themselves or on the social media networks that accept ads. But when Google or similar search engines accept payment for rankings, those websites are listed as sponsored links to underscore that their inclusion in the results is not organic. In the Internet era, to ignore the search engines risks jeopardizing a potential exponential expansion of the reach of the website. Early experience with the market potential of the Internet in the mid-1990s revealed that many webmasters were manipulating the key term algorithm to attract a wider market, often by using misleading key term indicators that ultimately wasted users' time, as search engines would simply download the site's initial screen and gather key terms (called "crawling") that would in turn be indexed. As the Internet expanded and the competition for traffic increased, search engines have by necessity grown far more protective and far more careful in sharing with webmasters how they gather the key terms and how those key terms are in turn used to rank the websites.

The strategy for webmasters is to design a website's initial page to attract the search engine's attention and, thus, to organically generate a high ranking in the results index that users are given on their screens. To do this, webmasters turn to professional search engine optimizers who are savvy in manipulating the search engine's indexing process. This has led to controversy among more traditional websites that

This is image of impression on search engine optimization.

believe the website's content should attract the audience rather than the deliberate manipulation of key terms and links to attract a better profile. Indeed, search engines themselves, recognizing the potential abuse of the system, have launched applications designed to single out websites using such manipulations with the goal of providing efficient and reliable user searches.

—*Joseph Dewey, MA, PhD*

BIBLIOGRAPHY

Auletta, Ken. *Googled: The End of the World As We Know It.* New York: Penguin, 2010. Print.

Croft, Bruce, and Trevor Strohman. *Search Engines: Information Retrieval in Practice.* Boston: Addison, 2009. Print.

Fleischner, Michael H. *SEO Made Simple: Strategies for Dominating the World's Largest Search Engine.* 4th ed. CreateSpace, 2014. Print.

Grappone, Jennifer, and Gradiva Couzin. *SEO: An Hour a Day.* New York: Sybex, 2011. Print.

Jones, Kristopher B. *SEO: Your Visual Blueprint for Effective Internet Marketing.* 3rd ed. Indianapolis: Wiley, 2013. Print.

Karthikeyan, K., and M. Sangeetha. "Page Rank Based Design and Implementation of Search Engine Optimization." *International Journal of Computer Applications* 40.4 (2012): 13–18. Print.

Kerdvilbulvech, Chutisant. "A New Method for Web Development Using Search Engine Optimization." *International Journal of Computer Science and Business Informatics* 3.1 (2013). Web. 26 Aug. 2013.

Levy, Steven. *In the Plex: How Google Thinks, Works, and Shapes Our Lives.* New York: Simon, 2011.

SEMANTIC MEMORY

Semantic memory is a system of memory involving the encoding, storage, and retrieval of facts, definitions, concepts, and other forms of cognitive knowledge. Semantic memory is often contrasted with episodic memory, the memory of what, where, and when a discrete event happened in one's own past experience. It is also distinguished from procedural memory, the knowledge of how to accomplish a task. Scientists study semantic memory and other types of memory to further our understanding of how the brain works and how humans and animals learn and retain knowledge. Studies of how semantic memory is formed, retained, and retrieved may also help researchers develop treatments and therapies for conditions that involve cognitive memory loss, such as Alzheimer's disease, dementia, amnesia, and traumatic brain injury.

OVERVIEW

Estonian Canadian psychologist and neuroscientist Endel Tulving first distinguished episodic memory from semantic memory in 1972 in "Episodic and Semantic Memory," a chapter in *Organization of Memory*, which he edited with Wayne Donaldson. At the time, Tulving wrote that he was making the distinction in order to facilitate discussion, not because he necessarily believed that the two systems were structurally or functionally separate.

According to Tulving, episodic and semantic memory differ in terms of the nature of the information they store, whether the reference is autobiographical or cognitive, the conditions and consequences of retrieval, how much they depend on each other, and their susceptibility to transformation and erasure of stored information by interference. Tulving theorized that semantic memory is essential for language use and likened it to a "mental thesaurus" because it deals with an individual's organized knowledge of words and symbols, what they mean, what they refer to, and how they are related and used. Unlike episodic memory, semantic memory encodes cognitive information rather than specific personal events; one can remember a fact such as a telephone number without necessarily remembering when and where one learned it. Semantic memory is less vulnerable than episodic memory to involuntary loss or change of information during the act of retrieval. The encoding and storage of information in sensory memory is independent of episodic memory.

Tulving further expounded on episodic and semantic memory in his much-cited 1983 book *Elements of Episodic Memory*. In this work, he asserts, among other things, the then-controversial theory that episodic memory and semantic memory are functionally distinct. Various experiments have supported Tulving's theory, demonstrating that semantic and episodic tasks are distinct. Since then, neuroimaging models have shown that episodic memory retrieval and semantic memory retrieval involve different parts of the brain. However, researchers remain divided as to whether semantic memory is still stored in some part of the hippocampal formation, as with episodic memory; involves a fully different area of the brain, such as the temporal neocortex; or is distributed among different regions.

There are a variety of disorders that impact semantic memory. Most notable are Alzheimer's disease and semantic dementia, both of which are studied in order to better understand semantic memory processes. Conversely, research into semantic memory may prove useful in treating or curing these and other memory disorders.

—Lisa U. Phillips, MFA

Bibliography

Binder, Jeffrey R., and Rutvik H. Desai. "The Neurobiology of Semantic Memory." *Trends in Cognitive Sciences* 15.11 (2011): 527–36. Print.

Fietta, Pierluigi, and Pieranna Fietta. "The Neurobiology of the Human Memory." *Theoretical Biology Forum* 104.1 (2011): 69–87. Print.

Foster, Jonathan K. "Memory." *New Scientist* 3 Dec. 2011: i–viii. Print

Hart, John, Jr., and Michael A. Kraut, eds. *Neural Basis of Semantic Memory*. Cambridge: Cambridge UP, 2007. Print.

Miller, Greg. "Making Memories." *Smithsonian* May 2010: 38–45. Print.

Shivde, Geeta, and Michael C. Anderson. "On the Existence of Semantic Working Memory: Evidence for Direct Semantic Maintenance." *Journal of Experimental Psychology: Learning, Memory, and Cognition* 37.6 (2011): 1342–70. Print.

Tulving, Endel. *Elements of Episodic Memory*. 1983. Oxford: Oxford UP, 2008. Print.

———. "Episodic and Semantic Memory." *Organization of Memory*. Ed. Tulving and Wayne Donaldson. New York: Academic, 1972. 381–403. Print.

———. "Episodic Memory: From Mind to Brain." *Annual Review of Psychology* 53 (2002): 1–25. Print.

Weisberg, Robert W., and Lauretta M. Reeves. *Cognition: From Memory to Creativity*. Hoboken: Wiley, 2013. Print.

SEMANTICS

Semantics is the study of meaning in language, whether natural (such as English) or artificial (such as that used in computer programming). Stemming from the Greek word *sēmainō*, which roughly translates as "to signify," semantics essentially examines the words and phrases that compose a language—referred to as "signifiers"—in terms of what they denote, or mean.

Key figures in twentieth-century semantics, when the field grew rapidly, include Richard Montague, Noam Chomsky, and Donald Davidson. Throughout the twentieth century, semantics was most prominently aligned with philosophy, but it is a field of study that also crosses over into mathematics, logic, computer science, and literary theory.

Overview

Beyond its general meaning, semantics, in philosophical circles, has several overlapping but distinct implications. Semiotics, a synonym of semantics, explores the basic causal relationship between words and symbols (or "signs") and what they denote. However, semantics also encompasses changes in the meaning of words and phrases over historical periods of time, and the term can be used to describe not only what words denote but also what they connote, or imply.

Furthermore, semantics may also indicate language used to sway opinion, such as that employed in an advertising campaign.

Despite its generalized definition and prevalence as a term in modern society, semantics can be a highly technical, often misunderstood field of linguistic study. Several of the theories that form the backbone of the discipline are rooted in logic and mathematics. For example, formal semantics, most often associated with Montague, employs a mathematical framework to dissect the construction of linguistic expression, using notation to diagram the elements of language. Truth-conditional semantics, pioneered by Davidson, is another complex subfield of semantics that asserts that the meaning of a sentence or phrase is linked to—is in fact the same as—its truthfulness. These and other similar theories, such as lexical semantics, illustrate the strong bond among the fields of linguistics, logic, and mathematics.

In the world of technology, semantics has become a key component of computer programming, sharing some theoretical similarities with linguistic semantics. Computer-science semantics also deals with the properties of language, albeit artificial ones that focus on programming. Specifically, semantics studies the process by which a computer language is executed. Thus, it has to do with executable commands—if the language used to process a command is incorrect, the command will not work for the specific software or hardware. Whether in linguistics or computer programming, semantics deals with deriving meaning from a string of words, symbols, or phrases.

A highly complex, interdisciplinary field of academic study, semantics is continually evolving, as new linguistic theories emerge about the connection between words and their meanings.

—*Christopher Rager, MA*

Bibliography

Davis, Steven, and Brendan S. Gillon. *Semantics: A Reader*. New York: Oxford UP, 2004. Print.

Kearns, Kate. *Semantics*. 2nd ed. New York: Palgrave, 2011. Print.

Li, Juanzi. *Semantic Web and Web Science*. New York: Springer, 2013. Digital file.

Löbner, Sebastian. *Understanding Semantics*. New York: Routledge, 2013. Digital file.

Nielson, Hanne Riis, and Flemming Nielson. *Semantics with Applications: A Formal Introduction*. New York: Wiley, 1992. Print.

Portner, Paul, Klaus von Heusinger, and Claudia Maienborn. *Semantics: An International Handbook of Natural Language Meaning*. Berlin: De Gruyter Mouton, 2013. Digital file.

Saeed, John I. *Semantics*. 4th ed. Hoboken: Wiley, 2015. Print.

Zimmermann, Thomas Ede, and Wolfgang Sternefeld. *Introduction to Semantics: An Essential Guide to the Composition of Meaning*. Boston: De Gruyter Mouton, 2013. Print.

SIGNAL PROCESSING

FIELDS OF STUDY

Algorithms; Information Technology; Digital Media

ABSTRACT

"Signal processing" refers to the various technologies by which analog or digital signals are received, modified, and interpreted. A signal, broadly defined, is data transmitted over time. Signals permeate everyday life, and many modern technologies operate by acquiring and processing these signals.

PRINCIPAL TERMS

- **analog signal:** a continuous signal whose values or quantities vary over time.
- **filter:** in signal processing, a device or procedure that takes in a signal, removes certain unwanted elements, and outputs a processed signal.
- **fixed point:** a type of digital signal processing in which numerical data is stored and represented with a fixed number of digits after (and sometimes before) the decimal point, using a minimum of sixteen bits.

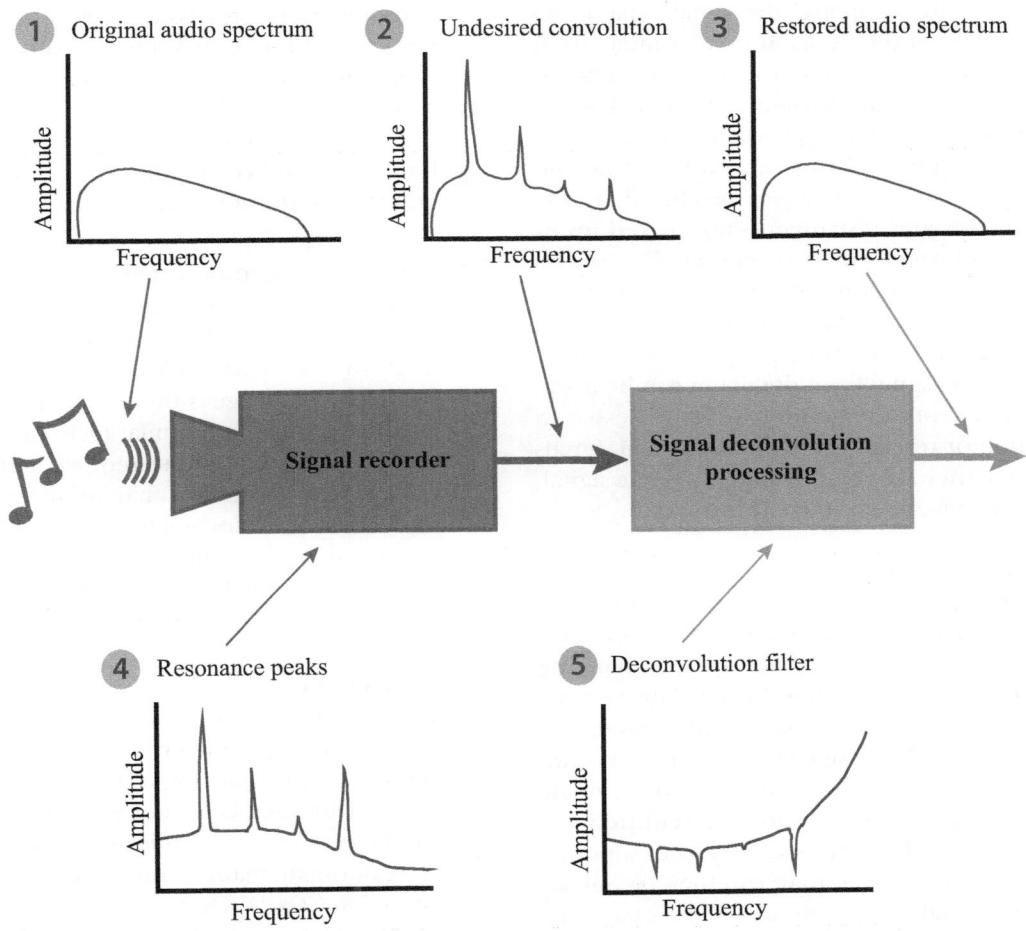

Signal processing involves sensors that receive a signal input (1). Programming to record the input may include deconvolution filters (5) to remove any undesired convolutions (2), such as harmonic resonance (4), so that the signal output (3) is restored.

- **floating point:** a type of digital signal processing in which numerical data is stored and represented in the form of a number (called the mantissa, or significand) multiplied by a base number (such as base-2) raised to an exponent, using a minimum of thirty-two bits.
- **Fourier transform:** a mathematical operator that decomposes a single function into the sum of multiple sinusoidal (wave) functions.
- **linear predictive coding:** a popular tool for digital speech processing that uses both past speech samples and a mathematical approximation of a human vocal tract to predict and then eliminate certain vocal frequencies that do not contribute to meaning. This allows speech to be processed at a faster bit rate without significant loss in comprehensibility.

Digital and Analog Signals

Signals can be analog or digital. Analog signals are continuous, meaning they can be divided infinitely into different values, much like mathematical functions. Digital signals, by contrast, are discrete. They consist of a set of values, and there is no information "between" two adjacent values. A digital signal is more like a list of numbers.

Typically, computers represent signals digitally, while devices such as microphones represent signals in an analog fashion. It is possible to convert a signal from analog to digital by "sampling" the signal—that is, recording the value of the signal at regular intervals. A digital signal can be converted to analog by generating an electrical signal that approximates the digital signal.

One example of analog-to-digital signal conversion is the setup used to record a music performance to CD. The sound waves produced by the musicians' voices and instruments are picked up by microphones plugged into a computer. The microphones convert the sound waves to analog electrical signals, which the computer then converts to digital ones. Each signal can then be processed separately. For example, the signal from the drum can be made quieter, or that from the guitar can be processed to add distortion or reverb. Once the individual signals are processed, they are combined into a single signal, which is then converted to the CD format.

PROCESSING SIGNALS

The core concept in signal processing is the Fourier transform. A transform is a mathematical operator that changes how data are expressed without changing the value of the data themselves. In signal processing, a transform changes how a signal is represented. The Fourier transform allows signals to be broken down into their constituent components. It turns a single signal into superimposed waves of different frequencies, just as a music chord consists of a set of superimposed musical notes. The signals can then be quickly and efficiently analyzed to suppress unwanted components ("noise"), extract features contained in an image, or filter out certain frequencies.

In signal processing, a filter is a device or process that takes in a signal, performs a predetermined operation on it (typically removing certain unwanted elements), and outputs a processed signal. For example, a low-pass filter removes high frequencies from a signal, such as the sound produced by a piccolo, and leaves only the low frequencies intact, such as those from a bass guitar. A high-pass filter does the opposite. It removes low frequencies from signals, keeping only the high frequencies.

When processing digital signals, one important consideration is how the data should be stored. The discrete values that make up a digital signal may be stored as either fixed point or floating point. Both formats have advantages and disadvantages. A digital signal processor (DSP) is optimized for one or the other. Fixed-point DSPs are typically cheaper and require less processing power but support a much smaller range of values. Floating-point DSPs offer greater precision and a wider value range, but they are costlier and consume more memory and power. Floating-point DSPs can also handle fixed-point values, but because the system is optimized for floating point, the advantages of fixed point are lost.

EXTRACTING INFORMATION FROM SIGNALS

Sometimes a signal contains encoded information that must be extracted. For example, a bar-code reader extracts the bar code from an acquired image. In this case, the image is the signal. It varies in space rather than in time, but similar techniques can be applied. Filters can be used to remove high spatial frequencies (such as sudden changes in pixel value) or low spatial frequencies (such as very gradual changes in pixel value). This allows the bar-code reader to find and process the bar code and determine the numerical value that it represents.

APPLICATIONS

Signal processing is everywhere. Modems take incoming analog signals from wires and turn them into digital signals that one computer can use to communicate with another. Cell phone towers work similarly to let cell phones communicate. Computer vision is used to automate manufacturing or tracking. Signals are stored in CDs, DVDs, and solid-state drives (SSDs) to represent audio, video, and text.

One particular application of signal processing is in speech synthesis and analysis. Speech signal processing relies heavily on linear predictive coding, which is based on the source-filter model of speech production. This model posits that humans produce speech through a combination of a sound source (vocal cords) and a linear acoustic filter (throat and mouth). The filter, in the rough shape of a tube, modifies the signal produced by the sound source. This modification produces resonant frequencies called "formants," which make speech sound natural but carry no inherent meaning. Linear predictive coding uses a mathematical model of a human vocal tract to predict these formants. They can then be removed for faster analysis of human speech or produced to make synthesized speech sound more natural.

—*Andrew Hoelscher, MEng, John Vines, and Daniel Horowitz*

Bibliography

Boashash, Boualem, ed. *Time-Frequency Signal Analysis and Processing: A Comprehensive Reference.* 2nd ed. San Diego: Academic, 2016. Print.

Lathi, B. P. *Linear Systems and Signals.* 2nd rev. ed. New York: Oxford UP, 2010. Print.

Owen, Mark. *Practical Signal Processing.* New York: Cambridge UP, 2012. Print.

Prandoni, Paolo, and Martin Vetterli. *Signal Processing for Communications.* Boca Raton: CRC, 2008. Print.

Proakis, John G., and Dimitris G. Manolakis. *Digital Signal Processing: Principles, Algorithms, and Applications.* 4th ed. Upper Saddle River: Prentice, 2007. Print.

Shenoi, Belle A. *Introduction to Digital Signal Processing and Filter Design.* Hoboken: Wiley, 2006. Print.

Vetterli, Martin, Jelena Kovačević, and Vivek K. Goyal. *Foundations of Signal Processing.* Cambridge: Cambridge UP, 2014. Print.

SOURCE CODE COMMENTS

FIELDS OF STUDY

Coding Techniques; Software Development; Software Engineering

ABSTRACT

Source code comments are annotations that are used to explain the purpose and structure of a computer program's code. The use of appropriate comments facilitates the efficient creation, debugging, and maintenance of computer programs and enables programmers to work together more effectively.

PRINCIPAL TERMS

- **annotation:** in computer programming, a comment or other documentation added to a program's source code that is not essential to the execution of the program but provides extra information for anybody reading the code.
- **software documentation:** materials that explain a software system's requirements, design, code, and functionality.
- **syntax:** in computer programming, the rules that govern how the elements of a programming language should be arranged.

Understanding Source Code Comments

Anyone familiar with the syntax of a computer language can read a computer program's source code to gain a rudimentary understanding what a statement or series of statements does. However, the code itself cannot reveal to a reader why a certain statement was used or the overall purpose of an entire program or section of code. This can make code difficult to understand, debug, and maintain, especially when it grows long and complex and is accessed by multiple people.

To address this problem, programmers use a technique called annotation. This is the process of adding comments to code to make programs easier to understand, modify, debug, and maintain. Computer code can be annotated in several ways. For example, a comment may be used to explain why a certain command or variable was used. Comments can also be used to describe functions, subroutines, classes, and methods. For this reason, comments are an important component of a project's software documentation. They are considered internal documentation, as they are included alongside the source code itself. In contrast, external documentation is kept separately, possibly as a digital file or a printed booklet.

Comments are written in plain language, not in the syntax of a computer language. Each computer language provides its own syntactical rules that allow the program, and anyone who is reading the program, to distinguish comments from executable lines of code. In many common programming languages, including C, C++, Java, and JavaScript, a comment is indicated by two forward slashes preceding the text of the comment:

    // This is a comment.

The forward slashes serve as a delimiter, which is one or more characters that indicate a boundary or separation. Other languages use different delimiters to indicate comments, such as an apostrophe or a hash symbol. A language may also use different delimiters to signify in-line comments, which begin with a delimiter and end at the next line break, and block comments, which both begin and end with a delimiter. Many languages that recognize forward slashes as delimiters also recognize slash-asterisk (/*) and asterisk-slash (*/) as beginning and ending delimiters for block comments. Additionally, individual programmers and companies often have preferences for how comments should be formatted or used. This can lead to variation in comments even within the same programming language.

All text contained within a comment is ignored by the compiler. Because of this, comments can be used to aid in testing and debugging code. For example, if a section of code is included in a comment and the error no longer occurs, the developer knows that the error is in that section of code.

Making Complex Programs Easy to Understand

Creating effective and useful comments is extremely important in real-world programming environments, which typically include teams of programmers working together on complex applications. As such projects are usually developed by multiple people over many weeks, months, or even years, proper use of comments allows programmers to quickly understand the purpose of code that was written by other programmers at other times. A lack of comments can make complex programs extremely difficult to understand. If appropriate comments are not provided, even experienced programmers will need to devote large amounts of time and effort to understanding the existing code before modifying it. Good comments form a basis from which all the components of a program and their interactions can be easily understood, greatly increasing the efficiency with which applications can be developed and updated.

Creating Effective Comments

The ability to create effective and useful comments is an important skill. Comments should not simply restate what a block of code does, but should focus on explaining why it does what it does and what its relationship is to the overall structure and function of the program. Developers must be committed to

---

SAMPLE PROBLEM

A developer is creating a procedure that prompts the user to enter a day of the week in order to see a restaurant's special for that day. The program connects to a database using a function called GetDailySpecial to retrieve the daily special, then displays it to the user. The variables are designated *DayOfWeek* and *DailySpecial*. The code for this program, written in the Visual Basic programming language, is as follows:

Dim DayOfWeek, DailySpecial As String
DayOfWeek = InputBox ("Enter the day of the week")
DailySpecial = GetDailySpecial(DayOfWeek)
MsgBox ("Today's special is " & DailySpecial)

What comments might be included so that anyone reading this code would understand why it was written? Note that Visual Basic uses a straight apostrophe and a space (or the keyword REM) as a delimiter for every line of comments.

Answer:

' The DayOfWeek variable stores the day of the week for which the user wishes
' to know the daily special. The DailySpecial variable holds the daily special for the
' day of the week the user has requested.
Dim DayOfWeek, DailySpecial As String
' Display in input box to allow the user to enter the day of the week for which they
' wish to know the special.
DayOfWeek = InputBox ("Enter the day of the week")
' Use the GetDailySpecial function to connect to the FoodService database and
' retrieve the daily special.
DailySpecial = GetDailySpecial(DayOfWeek)
' Display a message on the user's screen that informs the user of the daily special.
MsgBox ("Today's special is " & DailySpecial)

These comments explain why each line of code does what it does and what its relationship is to the overall structure and function of the program. The comments include information that cannot be determined from reading the code statement itself. For example, the fact that variables in the procedure are strings can be determined from the Dim statement.

---

including appropriate comments in their code, and development methodologies must support proper commenting. Testing and quality assurance efforts should include verification that useful comments are being used.

The Value of Comments

While some programmers believe well-written source code explains itself, most agree that comments are critical to nearly all projects. Programmers focus on writing effective and efficient code that is syntactically correct. However, the inherent limitations of programming languages make it difficult to describe code fully without using comments, especially on large projects. The consistent use of appropriate comments provides significant benefits to programmers and results in greater efficiency, reduced costs, and the development of more robust and effective code. It is therefore important for anyone interested in computer programming to understand the application of comments and to familiarize themselves with any rules for comment formatting that are specific to either the programming language or the project at hand.

—*Maura Valentino, MSLIS*

Bibliography

Friedman, Daniel P., and Mitchell Wand. *Essentials of Programming Languages*. 3rd ed., MIT P, 2008.

Haverbeke, Marijn. *Eloquent JavaScript: A Modern Introduction to Programming*. 2nd ed., No Starch Press, 2015.

MacLennan, Bruce J. *Principles of Programming Languages: Design, Evaluation, and Implementation*. 3rd ed., Oxford UP, 1999.

Schneider, David I. *An Introduction to Programming Using Visual Basic*. 10th ed., Pearson, 2017.

Scott, Michael L. *Programming Language Pragmatics*. 4th ed., Elsevier, 2016.

Van Roy, Peter, and Seif Haridi. *Concepts, Techniques, and Models of Computer Programming*. MIT P, 2004.

SPIRAL DEVELOPMENT

Fields of Study

Software Development; Software Engineering

Abstract

Spiral development is a software development methodology that combines elements of the waterfall, iterative, and incremental development methodologies while focusing on risk management. Spiral development is used to develop large, mission-critical systems that face high amounts of risk.

Principal Terms

- **proof of concept:** a prototype designed to evaluate the feasibility of a project, concept, or theory.
- **release:** the version of a software system or other product made available to consumers, or the act of making that version available.
- **requirement:** a necessary characteristic of a software system or other product, such as a service it must provide or a constraint under which it must operate.
- **software architecture:** the overall structure of a software system, including the components that make up the system, the relationships among them, and the properties of both those components and their relationships.

Understanding Spiral Development

Spiral development is a software development methodology that combines elements of several other development methodologies. It incorporates the linear structure of waterfall development with the iterative processes of iterative and incremental development. Elements of the prototyping methodology can also be used. The main emphasis of spiral development is on risk analysis and management in the design and implementation of software architecture.

At the core of the spiral development methodology are four phases. These phases are repeated in every iteration, or spiral, of a project. However, the

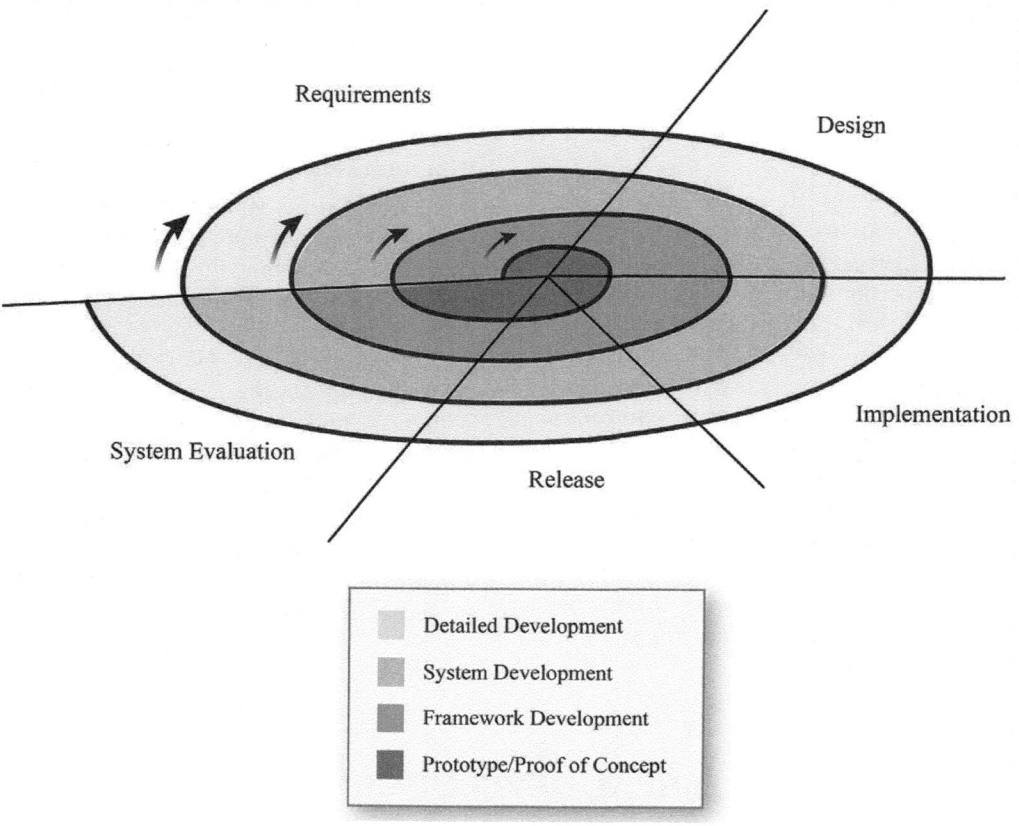

Developing software using the spiral model cycles through defining requirements, designing the software, implementing the code, and releasing the code for evaluation. In the initial cycle a proof of concept is tested, followed by iterations of prototype development until the product meets end user expectations and can be released without need for further requirements.

actual order and specific content of each phase can vary between projects, and they may be given various names. A common form of the spiral model specifies the four phases as planning, risk analysis, engineering, and evaluation. During the planning (or identification) phase, requirements are identified and refined. This is the stage in which system developers and analysts communicate closely with customers or other stakeholders.

During the risk analysis phase, potential risks are identified. Alternate solutions and other actions designed to minimize risks are developed. Types of risks considered include the risks of using new technologies and other technical aspects, risks that negatively affect the project schedule, financial risks, and market-related or other business-related risks. This phase often requires balancing various risk factors to achieve the greatest risk reduction possible in each situation.

Development and testing of the system occurs during the engineering phase. During the first spiral a proof of concept is developed. This allows system designers, developers, business analysts, and other stakeholders to confirm the feasibility of the proposed design. During subsequent spirals, new releases are developed depending on the specific needs of a given system. These may include prototypes, modules, and builds.

The fourth phase is the evaluation phase. During this step, customers, users, and other stakeholders evaluate the release and provide feedback to the development team. This input is then used to inform work conducted during the next spiral. The four phases are repeated in additional spirals until the final release fulfills all of the project's requirements.

The Benefits of Risk Management

Spiral development's greatest strength is its focus on risk management. Unlike in other software development methodologies, risk management is conducted in every iteration, increasing opportunities to minimize threats to a project's success. This makes the spiral model a good choice for medium to large mission-critical projects and those known to have a high amount of risk.

Spiral development also offers many of the same benefits as the methodologies it draws from. It retains the strong approval and documentation of the waterfall model. Like prototyping and iterative development, it provides deployable systems early in the development process. It can also adapt to changes in the project's requirements and greater end-user involvement relatively easily.

The main drawbacks of the spiral model are due to its complexity. Costs and overhead requirements tend to be high. The time involved may be lengthy, and a project's end may not be immediately clear. For this reason, small projects are usually not good candidates for spiral development. It is also important that project staff have experience in risk analysis and management, which may not be the case with all organizations. If risks are not properly identified and minimized, many of the model's benefits are lost.

Using Spiral Development for Large Systems

Spiral development is most suited for projects that are mission critical, are large in scope, and have a substantial amount of risk. For example, a department store chain with five hundred locations in forty-three states might need to develop a new point-of-sale system to control more than 7,500 cash registers. The system would be used by over twenty-five thousand sales associates daily. It would need to connect all stores to one another and to the company's seven regional offices. The system must be secure and meet all legal requirements for handling consumer credit information. In addition, the project will be developed using several technologies that are being used by the IT department for the first time.

This project would be an excellent candidate for spiral development. The project is large, involving a large IT department, hundreds of locations, and affecting many stakeholders. The system is mission critical, as any problem with the point-of-sale system will negatively affect the company's core business operations. The project also faces substantial risk, including the use of new technologies.

In addition, many elements of the project would benefit from the strengths of the methodologies encapsulated in spiral development. As thousands of employees will need to use the system a daily, involving end users in the development process will be important. A project of this size and scope will likely have to deal changing requirements. Multiple releases will most likely be required to develop, test, evaluated, and deploy the final system. These needs can be met by the spiral model's combination of sequential and iterative methods.

Risk Management Enables Project Success

All software development systems attempt to ensure that the development process is brought to a successful conclusion. Different methodologies, from traditional waterfall development to more recently developed models such as rapid application development (RAD) and extreme programming (XP), approach this problem in various ways. Spiral development is notable as a hybrid methodology incorporating elements from other approaches, with an added focus on risk management. Its strengths and flexibility mean it will remain relevant as new innovations in software development project management occur.

—*Maura Valentino, MSLIS*

Bibliography

Bell, Michael. *Incremental Software Architecture: A Method for Saving Failing IT Implementations*. John Wiley & Sons, 2016.

Friedman, Daniel P., and Mitchell Wand. *Essentials of Programming Languages*. 3rd ed., MIT P, 2008.

Jayaswal, Bijay K., and Peter C. Patton. *Design for Trustworthy Software: Tools, Techniques, and Methodology of Developing Robust Software*. Prentice Hall, 2007.

MacLennan, Bruce J. *Principles of Programming Languages: Design, Evaluation, and Implementation.* 3rd ed., Oxford UP, 1999.

Scott, Michael L. *Programming Language Pragmatics.* 4th ed., Elsevier, 2016.

Van Roy, Peter, and Seif Haridi. *Concepts, Techniques, and Models of Computer Programming.* MIT P, 2004.

Wysocki, Robert K. *Effective Project Management: Traditional, Agile, Extreme.* 7th ed., John Wiley & Sons, 2014.

STANDARD DEVIATION

Standard deviation is a mathematical value that is used to show the degree to which a given data point might deviate from the average, or mean, of the set to which it belongs. A small standard deviation means the data points are clustered close to the mean, while a larger value indicates that they are more spread out. The two main types of standard deviation are population standard deviation and sample standard deviation; which calculation to use depends on whether the data set being analyzed is complete or merely a representative sample of a larger set.

OVERVIEW

The standard deviation of a data set is the square root of the variance, which describes how far the data points are spread out from the mean. While the variance and the standard deviation are quite similar concepts, standard deviation is more useful in a real-world context, as it is expressed in the same units as the original data points, while the variance is expressed in those units squared. Thus, to determine the standard deviation of a data set, one must first determine the variance.

The first step in calculating the variance is to calculate the mean, which is done by adding all the values in the set together and then dividing them by the number of values in the set. Then subtract the mean from each individual value in the set and square each resulting difference. The goal of squaring the differences is to avoid dealing with negative numbers. Finally, calculate the mean of the squared differences by adding them all together and once again dividing by the number of values in the set. The number that results is the variance of the set. To determine the standard deviation, simply take the square root of the variance.

The above calculation is one of two basic formulas for standard deviation. It is often called the "population standard deviation" in order to differentiate it from the sample standard deviation. The population standard deviation is most accurate when the data points in the set represent the entirety of the data being analyzed. However, sometimes the data set is merely a sample of a larger population, and the results will be used to generalize about that larger population, such as when a fraction of a nation's residents are polled on a political issue and the results are extrapolated to represent the political attitudes of the entire nation. In these cases, the population standard deviation generally produces a value that is too low, so the sample standard deviation should be used instead. Though it still does not produce an entirely unbiased result, it is significantly more accurate.

To calculate the sample standard deviation, one must recalculate the variance to produce a sample variance. The calculation is the same except for the last step. Instead of dividing the sum of the squared differences by the number of values in the set, divide the sum by the number of values in the set *minus one*. This corrects for the tendency for the population standard deviation to be too low. Then, as before, simply take the square root of the sample variance to produce the sample standard deviation.

It is important to note that both the population standard deviation and the sample standard

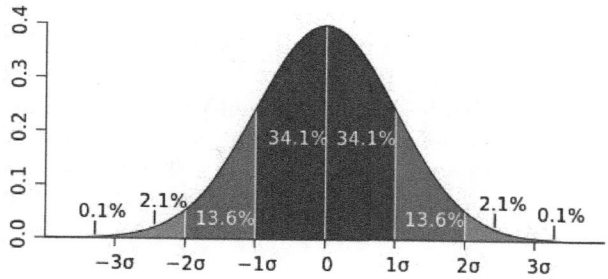

deviation assume a normal distribution of data, represented by a bell curve. When the distribution pattern deviates from this norm, further corrections may be necessary to accurately calculate the standard deviation. However, when the distribution does follow a bell curve, the standard deviation can communicate a great deal about the values within the data set. For example, the empirical rule, also known as the "68-95-99.7 rule," states that in a normal distribution, 68 percent of the data points in the set fall within one standard deviation of the mean, 95 percent fall within two standard deviations of the mean, and 99.7 percent fall within three standard deviations of the mean. Generally, outlying data points within a set are only considered statistically significant when they are more than one standard deviation from the mean, though the exact threshold of significance varies.

—*Randa Tantawi, PhD*

BIBLIOGRAPHY

Altman, Douglas G., and J. Martin Bland. "Standard Deviations and Standard Errors." *BMJ* 331.7521 (2005): 903. Print.

Hand, David J. *Statistics: A Very Short Introduction.* New York: Oxford UP, 2008. Print.

Kalla, Siddharth. "Calculate Standard Deviation." *Explorable.* Explorable.com, 27 Sept. 2009. Web. 4 Oct. 2013.

Lane, David M. "Measures of Variability." *Online Statistics Education: An Interactive Multimedia Course of Study.* Lane, n.d. Web. 4 Oct. 2013.

Orris, J. B. "A Visual Model for the Variance and Standard Deviation." *Teaching Statistics* 33.2 (2011): 43–45. Print.

Taylor, Jeremy J. "Confusing Stats Terms Explained: Standard Deviation." *Stats Make Me Cry.* Taylor, 1 Aug. 2010. Web. 4 Oct. 2013.

Urdan, Timothy C. *Statistics in Plain English.* 3rd ed. New York: Routledge, 2010. Print.

Weisstein, Eric W. "Standard Deviation." *Wolfram MathWorld.* Wolfram Research, n.d. Web. 4 Oct. 2013.

STANDPOINT THEORY

Standpoint theory is a means for understanding collective group discourse. The theory's most critical aspect is that a person's perspectives are created by his or her personal experiences in social groups. "Standpoint" refers to each individual's perspective of the world, which is usually based on position and outlook. These factors influence how people understand themselves and communicate with others. Standpoint theory assumes that social class provides a narrow viewpoint on interpersonal relationships, that dominating groups repress the inferior groups, and that dominating groups have disproportionate influence over inferior groups. Standpoint theory is based on essentialism, which refers members of a group (e.g., women) being labeled as having the same characteristics or all being very similar to each other. Critics of the theory state that its position between subjectivity and objectivity renders it unhelpful.

OVERVIEW

Standpoint theory was initially theory-based, but it is currently more communications-based. The theory originated from the work of German philosopher Georg Wilhelm Friedrich Hegel. Hegel was interested in the standpoints between masters and their slaves in the early 1800s. He noticed that the relationship between master and slave was related to a person's position and the group's influence on how

Patricia Hill Collins has used standpoint theory in her work regarding feminism and gender in the African American community.

Donna Jeanne Haraway (2007), known for writings that include "A Cyborg Manifesto: Science, Technology, and Socialist-Feminism in the Late Twentieth Century" (1985) and "Situated Knowledges: The Science Question in Feminism and the Privilege of Partial Perspective" (1988).

information and authority was dictated. In the 1980s, feminist theorist Nancy Hartsock focused on standpoint theory to understand differences between males and females. She based her work the Marxist assumption that a person's work influences his or her comprehension of the world. Hartsock created feminist standpoint theory, which emphasized women's social standpoints. However, many scholars note that Hartsock failed to include the viewpoints of non-Caucasian women. Standpoint theory does not propose that males and females are different, rather it argues that there are cultural and social factors that result in women constructing experiences that are different from their male counterparts.

One criticism of standpoint theory is the idea of dualism or double thinking. Many feminist scholars believe that much of human thought is structured around a group of dualisms or oppositions. For instance, people struggle with what is public or private, or what is objective or subjective. However, dualism usually infers that there is a hierarchical order between two opposites in which one is considered superior. In other words, people tend to devalue one and value the other. For instance, two terms associated with personality are rational and emotional. Women are typically associated with being more emotional than rational men, who are assumed to correlate with the superior element of the dualistic order. This, however, is a false dichotomy. All in all, standpoint theory allows a way to examine a social group's communication, experiences, and perspective.

—*Narissra Maria Punyanunt-Carter, MA, PhD*

BIBLIOGRAPHY

Allen, Brenda J. "Feminist Standpoint Theory: A Black Woman's (Re)view of Organizational Socialization." *Communication Studies* 47.4 (1996): 257–71. Print.

Edwards, Gail. "Standpoint Theory, Realism and the Search for Objectivity in the Sociology of Education." *British Journal of Sociology of Education* (2012): 1–18. Web. 22 July 2013.

Hekman, Susan. "Truth and Method: Feminist Standpoint Theory Revisited." *Signs* 22.2 (1997): 341–65. Print.

Longino, Helen E. "Feminist Standpoint Theory and the Problems of Knowledge." *Signs* 19.1 (1993): 201–12. Print.

Mahowald, Mary B. "On Treatment of Myopia: Feminist Standpoint Theory and Bioethics." *Feminism and Bioethics: Beyond Reproduction.* Ed. Susan M. Worf. New York: Oxford UP, 1996. Print.

Potter, Michael. "Loyalism, Women and Standpoint Theory." *Irish Political Studies* (2012): 1–17. Web. 22 July 2013.

Stoetzler, Marcel, and Nira Yuval-Davis. "Standpoint Theory, Situated Knowledge and the Situated Imagination." *Feminist Theory* 3.3 (2002): 315–33. Print.

Swigonski, Mary E. "The Logic of Feminist Standpoint Theory for Social Work Research." *Social Work* 39.4 (1994): 387–93. Print.

STATISTICAL INFERENCE

Statistical inference is the process of drawing conclusions from a piece of statistical analysis with a view to making recommendations for an organization, process, or policy. It is a process that requires both objective and subjective judgments, since statistics rarely indicate causation or provide definitive answers rather than indications. While the level of nuance such inference requires is understandable in an academic context, it does not always fit well in policy making or discourse for the public.

Overview

Statistical analysis is generally based on the concept of the confidence interval, or level of confidence in the results. A distribution of observed results is examined and compared to what would be expected. Calculations are then used to determine the level of significance of the difference between what was found and what was expected.

For general market research or academic research in business or management, a confidence interval of 95 percent is normally expected. This means that when a given result is declared to be significant, it can be assumed to be correct 95 percent of the time and incorrect for the remaining 5 percent. Researchers will most commonly conduct several different but related tests to try to reduce the likelihood of making a declaration that turns out to be false; for example, a questionnaire will have a number of questions about different aspects of a potential new product rather than just one. The level of confidence can be increased by adding more observations or respondents, but this must be balanced against the extra cost and time required, and uncertainty can never be eliminated completely. A confidence interval of 95 percent is a two-sigma approach, which means that it defines as nonsignificant any result that is within two standard deviations, or sigmas, from the mean. Even very rigorous testing, such as the search for the Higgs boson at the European Organization for Nuclear Research (CERN), which uses a five-sigma approach, will declare a false result once in every 3.5 million tests in which a significant result is observed.

Most forms of statistical analysis do not deal with causation. Cross-tabulations or correlation tests are used to indicate when an observed distribution of results is sufficiently far from what was expected to be declared significant, but such results do not actually explain what is happening. It is the responsibility of the researcher to interpret and provide all reasonable explanations for the results. In some cases, the results will be overwhelming and noncontroversial; for example, the relationship between certain demographic variations, such as level of income, ethnicity, and gender, and the propensity to vote for a particular political party is well established. In other cases, explaining the situation will involve much less concrete relationships and so will rely more on the conceptual framework that gave rise to the test in the first place and the degree of trust placed in the researcher.

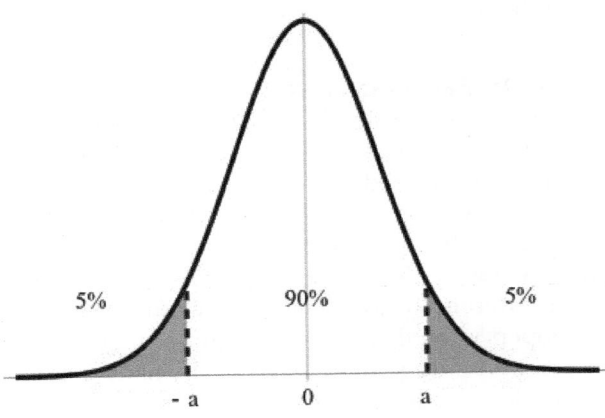

An illustration of a 90 percent confidence interval on a standard normal curve.

On occasion, statistical analysis does address causation to some degree. Tests based on linear regression aim to explain a portion of overall change in a measurement through particular variables; changes in labor productivity explain a proportion of the overall change in the gross domestic product of a country, for example. A similar approach is used in the case of factor analysis. Another technique, structural equation modelling (SEM), is intended to provide an explanatory causal path leading from variables to final results, such as showing how emotional antecedent states in combination with environmental variables can determine consumer behavior. In these cases, inference focuses on the degree of confidence that can be placed in a particular explanation of a complex phenomenon.

—*John Walsh, PhD*

BIBLIOGRAPHY

Boos, Dennis D., and Leonard A. Stefanski. *Essential Statistical Inference: Theory and Methods.* New York: Springer, 2013. Print.

Frankfort-Nachmias, Chava, and Anna Leon-Guerrero. *Social Statistics for a Diverse Society.* 7th ed. Thousand Oaks: Sage, 2014. Print.

Garthwaite, Paul H., Ian T. Jolliffe, and Byron Jones. *Statistical Inference.* 2nd ed. Oxford: Oxford UP, 2002. Print.

Kass, Robert E. "Statistical Inference: The Big Picture." *Statistical Science* 26.1 (2011): 1–9. Print.

Silver, Nate. *The Signal and the Noise: Why So Many Predictions Fail—but Some Don't.* New York: Penguin, 2012. Print.

Walker, Helen M., and Joseph Lev. *Statistical Inference.* New York: Holt, 1953. Print.

STRING-ORIENTED SYMBOLIC LANGUAGES (SNOBOL)

FIELDS OF STUDY

Programming Language; Coding Techniques; Programming Methodologies

ABSTRACT

String-oriented symbolic language (SNOBOL) refers to a family of programming languages developed in the 1960s by computer scientists working at Bell Laboratories. The final and most common version was SNOBOL4. Several later languages were developed as successors, and some of the language's unique aspects remain influential. As its name implies, SNOBOL was created for manipulation of symbolic string data. This allows programmers to easily search, change, and utilize string variables in just a few lines of code. SNOBOL and derivative languages are well-suited to pattern-matching tasks.

PRINCIPAL TERMS

- **function:** in computer programming, a self-contained section of code that carried out a certain task.
- **patterns:** in computer programming, repeated sequences or designs in code.
- **stack:** in computer science, a data structure that allows access only to the most recently added object.
- **statement:** a command that instructs a computer which action to perform.
- **string:** in computer programming, a data type consisting of characters arranged in a specific sequence.
- **variable:** in computer programming, a symbolic name that refers to data stored in a specific location in a computer's memory, the value of which can be changed.

SNOBOL PROGRAMMING

SNOBOL was developed by David Farber, Ralph Griswold, and Ivan Polonsky at Bell Labs beginning in 1962. They were driven to create a new language due to frustrations with other early programming languages. The result was an imperative programming language, using functions and statements that affect different kinds of variables. In particular, the developers wanted to create a language that made it easy to work with string variables. The name was settled on later, with the acronym chosen as a joke and worked out to stand for StriNg-Oriented symBOlic Language. Though designed for the specific needs of the team, the language quickly grew popular with others at Bell Labs and in the broader computing community.

The Bell Labs team designed the syntax and labels in SNOBOL to mimic speech, making it understandable even for beginning programmers. The language uses logic, operators, arrays, and standard loops. SNOBOL is dynamically typed, meaning that the programmer does not need to declare the type of variable being used. All variables are global, and SNOBOL programs often contain goto statements, which are common in older languages and machine languages. This linear way of programming stands in contrast to object-oriented languages that manipulate larger chunks of code called "objects."

SNOBOL Data Types

SNOBOL was designed to manipulate strings and match patterns. Since it was initially intended for use only at Bell Labs, the developers built only the features they needed into the original SNOBOL language. There was only one data type, no functions, and hardly any error control. As other programmers began expressing interest in a string-oriented language, the team added more features. SNOBOL2 was not widely released but added the ability to use functions. SNOBOL3 added the ability to create user-defined functions and became relatively popular. At that time, SNOBOL3 was only written for the IBM 7090 computer, which limited its use. Several programmers attempted to write equivalent versions for other computers, but these often contained bugs. SNOBOL4 fixed this compatibility problem by allowing itself to be run on a virtual machine, which could be used on many different computers. SNOBOL4 could also use arrays, tables, stacks, and other common features and data types.

Even with these additions, SNOBOL's main strength is its ability to powerfully manipulate strings. Of the various forms of string manipulation, SNOBOL relies on an approach called "pattern matching." Pattern matching is fast, versatile, and relatively easy to understand. Because SNOBOL makes it so easy to work with strings, it gained widespread usage among researchers in the humanities and other text-heavy fields. However, the language also has significant disadvantages. Its unique syntax and other properties can make it challenging to work with, particularly for programmers used to more common languages. It also lacks many of the control structures expected in programming languages, making it unsuited to many tasks.

String Manipulation Example

Consider a string named *Greeting* equal to "hello hello hello world, this is Bob." Perhaps a programmer wants to change each instance of "hello" to "goodbye." In SNOBOL, this conversion could be achieved with the following line of code:

Greeting "hello" = "goodbye"

The string Greeting would now equal "goodbye goodbye goodbye world, this is Bob." Such conversions are useful when editing a segment of text or when the programmer does not want a user-inputted phrase to contain a particular word or phrase. Other basic uses of SNOBOL include parsing strings, checking if a string contains a certain substring, and finding substrings based on the surrounding text.

SNOBOL syntax is somewhat different from many later languages. For example, an if-statement in many languages would appear something like the following:

```
IF (variable = "hello") THEN
    run statement
ELSE
    run other statement
```

A SNOBOL coder would not have to keep this code segment together. Instead, goto statements are used to direct the computer where to go next. Common SNOBOL syntax includes a label, a subject, and a goto command, each of which is optional. Here is an if-statement in SNOBOL:

```
variable "hello" :S(conditionTrue)
variable "hello" :F(conditionFalse)
(Any other code)
conditionTrue :(END)
conditionFalse :(END)
END
```

The :S on the first line tests whether the variable is equal to "hello." If so, it is a success, and proceeds to the line beginning with *conditionTrue*. If the variable does not equal "hello," the second line sends the computer to the *conditionFalse* line. In either case, the computer will follow the second goto command to the END line. Unlike an if-statement in most other languages, the SNOBOL code segment allows any amount of code to go in between any of the lines.

MODERN USAGE
The final official version of SNOBOL, SNOBOL4, was released in 1967. It enjoyed popularity into the 1980s, when object-oriented languages like Perl became favored for manipulating strings using regular expressions. In contrast with pattern matching, regular expressions use a collection of string-searching algorithms to find patterns within text. As object-oriented languages added more features, they became the preferred way to handle strings in all but a few specific cases.

Over the years, there have been attempts to revive SNOBOL. Original SNOBOL developer Ralph Griswold created a similar language called Icon in the late 1970s. It never enjoyed SNOBOL's widespread usage because it was too specialized. It did, however, receive attention from specialists for compilers, natural language recognition, and grammar analysis. Griswold also developed SL5, another language focused heavily on string processing. Decades later, SNOBOL and Icon served as inspirations for an object-oriented language called Unicon.

SNOBOL was a pioneer in the area of string manipulation and served as a template for other programming languages. Knowing the history of SNOBOL can give programmers a broader perspective on features of more modern programming languages.

—*Joshua Miller and Daniel Showalter, PhD*

BIBLIOGRAPHY
Farber, David J., et al. "SNOBOL, A String Manipulation Language." *Journal of the ACM*, vol. 11, no. 1, Jan. 1964, pp. 21–30, doi:10.1145/321203.321207.

Griswold, Ralph, and Madge T. Griswold. *The Icon Programming Language*. 3rd. ed., Peer-to-Peer Communications, 2002.

Jeffery, Clinton, et al. "Integrating Regular Expressions and SNOBOL Patterns into String Scanning: A Unifying Approach." *Proceedings of the 31st Annual ACM Symposium on Applied Computing*, 2016, pp. 1974–79, doi:10.1145/2851613.2851752.

Jeffery, Clinton, et al. "Pattern Matching in Unicon." *Unicon Project*, 16 Jan. 2017, unicon.org/utr/utr18.pdf. Accessed 4 May. 2017.

Koenig, Andrew. "The Snocone Programming Language." *Snobol4*, www.snobol4.com/report.htm. Accessed 4 May 2017.

Paine, Jocelyn. "Programs that Transform Their Own Source Code; or: the Snobol Foot Joke." *Dr. Dobb's*, UBM, 17 Jan. 2010, www.drdobbs.com/architecture-and-design/programs-that-transform-their-own-source/228701469. Accessed 4 May 2017.

STRUCTURAL EQUATION MODELING (SEM)

Structural equation modeling (SEM) is an advanced statistical analysis technique that is used by scientists in various fields. SEM diagrams look much like concept maps and allow readers to ascertain the essence of a study in a visual format. A single SEM diagram can often convey more information than multiple tables of results from linear-regression studies. SEM provided a breakthrough in theory testing by enabling researchers to thoroughly and efficiently examine the effects of complex constellations of variables on outcomes. Especially valuable in SEM is the ability to test how pivotal variables, called mediators, explain the effects of more distal variables on outcomes.

OVERVIEW
Prior to SEM's rise in popularity, researchers and statisticians relied more heavily on various forms of linear regression in order to predict outcomes. Linear regression is very valuable and still popular, but it does not as readily allow for testing the complex interrelationships among variables. Human brains are capable of both linear reasoning and parallel processing; both are valuable, but parallel processing is often necessary when analyzing rather complex sets of information. In studies, SEM figures resemble concept maps and often include the actual results of the study, making it easier to remember the researchers'

concepts and findings, especially for a person who favors visual-spatial thinking.

Latent variables are important in SEM. Represented by circles in SEM diagrams, they are composed of two or more directly measured variables, which are known as observed variables and represented in diagrams by squares. Latent variables are not directly measured by researchers; rather, they are statistically constructed composites of the theoretically related observed variables. For instance, a researcher could use four observed variables averaged across a neighborhood, such as levels of exercise, green space, positive social relationships, and safety, to compose a latent variable indicating the well-being of the neighborhood. Another scientist might use five different measures of how happy respondents feel in different aspects of their lives, each observed variables, to form the latent variable happiness.

Another key concept in SEM is the testing of mediators. Mediators are variables that exert their influence on an outcome on behalf of a variable that is otherwise not as closely connected with the outcome. For instance, parents' expectations that their young children will eventually graduate from college and earn an advanced degree promote various aspects of students' success during adolescence, but this effect is mediated by other important variables, such as children's expectations. Researchers who study parent expectations in a linear fashion may underestimate the effect of those expectations if they fail to account for important mediators. Likewise, SEM may help researchers in various fields capture a clearer picture of how a particular variable has an effect on outcomes.

—*John Mark Froiland, PhD*

Bibliography

Davison, Mark L., Yu-Feng Chang, and Ernest C. Davenport. "Modeling Configural Patterns in Latent Variable Profiles: Association with an Endogenous Variable." *Structural Equation Modeling* 21.1 (2014): 81–93. Print.

Hoyle, Rick H., ed. *Handbook of Structural Equation Modeling*. New York: Guilford, 2012. Print.

Hu, Li-tze, and Peter M. Bentler. "Cutoff Criteria for Fit Indexes in Covariance Structure Analysis: Conventional Criteria versus New Alternatives." *Structural Equation Modeling* 6.1 (1999): 1–55. Print.

Kenny, David A., Deborah A. Kashy, and William L. Cook. *Dyadic Data Analysis*. New York: Guilford, 2006. Print.

Kline, Rex B. *Principles and Practice of Structural Equation Modeling*. 3rd ed. New York: Guilford, 2011. Print.

Loehlin, John C. *Latent Variable Models: An Introduction to Factor, Path, and Structural Equation Analysis*. 4th ed. New York: Routledge, 2011. Print.

Shin, Tacksoo, Mark L. Davison, and Jeffrey D. Long. "Effects of Missing Data Methods in Structural Equation Modeling with Nonnormal Longitudinal Data." *Structural Equation Modeling* 16.1 (2009): 70–98. Print.

TECHNOLOGY IN EDUCATION

As the use of electronic teaching and learning tools has increased, technology in education has become a key focus for educators at all levels. Using electronic and digital tools is seen as a way to enhance learning and provide a beneficial experience for all students. Administration of educational programs has also benefited from the growth of technology, allowing student progress to be tracked and analyzed more successfully. This in turn permits a fine-tuning of learning objectives and corresponding curriculum units. It also allows teachers to more easily share student progress with parents. For example, teachers might produce charts on a laptop showing a particular student's successes and challenges in math, even breaking down the data by type of math problem. While many embrace these benefits, critics contend that an overuse of electronic devices in classrooms might detract from the learning experience for some students, making it too impersonal without enough of a social component.

BACKGROUND

A student in the late 1970s would have been aware of the new inventions entering society around them, from the personal computer (uncommon at the time) to early cable television. In classrooms, however,

Interactive whiteboards are tools in the classroom.

A high school computer lab.

technology was largely mechanical. For example, a teacher might use an overhead projector to show information on a screen to students. Alternatively, the teacher could show a science or history film on a 16 mm projector, which was wheeled from room to room. In the mid-1980s projectors gave way to videotapes and televisions, which continued to be wheeled from room to room.

By the late 1980s, personal computers had become common enough that many high schools had computer labs where students learned various skills. Most of the computers used DOS, and later the Windows operating system from Microsoft. In addition to word processing and printing, students could do advanced calculations on spreadsheets. Lotus 1-2-3, first released for MS-DOS systems in 1983, was a popular early spreadsheet program.

These software programs and the PCs they operated on represented the forefront of educational technology at the time. They were particularly beneficial to classes about business or math. Of course, students were excited to learn by using personal computers at their schools, which were still rare in the home environment.

By 2000 Windows-based PCs were common and the Apple iMac was taking off. Microsoft, Apple, and many other companies realized the benefits of getting their technology into schools with the hopes of creating lifelong customers. For that reason, they began to communicate to teachers and parents the benefits of their computing devices and software packages. That trend continues today, with software packages such as Microsoft Office having a home and student edition and Apple marketing aggressively to students. Because technology in education is big business, online outlets such as the Academic Superstore offering discounts on software for students who can verify their currently enrolled status. That is usually done by verifying a school-based e-mail.

Technology in Education Today

The range of electronic devices that can be used for teaching and learning has blossomed dramatically since the start of the twenty-first century. Touchscreen machines such as smartphones and tablet computers can fulfill a wide variety of educational functions. For example, a science student can physically manipulate a 3D model of an atom or molecule displayed on their tablet with a series of gestures, learning about the structure of each. A student in a music class can play the notes of a classical composition on his or her tablet by using it as a full piano, with visuals and sound to match.

Additionally, the newfound prevalence of wireless broadband Internet allows for the near-instant transfer of information in educational settings. A teacher leading a class can direct students to the website they established with course materials, or show them in real time how to do research on a site such as ERIC.gov.

New educational technology also allows classes to be delivered remotely. An American university can arrange for a guest lecture from a professor in Scotland, with the entirety of the presentation transmitted via a service such as Skype. Such presentations can even include active review and alteration of documents via free file-sharing services such as Google Drive. In these ways, modern technology in education has opened the doors for learning across borders. This includes the chance for students in poor areas to have access to learning resources that would previously have been inaccessible for them.

Companies view these new electronic avenues as opportunities for profit. In the key college textbook market, expensive and heavy books are beginning to be replaced with digital versions, which are not only easier to transport, but also easier to search. The hope, on the part of students at least, is that updates to digital textbooks will be less expensive to produce and the savings will be passed along to students. In some cases, downloaded digital textbooks come with automatic updates for future revisions, either for free or a modest fee.

These advantages, coupled with the popularity of using tablet computers as e-readers, is likely to make digital textbooks the default at most university by 2020. At the same time, the availability of digital textbooks is accelerating the development of e-learning programs. Many universities are using advances in educational technology to deliver full classes and complete degree programs to students around the country. This opens up the chance for students in different regions to attend the college of their choice, and it allows the universities to gain a new source of revenue from out-of-state students. Correspondingly, opportunities to learn and teach are expanded far beyond what would have been possible without the current revolution in technology.

—*Isaiah Flair, MA*

Bibliography

Benson, Susan N. Kushner, and Cheryl L. Ward. "Teaching with Technology: Using TPACK to Understand Teaching Expertise in Online Higher Education." *International Journal of Technology and Design Education* 23.2 (2013): 377–90. Print.

Edwards, Anthony David. *New Technology and Education.* New York: Continuum, 2012. Print.

Flanagan, Sara, Emily C. Bouck, and Jennifer Richardson. "Middle School Special Education Teachers' Perceptions and Use of Assistive Technology in Literacy Instruction." *Assistive Technology* 25.1 (2013): 24–30. Print.

Gillispie, Matthew. *From Notepad to iPad: Using Apps and Web Tools to Engage a New Generation of Students.* New York: Routledge, 2014. Print.

Lee, Silvia Wen-Yu, and Chin-Chung Tsai. "Technology-Supported Learning in Secondary and Undergraduate Biological Education: Observations from Literature Review." *Journal of Science Education and Technology* 22.2 (2013): 226–33. Print.

Maddux, Cleborne D., and D. Lamont Johnson. *Technology in Education: A Twenty-Year Retrospective.* Hoboken: Taylor, 2013. Print.

Norstrom, Per. "Engineers' Non-Scientific Models in Technology Education." *International Journal of Technology and Design Education* 23.2 (2013): 377–90. Print.

Periathiruvadi, Sita, and Anne N. Rinn. "Technology in Gifted Education: A Review of Best Practices and Empirical Research." *Journal of Research on Technology in Education* 45.2 (2012–2013): 153–69. Print.

Semela, Tesfaye, Thorsten Bohl, and Marc Kleinknecht. "Civic Education in Ethiopian Schools: Adopted Paradigms, Instructional Technology, and Democratic Citizenship in a Multicultural Context." *International Journal of Educational Development* 33.2 (2013): 156–64. Print.

Snape, Paul, and Wendy Fox-Turnbull. "Perspectives of Authenticity: Implementation in Technology Education." *International Journal of Technology and Design Education* 23.1 (2013): 51–68. Print.

Williams, P. John. "Research in Technology Education: Looking Back to Move Forward." *International Journal of Technology and Design Education* 23.1 (2013): 1–9. Print.

TEST DOUBLES

FIELDS OF STUDY
Software Development; Software Methodologies; Computer Science

ABSTRACT
Test doubles are pieces of code used to test other software components. Each of the five types of test double simulates functionality in order to test different capabilities, which can be useful during unit testing.

PRINCIPAL TERMS
- **dummy object:** code that simulates the existence of an object required by a component or system in order to test that component or system.
- **fake object:** code that mimics an object's expected functionality but implements it in another way; used to speed up testing; also called a "fake."
- **mock object:** code that implements the expected interface of an object needed by a component or system in order to test the behavior of that component or system; also called a "mock."
- **test spy:** code that simulates the functionality of a component required by the component or system under test in order to record the calls made by that component or system.
- **test stub:** a substitute interface that simulates an interface needed by the component or system under test; also called a "stub."

Test Doubles and Their Roles in Testing
Test doubles are pieces of code used to test components under test (CUTs), or systems under test (SUTs). A test double conforms to the interface (object methods and properties) required by the component being tested, and can help verify the SUT's correct functionality. Test doubles are useful when the other software components needed to test an SUT have not been developed or are otherwise unavailable. Some allow units of code to be tested when using the real components would be too slow for efficient testing. Each type provides different capabilities that can be useful during unit testing.

A dummy object implements an interface needed to test a component. They are used when an object must simply exist to provide the needed test functionality. For example, a unit test might be conducted to determine if a component can create an object or to provide objects that are needed to test code that adds objects to a data structure. Dummies supply the least functionality of the test double types.

A test stub can not only implement the needed interface (allowing it to serve the same functions as a dummy object), but also provide the functionality to supply inputs required by the SUT. For example, if the SUT requires the time of day be provided when a method in the test double is called, a stub could provide a time as requested. Stubs can be used to test for error and exception handling.

A test spy is typically used to record calls to a class that the SUT is using. For example, a test spy would count the number of times the SUT called the method that provides the time of day, unlike a stub, which would provide the time instead.

As the name implies, a fake object mimics the functionality expected of an object by the SUT but implements it differently from the actual object. Fakes are typically used to speed up testing when using the actual functionality would take too much time. For example, data could be stored in local memory to simulate the existence of a database during testing without connecting to a database server.

The last type of test double is the mock object. Unlike the other test doubles, which test the SUT's inputs, a mock tests the SUT's outputs. For example, a mock would be used to ensure that the SUT passes the correct parameters to the method that provides the SUT with the time of day.

Pros and Cons
Test doubles are a key element in a successful unit testing program. They allow components of a large software system to be tested individually and at any time in the development cycle without affecting those already tested and deployed or consuming valuable resources, such as network bandwidth and database storage. The various types of test doubles allow different testing approaches to be conducted with the least amount of effort spent on developing the test doubles. Test doubles are also well suited for testing applications developed using object-oriented

> **SAMPLE PROBLEM**
>
> Consider the following code sample in the Visual Basic language.
>
> Assuming that the getTime procedure is designed to store the time of day when the procedure is executed in the timeOfDay variable, what type of test double is being used to test myModule? What is the test attempting to verify?
>
> **Answer:**
>
> The test double being used is a stub. This is because it provides the input required so that the getTime function of myModule can be tested without generating an error. As the test double provides a static response of 9:00 p.m. to every call from the getTimeOfDay method, rather than mimicking the actual functionality of the getTimeOfDay method by returning the actual time of day, it cannot be considered a fake object.
>
> The test stub is being used to verify that the getTime function in myModule has successfully created an object of type Clock and has invoked the object's getTimeOfDay method.

```
Class Clock
    Dim time as String
    Public Function getTimeOfDay()
        timeNow = "9:00 pm"
        return timeNow
    End Function
End Class

Module myModule
    Sub getTime
        Dim myClock As Clock = New Clock()
        Dim timeOfDay As String
        timeOfDay = myClock.getTimeOfDay()
    End Sub
End Module
```

programming (OOP), which is widely used. Unit tests are often automated, too, saving valuable time in writing, debugging, and maintaining code.

Using test doubles requires developers with the knowledge and experience to create the appropriate test doubles and testing procedures. It also demands the financial and other resources needed to implement a robust testing regime. Therefore, it may not be an appropriate approach for small organizations with limited budgets and few IT staff.

Test Doubles in Action

Test doubles are useful in many real-world testing situations. They are particularly useful when testing large, complex, modular systems. For example, an international insurance corporation might be testing a customer relationship management application used by the global sales department. A large team of developers are building this large-scale application using an OOP language. The application features a modular design because it needs to be deployed quickly and be responsive to changes in the sales department's requirements. Such an application is an ideal candidate for the use of test doubles.

Doubling the Power of Unit Testing

Test doubles are a key component in a well-designed unit testing methodology. They allow flexible testing of any type of component in an array of real-world development situations. They work well with many development methodologies and programming languages. These include incremental and iterative development, agile development, extreme programming, and OOP languages. As long as these types of methodologies and technologies are used by developers, test doubles will be widely used.

—*Maura Valentino, MSLIS*

Bibliography

Haverbeke, Marijn. *Eloquent JavaScript: A Modern Introduction to Programming.* 2nd ed., No Starch Press, 2014.

MacLennan, Bruce J. *Principles of Programming Languages: Design, Evaluation, and Implementation.* 3rd ed., Oxford UP, 1999.

Meszaros, Gerard. *xUnit Test Patterns: Refactoring Test Code.* Pearson Education, 2009.

Schneider, David I. *An Introduction to Programming using Visual Basic.* 10th ed., Pearson, 2016.

Scott, Michael L. *Programming Language Pragmatics.* 4th ed., Morgan Kaufmann Publishers, 2016.

Van Roy, Peter, and Seif Haridi. *Concepts, Techniques, and Models of Computer Programming.* Massachusetts Institute of Technology, 2004.

THEORY OF MULTIPLE INTELLIGENCES

Date of original: 1 July 2013

In 1983 Harvard Howard Gardner advanced the theory of multiple intelligences (MI) in response to traditional intelligence tests that had previously been used as predictors of success in educational, occupational, and military settings. Rather than relying only upon standardized measures of language, mathematics, and logic as indicators of general intelligence, the theory of multiple intelligences proposes that humans are inherently intelligent nine distinct ways: visual/spatial; verbal/linguistic; mathematical/logical; bodily/kinesthetic; musical/rhythmic; intrapersonal; interpersonal; naturalistic; and existential. Gardner theorizes that while all humans possess each of these distinct intelligences, specific intelligences will be more prominent than others in each person.

Howard Gardner.

Overview

Gardner's theory of multiple intelligences challenged the long-standing use of standardized intelligence tests. The first widely used intelligence quotient (IQ) test was developed by French psychologist Alfred Binet in 1908 and revised in 1916 by Lewis Terman at Stanford University. This Stanford–Binet test, as it came to be called, was developed to predict school success for young children. Because the IQ tests were simple and inexpensive to administer, schools, employers, and the United States military adopted them to screen applicants for readiness and for future potential. The Scholastic Aptitude Test or SAT is an example of an aptitude test used to predict future college success; Gardner claims tests like SAT only measure intelligence in one dimension, thereby privileging a particular kind of intelligence over others.

Soon after Gardner's book, *Frames of Mind: The Theory of Multiple Intelligences,* was published in 1983, his new theory was quickly embraced by educators, though it was simultaneously roundly criticized by psychologists who claimed the theory of MI was not based in empirical science, but instead upon intuition and anecdotes. Those who embraced Gardner's theory of MI may have done so because it is easy to think of a person who is unusually gifted in some skill without having been successful in academics. For example, a gifted dancer exhibits extraordinary physical talent but might not perform well on standardized assessments. A car mechanic may intuitively know how to improve the performance of an engine but does not read well. Multiple intelligences has been used to explain the occurrence of seemingly innate, undeveloped abilities, such as four-year old Mozart's capacity to play and compose highly technical piano music that continues to challenge musicians today.

Though Gardner claims teachers sometimes mistake multiple intelligences for learning styles or learning preferences, many teachers have explored the implications of MI. In 1987 in Indianapolis, Indiana, the Key Learning Community was founded on Gardner's principles and with his direct involvement. Critics have accused teachers of having "dumbed down" their level of teaching and of poorly preparing students for traditional testing measures in their attempts to address the concept of MI in their classrooms. Nevertheless, teachers claim that recognizing the ways children are intelligent creates confidence and opportunities for meaningful learning.

Although schools continue to measure student learning through standardized testing, Gardner's widespread influence on the practice of teaching is undeniable. The acknowledgement of multiple intelligences has encouraged educators to include movement, music, visual support and graphic organizers, problem-solving, and social engagement to help students experience learning in a hands-on environment, rather than simply relying upon passive practices, such as listening and reading.

—*Susan R. Adams, MA, PhD*

BIBLIOGRAPHY

Armstrong, Thomas. *Multiple Intelligences in the Classroom*. 3rd ed. Alexandria: ASCD, 2009. Print.

Arnold, Ellen. *The MI Strategy Bank: 800+ Multiple Intelligence Ideas for the Elementary Classroom*. Chicago: Chicago Review P, 2012. Print.

Delgoshaei, Yalda, and Neda Delavari. "Applying Multiple-Intelligence Approach to Education and Analyzing Its Impact on Cognitive Development of Pre-school Children." *Procedia-Social and Behavioral Sciences* 32 (2012): 361–66. Print.

Gardner, Howard. *Frames of Mind: The Theory of Multiple Intelligences*. New York: Basic, 1983. Print.

———. *Multiple Intelligences: New Horizons in Theory and Practice*. Reprint. New York: Basic, 2006. Print.

———. "Reflections on My Works and Those of My Commentators." *MI at 25: Assessing the Impact and Future of Multiple Intelligences for Teaching and Learning*. Ed. Branton Shearer. New York: Teachers College P, 2009, 96–113. Print.

Humphrey, Neil. *Social and Emotional Learning: A Critical Appraisal*. Thousand Oaks: Sage, 2013. Print.

Schlinger, Henry D. "The Myth of Intelligence." *The Psychological Record* 53.1 (2012): 2. Print.

THEORY X AND THEORY Y

Theory X and Theory Y are two models of human motivation developed by Douglas McGregor while he was working as a professor at the Sloan School of Management at the Massachusetts Institute of Technology. McGregor presented and explained the two theories in what is considered a classic work of management science, his 1960 book *The Human Side of Enterprise*. Theory X and Theory Y represent two basic assumptions about the human capacity for and relationship to work.

Overview

Theory X hinges on the assumption that humans are inherently work-averse. The workplace is therefore authoritarian in nature, with top-down pressure serving as the primary mechanism of motivation. Under Theory X, employees are assumed to have little ambition and avoid responsibility, preferring a secure, base work environment. Therefore, they require constant supervision and coercion to maintain productivity. Workplaces that follow the Theory X model of management often use continual prodding, strict quotas, the threat of discipline, and rewards for performance to keep productivity at a desired level.

Theory Y, however, is participatory, in that it believes that humans will work of their own accord in an effort to meet their individual needs for self-respect and achievement. There is no need, in this scenario, for oppressive supervision or the looming threat of discipline. Theory Y teaches managers to focus an employee's self-sustaining motivation in a way that helps achieve company-wide goals. One of the ways this energy can be harnessed, according to Theory Y, is by strengthening an individual employee's commitment

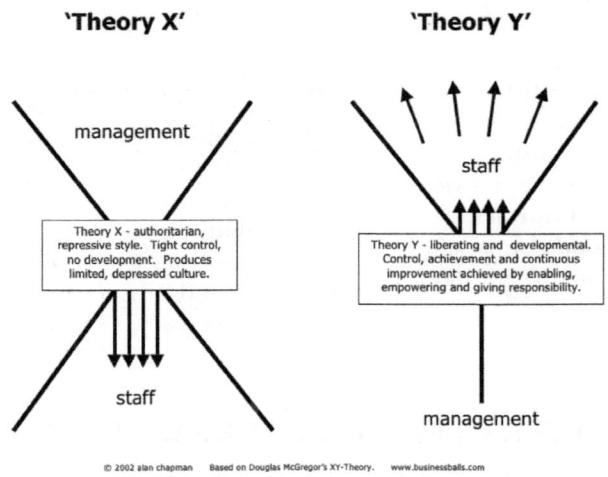

A comparison of Theory X and Theory Y.

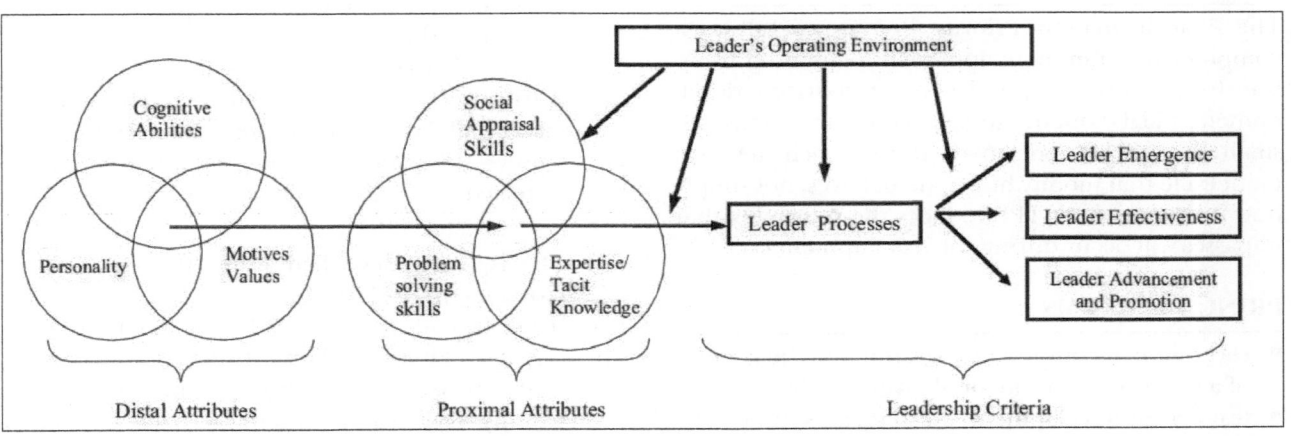

Theory X and Theory Y depend on the traits of management leadership.

to organizational objectives. People will work toward mutual objectives, Theory Y contends, only to the degree that the work fulfills their individual needs.

In Theory X, failure can be succinctly blamed on the inherent limitations of human resources. Theory Y spreads responsibility among a larger group of employees, and failure can only be blamed on a collective dysfunction and inability to overcome obstacles. McGregor believed Theory Y to be the model capable of motivating employees to reach their highest attainable level of achievement. Theory Y has emerged as the most popular model for management, particularly in the tech industry, where Internet giants such as Google and Facebook strike an employee-friendly, hands-off approach in the workplace that encourages individual creativity and self-realization.

—*Steven Miller*

Bibliography

"Douglas McGregor: Theory X and Theory Y." *Workforce* 81.1 (Jan. 2002): 32. Print.

"Guru: Douglas McGregor." *Economist*. Economist Newspaper, 3 Oct. 2008. Web. 18 Sept. 2013.

Haji Mohamed, Ramesh Kumar Moona, and Che Supian Mohamad Nor. "The Relationship between McGregor's X-Y Theory Management Style and Fulfillment of Psychological Contract: A Literature Review." *International Journal of Academic Research in Business and Social Sciences* 3.5 (May 2013): 715–20. Print.

Highhouse, Scott. "The Influence of Douglas McGregor." *TIP: The Industrial-Organizational Psychologist* 49.2 (Oct. 2011): 105–7. Print.

McGregor, Douglas. *The Human Side of Enterprise*. New York: McGraw-Hill, 1960. Print.

Neuliep, James W. "The Influence of Theory X and Y Management Style on the Perception of Ethical Behavior in Organizations." *Journal of Social Behavior & Personality* 11.2 (June 1996): 301–11. Print.

Stewart, Matthew. "Theories X and Y, Revisited." *Oxford Leadership Journal* 1.3 (June 2010): 1–5. Print.

"Theories X and Y." *Economist*. Economist Newspaper, 6 Oct. 2008. Web. 17 Sept. 2013.

TRANSFORMATION PRIORITY PREMISE (TPP)

FIELDS OF STUDY

Software Development; Coding Techniques; Computer Science

ABSTRACT

The Transformation Priority Premise (TPP) is a computer programming approach focused on using transformations to optimize the test-driven development (TDD) methodology. Transformations are small changes to code made during each development cycle that modify how code behaves, not simply how it is structured. TPP reduces the possibility that impasses will occur during the development cycle.

PRINCIPAL TERMS

- **recursion:** in computer programming, the process of a function, method, or algorithm calling itself.
- **test-driven development:** an iterative software development methodology based on frequent, short development cycles and unit testing; often abbreviated TDD.
- **unconditional statement:** statements in computer code that execute under all conditions.
- **unit test:** the process of testing an individual component of a larger software system.

Understanding the Transformation Priority Premise

The Transformation Priority Premise (TPP) is used to optimize computer code when using a test-driven development (TDD) methodology. TDD is an iterative software development methodology in which multiple unit tests are conducted based on the following steps:

> Create a test case for a specific required feature.
> Ensure that the test fails.
> Create code that causes the test to pass.
> Restructure (refactor) the code to remove duplication and make other changes as needed to improve the structure of the code that has been created.

TDD is often described using the red/green/refactor metaphor. First, a test is created that fails (red). Next, code is created that causes the test to pass (green). Finally, code is refactored as necessary. However, one of the problems developers encounter when using TDD is that an impasse may be reached during this cycle. An impasse occurs when the whole algorithm must be rewritten for the current test to pass. This can cause a long delay in development.

Software engineer Robert Cecil Martin (also known by the byname Uncle Bob) developed TPP as a way to reduce the possibility that impasses will occur during TDD. The TPP approach focuses on using transformations to modify code in order to make each new test case pass. Transformations are small changes to code that modify how the code behaves. This contrasts with refactoring, which simply changes how code is structured. In addition,

- ({} –>nil)
- (nil –>constant)
- (constant–>constant+)
- (constant–>scalar)
- (statement–>statements)
- (unconditional–>if)
- (scalar–>array)
- (array–>container)
- (statement–>recursion)
- (if–>while)
- (expression–>function)
- (variable–>assignment)

A transformation changes the behavior of code. This list of transformations is in order of priority when writing code in test-driven development (TDD). In some cases a description is given to explain the transformation.

transformations are directional in the sense that they change the behavior of code from more specific to more generic. Examples include changing a constant to a variable, changing an unconditional statement to an if-statement, and replacing statements with recursion.

Martin developed a hierarchy of transformations based on their complexity. For example, the transformation from a constant to a variable is less complex than the transformation from an unconditional statement to an if-statement. The core principle of TPP is that simple transformations have priority over more complex ones when changing code. The least complex transformation needed to pass the current test should be used. Future unit tests should be designed so that the simplest possible transformations can be used to make the code pass. This minimizes the risks inherent in making changes to code and reduces the likelihood of an impasse.

Considerations When Using TPP

TPP provides developers with an approach to developing code that is designed to address the problems that can arise during TDD. While the process has been shown to work effectively when developing many different algorithms and other software components, questions remain as to how widely it can be applied. The diverse requirements of modern computer programs and the differences in programming languages may provide challenges to the universality of TPP. Further study of the technique is needed to determine if all relevant transformations have been identified, if transformations have been prioritized correctly, and if these processes can be quantified and formalized.

TPP in Action

Consider a program that simulates a game of darts in which each player throws three darts per round. Each dart thrown can score 0 (missed the dartboard completely), 5, 10, 15, 20, 25, or 50 (hit the bull's-eye). Such a program might use a function to calculate a player's score after one round. To apply TPP in developing such a function, the developer might begin by creating a unit test for a simple test case: when all three darts miss the board completely.

First, a procedure is used to create the test. This procedure calls the computeScoreForRound function to compute the score for the round, passing it the values for each dart thrown. If the correct score is returned (0), the function returns 1, indicating success. If not, it returns 0, indicating failure. This function is shown below.

```
Function Test1 as Integer
   score = computeScoreForRound (0, 0, 0)
   IF score = 0 THEN
   Return 1
   ELSE
   Return 0
   END IF
End Function
```

Next, a function is created that returns 0. This will enable the test to pass.

```
Function computeScoreForRound (x as integer, y as integer, z as integer) as Integer
   Return 0
End Function
```

This completes the first unit test. Next, a new case is added in which two darts miss the board and one dart hits the board, scoring 5 points.

```
Function Test2 as Integer
  score = computeScoreForRound (0, 0, 5)
  IF score = 5 THEN
    Return 1
  ELSE
    Return 0
  END IF
End Function
```

In order to pass this test, the code is changed using a constant to variable transformation.

```
Function computeScoreForRound (x as integer, y as integer, z as integer) as Integer
  roundScore = x + y + z
  Return roundScore
End Function
```

Note how the new test is passed by transforming a constant value (0) in the first unit test to a variable, *roundScore*, in the second test.

The Power of Transformations

TPP promises to improve the efficiency of developers working in a TDD environment by helping to avoid impasses. It provides useful means of overcoming obstacles in designing and implementing unit tests. Though the specific list of transformations applied may vary among developers, the premise of prioritizing simplicity establishes a system that is easy to follow. The technique is likely to increase in importance and applicability as it is studied, improved, and formalized.

—*Maura Valentino, MSLIS*

Bibliography

Beck, Kent. *Test-Driven Development: By Example.* Pearson, 2014.

Farcic, Viktor, and Alex Garcia. *Test-Driven Java Development.* Packt Publishing, 2015.

Haverbeke, Marijn. *Eloquent JavaScript: A Modern Introduction to Programming.* 2nd ed., No Starch Press, 2014.

MacLennan, Bruce J. *Principles of Programming Languages: Design, Evaluation, and Implementation.* 3rd ed., Oxford UP, 1999.

Schneider, David I. *Introduction to Programming Using Visual Basic.* Pearson, 2016.

Scott, Michael L. *Programming Language Pragmatics.* 4th ed., Morgan Kaufmann Publishers, 2016.

Van Roy, Peter, and Seif Haridi. *Concepts, Techniques, and Models of Computer Programming.* Massachusetts Institute of Technology, 2004.

TREE STRUCTURES

FIELDS OF STUDY

Software Development; Coding Techniques; Computer Science

ABSTRACT

A tree structure is a way of representing how a hierarchical system is organized. As such, tree structures are ideal for storing hierarchical data. A binary search tree is a tree structure in which each nonterminal node has no more than two child nodes. Binary search trees are used in computer programming to store and manipulate data. They facilitate fast data retrieval, insertion, and deletion.

PRINCIPAL TERMS

- **binary:** something that consists of two parts; a binary tree structure is one in which every parent node has no more than two children.
- **child:** in a tree structure, a node that is directly related to another node that is one level above it in the hierarchy.
- **median:** the middle value in a sorted data set; in a balanced binary search tree, the median is the value stored in the root node.
- **node:** in a tree structure, a data structure that stores a unique value and, unless it is a terminal (leaf) node, contains references to one or more child nodes.

- **parent:** in a tree structure, a node that is directly related to one or more nodes that are one level below it in the hierarchy.
- **sibling:** in a tree structure, one of two or more nodes that share the same parent.

What Is a Tree Structure?

A tree structure is a method of representing a how a hierarchical system is organized. It resembles an upside-down tree, with the roots at the top and the branches growing downward, ending in leaves at the bottom. A classic example of a tree structure is an organizational chart.

The fundamental unit of a tree structure is called a node. Each node represents one of the entities in hierarchical system. The node at the top of the hierarchy is called the root node. In a company's organizational chart, for example, the root node might represent the president of the company. In most scenarios encountered in computer programming, tree structures have a single root node.

Relationships between nodes in a tree structure are described using kinship terminology. As in a family tree (another example of a tree structure), a node is the child of the node one level directly above it and the parent of any nodes one level directly below it. Children with the same parent are referred to as siblings. A node with no children is called a terminal node or a leaf node. Nonterminal nodes are ancestors to all nodes that descend directly from their children, which in turn are descendants of their ancestor nodes. The root node is at the top of the hierarchy and does not have a parent.

A binary search tree is a tree structure that is often used in computer programming to store data. A binary tree is one in which each node can have only two children, a left child and a right child. The value of the left-hand child is always lower than that of

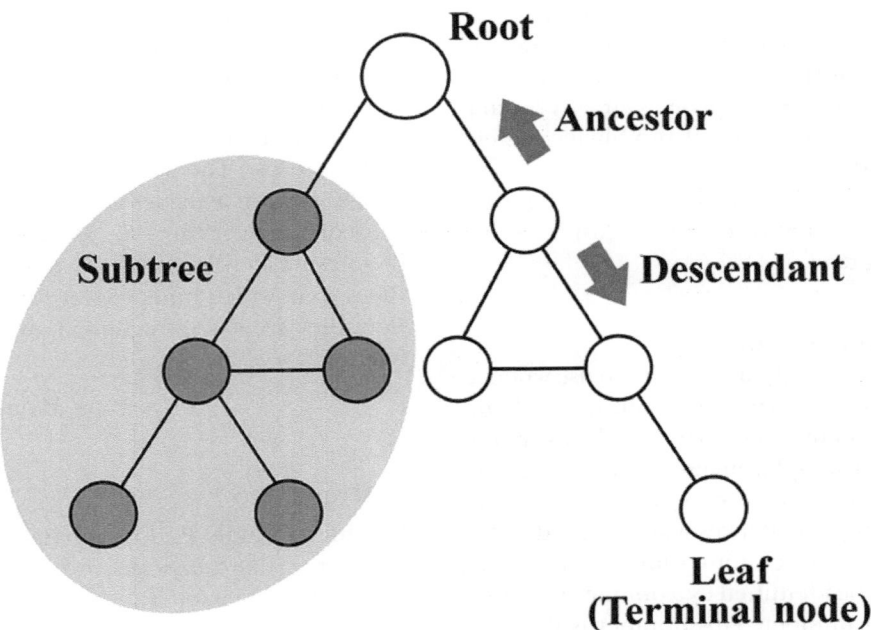

Tree structures are often used to organize, store, and quickly navigate through data. A binary tree is a tree structure in which a parent node can only have two children. Nodes are organized into multiple levels. Algorithms used to traverse these data structures will only ascend or descend through connected nodes. This example of a tree structure shows a root node with multiple subtrees descending from it. There are multiple descendants along the branches leading from the root. A nonterminal node is an ancestor to all lower-level nodes that descend from it. A terminal node, also called a leaf, is a node that has no children.

the right-hand child. Binary search trees are most efficient when they are balanced, meaning that both sides are the same height, or at most differ by only one level. The median value of the data set is stored in the root node, and all values lower than the median descend from the root's left child note, while all greater values descend from the right child node.

The Value of Tree Structures in Programming

Many types of data are organized in a hierarchical fashion. Examples include governmental entities (country, state or province, county, city) and biological taxonomy (domain, kingdom, phylum, class, order, family, genus, species). This makes the tree structure an excellent match for modeling such data.

Binary search trees offer additional advantages. Their design facilitates quick and flexible searches, insertions, and deletions. However, arrays or linked lists might offer improved performance over binary search trees, depending on the amount of data and the operation being performed. A properly balanced binary search tree can search for, insert, or delete data faster than an array, especially if there is a large number of data elements. Yet arrays can access data faster than binary search trees. If the binary search tree is unbalanced, the array may complete other operations faster as well. Linked lists take longer to search than binary search trees but are faster for inserting or deleting data.

Using a Tree Structure

Binary search trees are a good structure to use when a large amount of data needs to be frequently searched, and each data element is associated with a unique value, or key. For example, imagine an application used by customer service representatives for a large corporation. The representatives will frequently perform searches to retrieve customer information, and each customer can be identified by a unique customer identification (ID) number. This number is the key, and accessing it allows the program to retrieve all of the other data associated with the customer, such as their name and phone number.

The internal design of the binary search tree structure specifies that the value of right child node is always greater than the value of the left child node. So, the tree can be traversed efficiently in the following manner. First, the root node, which contains the median customer ID value, is examined. If the root node contains the customer ID value that the program is looking for, then the data has been located. If it does not, the value of the root node is compared to the search key. If the search key value is lower than the root node value, then the search algorithm checks the value in the left child node; if it is greater, the algorithm checks the right child node. If the child node value matches the search key, the correct customer ID has been located. If not, the same comparison procedure is conducted on the child node.

This process continues until the correct customer ID is located. This process is efficient because every time a node is checked, even if it does not match the search key, half of the remaining unchecked nodes are eliminated as possibilities.

Why Are Tree Structures Important?

Tree structures are an ideal construct to use when working with hierarchical data and with data that has unique keys. As so much of the information people deal with is hierarchically ordered, many people find tree structures easy to conceptualize and understand. It is an intuitive choice for modeling data objects and other hierarchical data structures in software architectures. The widespread use of hierarchical data in human activities across all areas of society has caused tree structures to become a key component in software development. As tree structures parallel the ways in which humans store and process information, they are like to remain an important tool in the future.

—*Maura Valentino, MSLIS*

Bibliography

Friedman, Daniel P., and Mitchell Wand. *Essentials of Programming Languages*. 3rd ed., MIT P, 2008.

Goodrich, Michael T., et al. *Data Structures and Algorithms in Java*. 6th ed., John Wiley & Sons, 2014.

Haverbeke, Marijn. *Eloquent JavaScript: A Modern Introduction to Programming*. 2nd ed., No Starch Press, 2015.

Lee, Kent D., and Steve Hubbard. *Data Structures and Algorithms with Python*. Springer, 2015.

MacLennan, Bruce J. *Principles of Programming Languages: Design, Evaluation, and Implementation.* 3rd ed., Oxford UP, 1999.

Scott, Michael L. *Programming Language Pragmatics.* 4th ed., Elsevier, 2016.

Schneider, David I. *An Introduction to Programming Using Visual Basic.* 10th ed., Pearson, 2017.

Van Roy, Peter, and Seif Haridi. *Concepts, Techniques, and Models of Computer Programming.* MIT P, 2004.

TURING TEST

FIELDS OF STUDY

Computer Science; Robotics

ABSTRACT

The Turing test is a game proposed by computer scientist and mathematical logician Alan Turing in 1950 to determine whether machines can think or at least simulate a thinking human well enough to fool a human observer. Despite criticism, the test has shaped the development and study of artificial intelligence ever since.

PRINCIPAL TERMS

- **artificial intelligence:** the intelligence exhibited by machines or computers, in contrast to human, organic, or animal intelligence.
- **automaton:** a machine that mimics a human but is generally considered to be unthinking.
- **chatterbot:** a computer program that mimics human conversation responses in order to interact with people through text; also called "talkbot," "chatbot," or simply "bot."
- **imitation game:** Alan Turing's name for his proposed test, in which a machine would attempt to respond to questions in such a way as to fool a human judge into thinking it was human.

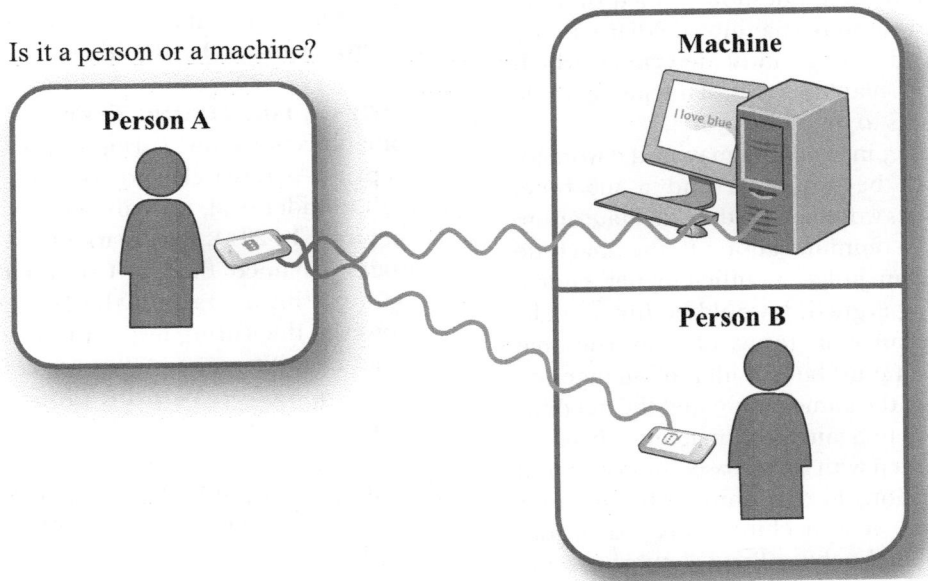

The Turing test of artificial intelligence should allow person A to determine whether they are corresponding via technology with a human or a machine entity.

Can Machines Think?

In 1950, British mathematician Alan Turing (1912–54) wrote a paper titled "Computing Machinery and Intelligence" in which he asked, "Can machines think?" The question was too difficult to answer directly, so instead he thought up a simple game to compare a machine's ability to respond in conversation to that of a human. If the machine could fool a human participant into believing it was human, then it could be considered functionally indistinguishable from a thinking entity (i.e., a person). This game later came to be known as the Turing test.

In Turing's time, digital computers and automata already existed. However, the notion of developing machines with programming sophisticated enough to engender something like consciousness was still very new. What Turing envisioned would, within the decade, become the field of artificial intelligence (AI): the quest for humanlike, or at least human-equivalent, intelligence in a machine.

The Imitation Game

To settle the question of whether machines can think, Turing proposed what he called an imitation game. He based it on a party game of the time, played by a man, a woman, and a third party (the judge). In the game, the judge stays in a separate room from the contestants and asks them questions, which they answer via writing or an impartial intermediary. Based on their answers, the judge must determine which player is which. The man tries to fool the judge, while the woman attempts to help him or her.

In Turing's game, instead of a man and a woman, the players would be a human and a machine. Computer terminals would be used to facilitate anonymous text-based communication. If the machine could fool a human judge a sufficient percentage of the time, Turing argued, it would be functionally equivalent to a human in terms of conversational ability. Therefore, it must be considered intelligent.

Later versions of the game eliminated the need for a human to compete against the machine. Instead, the judge is presented with a text-based interface and told to ask it questions to determine if he or she is talking to a person or a machine. This version has been the basis of the Loebner Prize chatterbot competition since 1995.

Natural Language Processing

Turing's challenge spurred the development of a subfield of AI known as "natural language processing" (NLP), devoted to creating programs that can understand and produce humanlike speech. The 1966 program ELIZA, written by German American computer scientist Joseph Weizenbaum (1923–2008), was very convincing in its ability to imitate a Rogerian psychotherapist ("How does that make you feel?"). Programs like ELIZA are called "chatterbots," or simply "bots." These programs became a feature of early text-based Internet communication. With the advent of smartphones, both Apple and Google have developed advanced NLP to underpin the voice-activated functions of their devices, most famously Apple's AI assistant Siri.

Criticism of the Turing Test

The Turing test has been a point of contention among AI researchers and computer scientists for decades. Objections fall into two main camps. Some object to the idea that the Turing test is a reasonable test of the presence of intelligence, while others object to the idea that machines are capable of intelligence at all. With regard to the former, some critics have argued that a machine without intelligence could pass the test if it had sufficient processing power and memory to automatically respond to questions with prewritten answers. Others feel that the test is too difficult and too narrow a goal to help guide AI research.

Legacy of the Turing Test

Despite objections, the Turing test has endured as a concept in AI research and robotics. Many similar, though not identical, tests are also informally called Turing tests. In 2014, sixty years after Turing's death, a program named Eugene Goostman made headlines for passing a version of the Turing test.

However, the Turing test's lasting contribution to AI is not its utility as a test, but the context in which Turing proposed it. With his proposal, Turing became the first major advocate for the idea that humanlike machine intelligence was indeed possible. Though he died before AI emerged as a proper field of study, his ideas shaped its development for decades thereafter.

—*Kenrick Vezina, MS*

BIBLIOGRAPHY

Ball, Phillip. "The Truth about the Turing Test." *Future*. BBC, 24 July 2015. Web. 18 Dec. 2015.

Myers, Courtney Boyd, ed. *The AI Report. Forbes*. Forbes.com, 22 June 2009. Web. 18 Dec. 2015.

Olley, Allan. "Can Machines Think Yet? A Brief History of the Turing Test." *Bubble Chamber*. U of Toronto's Science Policy Working Group, 23 June 2014. Web. 18 Dec. 2015.

Oppy, Graham, and David Dowe. "The Turing Test." *Stanford Encyclopedia of Philosophy (Spring 2011 Edition)*. Ed. Edward N. Zalta. Stanford U, 26 Jan. 2011. Web. 18 Dec. 2015.

Turing, Alan M. "Computing Machinery and Intelligence." *Mind* 59.236 (1950): 433–60. Web. 23 Dec. 2015.

"Turing Test." *Encyclopædia Britannica*. Encyclopædia Britannica, 23 Sept. 2013. Web. 18 Dec. 2015.

U

UNCERTAINTY REDUCTION THEORY (URT)

In 1975 communications researchers Charles Berger and Richard Calebrese developed the uncertainty reduction theory (URT). Their objective was to understand how two individuals communicate with each other during an initial encounter. They believed that when two strangers meet for the first time, the situation is fraught with uncertainty and vagueness. Hence, each individual attempts to lessen the ambiguity about the other person and their interpersonal relationship. Berger and Calebrese characterized two types of uncertainty that individuals encounter during an initial meeting: cognitive and behavioral. Cognitive uncertainty relates the observer's impression of the other person. Behavioral uncertainty relates to the actions a person will take. The researchers believed that the two types of uncertainties can be reduced with self-disclosure, especially as more frequent interactions occur between the individuals.

Mirroring, as shown in the openhanded gestures here, is the subconscious imitation of another. It can reduce uncertainty and produce a greater sense of engagement and belonging.

Overview

The uncertainty reduction theory consists of nine main axioms, as outlined in L. H. Turner and R. West's *Introducing Communication Theory*:

1. People experience uncertainty in initial interpersonal settings, which can be lessened as verbal communication increases.
2. Uncertainty is inversely correlated to nonverbal affiliative expressiveness.
3. Uncertainty is positively correlated with information-seeking strategies.
4. Intimacy and uncertainty are inversely correlated.
5. Reciprocity is positively correlated with uncertainty.
6. Similarities between individuals will reduce uncertainty.
7. Increased uncertainty results in decreased levels of liking.
8. Shared social networks reduce uncertainty and vice versa.
9. Uncertainty is inversely correlated to communication satisfaction.

URT provides strong predictions in many situations for determining whether strangers will become friends or not.

Berger and Calebrese noted that when individuals first meet someone, they have three antecedents that help them reduce uncertainty. These antecedents are whether or not the other person has the potential to reward or punish, whether or not the other person behaves contrary to normal expectations, and whether or not the person expects further encounters with the other person. Individuals will use three strategies in order to defeat their uncertainty. These include passive strategies (unobtrusive observation), active strategies (means other than by direct contact), and interactive strategies (engaging in conversation). Berger later found that in conversation, individuals tend not only to seek information from a new acquaintance but also to formulate a plan for navigating the situation and to "hedge," or couch their messages in humor, ambiguity, disclaiming, or discounting.

URT consists of seven basic assumptions. These assumptions are based on the idea that communication is the most important element of human behavior. The assumptions are that individuals feel uncertainty in interpersonal settings; uncertainty is an aversive state that produces cognitive stress; when individuals first meet their main concern is to reduce uncertainty or to increase predictability; interpersonal communication is a developmental process that happens in stages; interpersonal communication is the main process of uncertainty reduction; the quantity of data that individuals share changes through time; and that it is possible to predict individual's behavior. These assumptions provide an explanation for what people experience when they first meet with someone new and how those feelings will change as additional interactions occur in the future. URT can also be applied to shifting conditions in an established relationship as the individuals seek to understand how the other might behave in new situations.

Social scientists have examined URT in new relationships in such varied contexts as schools and businesses, romantic partnerships, and technology-mediated interactions including watching television or using a computer for communication. Some theorists maintain that uncertainty can be either positive or negative for the participants in an interaction or a relationship, particularly if reducing that uncertainty might harm one or the other individual.

—*Narissra Maria Punyanunt-Carter, MA, PhD*

Bibliography

Antheunis, Marjolijn L., Alexander P. Schouten, Patti M. Valkenburg, and Jochen Peter. "Interactive Uncertainty Reduction Strategies and Verbal Affection in Computer-Mediated Communication." *Communication Research* 39.6 (2012): 757–80. Print.

Berger, Charles R., and Michael Burgoon. *Communication and Social Influence Processes*. East Lansing: Michigan State UP, 1998. Print.

Berger, Charles R., and Richard J. Calabrese. "Some Explorations in Initial Interaction and Beyond: Toward a Developmental Theory of Interpersonal Communication." *Human Communication Research* 1.2 (1975): 99–112. Print.

Booth-Butterfield, Melanie, Steven Booth-Butterfield, and Jolene Koester. "The Function of Uncertainty Reduction in Alleviating Primary Tension in Small Groups." *Communication Research Reports* 5.2 (1998): 146–53. Print.

Deyo, Jessica, Price Walt, and Leah Davis. "Rapidly Recognizing Relationships: Observing Speed

Dating in the South." *Qualitative Research Reports in Communication* 12.1 (2011): 71–78. Print.

Haunani Solomon, Denise. "Uncertainty Reduction Theory." *The Concise Encyclopedia of Communication.* Ed. Wolfgang Donsbach. Malden: Wiley, 2015. Print.

Knobloch, Leanne K., and Laura E. Miller. "Uncertainty and Relationship Initiation." *Handbook of Relationship Initiation.* Ed. Susan Sprecher, Amy Wenzel, and John Harvey. New York: Psychology P, 2008. 121–34. Print.

Levine, Timothy R., Kim Sang-Yeon, and Merissa Ferrara. "Social Exchange, Uncertainty, and Communication Content as Factors Impacting the Relational Outcomes of Betrayals." *Human Communication* 13.4 (2010): 303–18. Print.

May, Amy, and Kelly E. Tenzek. "Seeking Mrs. Right: Uncertainty Reduction in Online Surrogacy Ads." *Qualitative Research Reports in Communication* 12.1 (2011): 27–33. Print.

Ramirez, Artemio. "The Effect of Interactivity on Initial Interactions: The Influence of Information Seeking Role on Computer-Mediated Interaction." *Western Journal of Communication* 73.3 (2009): 300–25. Print.

Turner, L. H., and R. West. *Introducing Communication Theory.* 4th ed. New York: McGraw, 2010. 147–65. Print.

UNICODE

FIELDS OF STUDY

Computer Science; Software Engineering; Information Technology

ABSTRACT

Unicode is a character-encoding system used by computer systems worldwide. It contains numeric codes for more than 120,000 characters from 129 languages. Unicode is designed for backward compatibility with older character-encoding standards, such as the American Standard Code for Information Interchange (ASCII). It is supported by most major web browsers, operating systems, and other software.

PRINCIPAL TERMS

- **glyph:** a specific representation of a grapheme, such as the letter *A* rendered in a particular typeface.
- **grapheme:** the *smallest* unit used by a writing system, such as alphabetic letters, punctuation marks, or Chinese characters.
- **hexadecimal:** a base-16 number system that uses the digits 0 through 9 and the letters *A, B, C, D, E,* and *F* as symbols to represent numbers.
- **normalization:** a process that ensures that different code points representing equivalent characters will be recognized as equal when processing text.
- **rendering:** the process of selecting and displaying glyphs.
- **script:** a group of written signs, such as Latin or Chinese characters, used to represent textual information in a writing system.
- **special character:** a character such as a symbol, emoji, or control character.

CHARACTER-ENCODING SYSTEMS

In order for computer systems to process text, the characters and other graphic symbols used in written languages must be converted to numbers that the computer can read. The process of converting these characters and symbols to numbers is called "character encoding." As the use of computer systems increased during the 1940s and 1950s, many different character encodings were developed.

To improve the ability of computer systems to interoperate, a standard encoding system was developed. Released in 1963 and revised in 1967, the American Standard Code for Information Interchange (ASCII) encoded ninety-five English language characters and thirty-three control characters into values ranging from 0 to 127. However, ASCII only provided support for the English language. Thus, there remained a need for a system that could encompass all of the world's languages.

Unicode was developed to provide a character encoding system that could encompass all of the scripts used by current and historic written languages. By 2016, Unicode provided character encoding for 129 scripts and more than 120,000 characters. These include special characters, such as control characters, symbols, and emoji.

Graphic character symbol	Hexadecimal character value																														
	0020	0	0030	@	0040	P	0050	`	0060	p	0070		00A0	°	00B0	À	00C0	Ð	00D0	à	00E0	ð	00F0
!	0021	1	0031	A	0041	Q	0051	a	0061	q	0071	¡	00A1	±	00B1	Á	00C1	Ñ	00D1	á	00E1	ñ	00F1
"	0022	2	0032	B	0042	R	0052	b	0062	r	0072	¢	00A2	²	00B2	Â	00C2	Ò	00D2	â	00E2	ò	00F2
#	0023	3	0033	C	0043	S	0053	c	0063	s	0073	£	00A3	³	00B3	Ã	00C3	Ó	00D3	ã	00E3	ó	00F3
$	0024	4	0034	D	0044	T	0054	d	0064	t	0074	¤	00A4	´	00B4	Ä	00C4	Ô	00D4	ä	00E4	ô	00F4
%	0025	5	0035	E	0045	U	0055	e	0065	u	0075	¥	00A5	µ	00B5	Å	00C5	Õ	00D5	å	00E5	õ	00F5
&	0026	6	0036	F	0046	V	0056	f	0066	v	0076	¦	00A6	¶	00B6	Æ	00C6	Ö	00D6	æ	00E6	ö	00F6
'	0027	7	0037	G	0047	W	0057	g	0067	w	0077	§	00A7	·	00B7	Ç	00C7	×	00D7	ç	00E7	÷	00F7
(	0028	8	0038	H	0048	X	0058	h	0068	x	0078	¨	00A8	¸	00B8	È	00C8	Ø	00D8	è	00E8	ø	00F8
)	0029	9	0039	I	0049	Y	0059	i	0069	y	0079	©	00A9	¹	00B9	É	00C9	Ù	00D9	é	00E9	ù	00F9
*	002A	:	003A	J	004A	Z	005A	j	006A	z	007A	ª	00AA	º	00BA	Ê	00CA	Ú	00DA	ê	00EA	ú	00FA
+	002B	;	003B	K	004B	[	005B	k	006B	{	007B	«	00AB	»	00BB	Ë	00CB	Û	00DB	ë	00EB	û	00FB
,	002C	<	003C	L	004C	\	005C	l	006C	|	007C	¬	00AC	¼	00BC	Ì	00CC	Ü	00DC	ì	00EC	ü	00FC
-	002D	=	003D	M	004D	]	005D	m	006D	}	007D	-	00AD	½	00BD	Í	00CD	Ý	00DD	í	00ED	ý	00FD
.	002E	>	003E	N	004E	^	005E	n	006E	~	007E	®	00AE	¾	00BE	Î	00CE	Þ	00DE	î	00EE	þ	00FE
/	002F	?	003F	O	004F	_	005F	o	006F		007F	¯	00AF	¿	00BF	Ï	00CF	ß	00DF	ï	00EF	ÿ	00FF

The Unicode Standard is a universally recognized coding system for more than 120,000 characters, using either 8-bit (UTF-8) or 16-bit (UTF-16) encoding. This chart shows the character symbol and the corresponding hexadecimal UTF-8 code. For the first 127 characters, UTF-8 and ASCII are identical.

Understanding the Unicode Standard

The Unicode standard encodes graphemes and not glyphs. A grapheme is the smallest unit used by a writing system, such as an alphabetic letter or Chinese character. A glyph is specific representation of a grapheme, such as the letter *A* rendered in a particular typeface and font size. The Unicode standard provides a code point, or number, to represent each grapheme. However, Unicode leaves the rendering of the glyph that matches the grapheme to software programs. For example, the Unicode value of U+0041 (which represents the grapheme for the letter *A*) might be provided to a web browser. The browser might then render the glyph of the letter *A* using the Times New Roman font.

Unicode defines 1,114,112 code points. Each code point is assigned a hexadecimal number ranging from 0 to 10FFFF. When written, these values are typically preceded by U+. For example, the letter *J* is assigned the hexadecimal number 004A and is written U+004A. The Unicode Consortium provides charts listing all defined graphemes and their associated code points. In order to allow organizations to define their own private characters without conflicting with assigned Unicode characters, ranges of code points are left undefined. One of these ranges includes all of the code points between U+E000 and U+F8FF. Organizations may assign undefined code points to their own private graphemes.

One inherent problem with Unicode is that certain graphemes have been assigned to multiple code points. In an ideal system, each grapheme would be assigned to a single code point to simplify text processing. However, in order to encourage the adoption of the Unicode standard, character encodings such as ASCII were supported in Unicode. This resulted in certain graphemes being assigned to more than one code point in the Unicode standard.

Unicode also provides support for normalization. Normalization ensures that different code points that represent equivalent characters will be recognized as

> **SAMPLE PROBLEM**
>
> Using a hexadecimal character chart as a reference, translate the following characters into their Unicode code point values:
> <, 9, ?, E, and @.
>
> Then select an undefined code point to store a private grapheme.
>
> **Answer:**
>
> Unicode uses the hexadecimal character code preceded by a U+ to indicate that the hexadecimal value refers to a Unicode character. Using the chart, <, 9, ?, E, and @ are associated with the following hexadecimal values: 003C, 0039, 003F, 0045, and 0040. Their Unicode code point values are therefore: U+003C, U+0039, U+003F, U+0045, and U+0040.
>
> A private grapheme may be assigned any code point value within the ranges U+E000 to U+F8FF, U+F0000 to U+FFFFF, and U+100000 to U+10FFFD. These code points are left undefined by the Unicode standards.

equal when processing text. For example, normalization ensures that the character é (U+00E9) and the combination of characters e (U+0065) and ? (U+0301) are treated as equivalent when processing text.

Using Unicode to Connect Systems Worldwide

Since its introduction in 1991, Unicode has been widely adopted. Unicode is supported by major operating systems and software companies including Microsoft and Apple. Unicode is also implemented on UNIX systems as well. Unicode has become an important encoding system for use on the Internet. It is widely supported by web browsers and other Internet-related technologies. While older systems such as ASCII are still used, Unicode's support for multiple languages makes it the most important character-encoding system in use. New languages, pictographs, and symbols are added regularly. Thus, Unicode remains poised for significant growth in the decades to come.

—*Maura Valentino, MSLIS*

Bibliography

Berry, John D. *Language Culture Type: International Type Design in the Age of Unicode.* New York: Graphis, 2002. Print.

Gillam, Richard. *Unicode Demystified: A Practical Programmer's Guide to the Encoding Standard.* Boston: Addison-Wesley, 2002. Print.

Graham, Tony. *Unicode: A Primer.* Foster City: M&T, 2000. Print.

Korpela, Jukka K. *Unicode Explained.* Sebastopol: O'Reilly Media, 2006. Print.

"The Unicode Standard: A Technical Introduction." *Unicode.org.* Unicode, 25 June 2015. Web. 3 Mar. 2016.

"Unicode 8.0.0." *Unicode.org.* Unicode, 17 June 2015. Web. 3 Mar. 2016.

"What Is Unicode?" *Unicode.org.* Unicode, 1 Dec. 2015. Web. 3 Mar. 2016.

UNIX

FIELDS OF STUDY

Operating Systems; Computer Science; Information Technology

ABSTRACT

UNIX is a computer operating system originally developed by researchers at Bell Laboratories in 1969. The term is also used to refer to later operating systems based in part on its source code. Several of the original UNIX operating system's features, such as its hierarchical file system and multiuser support, became standard in later systems.

PRINCIPAL TERMS

- **command-line interpreter:** an interface that interprets and carries out commands entered by the user.

The creators of UNIX were Dennis Ritchie (standing) and Ken Thompson (sitting).

- **hierarchical file system:** a directory with a treelike structure in which files are organized.
- **kernel:** the central component of an operating system.
- **multitasking:** capable of carrying out multiple tasks at once.
- **multiuser:** capable of being used by multiple users at once.
- **operating system (OS):** a specialized program that manages a computer's functions.

Origin of UNIX

As computer technology rapidly developed in the mid-twentieth century, programmers sought to create means of interfacing with computers and making use of their functions in a more straightforward, intuitive way. Chief among the goals of many programmers was the creation of an operating system (OS). OSs are specialized programs that manage all of computers' processes and functions. Although many different OSs were created over the decades, UNIX proved to be one of the most influential. UNIX inspired numerous later OSs and continues to be used in various forms into the twenty-first century.

Development of the original UNIX OS began in 1969 at Bell Laboratories, a research facility then owned by AT&T. Researchers at Bell had been working on the Multiplexed Information and Computing Service (Multics) project in collaboration with the Massachusetts Institute of Technology and General Electric. The group had focused on creating systems that allowed multiple users to access a computer at once. After AT&T left the project, Bell programmers Ken Thompson and Dennis Ritchie began work on an OS. Thompson coded the bulk of the system, which consisted of 4,200 lines of code, in the summer of 1969, running it on an outdated PDP-7 computer.

The OS was initially named the Unmultiplexed Information and Computing Service, a play on the name of the Multics project. That name was later shortened to UNIX. Thompson, Ritchie, and their colleagues continued to work on UNIX over the next several years. In 1973 they rewrote the OS in the new programming language C, created by Ritchie. UNIX gained popularity outside of Bell Laboratories in the mid-1970s. It subsequently inspired the creation of many UNIX variants and UNIX-like OSs.

Understanding UNIX

While different editions of UNIX vary, the majority of UNIX OSs have some common essential characteristics. UNIX's kernel is the core of the OS. The kernel is responsible for executing programs, allocating memory, and otherwise running the system. The user interacts with what is known as the "shell." The shell is an interface that transmits the user's commands to the kernel. The original UNIX shell was a command-line interpreter, a text-based interface into which the user types commands. Over time programmers developed a variety of different shells for UNIX OSs. Some of these shells were graphical user interfaces that enabled the user to operate the computer by interacting with icons and windows. Files saved to a computer running UNIX are stored in a hierarchical file system. This file system used a treelike structure that allowed folders to be saved within folders.

Using UNIX

In keeping with its origins as a project carried out by former Multics researchers, UNIX was designed to have multiuser capabilities. This enabled multiple people to use a single computer running UNIX at the same time. This was an especially important feature in the late 1960s and early 1970s. At this time, computers were not personal computers but large, expensive mainframes that took up a significant amount of space and power. Multiuser capabilities made it possible for the organization using the UNIX OS to maximize the functions of their computers.

UNIX is also a multitasking system, meaning it can carry out multiple operations at once. One of the first programs designed for UNIX was a text-editing program needed by the employees of Bell Laboratories. Over time, programmers wrote numerous programs compatible with the OS and its later variants, including games, web browsers, and design software.

UNIX Variants and UNIX-Like Operating Systems

At the time that UNIX was first developed, AT&T was prohibited from selling products in fields other than telecommunications. The company was therefore unable to sell its researchers' creation. Instead, the company licensed UNIX's source code to various institutions. Programmers at those many institutions rewrote portions of the OS's code, creating UNIX variants that suited their needs. Perhaps the most influential new form of UNIX was the Berkeley Software Distribution variant, developed at the University of California, Berkeley, in the late 1970s.

In the early twenty-first century, the multitude of UNIX variants and UNIX-derived OSs are divided into two main categories. Those that conform to standards established by an organization known as the Open Group, which holds the trademark to the UNIX name, may be referred to as "certified UNIX operating systems." Systems that are similar to UNIX but do not adhere to the Open Group's standards are typically known as "UNIX-like operating systems." The latter category includes Apple's OS X and the free, open-source system Linux. The mobile OSs Android and Apple iOS also fall under this category and account for hundreds of millions of users, arguably making UNIX-derived OSs the most widely used systems ever.

—*Joy Crelin*

Bibliography

Gancarz, Mike. *The UNIX Philosophy.* Woburn: Butterworth, 1995. Print.

"History and Timeline." *Open Group.* Open Group, n.d. Web. 28 Feb. 2016.

Raymond, Eric S. "Origins and History of Unix, 1969–1995." *The Art of UNIX Programming.* Boston: Pearson Education, 2004. Print.

Stonebank, M. "UNIX Introduction." *University of Surrey.* U of Surrey, 2000. Web. 28 Feb. 2016.

Toomey, Warren. "The Strange Birth and Long Life of Unix." *IEEE Spectrum.* IEEE, 28 Nov. 2011. Web. 28. Feb. 2016.

"What Is Unix?" *Knowledge Base.* Indiana U, 2015. Web. 28 Feb. 2016.

Worstall, Tim. "Is Unix Now the Most Successful Operating System of All Time?" *Forbes.* Forbes.com, 7 May 2013. Web. 7 Mar. 2016.

V

VARIABLES AND VALUES

FIELDS OF STUDY
Software Development; Coding Techniques; Computer Science

ABSTRACT
Computer programs use variables and values to refer to and manipulate information. The information stored in variables can be of different types, including numbers and strings of characters. Programs can perform mathematical, relational, and logical operations on the values stored in variables.

PRINCIPAL TERMS
- **character:** a symbol that represents a letter, number, punctuation, space, break, or other mark used in written language.
- **operation:** an action that a computer program performs on specified data, such as arithmetic calculation of the data contained in two variables.
- **string:** in computer programming, a data type consisting of characters arranged in a specific sequence.

STORING AND ACCESSING DATA
Computer programs need to store and manipulate vast amounts of different types of information. Without the ability to store and manipulate information, programs could not accomplish most tasks for which they are designed. Programmers use variables to refer to and use this information effectively. Variables can be thought of as containers that computer programs use to store and reference information in memory. The term "variable" is used because the information stored in a variable can change, or vary, during program execution.

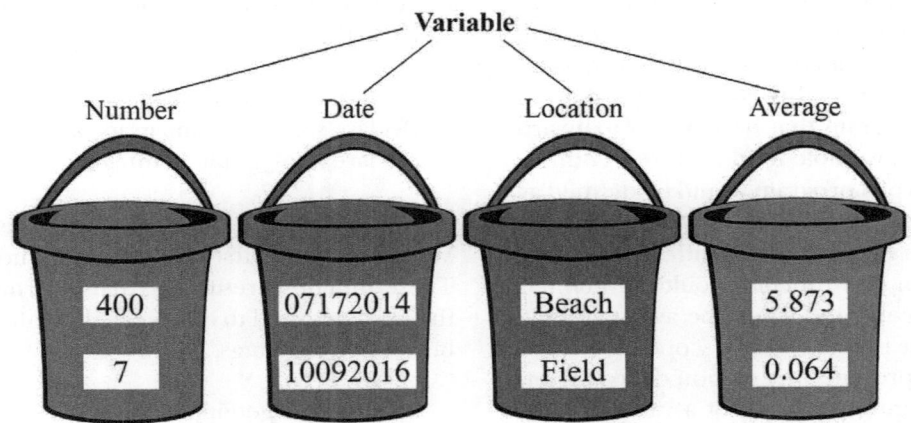

A variable stores values of a specified data type, such as integers, strings, or real numbers. Variables have names like Number, Date, Location, and Average. The data contained in each variable can be of one or more values, such as 7 and 400 for Number, 07/17/2014 and 10/09/2016 for Date, Beach and Field for Location, and 5.873 and 0.064 for Average.

The actual information or data referenced by a variable is called a "value." Variables can store various data types, including integers, real numbers, float numbers (decimals), Boolean (either-or), and strings. A string is a sequence of one or more characters. Characters are the symbols used in written language including letters, numbers, punctuation marks, and blank spaces.

Computer programs can perform operations on the values stored in variables. Typical operations include math calculations, logical comparisons, and relational operations, such as comparing two variables to see if they are equivalent.

Before a variable can be used in code, it must be declared, or given a unique name and data type. The name should describe the kind of information contained in the variable, not the value. The data type defines the possible values that the variable might contain. For example, a variable might be declared as holding any integer value from –2,147,483,648 to 2,147,483,648. Sometimes variables are assigned a starting value. Other variables can be used as part of a variable's starting value, as long as their own values are already declared in the code.

Important Considerations with Variables

Variables enable programmers to work with efficiently with large amounts of data of different types. However, potential errors can occur if variables are used incorrectly. To avoid these problems, programmers must consider a variable's scope and data type.

As variables are referenced by name, confusion and errors can result if two variables with the same name are used in unrelated sections of a program. Scope is used to address this issue. Scope defines the sections of a program that have access to a variable. For example, a variable used to store a counter used by all sections of a program would be defined as having a global scope, making the variable available to code throughout the program. A different variable that is used only in one function could be defined as having a local scope and would be available only to code within that process. Using scope to control access to variables prevents one section of code from inadvertently changing the value of a variable used by a different section of code. This also reduces the chance of naming conflicts.

Related to the concept of scope is lifetime. Lifetime indicates how long the variable will be available. A global variable is "brought to life" when the program runs and ends when the program ends. Local variables last as long as their section of the program runs.

The data type of a variable must be considered when performing operations on variables because operators behave differently when used with different data types. For example, if a program stores the value 5 in two variables defined as integers and the + operator is used to add the variables, the program sums them and gives the value 10. However, if the values are stored as strings and then added, the result is the value 55. This process of joining together is called "concatenation." Careful attention to using the appropriate data types can eliminate errors.

Variables in Practice

A computer program may need to calculate a student's final grade based on the score the student receives on two exams. First, the variables that will hold the values for the student's name must be declared. In the pseudocode below, the variables are declared as data type string, as they will hold text information.

```
Declare first_name as string
Declare last_name as string
Declare student_name as string
```

Next, the variables that will store the student's exam scores and final grade are declared. These variables are declared as integers, as they will hold numeric information. Failure to define the data type of these variables correctly could result in errors in calculating the student's final score.

```
Declare first_exam_grade as integer
Declare second_exam_grade as integer
Declare final_grade as integer
```

Next, functions are used to retrieve the first and last names of the first student in the student database. These functions return textual information, and so the variables used to reference that information must be defined as strings.

```
first_name = getFirstName(1)
last_name = getLastName(1)
```

In the next line of code, the student's first and last name are joined together and separated by a blank space using the + operator. A new variable, student_name, now contains the student's full name.

　　student_name = first_name + " " + last_name

This section of code uses a function to retrieve the first student's exam grades.

　　first_exam_grade = getExamGrade(1, 1)
　　second_exam_grade = getExamGrade(1, 2)

The student's final grade is then calculated using the addition and division operators. The calculation works correctly because both variables have been declared as integers. If the variables had been declared as strings, an error would have occurred.

　　final_grade = first_exam_grade + second_exam_grade / 2

Additional code can now use the values stored in these variables to display a message, create a report, or store the student's final score in a database.

The Importance of Variables and Values

Computer programs require access to a wide range of data of varying types. In fact, the ability to access, store, and manipulate large amounts of different types of data quickly and efficiently is one of the reasons computer programs are now used in almost every facet of modern life, from controlling railroad crossing gates to landing probes on the surface of Mars. Variables and values make it possible for other constructs and statements used in computer programming to function.

—*Maura Valentino, MSLIS*

Bibliography

Friedman, Daniel P., and Mitchell Wand. *Essentials of Programming Languages.* 3rd ed., MIT P, 2008.

Haverbeke, Marijn. *Eloquent JavaScript: A Modern Introduction to Programming.* 2nd ed., No Starch Press, 2014.

MacLennan, Bruce J. *Principles of Programming Languages: Design, Evaluation, and Implementation.* 3rd ed., Oxford UP, 1999.

Scott, Michael L. *Programming Language Pragmatics.* 4th ed., Morgan Kaufmann Publishers, 2016.

Schneider, David I. *An Introduction to Programming using Visual Basic.* 10th ed., Pearson, 2016.

Van Roy, Peter, and Seif Haridi. *Concepts, Techniques, and Models of Computer Programming.* MIT P, 2004.

WATERFALL DEVELOPMENT

FIELDS OF STUDY

Software Development; Programming Methodologies; Software Engineering

ABSTRACT

Waterfall development is a software development methodology based on a series of defined, ordered phases. Waterfall development is used to develop software systems where the system requirements are known at the start of the project and are unlikely to change during development. Waterfall development works best when used with mature, stable technologies and experienced teams.

PRINCIPAL TERMS

- **implementation:** a step in software development in which project requirements and specifications are converted to code and working systems are developed.
- **release:** the version of a software system or other product made available to consumers, or the act of making that version available.
- **requirement:** a necessary characteristic of a software system or other product, such as a service it must provide or a constraint under which it must operate.
- **unit test:** the process of testing an individual component of a larger software system.

Understanding the Waterfall Methodology

Waterfall development is a software methodology based on a series of defined, ordered phases. Each phase is completed before the next phase is begun. In the model's most basic form, no phase is repeated. Outputs from each phase usually serve as inputs for the phase that follows. The waterfall model is simple and straightforward. It is also one of the oldest and most widely used software development methodologies. It grew out of manufacturing and other industries that used a similar linear, step-by-step methodology for business systems and models. Many later software development models were designed to address perceived weaknesses in the standard waterfall model.

Different versions of the model have varying numbers of phases or names for the phases. At its most basic, the methodology has five phases: requirements analysis, system design, implementation, testing (verification), and maintenance. Many versions further include a deployment phase between verification and maintenance.

During requirements analysis, all possible requirements for the system are identified, documented, and approved by the project's stakeholders. Some versions break this phase into separate planning and analysis phases. Identifying all needed requirements at this stage is crucial, as the model does not easily incorporate changes to requirements and specifications in later phases.

During system design, those defined requirements and specifications are used to develop a detailed system design. This design specifies the overall system architecture and the hardware and software technologies that will be used to implement the system.

The system design is used to create working systems during implementation. This is the phase when code is written. The components that will make up the system are developed and go through unit testing.

Once the components have been developed, they are integrated into a single system for verification, or the integration and testing phase. After integration, components are tested to ensure the entire system interoperates correctly and is error free.

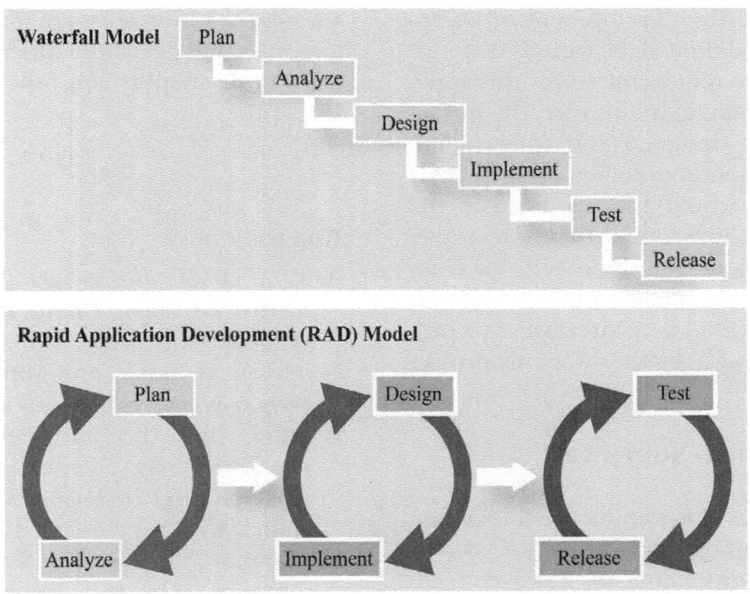

Traditional software development follows a waterfall model of completing one phase before starting the next. It flows from planning to analyzing, to implementing, to testing, to release. This is different from the rapid application development model, which includes cycles of planning and analysis, cycles of design and implementation, and cycles of testing and release to incrementally build the final product.

During deployment, the integrated and tested final system, or release, is delivered to the customer or made available for use or sale. The final phase is maintenance. During this phase, any errors or other problems that are detected once the system is in use are identified and corrected. Further releases designed to improve the system may be developed and deployed during this phase.

The Pros and Cons of Waterfall Development

Waterfall development has been used successfully for decades. The methodology is easy to understand and manage as each phase has agreed upon deliverables, a formal review process, and is completed before the next phase is begun. Projects that use this methodology are usually well documented. There is minimal customer and end-user involvement after the requirements phase. Because the design is finalized early on and changes little in development, waterfall development works well for projects where systems are integrated with other systems. The method also minimizes the potential for the "piecemeal effect" to affect a project. No new components that might introduce errors and other problems into the system later in development are added after the design phase.

However, waterfall development is not well suited for all scenarios. Iterative development models such as rapid application development (RAD) and spiral development are better for projects with loosely defined requirements that will likely change in development and testing. Similarly, systems and applications designed for release in a fast-moving, competitive environments are more suited for agile and iterative development methodologies that focus on producing working systems in the shortest time.

Waterfall Development in Practice

Waterfall development is most effective for development projects with well-defined requirements. For example, a manufacturer of precision scientific equipment may require the development of an application to display real-time data captured by an existing instrument. The requirements for this application are unlikely to change. In this case, the instrument has already been designed and constructed. Iterative development methodologies such as RAD and the spiral method would add overhead to the project and likely result in a longer time to completion than the simpler waterfall approach. Such an application relies only minimally on user input and does not need a complicated user interface, so a prototyping model would be less suited for this project as well.

A Proven Approach to Software Development

Waterfall development is a proven methodology that works well when developing straightforward systems with clearly defined requirements using proven technologies. It is well suited for use in industries such as construction and manufacturing that use linear, step-by-step business and production models. The simplicity of the model, along with its easy-to-understand linear process, set phases and deliverables, and robust documentation, makes it well suited for organizations that lack software development resources and experienced project managers. Despite being among the oldest software development methodologies, waterfall development continues to offer advantages to developers and other stakeholders in many scenarios.

—Maura Valentino, MSLIS

Bibliography

Bell, Michael. *Incremental Software Architecture: A Method for Saving Failing IT Implementations*. John Wiley & Sons, 2016.

Friedman, Daniel P., and Mitchell Wand. *Essentials of Programming Languages*. 3rd ed., MIT P, 2008.

Jayaswal, Bijay K., and Peter C. Patton. *Design for Trustworthy Software: Tools, Techniques, and Methodology of Developing Robust Software*. Prentice Hall, 2007.

MacLennan, Bruce J. *Principles of Programming Languages: Design, Evaluation, and Implementation*. 3rd ed., Oxford UP, 1999.

Scott, Michael L. *Programming Language Pragmatics*. 4th ed., Morgan Kaufmann Publishers, 2016.

Van Roy, Peter, and Seif Haridi. *Concepts, Techniques, and Models of Computer Programming*. MIT P, 2004.

Wysocki, Robert K. *Effective Project Management: Traditional, Agile, Extreme*. 7th ed., John Wiley & Sons, 2014.

WEB DESIGN

Web design includes the planning, creation, and maintenance of a website and associated web pages intended for use on an online, web-based platform. Since the World Wide Web gained popularity in the early 1990s, companies, organizations, academic institutions, and government agencies have recognized the importance of creating and maintaining a user-friendly online presence to provide users with a variety of information, services, and products. Users' expectations have resulted in a shift in web design that has transformed an organization's web content from a static environment to a more dynamic one, driven by users' demand for data at the touch of the fingertips. In a swiftly changing technology environment, web designers must be able to provide users with access to data across platforms that run on computers, tablets, smartphones, and any other device that may hit the market at a moment's notice.

Brief History

In its earliest stages, web design focused on conveying static information viewed through a simple web browser, such as Mosaic. Web designers worked with computer languages and markup tools such as JavaScript, Dynamic HTML, and Cascading Style Sheets. However, while these initial tools were limited in scope and capability, companies such as Microsoft quickly realized the potential for increased capability and options in such applications and invested in research and development to improve them. Web

Web & Graphic Design advertisement for *thewitchez*.

design grew quickly in capability in response to users' ever-increasing demands.

Web designers use particular computer coding languages, such as C++, JavaScript, Python, HTML, ColdFusion, Perl, Ruby, SQL, and PHP, to name only a few of the thousand computer languages in existence. These languages are based on a system of syntax and semantics that follow a sets of rules or sequences that drive computer programs and/or applications. These languages are used to develop new functions used in web design.

In 1996, web design took a leap forward when computer programmers developed the application known as Flash, which acted as a simple content development tool that incorporated basic layout principles, drawing tools, a limited precursor to ActionScript, and a time line. Additionally, Flash gave many web designers a new set of capabilities for animation and images, and soon found popularity among web designers for specific markets to develop complete websites. However, Flash required a plug-in and had issues with compatibility across operating systems, which made some web designers shy away from it.

Modern web design plays a critical role in the everyday lives of many individuals, and it incorporates elements of good graphic design and information technology. From a technological standpoint, good web design leverages graphic design to create a dynamic web environment through which users can move easily and trouble-free.

Impact

Web users typically scan a website expecting to instantly find the information they are seeking. However, good graphics and visual appeal, while important, are not considered the top elements of good web design. Instead, most web designers credit usability as their number-one priority. If a visitor to a website can easily access and discover information or products, he or she is likely to revisit the site.

The users' ability to understand and interact with the content of a website often depends on their understanding of how the website works. This is often referred to as the user experience, and it is critical to designing a functional and successful web page. The user experience can be related to layout, but it also includes design elements that provide clear and intuitive instructions on how to use the website. This principle drives contemporary web design, which often centers on duplicating features found on other websites in an effort to create a more universal user experience across multiple websites, regardless of organization, business, or product.

Other key elements in a good web design include: (1) positioning information and graphics so that the user will quickly identify what he or she needs; (2) using color to guide users' actions; (3) using contrast to make some page elements stand out while relegating others to a secondary position; and (4) ease of navigation so that users can move between elements of the site.

Web designers are responsible for the visual aspect of a website, which includes the layout, color scheme, and typography of each page, while web developers often build the web platform and maintain the server space in which a website is housed. Both professionals must have extensive knowledge of computer languages used in web development and design.

Additionally, content management tools have allowed many companies and organizations a simple way to create and maintain an online web presence, reducing overall design and production costs while allowing them to exist in an e-commerce or online information environment, increasing visibility of missions or products to a ever growing number of online users.

As Internet-based business continues to grow, the field of web design continues to change. Search Engine Optimization (SEO), in which key words are embedded in a website's text in order to drive traffic from search engines such as Google, has become an integral aspect of web design, as SEO can help determine a website's placement on a search-results page.

Another trend is responsive web design, meaning web pages are designed to flow fluidly no matter the platform on which users view a website. Thus, responsive design ensures that visual quality is not lost if a page is viewed on a portable device such as a tablet. In the same vein, because of the increasing use of portable devices, web design has incorporated enhanced scrolling and is moving away from hyperlinking text, as such can become cumbersome when viewing a website on a smartphone, for example. Furthermore, web designers increasingly make use of visuals and animation to capture the attention of clients.

—*Laura L. Lundin, MA*

BIBLIOGRAPHY

Blijlevens, Janneke, et al. "The Influence of Product Exposure on Trendiness and Aesthetic Appraisal." *International Journal of Design* 7.1 (2013): 55–67. Print.

Chisholm, Wendy, and Matt May. *Universal Design for Web Applications: Web Applications that Reach Everyone.* Cambridge: O'Reilly, 2008. Print.

Connor, Joshua. *ProHTML.5 Accessibility.* New York: Apress, 2012. Print.

Harper, Simon, and Yeliz Yesilada, eds. *Web Accessibility: A Foundation for Research.* New York: Springer, 2008. Print.

Horton, Sarah, and Whitney Quesenbery. *Universal Design for Web Accessibility.* New York: Rosenfeld Media, 2013. Print.

Hricko, Mary. *Design and Implementation of Web-Enabled Teaching Tools.* Hershey: Idea Group, 2002. Print.

Kim, Sung Yeon, and Kwangsu Cho. "Usability and Design Guidelines of Smart Canes for Users with Visual Impairments." *International Journal of Design* 7.1 (2013): 99–110. Print.

Krug, Steve. *Don't Make Me Think: A Common Sense Approach to Web Usability.* 3rd ed. Berkeley: New Riders, 2014. Print.

Miller, Luke. *The Practitioner's Guide to User Experience Design.* New York: Grand Central, 2015. Print.

O'Toole, Greg. *Sustainable Web Ecosystem Design.* New York: Springer, 2013. Print.

Sklar, Joel. *Principles of Web Design.* 6th ed. Boston: Cengage, 2015. Print.

WEB GRAPHIC DESIGN

FIELDS OF STUDY
Graphic Design; Digital Media; Programming Language

ABSTRACT
Web graphic design is the use of graphic design techniques in designing websites. Web graphic designers must balance the marketing aspects of a website with aesthetic design criteria. They also attempt to increase the likelihood that the website will be found in search results and therefore be an effective advertising tool.

PRINCIPAL TERMS

- **logotype:** a company or brand name rendered in a unique, distinct style and font; also called a "wordmark."
- **search engine optimization (SEO):** techniques used to increase the likelihood that a website will appear among the top results in certain search engines.
- **tableless web design:** the use of style sheets rather than HTML tables to control the layout of a web page.
- **typography:** the art and technique of arranging type to make language readable and appealing.
- **wireframe:** a schematic or blueprint that represents the visual layout of a web page, without any interactive elements.

DESIGNING FOR THE WEB
Web graphic design is a subfield of graphic design that focuses on designing for the web. It typically involves a blend of graphic design techniques and computer programming. Many websites are used as marketing materials for businesses and organizations.

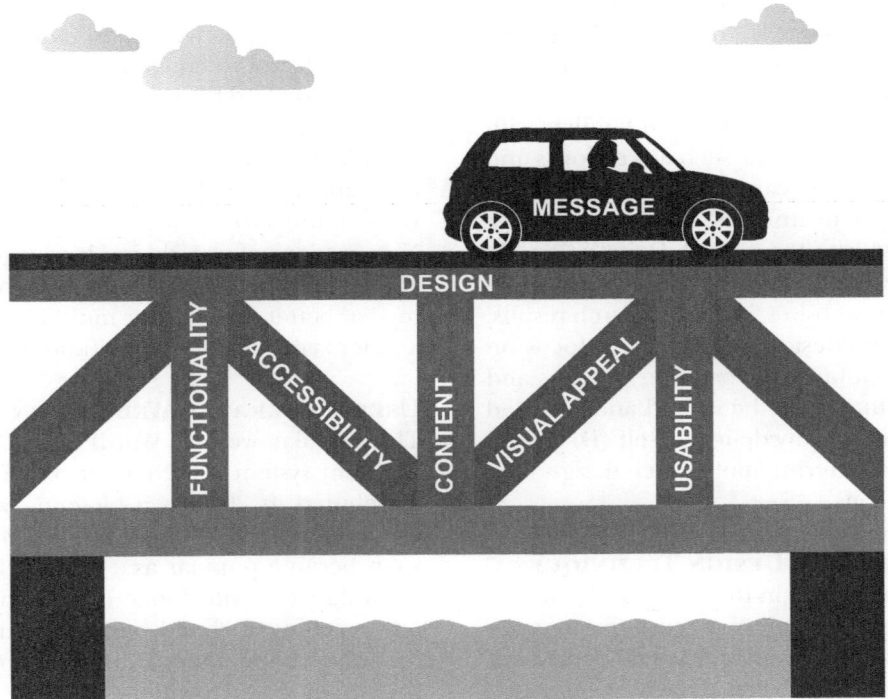

A quality website design ensures that the message gets across. Good web design incorporates accessible content, provided in a visually appealing way, through strong website back-end functionality and front-end usability.

Consequently, web designers often incorporate business logos and other promotional materials. They also use search engine optimization (SEO) techniques to increase the likelihood that the web page will be found by search engines.

WEB DESIGN FUNCTIONALITY

Most websites, whether business, advocacy, news, or personal sites, serve as both marketing and informational tools. A website that represents a specific brand will incorporate iconic logos, logotypes, or combination marks to aid brand recognition. An iconic logo is a symbol or emblem that represents a person, business, or organization. A logotype is a company or brand name rendered in a unique or proprietary font and style. Combination marks combine icons with logotypes.

In addition to being informative, a website should be visually appealing and easy to understand. Skilled use of typography helps web designers achieve these goals. For example, while logotypes are designed to catch the eye, important information should be presented in a font that is aesthetically pleasing yet unobtrusive. A font that draws attention to itself will detract from the message of the text. Typography techniques such as this help web designers make websites easy to read and navigate.

In addition, web designers must be familiar with SEO, which relies on elements such as keywords and links to ensure that users can find a website using a search engine. SEO techniques are most effective when incorporated into the overall design of a website. As such, many web graphic designers also help users optimize their websites for better search results.

By the 2010s, web design had begun to focus on designing for the mobile web, creating websites and e-commerce sites that could be viewed and accessed using mobile devices. Many do-it-yourself (DIY) website builders started offering mobile web design templates and conversions.

DEVELOPMENT OF WEB DESIGN TECHNIQUES

Modern web design began in the 1990s, with the creation and adoption of hypertext markup language (HTML). This markup language specifies the location and appearance of objects displayed on a web page. In the mid-1990s, web designers began using HTML tables. These consist of static cells that can be arranged on a page to specify the location of text or objects. By placing tables within tables, designers could create a richer experience for websites.

The programming language JavaScript was developed in 1995 to code website behavior. This made individual web pages interactive for the first time. With JavaScript, web designers could embed image galleries, add functions such as drop-down menus, and make websites respond visually when a user clicked on text or images. Flash, developed in 1996, allowed designers to add animated elements to websites. However, not all browsers supported Flash, so it was not as universally useful as JavaScript.

The next major step in web graphic design was the introduction of tableless web design. This approach to web design does not rely on HTML tables. Instead, it uses style sheets to format pages. The Cascading Style Sheets (CSS) language, first introduced in 1996, allows designers to separate the visual elements of the design from the content. CSS can determine the appearance and spacing, while individual elements are still described in HTML. Personal Home Page (PHP), another language used in web design, was developed in 1994 but gained in popularity during the late 1990s. It provides further options for making web pages interactive. When used with HTML, PHP allows websites to create new content based on user information and to collect information from visitors.

In the 2010s, most web design is still based on a combination of CSS and HTML. Programming in PHP and the database server software MySQL provides options for greater flexibility and interactivity. New versions of web design languages, such as CSS4 and HTML5, offer more ways of incorporating multimedia, better support for multilingual websites, and a wider array of aesthetic options for designers.

USER-GENERATED WEB DESIGN

The popular website WordPress is a content management system (CMS) built on PHP and MySQL. It debuted in 2003 as a blogging website, offering users templates and tools to design blogs. The site soon became popular as a basic web design system, allowing users with limited knowledge of HTML or other coding to build basic and functional, if aesthetically rudimentary, websites. Soon web designers began tailoring WordPress sites for customers. This has essentially served as a shortcut for the web design industry. As of 2015, WordPress remained among the most popular web design systems in the world, accounting for about half of all CMS-based websites on the Internet.

The debut of WordPress launched a new era in DIY web graphic design. Other websites such as Squarespace, Wix, and Weebly soon began offering users the ability to design websites quickly and easily, without the need to understand programming languages. Most such sites provide users templates in the form of wireframes. Wireframes are structural layouts that specify the location of images, text, and interactive elements on a web page but are not themselves interactive. Users can then insert their own images and text and, depending on the underlying program, also rearrange the layout of the wireframe. Like most elements of online commerce and marketing, web graphic design is moving toward user-generated and user-influenced design in order to open up to broader audiences.

—*Micah L. Issitt*

BIBLIOGRAPHY

Allanwood, Gavin, and Peter Beare. *User Experience Design: Creating Designs Users Really Love.* New York: Fairchild, 2014. Print.

Cezzar, Juliette. "What Is Graphic Design?" *AIGA.* Amer. Inst. of Graphic Arts, 2016. Web. 16 Mar. 2016.

Hagen, Rebecca, and Kim Golombisky. *White Space Is Not Your Enemy: A Beginner's Guide to Communicating Visually through Graphic, Web & Multimedia Design.* 2nd ed. Burlington: Focal, 2013. Print.

Malvik, Callie. "Graphic Design vs. Web Design: Which Career Is Right for You?" *Rasmussen College.* Rasmussen Coll., 25 July 2013. Web. 16 Mar. 2016.

Schmitt, Christopher. *Designing Web & Mobile Graphics: Fundamental Concepts for Web and Interactive Projects.* Berkeley: New Riders, 2013. Print.

Williams, Brad, David Damstra, and Hal Stern. *Professional WordPress: Design and Development.* 3rd ed. Indianapolis: Wiley, 2015. Print.

WORKING MEMORY

Working memory is the brain system that temporarily stores and processes information used in complex cognitive functions such as learning, reasoning, language comprehension, long-term memory retrieval, information filtering, and problem solving. It is sometimes conflated with short-term memory because it is temporary in nature and limited in capacity. The difference is that short-term memory is concerned with the passive storage and retrieval of the recent past, while working memory deals with the active processing of the same information. For example, short-term memory allows an individual to remember a message he or she just heard, while working memory allows the person to understand the meaning of the message, determine the relevant information, and formulate a response.

Researchers study working memory to gain a better understanding of how people learn, reason, understand, focus, and solve problems. Such work has contributed to advances in medical knowledge regarding conditions such as Alzheimer's disease and learning disabilities.

OVERVIEW

Stanford University researchers Richard C. Atkinson and Richard M. Shiffrin proposed a model of short-term memory in 1968. In their model, short-term memory holds a limited amount of information for a brief period of time with minimal processing. The model is described as multistore or dual-store because it stores all information, whether visual, spatial, or auditory, via the same system. It is also considered a unitary system, or a single system without any subsystems. This model generated a lot of research, which in turn exposed the flaws in the model.

In 1974, Alan Baddeley and Graham Hitch devised an alternative to the Atkinson-Shiffrin model. Called working memory, their model supposes that there are different stores for different types of information, rather than a single store for everything. Baddeley and Hitch also proposed that working memory has a central executive that manages two subsystems. These two subsystems are the phonological loop, which stores auditory information, and the visual-spatial "sketch pad"—also sometimes referred to

Richard C. Atkinson, whose research had an impact on advances in computer-assisted instruction and methods for optimizing the learning process.

as the "mind's eye"—which stores visual and spatial information.

Various studies suggest different functions of certain brain regions that contribute to working memory. Experiments dating back to the 1930s identified the prefrontal cortex as a key part of the brain contributing to spatial working memory, and later use of neuroimaging confirmed its importance. However, advanced brain imaging has also shown that working memory involves areas other than the prefrontal cortex, and that most tasks of working memory may in fact use a complex network of different areas. Other areas seen to be activated during certain working memory tasks include the posterior parietal cortex, the thalamus, the anterior cingulate cortex, and the superior frontal sulcus.

People use both visual-spatial and phonological working memory subsystems every day. Together, these subsystems form the basis of the brain's executive function, allowing an individual to pay attention, organize, solve problems, and plan ahead—all of which are key to being able to learn in the classroom, work, and complete numerous other tasks. Both types of working memory usually develop at about the same rate in childhood and reach their peak in early adulthood. The fact that working memory develops over time means that, for example, as students grow older, they are able to follow longer sets of instructions.

Ongoing research has revealed significant factors that may impair working memory, leading to insights into the functional pathways of memory and potential avenues of prevention or treatment of certain disorders. A 2009 study found that strong psychological stress can lower the functionality of the prefrontal cortex, inhibiting working memory. A 2011 paper found that the frequent abuse of alcohol damaged working memory and associated visuospatial capabilities. Conditions such as developmental coordination disorder, attention deficit hyperactivity disorder (ADHD), and autism spectrum disorders have been investigated in regards to the influence of working memory.

People with a weakened working memory are more likely to have learning disorders, and working memory has been shown to be a more dependable indicator of a subject's ability to learn than his or her intelligence quotient (IQ). Research has linked poor working memory to a greater likelihood of risk taking, which in turn has been connected to various social problems; a 2015 study found that adolescents with weaker working memories were more likely to become sexually active earlier and engage in unprotected sex. Researchers have also found that because visual-spatial working memory and phonological working memory can operate independently of each other, a person's working memory might be impaired in one area but not in the other. Thus, students with weak visual-spatial working memory might learn to compensate with their stronger auditory working memory and effectively conceal their disability. Testing both kinds of working memory is a critical step toward helping students with learning disabilities devise successful learning plans.

—*Lisa U. Phillips, MFA*

Bibliography

Atkinson, R. C., and R. M. Shiffrin. "Human Memory: A Proposed System and Its Control Processes." *The Psychology of Learning and Motivation*. Ed. Kenneth W. Spence and Janet Taylor Spence. Vol. 2. New York: Academic, 1968. 89–195. Print.

Baddeley, Alan D., and Graham Hitch. "Working Memory." *The Psychology of Learning and Motivation*. Ed. Gordon H. Bower. Vol. 8. New York: Academic, 1974. 47–89. Print.

Bernecker, Sven. *Memory: A Philosophical Study*. Oxford: Oxford UP, 2010. Print.

Hendrix, Dannie M., and Orval Holcomb, eds. *Psychology of Memory*. New York: Nova, 2012. Print.

Khurana, Atika, et al. "Stronger Working Memory Reduces Sexual Risk Taking in Adolescents, Even After Controlling for Parental Influences." *Child Development* (2015). n. pag. Web. 24 Jun. 2015.

Levin, Eden S., ed. *Working Memory: Capacity, Developments and Improvement Techniques*. New York: Nova, 2011. Print.

Morin, Amanda. "5 Ways Kids Use Working Memory to Learn." *Understood*. Understood.org, 16 Dec. 2013. Web. 24 Jun. 2015.

Sanders, Laura. "Brain Separates Working Memory: Hemispheres Independent in Mental Version of RAM." *Science News* 30 July 2011: 10–11. Print.

St. Clair-Thompson, Helen, ed. *Working Memory: Developmental Differences, Component Processes and Improvement Mechanisms*. New York: Nova, 2013. Print.

Surprenant, Aimée M., and Ian Neath. *Principles of Memory*. New York: Taylor, 2009. Print.

Weisberg, Robert W., and Lauretta M. Reeves. *Cognition: From Memory to Creativity*. Hoboken: Wiley, 2013. Print.

WORSE-IS-BETTER

FIELDS OF STUDY

Software Development; Software Methodologies; Computer Science

ABSTRACT

The worse-is-better approach to software development was created in response to the Massachusetts Institute of Technology (MIT) approach. It is based on the idea that the quality of a software solution does not improve with added functionality (better), but that a simpler solution (worse) might be more practical and usable. Worse-is-better designs take less time and effort to create and implement. They are also more adaptable to changing requirements.

PRINCIPAL TERMS

- **completeness:** the extent to which all user requirements have been implemented and verified.
- **consistency:** the extent to which a solution contains no conflicting requirements, assertions, constraints, functions, or components.
- **correctness:** the extent to which a solution meets its requirements and specifications and is free of defects.
- **simplicity:** the quality of being easily understood and lacking complexity.
- **usability:** a measure of the degree to which a user can deploy a specified solution meet their requirements.

FEWER FUNCTIONS, MORE USABLE

Worse-is-better is an approach to software development based on the idea that the quality of a software solution does not necessarily increase with added functionality (better), but that a simpler solution (worse) might be more practical and provide enough usability. The worse-is-better approach, or New Jersey approach, was developed in the 1980s. Developer Richard P. Gabriel, who outlined the approach in a 1991 paper on the Lisp programming language, saw the C programming language and UNIX operating systems as successfully following worse-is-better. Meanwhile, Lisp and the Common Lisp Object System (CLOS), created using the well-known MIT approach (or right-thing approach), lacked features found in C.

According to Gabriel, four characteristics define the worse-is-better approach: simplicity, correctness, consistency, and completeness. Simplicity is the primary factor. A worse-is-better design must have simple implementations and interfaces. However, this is imperative for the implementation. In contrast, in the MIT approach, simplicity is more important for the interface.

Correctness is the extent to which a solution meets the user's requirements and system specifications and lacks defects. A worse-is-better design should be "correct in all observable aspects." Simplicity remains more desirable than correctness, however. When following the MIT approach, a design must be "correct in all observable aspects" and incorrectness is unacceptable.

Consistency is the extent to which a solution contains no internal conflicts. A worse-is-better design must be reasonably consistent, but consistency may be sacrificed if needed. The preferred approach is to simplify the design by removing parts that introduce complexity or inconsistency. In contrast, in the MIT approach, a design must be consistent, and consistency and correctness are equally essential.

Completeness is the extent to which all requirements have been implemented and verified. A worse-is-better design must meet as many requirements as practical, including "all reasonably expected use cases." However, completeness can be sacrificed in favor of correctness and consistency, and must be sacrificed to preserve simplicity. The difference when following the MIT approach is that simplicity cannot "overly reduce completeness."

Strengths and Weaknesses

The worse-is-better approach does not necessarily result in a design that is more or less complete, correct, and consistent than software designed using other methodologies. The design does, however, take less time and effort to create and implement and is more adaptable. Because solutions developed using a worse-is-better approach are implemented with faster, they offer the "first-mover advantage." This is the idea that the first occupant of a certain market segment can gain greater market share or profits. After allowing a solution to gain acceptance and market share in an incomplete but useful state, revisions can be deployed until the solution reaches a high level of completeness. Gathering user input is critical to prioritizing the development of new features to meet their needs.

This can be preferable to the approach taken when following the MIT approach. When a solution is not deployed until it is complete, there is a chance that the need that the solution was designed to fill will already have been met by a competing solution. This would result in lost time, effort, and potential profit for the organization.

The worse-is-better approach may not be preferable to the MIT approach in all cases. For example, if the project requirements can be clearly and completely defined at the outset and are unlikely to change during the software life cycle, then a linear MIT approach may produce a better design.

Worse-Is-Better in Practice

An Internet start-up is developing the design for a new social media application (app). The concept offers a unique, innovative approach to connecting businesses with their potential customers through social media. Worse-is-better would be a good approach to follow while designing this solution for the several reasons.

Perhaps the main factor in this case is the need to release the app into the market in the least amount of time. Social media and marketing apps are a highly competitive marketplace, with many companies releasing new apps on frequent basis. If a competitor releases an app with similar functionality, it may be impossible for the design to penetrate the market even if the design is superior to the competing app in completeness and correctness. As the company is a start-up with no other lines of business or source of revenue, this could result in the company's failure. The worse-is-better approach is also suitable in this case because app users expect to wait for improved functionality to be delivered in product updates. Thus, the incompleteness of the solution would not hinder its adoption.

Success with a Worse Design

The worse-is-better approach reverses traditional thinking by positing that incomplete software that can be developed and released faster might be more successful than designs that offer all possible functionality upon release. Worse-is-better programming is ideally suited for use in contemporary software development, with competition growing and software users expecting added functionality in later version releases.

—*Maura Valentino, MSLIS*

BIBLIOGRAPHY

Bell, Michael. *Incremental Software Architecture: A Method for Saving Failing IT Implementations.* John Wiley & Sons, 2016.

Gabriel, Richard P. "Lisp: Good News, Bad News, How to Win Big." *AI Expert*, vol. 6, no. 6, 1991, pp. 30–39.

Goldman, Ron, and Richard P. Gabriel. *Innovation Happens Elsewhere: Open Source as Business Strategy.* Morgan Kaufmann Publishers, 2005.

Jayaswal, Bijay K., and Peter C. Patton. *Design for Trustworthy Software: Tools, Techniques, and Methodology of Developing Robust Software.* Pearson Education, 2006.

MacLennan, Bruce J. *Principles of Programming Languages: Design, Evaluation, and Implementation.* 3rd ed., Oxford UP, 1999.

Wysocki, Robert K. *Effective Project Management: Traditional, Agile, Extreme.* 7th ed., John Wiley & Sons, 2014.

TIME LINE OF INVENTIONS AND ADVANCEMENTS IN PROGRAMMING AND CODING

1948	Plankalkül (Plan Calculus)	Konrad Zuse begins work on Plankalkül (Plan Calculus), the first algorithmic programming language, with the goal of creating the theoretical preconditions for the solution of general problems. Seven years earlier, Zuse had developed and built the world's first binary digital computer, the Z1. He completed the first fully functional program-controlled electromechanical digital computer, the Z3, in 1941. Only the Z4 — the most sophisticated of his creations — survived World War II.
	The Mathematical Theory of Communication	American mathematician Claude Shannon writes *The Mathematical Theory of Communication*, laying the groundwork for understanding the theoretical limits of communication between people and machines. As part of this work Shannon identified the bit as a fundamental unit of information and, coincidentally, the basic unit of computation.
1952	Grace Hopper completes A-0	Mathematician Grace Hopper completes A-0, a program that allows a computer user to use English-like words instead of numbers to give the computer instructions. It possessed several features of a modern-day compiler and was written for the UNIVAC I computer, the first commercial business computer system in the United States.
1953	John Backus completes Speedcode	John Backus completes Speedcode for IBM's first large-scale scientific computer, the IBM 701. Although using Speedcode demanded a significant amount of scarce memory, it greatly reduced the time required to write a program. In 1957, Backus became project leader of the IBM FORTRAN project, which became the most popular scientific programming language in history and is still in use today.
1957	FORTRAN	An IBM team led by John Backus develops FORTRAN, a powerful scientific computing language that uses English-like statements. Some programmers were skeptical that FORTRAN could be as efficient as hand coding, but that sentiment disappeared when FORTRAN proved it could generate efficient code. Over the ensuing decades, FORTRAN became the most often used language for scientific and technical computing. FORTRAN is still in use today.
	MATH-MATIC	Sperry Rand releases a commercial compiler for its UNIVAC I computer. Developed by programmer Grace Hopper as a refinement of her earlier innovation, the A-0 compiler, the new version was called MATH-MATIC. Earlier work on the A-0 and A-2 compilers led to the development of the first English-language business data processing compiler, B-0 (FLOW-MATIC), also completed in 1957.

1959	SRI designs ERMA	SRI International designs ERMA (Electronic Recording Machine, Accounting), for Bank of America. At the time, accounts were posted manually, a method that would quickly be outstripped by the growth in check writing after World War II. The ERMA project digitized checking by creating a computer-readable font. A special scanner read account numbers preprinted on checks using magnetic ink character recognition. In just one hour, ERMA could process the number of accounts that would have taken a well-trained banker nearly 17 workdays to complete.
1960	COBOL (Common Business-Oriented Language)	A team drawn from several computer manufacturers and the Pentagon develop COBOL—an acronym for Common Business-Oriented Language. Many of its specifications borrow heavily from the earlier FLOW-MATIC language. Designed for business use, early COBOL efforts aimed for easy readability of computer programs and as much machine independence as possible. Designers hoped a COBOL program would run on any computer for which a compiler existed with only minimal modifications.
1961	Compatible Time-Sharing System (CTSS) is Demonstrated	The increasing number of users needing access to computers in the early 1960s leads to experiments in timesharing computer systems. Timesharing systems can support many users – sometimes hundreds – by sharing the computer with each user. CTSS was developed by the MIT Computation Center under the direction of Fernando Corbató and was based on a modified IBM 7094 mainframe computer. Programs created for CTSS included RUNOFF, an early text formatting utility, and an early inter-user messaging system that presaged email. CTSS operated until 1973.
1962	Kenneth Iverson writes *A Programming Language*	Kenneth Iverson's book *A Programming Language* details a form of mathematical notation that he had developed in the late 1950s while an assistant professor at Harvard University. IBM hired Iverson and it was there that APL evolved into a practical programming language. APL was widely used in scientific, financial, and especially actuarial applications. Powerful functions and operators in APL are expressed with special characters, resulting in very concise programs.
1963	ASCII	ASCII — American Standard Code for Information Interchange — permits machines from different manufacturers to exchange data. The ASCII code consisted of 128 unique strings of ones and zeros. Each sequence represented a letter of the English alphabet, an Arabic numeral, an assortment of punctuation marks and symbols, or a function such as a carriage return. ASCII can only represent up to 256 symbols, and for this reason many other languages are better supported by Unicode, which has the ability to represent over 100,000 symbols.
	Ivan Sutherland publishes Sketchpad	Ivan Sutherland publishes Sketchpad, an interactive, real-time computer drawing system, as his MIT doctoral thesis. Using a light

		pen and Sketchpad, a designer could draw and manipulate geometric figures on a computer screen. Blossoming into the best known of the early drawing applications, Sketchpad influenced a generation of design and drafting programs. Although used mostly for engineering drawings, it had some artistic applications, including a famous drawing of Nefertiti that could be animated to a limited extent.
1964	IBM introduces SABRE	IBM introduces the SABRE reservation system for American Airlines. First tested in 1960, the system took over American's reservations four years later. Running on dual IBM 7090 mainframes, SABRE was inspired by IBM's work on the SAGE air-defense system. SABRE, which became a separate travel-services company in 2000, owns the Travelocity website.
	Thomas Kurtz and John Kemeny create BASIC	Thomas Kurtz and John Kemeny create BASIC (Beginner's All-purpose Symbolic Instruction Code), an easy-to-learn programming language, for their students at Dartmouth College who had no prior programming experience. Its use spread widely to schools all over the world. Over a decade later, most early personal computers were shipped with a version of BASIC embedded in their system, which opened up programming to an entirely new audience.
1965	Simula is written by Kristen Nygaard and Ole-John Dahl	Simula, an object-oriented language, is written by Kristen Nygaard and Ole-John Dahl at the Norwegian Computing Center. Based largely on the Algol 60 programming language, Simula grouped data and instructions into blocks called objects, each representing one facet of a system intended for simulation. In addition to simulation, Simula also has applications in computer graphics, process control, scientific data processing and other fields.
1967	Seymour Papert designs LOGO	Seymour Papert designs LOGO as a computer language for children. Initially a drawing program, LOGO controlled the actions of a mechanical "turtle," which traced its path with pen on paper. Electronic turtles made their designs on a video display monitor. Papert emphasized creative exploration over memorization of facts: "People give lip service to learning to learn, but if you look at curriculum in schools, most of it is about dates, fractions, and science facts; very little of it is about learning. I like to think of learning as an expertise that every one of us can acquire."
1968	"GO TO considered harmful" letter is published	Edsger Dijkstra's "GO TO considered harmful" letter is published in Communications of the ACM, fires the first salvo in the structured programming wars. He called for abolishing the unrestricted GOTO statements used in higher-level languages, and argued that they complicated programming. The ACM considered the resulting acrimony sufficiently harmful that it established a policy of no longer printing articles taking such an assertive position against a coding practice.

	CICS is released	CICS (Customer Information Control System), an IBM transaction processing system, is released. Before CICS was introduced, many industries used punched card batch processing for high-volume customer transactions. As it allowed online transaction processing, CICS was able to replace this method and greatly sped up the way that companies interacted with their customers. It was first used in the public utility industry for access to customer information and transactions, but soon after its release it was quickly adopted by a wide spectrum of industries including banking, oil, insurance and even smaller companies. Although it was originally intended to only last a few years, CICS is still in use today.
1969	Kenneth Thompson and Dennis Ritchie develop UNIX	AT&T Bell Labs programmers Kenneth Thompson and Dennis Ritchie develop the UNIX operating system on a spare DEC mini-computer. UNIX combined many of the timesharing and file management features offered by Multics, from which it took its name. (Multics, a project of the mid-1960s, represented one of the earliest efforts at creating a multi-user, multi-tasking operating system.) The UNIX operating system quickly secured a wide following, particularly among engineers and scientists, and today is the basis of much of our world's computing infrastructure.
	The RS-232-C standard is adopted	The RS-232-C standard for communications is adopted by the Electronic Industries Association. The standard permits computers and peripheral devices to transmit information serially — that is, one bit at a time. RS-232-C compatible ports were widely used for equipment like printers and modems. Compared to more modern interfaces, serial connections had slow transmission speeds, were bulky and have been largely replaced by USB ports on new PCs and peripheral equipment.
1970	Pascal is introduced	The Pascal programming language, named after Blaise Pascal, a French physicist, mathematician and inventor turned philosopher, is introduced by Professor Niklaus Wirth. His aim with Pascal was to develop a programming language applicable to both commercial and scientific applications, and which could also be used to teach programming techniques to college students. It was closely based on ALGOL 60, which Wirth had also helped to develop.
1972	C programming language is released	The C programming language is released. Dennis Ritchie and his team created C based on the earlier language BCPL (Basic Combined Programming Language) and soon after re-wrote the source code for Unix in C. As such, Unix was easily ported to other computers and spread swiftly. C is still widely used today.
1976	CP/M is developed	Gary Kildall develops the first commercially successful operating system for microcomputers, CP/M. He and his wife established Intergalactic Digital Research (modestly dropping "Intergalactic" later) to market it. CP/M made it possible for one version of a program to run on a variety of computers built around eight-bit microprocessors. At one point Digital Research and Microsoft were

		approached by IBM about providing an operating system for its PC. Microsoft won the competition with its own operating system, called MS-DOS.
1978	WordStar is created	Rob Barnaby creates WordStar while at MicroPro International. Among the first popular word processing systems for personal computers, WordStar originally ran on the CP/M operating system, and later on DOS and Windows. In 1981, it had significant market share, in part because it came bundled with the Osborne 1 computer. WordStar retained a loyal following well after Microsoft Word surpassed it in sales.
1979	Visicalc is developed	Harvard MBA candidate Dan Bricklin and programmer Bob Frankston develop VisiCalc, the program that turned the personal computer into a business machine. Initially developed for the Apple II, whose sales it boosted, VisiCalc automated the recalculation of spreadsheets, allowing users to ask "What if?" questions of their financial information.
1981	MS-DOS released with the IBM PC	MS-DOS, or Microsoft Disk Operating System, the basic software for the newly released IBM PC, is the start of a long partnership between IBM and Microsoft, which Bill Gates and Paul Allen had founded only six years earlier. IBM's PC inspired hardware imitators in the 1980s, but for software, most licensed MS-DOS. MS-DOS was eventually supplanted by Microsoft's Windows operating system.
1982	Mitch Kapor develops Lotus 1-2-3	Mitch Kapor develops Lotus 1-2-3, a software suite for the IBM PC based on a word processor, spreadsheet, and database. It quickly became the first "killer application" for the IBM PC, and contributed to the success of the PC in business. IBM purchased Lotus in 1995.
1983	Microsoft introduces Word	Microsoft announces Word, originally called Multi-Tool Word. In a marketing blitz, Microsoft distributed 450,000 disks containing a demonstration version of its Word program in the November issue of PC World magazine, giving readers a chance to try the program for free. It competed with WordPerfect for market share as a word processing program, and it was not until Microsoft Word for Windows was introduced in 1989 that it became a global standard.
	Richard Stallman develops GNU	Richard Stallman, a programmer at MIT's Artificial Intelligence Lab, set out to develop a free alternative to the popular Unix operating system. This operating system called GNU (for Gnu's Not Unix) was going to be free of charge but also allow users the freedom to change and share it. Stallman founded the Free Software Foundation (FSF) based on this philosophy in 1985. While the GNU work did not immediately result in a full operating system, it provided the necessary tools for creating another Unix-type system known as Linux. The software developed as part of the GNU project continues to form a large part of Linux, which is why the FSF asks for it to be called GNU/Linux.

1984	Matlab is released	Matlab (Matrix Laboratory), a high-level programming language, is released. It was designed by Professor Cleve Moler of the University of New Mexico and was initially intended to help students use mathematical software libraries without requiring knowledge of the scientific programming language FORTRAN. Its roots began in the academic community, but it spread quickly to many other areas of technical computing and is widely used today.
	Verilog is created	Phil Moorby and Prabhu Goel of Gateway Design Automation create Verilog, a hardware description language that is used in the design of digital circuitry. Initially designed for Gateway's Verilog XL Design Logic Simulator, it was a vast improvement over methods being used by circuit designers at the time. Gateway Design Automation was acquired in 1989 by Cadence Design, which released the Verilog Hardware Description Language (HDL) into the public domain the following year. Verilog is now one of two hardware description languages used in the world today to design complex digital systems.
1985	The C++ Programming Language is published	The C++ programming language emerges as the dominant object-oriented language in the computer industry when Bjarne Stroustrup publishes the book The C++ Programming Language. Stroustrup, from AT&T Bell Labs, said his motivation stemmed from a desire to create a language that would allow for more complex programs and which combined the low-level features of BCPL with the high-level structures of Simula. According to Stroustrup: "C++ is a general purpose programming language designed to make programming more enjoyable for the serious programmer."
	Aldus announces PageMaker	Aldus announces its PageMaker program for use on Macintosh computers, launching the desktop publishing revolution. Two years later, Aldus released a version for the IBM PC. Developed by Paul Brainerd, PageMaker allowed users to combine graphics and text easily into professional quality documents. Pagemaker was one of three components to the desktop publishing revolution. The other two were the invention of Postscript by Adobe and the LaserWriter laser printer from Apple. All three were necessary to create a desktop publishing environment.
1987	Perl is written by Larry Wall	Perl (Practical Extraction and Report Language) is written by Larry Wall. It was intended to facilitate report processing and could scan and extract information from text files and ultimately create reports generated from that information. It was designed for ease of use and quick programming and has found multiple applications in every branch of computing. It is very useful in making other programs work together and has been called "the duct tape of the Internet."

	William Atkinson designs HyperCard	Apple engineer William Atkinson designs HyperCard, a software tool that simplifies development of in-house applications. In HyperCard, programmers built "stacks" of information with the concept of hypertext links between stacks of pages. As a stack author, a programmer employed various tools to create his own stacks, linked together as a sort of slide show. Apple distributed the program free with Macintosh computers until 1992. Hypercard influenced the creation on the Internet protocol HTTP and JavaScript.
1988	Mathematica is created	Mathematica is created by Stephen Wolfram, a British scientist. It was a symbolic mathematical programming language used in mathematical, scientific, academic, and engineering fields. Mathematica was a complete ecosystem for computing that allowed symbolic entry of mathematical functions and equations as well as graphical display of the results.
1990	Microsoft ships Windows 3.0	Microsoft ships Windows 3.0. Compatible with DOS programs, the first successful version of Windows finally offered good enough performance to satisfy PC users. For the new version, Microsoft updated the interface and created a design that allowed PCs to support large graphical applications for the first time. It also allowed multiple programs to run simultaneously on its Intel 80386 microprocessor. Microsoft lined up a number of other applications ahead of time that ran under Windows 3.0, including versions of Microsoft Word and Microsoft Excel. As a result, PC users were exposed to the user-friendly concepts of the Apple Macintosh, making the IBM PC more popular.
	Photoshop is released	Photoshop is released. Created by brothers John and Thomas Knoll, Photoshop was an image editing program and the most popular software program published by Adobe Systems. Thomas, while earning a PhD at the University of Michigan, had created an early version of the program in 1987, and John saw a practical use for it as a special effects staff member at Industrial Light & Magic. It was then used for image editing in the "pseudopod" scene in the movie The Abyss. When Adobe saw potential in the project they bought a license for distribution in 1989 and released the product on February 19, 1990.
1991	Linus Torvalds releases the Linux kernel	Designed by Finnish university student Linus Torvalds, the Linux kernel is released to several Usenet newsgroups. Almost immediately, enthusiasts began developing and improving it, such as adding support for peripherals and improving its stability. In February 1992, Linux became free software or, as its developers preferred to say after 1998, "open source." Linux also incorporated some elements of the GNU operating system and is used today in devices ranging from smartphones to supercomputers.

	PGP is introduced	Pretty Good Privacy, or PGP, a public-key encryption program, is introduced and is used for securing texts, emails and files. Its inventor, software engineer Phil Zimmermann, created it as a tool for people to protect themselves from intrusive governments, businesses, and institutions around the world. Zimmermann posted PGP on the Internet in 1991 where it was available as a free download. The United States government, concerned about the strength of PGP, which rivaled some of the best secret codes in use at the time, prosecuted Zimmermann but dropped its investigation in 1996.
1993	FreeBSD is launched	FreeBSD, a complete Unix-like operating system is launched. It was the most widely used open-source BSD (Berkeley Software Distribution) variant. After its initial release, the software was significantly re-engineered due to a lawsuit between Unix copyright holder Unix Systems Laboratories and the University of California, Berkeley. The lawsuit revolved around source code in Berkeley's 4.3BSD-Lite which was the basis of the FreeBSD operating system. FreeBSD incorporated features including networking, storage, security, portability and Linux compatibility.
	Microsoft Windows NT is released	Microsoft Windows NT is released. Work on the project began in the late 1980s in an effort spearheaded by a group of former Digital Equipment Corporation employees led by Dave Cutler. It was the first truly 32-bit version of Windows from Microsoft, which made it appealing to high-end engineering and scientific users that required better performance. A number of subsequent versions of Windows were based on NT technology.
1995	Java 1.0 is introduced	Java 1.0 is introduced by Sun Microsystems. The Java platform's "Write Once, Run Anywhere" functionality let a program run on any system, offering users independence from traditional large software vendors like Microsoft or Apple. The project was a successor to the Oak programming language created by James Gosling in 1991.
	JavaScript is developed	JavaScript, an object-based scripting language, is developed at Netscape Communications by Brendan Eich. It was used extensively across the Internet on both client and server sides. Although it shared its name with the Java programming language, the two are completely different.
1997	Microsoft introduces Visual Studio	Microsoft introduces Visual Studio. Bundled within Visual Studio were a number of programming tools, as Microsoft's intent was to create a single environment where developers could use different programming languages. The idea of visual programming is to allow programmers to develop software using built-in visual elements (like in a block diagram) instead of text.
2000	Y2K bug	During the late 1990s, the impending Year 2000 (Y2K) bug fuels news reports that the onset of the year 2000 will cripple telecommunications, the financial sector and other vital infrastructure. The issue was rooted in the fact that date stamps in most previously

		written software used only two digits to represent year information. This meant that some computers might not be able to distinguish the year 1900 from the year 2000. Although there were some minor glitches on New Year's Day in 2000, no major problems occurred, in part due to a massive effort by business, government and industry to repair their code beforehand.
2001	BitTorrent is launched	BitTorrent, a peer-to-peer file sharing service, is launched by BitTorrent, Inc. It was developed by Bram Cohen and was initially an open source program, but became closed source in 2005. BitTorrent enabled users to upload and download files, typically music and movies. It came under scrutiny of copyright holders – such as the music and motion picture industries – which claimed BitTorrent facilitated theft of their intellectual property.
	Mac OS X is released	Mac OS X is released. It was a significant departure from the classic Mac OS as it was based on the Unix-like operating systems FreeBSD, NetBSD and NeXTSTEP/OPENSTEP. OS X introduced a more stable and reliable platform and multiple applications could more efficiently be run at the same time. Mac OS X 10.7 ("Lion") was the first version to support 64-bit Intel processors. It came pre-installed on all Macs beginning in 2011.
	Windows XP is released	The Windows XP operating system is released. Based on the Windows NT kernel, XP was considered more stable than previous versions of the operating system. XP was widely adopted by industry and persisted much longer than Microsoft planned. For example, in 2014, 95% of the world's automated teller machines ran XP. Microsoft support for XP ended on April 8, 2014.
	iTunes is released	Apple's iTunes is released. It was based on Bill Kincaid's SoundJam MP software, the rights to which Apple purchased. Initially, iTunes was only supported on the Mac operating system and functioned as a media player and media management tool. iTunes allowed users to record music from CDs, bring it into iTunes, mix it with other songs and then burn a custom CD. When the Apple iTunes music store was launched in 2003, it transformed music distribution and the entire music industry. Less than a week after its launch, over one million songs were downloaded. By 2013, over 25 billion songs had been downloaded from the iTunes store.
2005	Hadoop is developed	Hadoop is an open source software project initially developed by Google as a means of extracting search results from large amounts of unstructured data, such as data found on the web. It was used by many large corporations where networked scalability, cost effectiveness and fault tolerance were critical to their business models. Companies such as Google, Yahoo, American Airlines, IBM and Twitter all used Hadoop, and it could be scaled from a single server to thousands. With Hadoop different types of data could be seamlessly integrated and Hadoop could redirect work to another system if a node failed in the cluster.

2007	Scratch is publicly released	Scratch is released to the public. A free programming language that focused on education, it was designed by a team led by Mitchel Resnick at the MIT Media Lab Lifelong Kindergarten Group. Intended to be used by educators, students and parents as a teaching language, it had a number of applications in educational settings. These included math, computer science, language arts and social studies. Its interface allowed novice users to stack and organize block commands to write programs. Scratch has millions of users worldwide and is available in more than 40 languages.
2010	Reports of the Stuxnet virus surface	The Stuxnet virus is widely reported in the media due to attacks centered in Iran. The virus attempted to damage uranium enrichment centrifuges used in Iran's nuclear development program by causing damaging speed variations. Although it was recognized that some centrifuges were rendered inoperable by the virus, the full extent of the damage remained unknown. Stuxnet brought attention to the fragile nature of global infrastructure in a networked world.
2011	Adobe Creative Cloud is Announced	Adobe Creative Cloud is announced as a subscription and cloud-based model of distribution for its major software products. Adobe Acrobat, Illustrator, Dreamweaver, Photoshop, and others, could be subscribed to either as a complete package or individually to suit user needs. This model also allowed Adobe to begin releasing continuous updates to their products, shortening the development cycle and the time need to incorporate new features.
2012	Facebook Acquires Instagram	Instagram, an image-sharing and social networking application, is purchased by Facebook for nearly $1 billion. It was initially launched in October 2010 by founders Kevin Systrom and Mike Krieger and became an instant hit, with over 100 million active users by early 2013. Photos and videos (with 15 second maximum length) could be shared among users, who could then annotate these images with specific hash tags to enable them to be easily shared among other social media platforms such as Twitter and Facebook. Instagram also allowed users to manipulate their photos with a variety of digital filters such as "Slumber," "Kelvin," "1977," "Sierra," and "Inkwell."
2013	The Stable Release of Microsoft Office 365 is Unveiled	An updated Microsoft Office 365 is announced. It was a subscription-based software product. Microsoft's Word, Excel, OneNote, PowerPoint, Outlook, Access, and Publisher were all available in packages for a monthly or annual subscription. Also included with a subscription was 1 TB of cloud storage on Microsoft's One Drive (formerly Skydrive). Home, personal, university, business, and enterprise subscription plans were made available for a wide range of users. Microsoft's change to a subscription model was not unique: Apple, Adobe, IBM and many other large software and technology companies adopted this model as well.

2014	Apple Pay is Released	The Apple Pay mobile payment system is introduced into Apple's product ecosystem. Initially only available for the iPhone 6 and 6 Plus, iWatch, iPad Air 2, and iPad Mini 3, many major banks and credit card companies participated in the Apple Pay system. The device's near field communications (NFC) interface, Passbook app, and Apple's Touch ID system worked in tandem with point-of-sale systems in retail outlets to complete transactions. Apple Pay could also be used for online purchases.
	HTML 5 is Announced	HTML 5 is announced as the successor to HTML 4, which had become the standard for web markup languages in 1997. Markup languages describe how web pages will look and function. Work on HTML 5 had begun in 2004 under the auspices of the Web Hypertext Application Technology Working Group. It was simplified compared to its predecessors and was intended to be human-readable. HLTML 5 also offered a number of improvements for multimedia, such as simplifying the embedding of content such as streaming video and games into web pages.
	Heartbleed Bug Discovered	The Heartbleed bug is uncovered as a dangerous security flaw in the code base of the OpenSSL cryptographic software library. OpenSSL protected a significant portion of the world's web servers, and nearly 20% of them were found to be vulnerable to attack from this particular security bug, which allowed hackers to eavesdrop on the communications of unsuspecting victims and steal sensitive information such as user names and passwords, emails, instant messages, and even confidential files and documents. Although it was a dangerous and widespread bug, installation of the "Fixed OpenSSL" library by service providers and users greatly reduced its effectiveness.

GLOSSARY

~a~

abstract: [Mathematical] algebras and logics that describe several different concrete algebras and logics; Software engineering] descriptions that do not swamp you with unnecessary detail; they provide enough information to use something without knowing its detailed construction.

abstract class: A class that cannot be directly constructed, one that can be constructed only through construction of some of its subclasses.

abstract type: A type in a nominative type system that cannot be instantiated.

accessor: A method or member function that does not change the object to which it is applied, also known as a "const" function.

actual argument: A value, or reference to a value, passed to a function.

actual parameter: Any parameter in the call of a subprogram.

algorithm: A description in precise but natural language plus mathematical notation of how a problem is solved.

alias: Two names or identifiers are aliases if they name or identify the same thing.

app: An application that executes on a small, hand-held device.

application: A program or integrated suite of programs that has a defined function.

argument: A value, or reference to a value, passed to a function; an actual argument.

arithmetic operations: Addition, subtraction, multiplication, and division ideally forming an abstract data type (ADT) with the algebraic properties of a ring or field.

array: An ordered sequence of same-typed values whose elements are fast to access by their numerical index in the array.

ASCII: The original common character code for computers using 8 bits.

assignment: [Statement] A statement with an expression and a variable. The expression is evaluated and the result is stored in the variables.

associativity: Rules for determining which of two identical infix operators should be evaluated first.

~b~

BASIC: Beginners All-purpose Symbolic Instruction Code, a family of languages developed for teaching programming and given away with early IBM PCs.

binary: Pertaining to 2. Binary operators have two operands. Binary numbers have base 2 and use 2 symbols.

binding: A relationship between two things, typically an identifier and some one of its properties or attributes. For example a variable is an identifier bound to a piece of storage in the main memory of the computer.

bit: Binary digit. A unit of information introduced by Shannon in the 1940's.

block: [Program structure] A piece of source code that has one or more declarations in it.

Boolean: [Adjective] Any data type that follows George Boole's algebraic axioms. The commonest Boolean data has two values {true, false} and the operations of and, or, and not.

byte: Eight bits.

~c~

C: Programming Language invented to help develop operating systems.

C++: Hybrid child of C with object oriented features and generic functions and classes.

call: To make use of something by writing its name and the correct protocol; A piece of code that transfers control, temporarily, to a subprogram and suspends the original code until the subprogram returns to the following statement etc.

call by(X): Old-fashioned way of saying: pass by(X).

chain: [Data structure] Any kind of linked list, a set of records where each record identifies the next record in some sequence or other.

class: A description of a collection of objects that have similar structures and behaviors.

CLOS: "Common LISP Object System," a modern LISP.

COBOL: COmmon Business Oriented Language.

code: [Noun] A piece of text that can not be understood without a key, hence the source code for a program

coercion: An implicit type conversion that lets a smart compiler work out the wrong meaning for a programmers typing mistake.

compile: [Verb] Translate source code into executable object code.

compiler: A computer program which transforms source code into object code.

component: [Technology] A unit of composition with contractually specified interfaces and only explicit context dependencies; components can be deployed and composed by third parties, often a collection of objects with a given set of methods for handling them and abstract classes that can be defined by other people.

compound: A single statement or object that can have any number of other statements as its parts.

conditional: An expression or statement that selects one out of a number of alternative subexpressions.

constant: An identifier that is bound to an invariant value.

constructor: A method or function in a class that creates a new object of that class.

control statement: Statements that permit a processor to select the next of several possible computations according to various conditions.

~d~

data type: A collection of values together with the operations that use them and produce them, plus the assumptions that can be made about the operations and values.

declaration: A piece of source code that adds a name to the program's environment and binds it to a class of meanings, and may also define the name.

default: An item provided in place of an omitted item.

default constructor: Actions carried out to create an object of a given class when no other data is provided about the object.

definition: A piece of source code or text that binds a name to a precise "definite" meaning. A definition may implicitly also declare the name at the same time or bind more information to an already defined name.

destructor: In object-oriented programming, the command sequence that is launched when the execution of an object is finished.

dump: A formatted listing of the contents of program storage, especially when produced automatically by a failing program.

dynamic: Something that is done as the program runs rather than by the compiler before the program runs.

dynamic binding: A binding that can be made at any time as a program runs and may change as the program runs.

dynamic polymorphism: A kind of polymorphism where the current type of an object determines which of several alternate subprograms is invoked as the program runs.

dynamic scoping: Determining the global environment of a subprogram as that which surrounds its call.

dynamically scoped: Something that uses dynamic scoping.

~e~

EBNF: Extended BNF. A popular way to define syntax as a dictionary of terms defined by using iteration, options, alternatives, etc.

encapsulated: [Programming] Coding that can be changed with out breaking client code; Being able to place all relevant information in the same piece of code; for example data and the operations that manipulate it in a C++ class.

encapsulation: [Programming] The ability to hide unwanted details inside an interface so that the result works like a black box or vending machine, providing useful services to many clients (programs or people).

enumeration: A data type whose values are a set of mutually exclusive named constants.

exception: An interruption in normal processing, especially as caused by an error condition.

expression: A shorthand description of a calculation.

~f~

fixed point: A form of arithmetic that always has the same number of places on either side of the decimal point giving bounded rounding errors, speed, simplicity, and a comparatively small range. Contrast with floating point.

floating point: A form of arithmetic that always preserves the same number of digits but allows the decimal point to be placed anywhere among them. This gives unbounded errors, a wider range, and a more complex processor.

flow chart: A schematic representation of the logic that defines the flow of control through a program.

formal argument: A parameter in a function definition.

formal parameter: The symbol used inside a subprogram in place of the actual parameter provided when the subprogram is called.

FORTRAN: FORmula TRANslation. There have been many FORTRANs. The series includes: I, II, IV, 66, 77, and 90 so far. Its author has said that he doesn't know what the programming language used in the next millennium will look like but he's sure it will be called FORTRAN.

function: A subprogram that returns a value but does not change its parameters or have side effects; any subprogram; [Mathematics] A total many to one relation between a domain and a co-domain; A routine that receives zero or more arguments and may return a result.

functional programming: A programming paradigm that treats computation as the evaluation of mathematical functions, avoids state and mutable data, and makes it easy to construct functions as if they were data objects.

fundamental data type: A type of data that is not defined by a class, struct, or union declaration.

~g~

garbage: A piece of storage that has been allocated but can no longer be accessed by a program. If not collected and recycled garbage can cause a memory leak.

generic: A package or subprogram that can generate a large number of similar yet different packages or subprograms.

global: Something that can be used in all parts of program.

goto: A 4-letter word no longer considered correct that is still usable in all practical languages to indicate an unconditional jump.

grammar: [Math] A set of definitions that define the syntax of a language. A grammar generates the strings in the language and so implicitly describes how to recognize and parse strings in the language.

~h~

header: [C++] The first part of a function definition that describes how to call the function but does not describe what it does. The header defines the function's signature. A function header can be separated from its function when the body of the function is replaced by a semicolon. This allows information hiding and separate development.

header file: [C++] A collection of function headers, class interfaces, constants and definitions that is read by a compiler and changes the interpretation of the rest of the program by (for example) defining operation for handling strings.

heap: An area of memory reserved for dynamically allocated data objects, contrasted to the stack.

HTML: HyperText Markup Language; used to define pages on the WWW.

~i~

identifier: A formal name used in source code to refer to a variable, function, procedure, package, etc.

identity: [Mathematics] An equation that is true for all values of its variables.

identity operation: An operation that returns its arguments unchanged.

implementation: The way something is made to work. There are usually many ways to implement something.

in: A way of handling parameters that gives a subprogram access to the value of an actual parameter without permitting the subprogram to change the actual parameter. Often implemented by pass by value.

infix: An operator that is placed between operands. Infix notation dates back to the invention of algebra four hundred years ago.

information hiding: The doctrine that design choices should be hidden by the modules in which they are implemented

inheritance: [Objects] The ability to easily construct new data types or classes by extending existing structures, data types, or classes.

inout: A way of handling parameters that lets a subprogram both use and change the values of an actual parameter. It can be implemented by pass by reference, pass by name, or pass by value result.

input: Data supplied to some program, subprogram, OS, machine, system, or abstraction.

int: [Integer data type] Fixed point data representing a subset of the whole numbers.

integer: A data type for integer values.

interpreter: A program which executes another program written in a programming language other than machine code.

item field: A component in a compound data structure.

iteration: Repetition of a mathematical or computational procedure; new version of a piece of computer hardware or software.

iterator: An object that is responsible for tracking progress through a collection of other objects. Often it is implemented as a reference or pointer plus methods for navigating the set of objects. The C++ STL provides many iterators Java has an Enumeration class for iterators.

~j~

Java: Object oriented language that has a C-like syntax.

~l~

link: [Verb] to connect to things together. In computing: to place addresses in one part of memory so that they identify other parts of memory.

link editor: Link loader.

link loader: A program that carries out the last stage of compilation by binding together the different uses of identifiers in different files.

LISP: LISt Processing language, The key versions are LISP1.5, CLOS, and Scheme.

loader: Nowadays a link loader, in the past any program that placed an executable program and placed it into memory.

local: Related to the current instruction rather than a larger context.

logic programming: A style or paradigm of computer programming exemplified by the language Prolog.

long: Fixed point data type that may have more bits than integers.

~m~

machine code: System of instructions and data directly understandable by a computer's central processing unit.

mapping: [Mathematics] A relationship that takes something and turns it into something uniquely determined by the relationship.

Matrix: [Mathematics] An ADT that can be implemented by rectangular arrays and has many of the arithmetic operations defined on them. A matrix abstracts the structure and behavior of linear maps.

method: In object-oriented languages, a subroutine or function belonging to a class or object.

module: A program that is linked with others to form a functioning application; one method of implementing a subroutine

mutator: A method that is permitted to change the state of the object to which it is applied.

~n~

narrowing: A conversion that converts an object to a type that cannot include all the possible values of the object.

natural numbers: The numbers 1,2,3,4...

~o~

object: An instance of a class; [Analysis] Something that is uniquely identifiable by the user; [Code] A piece identifiable storage that can suffer and/or perform various operations defined by the objects type; [Design] A module that encapsulates a degree of intelligence and know how and has specialized responsibilities.

object code: The output of a compiler or assembler, not necessarily executable directly without linking to other modules.

object file: A piece of compiled code that is linked into a compiled program after compilation and either during loading or when the program is running. Do not confuse this use of object with the later use in programming, analysis, and design.

object-oriented: Using entities called objects that can process data and exchange messages with other objects.

operation: One of a set of functions with special syntax and semantics that can be used to construct an expression.

operator: [Lexeme] A symbol for an operation. Operators can infix, prefix, or postfix.

operator associativity: Rules that help define the order in which an expression is evaluated when two adjacent infix operators are identical.

operator precedence: Rules that help define the order in which an expression is evaluated when two infix operators can be done next.

out: Any mode of passing parameters that permits the subprogram to give a value to an actual parameter without letting the subprogram no what the original value of the subprogram. Only available for general parameters in Ada, it can be implemented by pass by result.

output: A means whereby data or objects are passed from a part to a wider context; for example a program sending data to the operating system so that you can see it on the screen.

overload: To provide multiple context dependent meanings for a symbol in a language.

overloading: Giving multiple meanings to a symbol depending on its context.

~p~

paradigm: A fundamental style of computer programming to which the design of a programming language typically has to cater, such as imperative programming, declarative programming, or, on a finer level, functional programming, logic programming or object-oriented programming.

parameter: A name in a function or subroutine definition that is replaced by, or bound to, the corresponding actual argument when the function or subroutine is called; [Mathematics] A variable constant or perhaps a constant variable; [Programming] Something that is used in a subprogram that can be changed when the subprogram is called.

parameter passing: The means by which the actual parameters in a call of a subprogram are connected with the formal parameters in the definition of the subprogram.

parametric polymorphism: A piece of code describes a general form of some code by using a parameter. Different instances or special cases are created by replace these parameters by actual parameters. Templates in C++, Generics in Ada, and Functors in SML are particular implementations of this idea.

parse: To convert a sequence of tokens into a data structures (typically a tree and a name table) that can be used to interpret or translate the sequence.

pass by reference: Parameter passing where the parameter is implemented by providing an access path to the actual parameter from the formal parameter. Actions written as if they use or change the formal parameter use or change the actual parameter instead.

pass by value: Parameter passing where the actual parameter is evaluated (if necessary) and the value placed in a location bound to the formal parameter.

PL: Programming Language.

pointer: Data type with values that are addresses of other items of data.

polymorphism: [Objects] The ability of a function to apply to more than one type of object or data.

positional parameter: A parameter that is bound by its position.

postfix: [Operator] An operator that is placed after its single operand.

predicate: [Logic] A formula that may contain variables, that when evaluated should be either true or false; [Prolog] A procedure that can fail or succeed to achieve a goal. Success is finding an instance of a formula that is true and failure means failing to find such an instance. It is assumed that failing to find a solution is proof that the predicate is false. In fact the definition of the predicate may be incomplete or some infinite instance is needed to fit the predicate.

prefix: [Operator] An operator that is placed in front of its single operand.

primitive: Something that does not need to be defined.

procedure: A subroutine or function coded to perform a specific task.

program: A software application, or a collection of software applications, designed to perform a specific task.

Prolog: PROgramable LOGic.

protocol: [Networking] Rules for sending and receiving data and commands over the network; [Subprogram] Rules for calling a subprogram.

prototype: [Software engineering] A piece of software that requires more work before it is finished, but is complete enough for the value of the finished product to be evaluated or the currant version improved.

~q~

quicksort: [Algorithm] Split the data into two roughly equal parts with all the lesser elements in one and the greater ones in the other and then sort each part; Prof. C. A. R. Hoare wrote this as a young programmer and team leader. His future career started with the publication of this elegant recursion for placing an array of numbers into order.

~r~

real: a number containing a decimal point, e.g. the number *pi* is a real number with a value of approximately 3.14159268.

relational: Pertaining to a relation.

relational expression: An infix expression in which two non-Boolean values are compared and a Boolean value returned.

relational operator: An infix operator that returns a Boolean value when given non-Boolean operands.

run time: The time during which a program is executing, as oppose to the compile time.

~s~

Scheme: A modern statically scoped version of LISP.

scope: The parts of a program where a particular identifier has a particular meaning (set of bindings).

scoped: Pertains to languages with particular scoping rules.

scoping: The rules used to determine an identifier's scope in a language.

selection: A statement that chooses among several possible executions paths in a program.

semantics: A description of how the meaning of a valid statement or sentence can be worked out from its parsed form.

set: A collection of objects, usually of the same type, described either by enumerating the elements or by stating a common property, or by describing rules for constructing items in the set.

side effect: A function or expression has a side effect if executing it changes the values of global variables or its arguments.

source code: Human-readable instructions in a programming language, to be transformed into machine instructions by a compiler, interpreter, assembler or other such system.

stack: A collection of data items where new items are added and old items retrieved at the same place, so that the last item added is always the first item retrieved, and so on. An important part of compilers, interpreters, processors, and programs.

static: [C] A keyword with too many different meanings pertaining to the life history and scope of variables.

string: A data type for a sequence of characters such as letters of English alphabet.

structure: [Data type] A finite collection of named items of data of different types.

subclass: In object-oriented programming, an object class derived from another class (its superclass) from which it inherits a base set of properties and methods.

subprogram: A piece of code that has been named and can be referred to by that name (called) as many times as is needed. Either a procedure or a function.

subprogram header: The part of a subprogram definition that describes how the subprogram can be called without defining how it works.

subroutine: A section of code that implements a task. While it may be used at more than one point in a program, it need not be.

subtype: A type S is a subtype of type T if every valid operation on an object of type T is also a valid operation of type S.

superclass: A class that passes attributes and methods down the hierarchy to subclasses.

syntax: A description of the rules that determine the validity and parsing of sentences or statements in a language.

~t~

ternary: Pertaining to 3. Ternary operators have two operands. Ternary numbers have base 3 and use 3 symbols.

token: A particular representation of lexemes.

tree: A collection of connected objects called nodes with all nodes connected indirectly by precisely one path. An ordered tree has a root and the connections lead from this root to all other nodes. Nodes at the end of the paths are called leaves. The connections are called branches. All computer science tress are drawn upside-down with the root at the top and the leaves at the bottom.

type: A tag attached to variables and values used in determining what values may be assigned to what variables; A collection of similar objects, See ADT and data type. Objects can be fundamental, pointers, or have their type determined by their class.

~u~

UML: Unified Modeling Language.

unary: Pertaining to 1. Unary operators have one operand, unary numbers use base 1 and one symbol.

UNICODE: A new 16 bit International code for characters. Used in Java.

~v~

variable: A named memory location in which a program can store intermediate results and from which it can read them.

virtual: [C++] A member function or method is virtual if when applied to a pointer the class of the object pointed at is used rather than the class of the pointer. Virtual inheritance means that when a class in inherited by two different path only one single parent object is stored for both paths.

void function: [C] A procedure.

void pointer: [C] A pointer to an object of unknown type and size.

~w~

widening: A conversion that places an object in a type that includes all the possible values of the type.

~x~

XBNF: [MATHS] An extension to EBNF invented by Dr. Botting so that ASCII can be used to describe formal syntax and semantics.

BIBLIOGRAPHY

3D Printing Processes Industry, 2015. Web. 6 Jan. 2016.

"About Additive Manufacturing." *Additive Manufacturing Research Group.* Loughborough U, 2015. Web. 6 Jan. 2016.

Abraham, Prabhakaran, Mustafa Almahdi Algaet, and Ali Ahmad Milad. "Performance and Efficient Allocation of Virtual Internet Protocol Addressing in Next Generation Network Environment." *Australian Journal of Basic & Applied Sciences* 7.7 (2013): 827–32. Print.

Abramovich, Sergei, ed. *Computers in Education.* 2 vols. New York: Nova, 2012. Print.

Abu-Mostafa, Yaser S. "Machines That Think for Themselves." *Scientific American* July 2012: 78-81. Print.

Adamczyk, Maria. "Forum for the Ugly People—Study of an Imagined Community." *Sociological Review* 58 (Dec. 2010): 97–113. Print.

Adee, Sally. "Thanks for the Memories." *IEEE Spectrum.* IEEE, 1 May 2009. Web. 10 Mar. 2016.

Agazzi, Evandro. *Scientific Objectivity and Its Contexts.* Springer, 2014.

Agrawal, Manindra, S. Barry Cooper, and Angsheng Li, eds. *Theory and Applications of Models of Computation: 9th Annual Conference, TAMC 2012, Beijing, China, May 16–21, 2012.* Berlin: Springer, 2012. Print.

Ahmad, Janice, and Traqina Q. Emeka. "Rational Choice Theory." *Encyclopedia of Criminology and Criminal Justice* (Jan. 2014): 1–5. Web. 5 July 2015.

Alderman, Ellen and Caroline Kennedy. *The Right to Privacy.* New York: Knopf, 1995. Print.

Alkire, Brien. *Applications for Navy Unmanned Aircraft Systems.* Santa Monica: Rand, 2010. Print.

Allanwood, Gavin, and Peter Beare. *User Experience Design: Creating Designs Users Really Love.* New York: Fairchild, 2014. Print.

Allen, Anita L. *Unpopular Privacy: What Must We Do?* New York: Oxford UP, 2011. Print.

Allen, Brenda J. "Feminist Standpoint Theory: A Black Woman's (Re)view of Organizational Socialization." *Communication Studies* 47.4 (1996): 257–71. Print.

Altman, Douglas G., and J. Martin Bland. "Standard Deviations and Standard Errors." *BMJ* 331.7521 (2005): 903. Print.

Amadeo, Ron. "The History of Android." *Ars Technica.* Condé Nast, 15 June 2014. Web. 2 Jan. 2016.

Ambainis, Andris. "What Can We Do with a Quantum Computer?" *Institute Letter* Spring 2014: 6–7. *Institute for Advanced Study.* Web. 24 Mar. 2016.

Ambinder, Marc. "What's Really Limiting Advances in Computer Tech." *Week.* The Week, 2 Sept. 2014. Web. 4 Mar. 2016.

Amer. Standards Assn. *American Standard Code for Information Interchange.* Amer. Standards Assn., 17 June 1963. Digital file.

Ammori, Marvin. "The Case for Net Neutrality." *Foreign Affairs* 93.4 (2014): 62–73. Print.

Anderson, Chris. "How I Accidentally Kickstarted the Domestic Drone Boom." *Wired* (June, 2012). http://www.wired.com/2012/06/ff_drones/. This article describes a number of drone uses as well as the author's recollection of building his own drone.

Anderson, Deborah. "Global Linguistic Diversity for the Internet." *Communications of the ACM* Jan. 2005: 27. PDF file.

Anderson, Mike, and Sergio Della Salla. *Neuroscience in Education: The Good, the Bad, and the Ugly.* New York: Oxford UP, 2012. Print.

Anderson, Steve. "Net Neutrality: The View from Canada." *Media Development* 56.1 (2009): 8–11. Print.

Anderson, Thomas, and Michael Dahlin. *Operating Systems: Principles and Practice.* West Lake Hills: Recursive, 2014. Print.

Andrews, Jean. *A+ Guide to Hardware: Managing, Maintaining, and Troubleshooting.* 6th ed. Boston: Course Tech., 2014. Print.

———. *A+ Guide to Managing and Maintaining Your PC.* 8th ed. Boston: Course Tech., 2014. Print.

Android: A Visual "Android: A Visual History." *Verge.* Vox Media, 7 Dec. 2011. Web. 2 Jan. 2016.

Ang, Tom. *Digital Photography Essentials.* New York: DK, 2011. Print.

———. *Digital Photography Masterclass.* New York: DK, 2013. Print.

———. *How to Photograph Absolutely Everything: Successful Pictures from Your Digital Camera.* New York: DK, 2009. Print.

Anthes, Gary. "Back to Basics: Algorithms." *Computerworld.* Computerworld, 24 Mar. 2008. Web. 19 Jan. 2016.

Antheunis, Marjolijn L., Alexander P. Schouten, Patti M. Valkenburg, and Jochen Peter. "Interactive Uncertainty Reduction Strategies and Verbal Affection in Computer-Mediated Communication." *Communication Research* 39.6 (2012): 757–80. Print.

Antoy, Sergio, and Michael Hanus. "Functional Logic Programming." *Communications of the ACM* 53 (2010): 74–85. Print.

Anzulewicz, Anna, et al. "Does Level of Processing Affect the Transition from Unconscious to Conscious Perception." *Consciousness and Cognition* 36 (2015): 1–11. Print.

Ao, Sio-Iong, and Len Gelman, eds. *Electrical Engineering and Intelligent Systems.* New York: Springer, 2013. Print.

Ariely, Dan, and Gregory S. Berns. "Neuromarketing: The Hope and Hype of Neuroimaging in Business." *Nature Reviews Neuroscience* 11.4 (2010): 284–92. Print.

Armstrong, Thomas. *Multiple Intelligences in the Classroom.* 3rd ed. Alexandria: ASCD, 2009. Print.

Arnold, Ellen. *The MI Strategy Bank: 800+ Multiple Intelligence Ideas for the Elementary Classroom.* Chicago: Chicago Review P, 2012. Print.

Atkinson, R. C., and R. M. Shiffrin. "Human Memory: A Proposed System and Its Control Processes." *The Psychology of Learning and Motivation.* Ed. Kenneth W. Spence and Janet Taylor Spence. Vol. 2. New York: Academic, 1968. 89–195. Print.

"The Attack of the MOOCs." *The Economist.* Economist, 20 July 2013. Web. 29 July 2013.

Auletta, Ken. *Googled: The End of the World As We Know It.* New York: Penguin, 2010. Print.

Australian National University. *Binary Representation and Computer Arithmetic.* Australian National U, n.d. Digital file.

Baca, Murtha, ed. *Introduction to Metadata.* 2nd ed. Los Angeles: Getty Research Inst., 2008. Print.

Baddeley, Alan D., and Graham Hitch. "Working Memory." *The Psychology of Learning and Motivation.* Ed. Gordon H. Bower. Vol. 8. New York: Academic, 1974. 47–89. Print.

Badia, Antonio, and Daniel Lemire. "A Call to Arms: Revisiting Database Design." *ACM SIGMOD Record* 40.3 (2009): 61–69. Print.

Bajarin, Tim. "Google Is at a Major Crossroads with Android and Chrome OS." *PCMag.* Ziff Davis, 21 Dec. 2015. Web. 4 Jan. 2016.

Bajo, Javier, et al., eds. *Highlights of Practical Applications of Agents, Multi-Agent Systems, and Sustainability.* Proc. of the International Workshops of PAAMS 2015, June 3–4, 2015, Salamanca, Spain. Cham: Springer, 2015. Print.

Baker-Eveleth, L., and R. Stone. "Expectancy Theory and Behavioral Intentions to Use Computer Applications." *Interdisciplinary Journal of Information* 3 (2008): 135–46. Print.

Ball, Phillip. "The Truth about the Turing Test." *Future.* BBC, 24 July 2015. Web. 18 Dec. 2015.

Bandler, Richard, and John T. Grinder, *Frogs into Princes: Neuro Linguistic Programming*, Moab: Real People P, 1979. Print.

Banga, Cameron, and Josh Weinhold. *Essential Mobile Interaction Design: Perfecting Interface Design in Mobile Apps.* Upper Saddle River: Addison-Wesley, 2014. Print.

Baptiste, Philippe, Claude Le Pape, and Wim Nuijten. *Constraint-Based Scheduling: Applying Constraint Programming to Scheduling Problems.* New York: Springer, 2013. Print.

Barkin, Eric. "The Prospects and Limitations of Neuromarketing." *CRM* July 2013: 46–50. Print.

Barnett, Chance. "Top 10 Crowdfunding Sites for Fundraising." *Forbes.com.* Forbes.com LLC, 8 May 2013. Web. 7 Aug. 2013.

Barnhardt, Richard K., and Eric Shappee. *Introduction to Unmanned Aircraft Systems.* New York: CRC, 2012. Print.

Beattie, Andrew. "Cloud Computing: Why the Buzz?" *Techopedia.* Techopedia, 30 Nov. 2011. Web. 21 Jan. 2016.

Beck, Kent. *Test-Driven Development: By Example.* Pearson, 2014.

Beck, Ulrich. "Cosmopolitanism as Imagined Communities of Global Risk." *American Behavioral Scientist* 55.10 (Oct. 2011): 1346–61. Print.

Becker, Lee A. "VIII. The Internal Validity of Research." *Effect Size Calculators*. U of Colorado: Colorado Springs, 16 Mar. 1998. Web. 28 July 2015.

Beelders, Tanya R., and Jean-Pierre L. Du Plessis. "Syntax Highlighting as an Influencing Factor When Reading and Comprehending Source Code." *Journal of Eye Movement Research*, vol. 9, no. 1, 2016, pp. 2207–19.

Beeler, Robert A., *How to Count: An Introduction to Combinatorics*. New York: Springer, 2015. Print.

Bell, Michael. *Incremental Software Architecture: A Method for Saving Failing IT Implementations*. John Wiley & Sons, 2016.

Bell, Tim, et al. "Algorithms." *Computer Science Field Guide*. U of Canterbury, 3 Feb. 2015. Web. 19 Jan. 2016.

Bell, Tom. *Programming: A Primer; Coding for Beginners*. London: Imperial Coll. P, 2016. Print.

Belton, Padraig. "Coding the Future: What Will the Future of Computing Look Like?" *BBC News*, BBC, 15 May 2015, www.bbc.com/news/business-32743770. Accessed 24 Feb. 2016.

Bembenik, Robert, Łukasz Skonieczny, Henryk Rybiński, Marzena Kryszkiewicz, and Marek Niezgódka, eds. *Intelligent Tools for Building a Scientific Information Platform: Advanced Architectures and Solutions*. New York: Springer, 2013. Print.

Benjamin, Medea. *Drone Warfare: Killing by Remote Control*. London: Verso. 2013. Print.

Bennett, Sue, and K. Maton. "Beyond the 'Digital Natives' Debate: Towards a More Nuanced Understanding of Students' Technology Experiences." *Journal of Computer Assisted Learning* 26.5 (2010): 321–31. Print.

Benson, Susan N. Kushner, and Cheryl L. Ward. "Teaching with Technology: Using TPACK to Understand Teaching Expertise in Online Higher Education." *International Journal of Technology and Design Education* 23.2 (2013): 377–90. Print.

Berge, Zane L., and Lin Muilenburg. *Handbook of Mobile Learning*. New York: Routledge, 2013. Print.

Bergen, Peter, and Jennifer Rowland. "Did Obama Keep His Drone Promises?" *CNN*. Cable News Network, 25 Oct. 2013. Web. 22 May 2015.

Berger, Charles R., and Michael Burgoon. *Communication and Social Influence Processes*. East Lansing: Michigan State UP, 1998. Print.

_____., and Richard J. Calabrese. "Some Explorations in Initial Interaction and Beyond: Toward a Developmental Theory of Interpersonal Communication." *Human Communication Research* 1.2 (1975): 99–112. Print.

Bergman, Manfred Max. *Advances in Mixed Methods Research: Theories and Applications*. Thousand Oaks: Sage, 2008, Print.

Bernal, Paul. *Internet Privacy Rights: Right to Protect Autonomy*. New York: Cambridge UP, 2014. Print.

Bernecker, Sven. *Memory: A Philosophical Study*. Oxford: Oxford UP, 2010. Print.

Berns, Andrew, and Sukumar Ghosh. "Dissecting Self-* Properties." *SASO 2009: Third IEEE International Conference on Self-Adaptive and Self-Organizing Systems*. Los Alamitos: IEEE, 2009. 10–19. *Andrew Berns: Homepage*. Web. 20 Jan. 2016.

Berry, John D. *Language Culture Type: International Type Design in the Age of Unicode*. New York: Graphis, 2002. Print.

Berry, M. A. J., and G. Linoff. *Data Mining Techniques For Marketing, Sales and Customer Support*. Hoboken, NJ: Wiley, 1997.

Bethune, James. *Engineering Graphics with AutoCAD 2014*. San Francisco: Peachpit, 2013. Print.

Biere, Armin, Amir Nahir, and Tanja Vos, eds. *Hardware and Software: Verification and Testing*. New York: Springer, 2013. Print.

Binder, Jeffrey R., and Rutvik H. Desai. "The Neurobiology of Semantic Memory." *Trends in Cognitive Sciences* 15.11 (2011): 527–36. Print.

Binh, Le Nguyen. *Digital Processing: Optical Transmission and Coherent Receiving Techniques*. Boca Raton: CRC, 2013. Print.

Bittman, Michael, Leonie Rutherford, Jude Brown, and Lens Unsworth. "Digital Natives? New and Old Media and Children's Outcomes." *Australian Journal of Education* 55.2 (2011): 161–75. Print.

Black, Jeremy. *The Power of Knowledge: How Information and Technology Made the Modern World*. New Haven: Yale UP, 2014. Print.

Blank, Andrew G. *TCP/IP Jumpstart: Internet Protocol Basics*. 2nd ed. San Francisco: Sybex, 2002. Print.

Blijlevens, Janneke, et al. "The Influence of Product Exposure on Trendiness and Aesthetic Appraisal." *International Journal of Design* 7.1 (2013): 55–67. Print.

Boashash, Boualem, ed. *Time-Frequency Signal Analysis and Processing: A Comprehensive Reference*. 2nd ed. San Diego: Academic, 2016. Print.

Bone, Simon, and Matias Castro. "A Brief History of Quantum Computing." *SURPRISE* May–June 1997: n. pag. *Department of Computing, Imperial College London*. Web. 24 Mar. 2016.

Booch, Grady, et al. *Object-Oriented Analysis and Design with Applications*. 3rd ed. Upper Saddle River: Addison, 2007. Print.

Boos, Dennis D., and Leonard A. Stefanski. *Essential Statistical Inference: Theory and Methods*. New York: Springer, 2013. Print.

Booth-Butterfield, Melanie, Steven Booth-Butterfield, and Jolene Koester. "The Function of Uncertainty Reduction in Alleviating Primary Tension in Small Groups." *Communication Research Reports* 5.2 (1998): 146–53. Print.

Borkar, Shekhar, and Andrew A. Chien. "The Future of Microprocessors." *Communications of the ACM*. ACM, May 2011. Web. 3 Mar. 2016.

Bowden, Mark. "How the Predator Drone Changed the Character of War." *Smithsonian* (November, 2013). http://www.smithsonianmag.com/history/how-the-predator-drone-changed-the-character-of-war-3794671/?no-ist. This article discusses the use of drones, particularly Predator drones, in war.

Boyle, Randall, and Raymond R. Panko. *Corporate Computer Security*. 4th ed. Boston: Pearson, 2015. Print.

Brabham, Daren C. *Crowdsourcing*. Cambridge: MIT Press, 2013. Print.

Bradley, Tony. "Experts Pick the Top 5 Security Threats for 2015." *PCWorld*. IDG Consumer & SMB, 14 Jan. 2015. Web. 12 Mar. 2016.

Brandom, Russell. "Google Survey Finds More than Five Million Users Infected with Adware." *The Verge*. Vox Media, 6 May 2015. Web. 12 Mar. 2016.

Brenner, Susan W. *Cybercrime and the Law: Challengers, Issues, and Outcomes*. Boston: Northeastern UP, 2012. Print.

Bright, Peter. "Locking the Bad Guys Out with Asymmetric Encryption." *Ars Technica*. Condé Nast, 12 Feb. 2013. Web. 23 Feb. 2016.

Brill, Steven. *After: How America Confronted the September 12 Era*. New York: Simon, 2003. Print.

Brodley, Carla E. "Challenges and Opportunities in Applied Machine Learning." *AI Magazine* 33.1 (2012): 11–24. Print.

Brooks, R. R. *Introduction to Computer and Network Security: Navigating Shades of Gray*. Boca Raton: CRC, 2014. Digital file.

Brown, Adrian. *Graphics File Formats*. Kew: Natl. Archives, 2008. PDF file. Digital Preservation Guidance Note 4.

Bryden, Douglas. *CAD and Rapid Prototyping for Product Design*. London: King, 2014. Print.

Bucchi, Massimiano, and Brian Trench, eds. *Routledge Handbook of Public Communication of Science and Technology*. 2nd ed. New York: Routledge, 2014. Print.

Burger, John R. *Brain Theory from a Circuits and Systems Perspective: How Electrical Science Explains Neuro-Circuits, Neuro-Systems, and Qubits*. New York: Springer, 2013. Print.

Burke, Brian. "The Gamification of Business." *Forbes*. Forbes.com, 21 Jan. 2013. Web. 20 Aug. 2013.

———. *Gamify: How Gamification Motivates People to Do Extraordinary Things*. Brookline: Bibliomotion, 2014. Print.

Butler, Brandon. "Massive Open Online Courses: Legal and Policy Issues for Research Libraries." *Association of Research Libraries*. Assoc. of Research Libraries, 22 Oct. 2012. Web. 5 Aug. 2013.

Buxton, Stephen. *Database Design: Know it All*. Boston: Morgan Kaufmann, 2009. Print.

Bwalya, Kelvin J., Nathan M. Mnjama, and Peter M. I. I. M. Sebina. *Concepts and Advances in Information Knowledge Management: Studies from Developing and Emerging Economies*. Boston: Elsevier, 2014. Print.

Calude, Cristian S., ed. *The Human Face of Computing*. London: Imperial Coll. P, 2016. Print.

Campbell, Joyce M. "General Considerations in the Clinical Application of Electrical Stimulation." *International Functional Electrical Stimulation Society*. IFESS, n.d. Web. 7 Oct. 2013.

Campbell-Kelly, Martin, William Aspray, Nathan Ensmenger, and Jeffrey R. Yost. *Computer: A History of the Information Machine*. Boulder: Westview, 2014. Print.

Candela, Leonardo, Donatella Castelli, and Pasquale Pagano. "History, Evolution, and Impact of Digital Libraries." *E-Publishing and Digital Libraries: Legal and Organizational Issues*. Ed. Ioannis Iglezakis, Tatiana-Eleni Synodinou, and Sarantos Kapidakis. Hershey: IGI Global, 2011. 1–30. Print.

Card, Noel A. *Applied Meta-analysis for Social Science Research*. New York: Guilford, 2012. Print.

Carlson, Wayne. "A Critical History of Computer Graphics and Animation." *Ohio State University*. Ohio State U, 2003. Web. 31 Jan. 2016.

Cassar, G., and H. Friedman. "Does Self-Efficacy Affect Entrepreneurial Investment?" *Strategic Entrepreneurial Journal* 3.3 (2009): 241–60. Print.

Cavusgil, S. Tamer, et al. *International Business: The New Realities*. Frenchs Forest, 2015. Print.

Cazzola, Walter, and Edoardo Vacchi. "@Java: Bringing a Richer Annotation Model to Java." *Computer Languages, Systems & Structures* 40.1 (2014): 2–18. Print.

Ceberio, Martine, and Vladik Kreinovich. *Constraint Programming and Decision Making*. New York: Springer, 2014. Print.

Celada, Laura. "What Are the Most Common Graphics File Formats." *FESPA*. FESPA, 27 Mar. 2015. Web. 11 Feb. 2016.

Cezzar, Juliette. "What Is Graphic Design?" *AIGA*. Amer. Inst. of Graphic Arts, 2016. Web. 16 Mar. 2016.

Chan, Melanie. *Virtual Reality: Representations in Contemporary Media*. New York: Bloomsbury, 2014. Print.

Chandran, Ravi. "DW-on-Demand: The Data Warehouse Redefined in the Cloud." *Business Intelligence Journal* 20.1 (2015): 8–13. *Business Source Complete*. Web. 8 June 2015.

Chao, Loretta. "Tech Partnership Looks beyond the Bar Code with Digital Watermarks." *Wall Street Journal*. Dow Jones, 12 Jan. 2016. Web. 14 Mar. 2016.

Chau, S. C., and M. T. Lu. "Understanding Internet Banking Adoption and Use Behavior: A Hong Kong Perspective." *Journal of Global Information Management* 12.3 (2009): 21–43. Print.

Cheever, Erik. "Representation of Numbers." *Swarthmore College*. Swarthmore College, n.d. Web. 20 Feb. 2016.

Chen, Ding-Geng, and Karl E. Peace. *Applied Meta-analysis with R*. Boca Raton: CRC, 2013. Print.

Chen, Yufeng, and Zhiwu Li. *Optimal Supervisory Control of Automated Manufacturing Systems*. Boca Raton: CRC, 2013. Print.

Cheung, Mike W. L. *Meta-Analysis: A Structural Equation Modeling Approach*. Malden: Wiley, 2015. Print.

Chisari, Carmelo, et al. "Chronic Muscle Stimulation Improves Muscle Function and Reverts the Abnormal Surface EMG Pattern in Myotonic Dystrophy: A Pilot Study." *Journal of NeuroEngineering and Rehabilitation* 10.1 (2013): 94. Print.

Chisholm, Wendy, and Matt May. *Universal Design for Web Applications: Web Applications that Reach Everyone*. Cambridge: O'Reilly, 2008. Print.

Chowdhury, G. G., and Schubert Foo, eds. *Digital Libraries and Information Access: Research Perspectives*. New York: Neal, 2012. Print.

Chrząszcz, Jacek, Patryk Czarnik, and Aleksy Schubert. "A Dozen Instructions Make Java Bytecode." *Electronic Notes in Theoretical Computer Science* 264.4 (2011): 19–34. Print.

Chua, Chee Kai, Kah Fai Leong, and Chu Sing Lim. *Rapid Prototyping: Principles and Applications*. Hackensack: World Scientific, 2010. Print.

Churcher, Clare. *Beginning Database Design: From Novice to Professional*. 2nd ed. Berkeley: Apress, 2012. Print.

Churchhouse, Robert. *Codes and Ciphers*. Cambridge, England: Cambridge University Press, 2002.

Ciletti, Michael D. *Advanced Digital Design with the Verilog HDL*. 2nd ed. Upper Saddle River: Prentice, 2010. Print.

Cirani, Simone, Gianluigi Ferrari, and Luca Veltri. "Enforcing Security Mechanism in the IP-Based Internet of Things: An Algorithmic Overview." *Algorithms* 6.2 (2013): 197–226. Print.

Clark, Martin P. *Data Networks, IP, and the Internet*. Hoboken: Wiley, 2003. Print.

Clifford, Catherine. "Crowdfunding Generates More Than $60,000 an Hour." *Entrepreneur*. Entrepreneur Media, 19 May 2014. Web. 22 May 2015.

Cline, Hugh F. *Information Communication Technology and Social Transformation: A Social and Historical Perspective*. New York: Routledge, 2014. Print.

Coker, Christopher. *Warrior Geeks: How 21st Century Technology Is Changing the Way We Fight and Think about War*. London: Hurst, 2013. Print.

Coleman, Liv. "'We Reject: Kings, Presidents, and Voting': Internet Community Autonomy in Managing the Growth of the Internet." *Journal of Information Technology & Politics* 10.2 (2013): 171–89. Print.

Coll, Steve. "The Unblinking Stare." *New Yorker*. Condé Nast, 24 Nov. 2014. Web. 22 May 2015.

Colman, Andrew M. *A Dictionary of Psychology*. 4th ed. New York: Oxford UP, 2015. Print.

"Combinatorics." *Mathigon*. Mathigon, 2015. Web. 10 Feb. 2016.

Computer Security Institute. *15th Annual 2010/2011 Computer Crime and Security Survey*. New York: Computer Security Inst., n.d. PDF file.

"Computer-Aided Design (CAD) and Computer-Aided Manufacturing (CAM)." *Inc.* Mansueto Ventures, n.d. Web. 31 Jan. 2016.

Connor, Joshue. *ProHTML.5 Accessibility*. New York: Apress, 2012. Print.

Cook, Vickie S. "Net Neutrality: What Is It and Why Should Educators Care?" *Delta Kappa Gamma Bulletin* 80.4 (2014): 46–49. Print.

Cookies: Leaving a Trail on the Web "Cookies: Leaving a Trail on the Web." *OnGuard Online*. US Federal Trade Commission, Nov. 2011. Web. 25 Sept. 2013.

Cooper, Stephen. "Motherboard Design Process." *MBReview.com*. Author, 4 Sept. 2009. Web. 14 Mar. 2016.

Cooper, Steve, and Mehran Salami. "Reflections on Stanford's MOOCs." *Communications of the ACM* 56.2 (2013): 28-30. *Association of Computing Machinery*. Web. 5 Aug. 2013.

Corbet, Jonathan, Alessandro Rubini, and Greg Kroah-Hartman. *Linux Device Drivers*. 3rd ed. Cambridge: O'Reilly, 2005. Print.

Cormen, Thomas H. *Algorithms Unlocked*. Cambridge: MIT P, 2013. Print.

_____., et al. *Introduction to Algorithms*. 3rd ed. Cambridge: MIT P, 2009. Print.

Costello, Vic, Susan Youngblood, and Norman E. Youngblood. *Multimedia Foundations: Core Concepts for Digital Design*. New York: Focal, 2012. Print.

Cotton, Charlotte. *The Photograph as Contemporary Art*. London: Thames, 2009. Print.

Cox, Mike, Ellen Mulder, and Linda Tadic. *Descriptive Metadata for Television*. Burlington: Focal, 2006. Print.

Coyle, John J., et al. *Supply Chain Management: A Logistics Perspective*. 9th ed. Mason: South-Western, 2013. Print.

Craik, Fergus I. M., and Endel Tulving, "Depth of Processing and the Retention of Words in Episodic Memory," *Journal of Experimental Psychology*, 1975, 104(3), 268–94. Print.

Crain, Rance. "Neuromarketing Threat Seems Quaint in Today's Ad Landscape." *Advertising Age* 8 July 2013: 22. Print.

Creeber, Glen, and Royston Martin, eds. Digital Cultures: Understanding New Media. Berkshire: Open UP, 2009. Print.

Cresswell, John, W. *Research Design: Qualitative, Quantitative, and Mixed Methods Approaches*. Thousand Oaks: Sage, 2013. Print

_____., and Vicki Lynn Plano Clark. *Designing and Conducting Mixed Methods Research* Thousand Oaks: Sage, 2010. Print

_____. *Research Design: Qualitative, Quantitative, and Mixed Methods Approaches*. Thousand Oaks: Sage, 2013. Print.

Croft, Bruce, and Trevor Strohman. *Search Engines: Information Retrieval in Practice*. Boston: Addison, 2009. Print.

Crothers, Brooke. "Microsoft Explains Quantum Computing So Even You Can Understand." *CNET*. CBS Interactive, 25 July 2014. Web. 24 Mar. 2016.

Cunningham, William Michael. *The JOBS Act: Crowdfunding for Small Businesses and Startups*. New York: Springer, 2012. Print.

Currim, Sabah, et al. Using a Knowledge Learning Framework to Predict Errors in Database Design." *Elsevier* 40 (Mar. 2014): 11–31. Print.

Dahlberg, T., et al. "Past, Present, and Future of Mobile Payments Research: A Literature Review." *Electronic Commerce Research and Applications* 7.2 (2008): 165–81. Print.

Dale, Nell, and John Lewis. *Computer Science Illuminated*. 6th ed. Burlington: Jones, 2016. Print.

Daniel, John. "Making Sense of MOOCs: Musings in a Maze of Myth, Paradox and Possibility." *JIME Journal of Interactive Media in Education*. JIME, Dec. 2012. Web. 5 Aug. 2013.

Davidoff, Sherri, and Jonathan Ham. *Network Forensics: Tracking Hackers through Cyberspace*. Upper Saddle River: Prentice, 2012. Print.

Davis, Harold. *The Way of the Digital Photographer: Walking the Photoshop Post-Production Path to More Creative Photography*. San Francisco: Peachpit, 2013. Print.

Davis, Steven, and Brendan S. Gillon. *Semantics: A Reader*. New York: Oxford UP, 2004. Print.

Davison, Mark L., Yu-Feng Chang, and Ernest C. Davenport. "Modeling Configural Patterns in Latent Variable Profiles: Association with an Endogenous Variable." *Structural Equation Modeling* 21.1 (2014): 81–93. Print.

Dawson, Ross, and Steve Bynghall. *Getting Results from Crowds: The Definitive Guide to Using Crowdsourcing to Grow Your Business*. San Francisco: Advanced Human Technologies, 2011. Print.

de Jong, Gerard, and Moshe Ben-Akiva. "Transportation and Logistics in Supply Chains." *Supply Chain Management, Marketing and Advertising, and Global Management*. Ed. Hossein Bidgoli. Hoboken: Wiley, 2010. 146–58. Print. Vol. 2 of *The Handbook of Technology Management*. 3 vols.

De Micheli, Giovanni. *Synthesis and Optimization of Digital Circuits*. New York: McGraw, 1994. Print.

De Veaux, R. D. "Data Mining: A View From Down in the Pit." Stats 34 (2002).

_____, and H. Edelstein. "Reducing Junk Mail Using Data Mining Techniques." In *Statistics: A Guide to the Unknown*. 4th ed. Belmont, CA: Thomson, Brooks-Cole, 2006.

Dehn, Milton J. *Working Memory and Academic Learning: Assessment and Intervention*. Hoboken: Wiley, 2011. Print.

Deitel, Paul J., et al. Preface. *Android: How to Program, Global Edition*. 2nd ed., Pearson, 2015, pp. 19–30.

Delforge, Pierre. "America's Data Centers Consuming and Wasting Growing Amounts of Energy." *NRDC*. Natural Resources Defense Council, 6 Feb. 2015. Web. 17 Mar. 2016.

Delfs, Hans, and Helmut Knebl. *Introduction to Cryptography: Principles and Applications*. 3rd ed. Berlin: Springer, 2015. Print.

Delgoshaei, Yalda, and Neda Delavari. "Applying Multiple-Intelligence Approach to Education and Analyzing Its Impact on Cognitive Development of Pre-school Children." *Procedia-Social and Behavioral Sciences* 32 (2012): 361–66. Print.

Demirkan, H., and R. J. Kauffman. "Service-Oriented Technology and Management: Perspectives on Research and Practice for the Coming Decade." *Electronic Commerce Research and Applications* 7.4 (2008): 356–76. Print.

Dennis, Alan, Barbara Haley Wixom, and David Tegarden. *Systems Analysis and Design: An Object-Oriented Approach with UML*. 5th ed. Hoboken: Wiley, 2015. Print.

"Design and Technology: Manufacturing Processes." *GCSE Bitesize*. BBC, 2014. Web. 31 Jan. 2016.

Devadas, Srinivas, Abhijit Ghosh, and Kurt Keutzer. *Logic Synthesis*. New York: McGraw, 1994. Print.

Devlin, Barry A., and Paul T. Murphy. "An Architecture for a Business and Information System." *IBM Systems Journal* 27.1 (1988): 60–80. Print.

Dey, Pradip, and Manas Ghosh. *Computer Fundamentals and Programming in C*. 2nd ed. New Delhi: Oxford UP, 2013. Print.

Deyo, Jessica, Price Walt, and Leah Davis. "Rapidly Recognizing Relationships: Observing Speed Dating in the South." *Qualitative Research Reports in Communication* 12.1 (2011): 71–78. Print.

Dice, Pete. *Quick Boot: A Guide for Embedded Firmware Developers*. Hillsboro: Intel, 2012. Print.

"Digital Evidence and Forensics." *National Institute of Justice*. Office of Justice Programs, 28 Oct. 2015. Web. 12 Feb. 2016.

Dikkers, Seann, John Martin, and Bob Coulter. *Mobile Media Learning: Amazing Uses of Mobile Devices for Learning*. Pittsburgh: ETC P, Carnegie Mellon U, 2011. Print.

Dimitri, Giovanna Maria. "The Impact of Syntax Highlighting in Sonic Pi." *Psychology of Programming Interest Group*, 2015, www.ppig.org/sites/default/files/2015-PPIG-26th-Dimitri.pdf. Accessed 22 Feb. 2017.

Dingli, Alexiei, and Dylan Seychell. *The New Digital Natives: Cutting the Chord*. Heidelberg: Springer, 2015. eBook Collection (EBSCOhost). Web. 19 June 2015.

Doeppner, Thomas W. *Operating Systems in Depth*. Hoboken: Wiley, 2011. Print.

Domingos, Pedro. "A Few Useful Things to Know about Machine Learning." *Communications of the ACM* 55.10 (2012): 78-87. Print.

Donahue, Gary A. *Network Warrior*. Sebastopol: O'Reilly, 2011. Print.

Douglas McGregor: Theory X "Douglas McGregor: Theory X and Theory Y." *Workforce* 81.1 (Jan. 2002): 32. Print.

Downes, Larry. "Unscrambling the FCC's Net Neutrality Order: Preserving the Open Internet—But Which One?" *CommLaw Conspectus* 20.1 (2011): 83–128. Print.

Dredge, Stuart. "Kickstarter's Biggest Hits: Why Crowdfunding Now Sets the Trends." *Guardian*. Guardian News and Media, 17 Apr. 2014. Web. 22 May 2015.

Drucker, Peter F. "Knowledge-Worker Productivity: The Biggest Challenge." *California Management Review* 41.2 (1999): 79–94. Print.

Dushnitsky, Gary. "What If Crowdfunding Becomes the Leading Source of Finance for Entrepreneurs or Growing Companies?" *Forbes*. Forbes, 20 May 2015. Web. 22 May 2015.

Dyson, George. *Turing's Cathedral: The Origins of the Digital Universe*. London: Penguin Books, 2013. Print.

Eadicicco, Lisa. "Here's Why Drone Delivery Won't Be a Reality Any Time Soon." *Time* (November, 2015). http://time.com/4098369/amazon-google-drone-delivery/. This article identifies a number of potential issues related to drone delivery services.

Edmonds, W. Alex, and Thomas D. Kennedy. *An Applied Reference Guide to Research Designs: Quantitative, Qualitative, and Mixed Methods* Thousand Oaks: Sage, 2012. Print.

Edwards, Anthony David. *New Technology and Education*. New York: Continuum, 2012. Print.

Edwards, Gail. "Standpoint Theory, Realism and the Search for Objectivity in the Sociology of Education." *British Journal of Sociology of Education* (2012): 1–18. Web. 22 July 2013.

Edwards, Jim. "Proof That Android Really Is for the Poor." *Business Insider*. Business Insider, 27 June 2014. Web. 4 Jan. 2016.

Edwards, Paul N. *A Vast Machine: Computer Models, Climate Data, and the Politics of Global Warming*. Cambridge: MIT P, 2010. Print.

Eisner, Donald A. *The Death of Psychotherapy: From Freud to Alien Abductions*. Westport: Greenwood, 2000. Print.

Elliott, Larry, and Dan Atkinson. *Fantasy Island: Waking Up to the Incredible Economic, Political and Social Illusions of the Blair Legacy*. London: Constable, 2007. Print.

Emurian, Henry H., and Peng Zheng. "Programmed Instruction and Interteaching Applications to Teaching Java: A Systematic Replication." *Computers in Human Behavior* 26.5 (2010): 1166–75. Print.

Englander, Irv. *The Architecture of Computer Hardware and System Software: An Information Technology Approach*. 5th ed. Hoboken: Wiley, 2014. Print.

Epstein, Lee, and Thomas G. Walker. *Constitutional Law for a Changing America: A Short Course*. 6th ed. Los Angeles: Sage, 2015. Print.

Erben, Tony, Ruth Ban, and Martha E. Castañeda. *Teaching English Language Learners through Technology*. New York: Routledge, 2009. Print.

Esslinger, Bernhard, et al. *The CrypTool Script: Cryptography, Mathematics, and More*. 11th ed. Frankfurt: CrypTool, 2013. *CrypTool Portal*. Web. 2 Mar. 2016.

Estelles-Miguel, Sofia, Ignacio Gil-Pechuán, and Fernando J. Garrigos-Simon. *Advances in Crowdsourcing*. Cham: Springer, 2015. *eBook Collection (EBSCOhost)*. Web. 19 June 2015.

Estes, Brent, and Barbara Polnick. "Examining Motivation Theory in Higher Education: An Expectancy Theory Analysis of Tenured Faculty Productivity." *International Journal of Management, Business and Administration* 15.1 (2012): n.p. Print.

———. "Predicting Productivity in a Complex Labor Market: A Sabermetric Assessment of Free Agency on Major League Baseball Performance." *Business Studies Journal* 3.1 (2011): 23–58. Print.

Estopace, Eden. "Event-Driven Marketing: Retail's Next Big Step?" *Enterprise Innovation*. Questex Media, 25 Mar. 2013. Web. 7 Oct. 2013.

Everitt, Paul. "Make Sense of Your Variables at a Glance with Semantic Highlighting." *PyCharm Blog*, JetBrains, 19 Jan. 2017, blog.jetbrains.com/pycharm/2017/01/make-sense-of-your-variables-at-a-glance-with-semantic-highlighting/. Accessed 17 Feb. 2017.

"An Explanation of Event Driven Marketing." *Eventricity*. Eventricity Ltd, n.d. Web. 7 Oct. 2013.

Fain, Paul. "Only Sometimes for Online." *Inside Higher Ed*. Inside Higher Ed., 26 Apr. 2013. Web. 14 Sept. 2014.

Farahani, Reza Zanjirani, Shabnam Rezapour, and Laleh Kardar, eds. *Logistics Operations and Management: Concepts and Models*. Waltham: Elsevier, 2011. Print.

Farber, David J., et al. "SNOBOL, A String Manipulation Language." *Journal of the ACM*, vol. 11, no. 1, Jan. 1964, pp. 21–30, doi:10.1145/321203.321207.

Farcic, Viktor, and Alex Garcia. *Test-Driven Java Development*. Packt Publishing, 2015.

Farin, G. E. Curves and Surfaces for Computer-Aided Geometric Design: A Practical Guide. 4th ed. San Diego: Academic, 1997. Print.

Faticoni, Theodore G., Combinatorics: An Introduction. New York: Wiley, 2014. Digital file.

Feder, Samuel L., and Luke C. Platzer. "FCC Open Internet Order: Is Net Neutrality Itself Problematic for Free Speech?" Communications Lawyer 28.1 (2011): 1–26. Print.

Federal Aviation Administration. https://www.faa.gov/uas/model_aircraft/. "Model Aircraft Operations." Accessed February, 2016. This site identifies the rules and regulations that the Federal Aviation Administration has established for people who fly drones as a hobby.

Federal Bureau of Investigation. "Cyber Crime." FBI.gov. Department of Justice, n.d. Web 30 July 2013.

Feigenbaum, Edward Albert., and Julian Feldman. Computers and Thought. AAAI Press, 1995.

Fietta, Pierluigi, and Pieranna Fietta. "The Neurobiology of the Human Memory." Theoretical Biology Forum 104.1 (2011): 69–87. Print.

File, Thom. "Computer and Internet Use in the United States: Population Characteristics." Census.gov. US Dept. of Commerce, 2013. PDF file.

Finn, Patrick. "Primer on Research: Bias and Blinding; Self-Fulfilling Prophecies and Intentional Ignorance." ASHA Leader June 2006: 16–22. Web. 3 Oct. 2013.

Fischer, Eric. The Evolution of Character Codes, 1874–1968. N.p.: Fischer, n.d. Trafficways.org. Web. 22 Feb. 2016.

Fisher, Carl Erik, Lisa Chin, and Robert Klitzman. "Defining Neuromarketing: Practices and Professional Challenges." Harvard Review of Psychiatry 18.4 (2010): 230–37. Print.

Flanagan, Sara, Emily C. Bouck, and Jennifer Richardson. "Middle School Special Education Teachers' Perceptions and Use of Assistive Technology in Literacy Instruction." Assistive Technology 25.1 (2013): 24–30. Print.

Fleischner, Michael H. SEO Made Simple: Strategies for Dominating the World's Largest Search Engine. 4th ed. CreateSpace, 2014. Print.

Follin, Steve. "Preparing for IT Infrastructure Autonomics." IndustryWeek. Penton, 19 Nov. 2015. Web. 20 Jan. 2016.

Foote, Steven. Learning to Program. Upper Saddle River: Pearson, 2015. Print.

Foster, Jonathan K. "Memory." New Scientist 3 Dec. 2011: i–viii. Print

Fowler, Geoffrey A. "An Early Report Card on Massive Open Online Courses." Wall Street Journal. Dow Jones, 8 Oct. 2013. Web. 18 Jun. 2015.

Fox, Richard. Information Technology: An Introduction for Today's Digital World. Boca Raton: CRC, 2013. Print.

Fox, Susannah. "51% of U.S. Adults Bank Online." Pew Internet: Pew Internet & American Life Project. Pew Research Center. 7 Aug. 2013. Web. 20 Aug. 2013.

Franceschi-Bicchierai, Lorenzo. "Love Bug: The Virus That Hit 50 Million People Turns 15." Motherboard. Vice Media, 4 May 2015. Web. 16 Mar. 2016.

Frankfort-Nachmias, Chava, and Anna Leon-Guerrero. Social Statistics for a Diverse Society. 7th ed. Thousand Oaks: Sage, 2014. Print.

Franklin, Stan. Artificial Minds. MIT Press, 2001.

Freeman, Eric, and David Gelernter. "Lifestreams: A Storage Model for Personal Data." SIGMOD Record 25.1 (1996): 80–86. Print.

Frenzel, Louis E., Jr. Electronics Explained: The New Systems Approach to Learning Electronics. Burlington: Elsevier, 2010. Print.

"Frequently Asked Questions." Digital Watermarking Alliance. DWA, n.d. Web. 11 Mar. 2016.

Friedman, Daniel P., and Mitchell Wand. Essentials of Programming Languages. Cambridge: MIT P, 2006. Print.

Friedman, Debra, and Michael Hechter. "The Contribution of Rational Choice Theory to Macrosociological Research." Sociological Theory 6.2 (1988): 201–18. Print.

"Functional Electrical Stimulation (FES) Factsheet." Multiple Sclerosis Trust. Multiple Sclerosis Trust, Dec. 2012. Web. 7 Oct. 2013.

Gabriel, Richard P. "Lisp: Good News, Bad News, How to Win Big." AI Expert, vol. 6, no. 6, 1991, pp. 30–39.

Gaffin, Julie C. Internet Protocol 6. New York: Novinka, 2007. Print.

Gais, Hannah. "Is the Developing World 'MOOC'd Out'?" Al Jazeera America. Al Jazeera America, 17 July 2014. Web. 18 Jun. 2015.

Gallagher, Sean. "'Locky' Crypto-Ransomware Rides In on Malicious Word Document Macro." Ars Technica. Condé Nast, 17 Feb. 2016. Web. 16 Mar. 2016.

Gallagher, Sean. "Though 'Barely an Operating System,' DOS Still Matters (to Some People)." *Ars Technica*. Condé Nast, 14 July 2014. Web. 31 Jan. 2016.

Gancarz, Mike. *The UNIX Philosophy*. Woburn: Butterworth, 1995. Print.

García-Barriocanal, Elena, et al., eds. *Metadata and Semantic Research*. New York: Springer, 2011. Print.

Gardner, Howard. "Reflections on My Works and Those of My Commentators." *MI at 25: Assessing the Impact and Future of Multiple Intelligences for Teaching and Learning*. Ed. Branton Shearer. New York: Teachers College P, 2009, 96–113. Print.

———. *Frames of Mind: The Theory of Multiple Intelligences*. New York: Basic, 1983. Print.

———. *Multiple Intelligences: New Horizons in Theory and Practice*. Reprint. New York: Basic, 2006. Print.

Garrido, José M., Richard Schlesinger, and Kenneth E. Hoganson. *Principles of Modern Operating Systems*. 2nd ed. Burlington: Jones, 2013. Print.

Garthwaite, Paul H., Ian T. Jolliffe, and Byron Jones. *Statistical Inference*. 2nd ed. Oxford: Oxford UP, 2002. Print.

Garza, George. "Working with the Cons of Object Oriented Programming." Ed. Linda Richter. *Bright Hub*. Bright Hub, 19 May 2011. Web. 6 Feb. 2016.

Gaudin, Sharon. "Quantum Computing May Be Moving out of Science Fiction." *Computerworld*. Computerworld, 15 Dec. 2015. Web. 24 Mar. 2016.

Gee, James Paul. *Unified Discourse Analysis: Language, Reality, Virtual Worlds, and Video Games*. New York: Routledge, 2015. Print.

Gibbs, W. Wayt. "Autonomic Computing." *Scientific American*. Nature Amer., 6 May 2002. Web. 20 Jan. 2016.

Gibson, Barbara P. *The Complete Guide to Understanding and Using NLP*. Print. Ocala: Atlantic, 2011. Print.

Gibson, Jerry D., ed. *Mobile Communications Handbook*. 3rd ed. Boca Raton: CRC, 2012. Print.

Gillam, Richard. *Unicode Demystified: A Practical Programmer's Guide to the Encoding Standard*. Boston: Addison-Wesley, 2002. Print.

Gillespie, Tarleton, Pablo J. Boczkowski, and Kirsten A. Foot, eds. *Media Technologies: Essays on Communication, Materiality, and Society*. Cambridge: MIT P, 2014. Print.

Gillispie, Matthew. *From Notepad to iPad: Using Apps and Web Tools to Engage a New Generation of Students*. New York: Routledge, 2014. Print.

Gkoutzinis, Apostolos. *Internet Banking and the Law in Europe: Regulation, Financial Integration and Electronic Commerce*. Cambridge UP, 2010. Print.

Glaser, Anton. *History of Binary and Other Nondecimal Numeration*. Rev. ed. Los Angeles: Tomash, 1981. Print.

Glaser, J. D. *Secure Development for Mobile Apps: How to Design and Code Secure Mobile Applications with PHP and JavaScript*. Boca Raton: CRC, 2015. Print.

Gleason, Ann Whitney. *Mobile Technologies for Every Library*. Lanham: Rowman, 2015. eBook Collection (EBSCOhost). Web. 1 July 2015.

Goggin, Gerard. *Cell Phone Culture: Mobile Technology in Everyday Life*. New York: Routledge, 2006. Print.

Gogolin, Greg. *Digital Forensics Explained*. Boca Raton: CRC, 2013. Print.

Goldman, Ron, and Richard P. Gabriel. *Innovation Happens Elsewhere: Open Source as Business Strategy*. Morgan Kaufmann Publishers, 2005.

Goldsborough, Reid. "Android on the Rise." *Tech Directions* May 2014: 12. *Academic Search Complete*. Web. 2 Jan. 2016.

Gollakota, Kamala, and Kokila Doshi. "Diffusion of Technological Innovations in Rural Areas." *Journal of Corporate Citizenship* 41 (2011): 69–82. Print.

Gong, Yihong. *Intelligent Image Databases: Towards Advanced Image Retrieval*. Hingham: Kluwer, 1998. Print.

Gonzalez, Teofilo, and Jorge Díaz-Herrera, eds. *Computing Handbook: Computer Science and Software Engineering*. 3rd ed. Boca Raton: CRC, 2014. Print.

Goodchild, Michael F. "The Alexandria Digital Library Project: Review, Assessment, and Prospects." *Trends in Information Management* 1.1 (2005): 20–25. Web. 21 Aug. 2013.

Goodman, Danny, Michael Morrison, and Brendan Eich. *Javascript Bible*. New York: Wiley, 2007. Print.

Goodrich, Michael T., et al. *Data Structures and Algorithms in Java*. 6th ed., John Wiley & Sons, 2014.

Goriunova, Olga, ed. *Fun and Software: Exploring Pleasure, Paradox, and Pain in Computing*. New York: Bloomsbury, 2014. Print.

Gould, Jay E. *Concise Handbook of Experimental Methods for the Behavioral and Biological Sciences*. Boca Raton: CRC, 2002. Print.

Gourley, David, et al. "Client Identification and Cookies." *HTTP: The Definitive Guide*. Sebastopol, CA: O'Reilly Media, 2002. 257–76. Print.

Govindjee, S. *Internal Representation of Numbers*. Dept. of Civil and Environmental Engineering, U of California Berkeley, Spring 2013. Digital File.

Grad, Burton, and Thomas J. Bergin. "History of Database Management Systems." *IEEE Annals of the History of Computing* 31.4 (2009): 3–5. Print.

Graham, Mark. "The Knowledge Based Economy and Digital Divisions of Labour." *The Companion to Development Studies*. Ed. V. Desai and R. Potter. 3rd ed. London: Routledge, 2014. 189–95. Print.

Graham, Tony. *Unicode: A Primer*. Foster City: M&T, 2000. Print.

Grant, August E., and Jennifer H. Meadows, eds. *Communication Technology Update and Fundamentals*. 13th ed. Waltham: Focal, 2012. Print.

"Graphical User Interface (GUI)." *Techopedia*. Techopedia, n.d. Web. 5 Feb. 2016.

Grappone, Jennifer, and Gradiva Couzin. *SEO: An Hour a Day*. New York: Sybex, 2011. Print.

Grassl, Wolfgang. "Aquinas on Management and Its Development." *Journal of Management Development* 29.7–8 (2010): 706–15. Print.

Greenberg, Jerald, and Robert Folger. *Controversial Issues in Social Research Methods*. New York: Springer, 1988. Print.

Greenfield, Larry. *Data Warehousing Information Center*. LGI Systems, 1995. Web. 8 Oct. 2013.

Griffith, Eric. "What Is Cloud Computing?" *PC Magazine*. Ziff Davis, PCMag Digital Group, 17 Apr. 2015. Web. 8 June 2015.

Griffiths, Devin C. *Virtual Ascendance: Video Games and the Remaking of Reality*. Lanham: Rowman, 2013. Print.

Grinnell, Frederick. "Research Integrity and Everyday Practice of Science." *Science and Engineering Ethics*, vol. 19, no. 3, 2013, pp. 685–701. *Academic Search Complete*, search.ebscohost.com/. Accessed 19 Apr. 2017.

Griswold, Ralph, and Madge T. Griswold. *The Icon Programming Language*. 3rd. ed., Peer-to-Peer Communications, 2002.

Groux, Catherine. "Study Analyzes Characteristics of Online Learners." *Learning House*. Learning House, 27 July 2012. Web. 14 Sept. 2014.

Gruzd, Anatoliy, et al. "Imagining Twitter as an Imagined Community." *American Behavioral Scientist* 55.10 (Oct. 2011): 1294–318. Print.

Guichard, David. "An Introduction to Combinatorics and Graph Theory." *Whitman*. Whitman Coll., 4 Jan 2016. Web. 10 Feb. 2016.

Gupta, Siddarth, and Vagesh Porwal. "Recent Digital Watermarking Approaches, Protecting Multimedia Data Ownership." *Advances in Computer Science* 4.2 (2015): 21–30. Web. 14 Mar. 2016.

"Guru: Douglas McGregor." *Economist*. Economist Newspaper, 3 Oct. 2008. Web. 18 Sept. 2013.

Hachtel, Gary D., and Fabio Somenzi. *Logic Synthesis and Verification Algorithms*. New York: Springer, 2006. Print.

Hagen, Rebecca, and Kim Golombisky. *White Space Is Not Your Enemy: A Beginner's Guide to Communicating Visually through Graphic, Web & Multimedia Design*. 2nd ed. Burlington: Focal, 2013. Print.

Haji Mohamed, Ramesh Kumar Moona, and Che Supian Mohamad Nor. "The Relationship between McGregor's X-Y Theory Management Style and Fulfillment of Psychological Contract: A Literature Review." *International Journal of Academic Research in Business and Social Sciences* 3.5 (May 2013): 715–20. Print.

Hakala, Tuomas, et al. "An Experiment on the Effects of Program Code Highlighting on Visual Search for Local Patterns." *18th Workshop of the Psychology of Programming Interest Group*, U of Sussex, Sept. 2006, www.ppig.org/papers/18th-hakala.pdf. Accessed 7 Mar. 2017.

Hall, Gene E., R. C. Wallace, and W. A. Dossett. *A Developmental Conceptualization of the Adoption Process within Educational Institutions*. Austin: U of Texas, 1973. Print.

Halperin, Sandra, and Oliver Heath. *Political Research: Methods and Practical Skills*. New York: Oxford UP, 2012. Print.

Hand, David J. *Statistics: A Very Short Introduction*. New York: Oxford UP, 2008. Print.

Hansen, V. "Predator Drone Attacks." *New England Law Review* 46 (2011): 27–36. Print.

Harbour, Jonathan S. *Beginning Game Programming*. 4th ed. Boston: Cengage, 2015. Print.

Harding, Sandra. *Objectivity and Diversity: Another Logic of Scientific Research*. U of Chicago P, 2015.

Harel, Jacob. "SynthOS and Task-Oriented Programming." *Embedded Computing Design*. Embedded Computing Design, 2 Feb. 2016. Web. 7 Feb. 2016.

Harper, Robert. *Practical Foundations for Programming Languages*. Cambridge: Cambridge UP, 2013. Print.

Harper, Simon, and Yeliz Yesilada, eds. *Web Accessibility: A Foundation for Research*. New York: Springer, 2008. Print.

Harrington, Jan L. *Relational Database Design and Implementation*. 4th ed., Elsevier, 2016.

Harris, David Money, and Sarah L. Harris. *Digital Design and Computer Architecture*. 2nd ed. Waltham: Morgan, 2013. Print.

Harrison, Virginia, and Jose Pagliery. "Nearly 1 Million New Malware Threats Released Every Day." *CNNMoney*. Cable News Network, 14 Apr. 2015. Web. 16 Mar. 2016.

Hart, Archibald D., and Sylvia Hart Frejd. *The Digital Invasion: How Technology Is Shaping You and Your Relationships*. Grand Rapids: Baker, 2013. Print.

Hart, John, Jr., and Michael A. Kraut, eds. *Neural Basis of Semantic Memory*. Cambridge: Cambridge UP, 2007. Print.

Hassoun, Soha, and Tsutomu Sasao, eds. *Logic Synthesis and Verification*. Norwell: Kluwer, 2002. Print.

Haunani Solomon, Denise. "Uncertainty Reduction Theory." *The Concise Encyclopedia of Communication*. Ed. Wolfgang Donsbach. Malden: Wiley, 2015. Print.

Hauptmann, Alexander G., Michael J. Witbrock, and Michael G. Christel. "News-on-Demand: An Application of Informedia Technology." *D-Lib Magazine*. Corp. for Natl. Research Initiatives, Sept. 1995. Web. 24 Feb. 2014.

Haverbeke, Marijn. *Eloquent JavaScript: A Modern Introduction to Programming*. 2nd ed., No Starch Press, 2015.

Heaven, Douglas. "Higher State of Mind." *New Scientist* 10 Aug. 2013: 32?35. Print.

Hebert, Dustin M. "Innovation Diffusion Factors Affecting Electronic Assessment System Adoption in Teacher Education." *National Teacher Educational Journal* 5.2 (2012): 35–44. Print.

Heisler, Yoni. "The History and Evolution of iOS, from the Original iPhone to iOS 9." *BGR*. BGR Media, 12 Feb. 2016. Web. 26 Feb. 2016.

Hekman, Susan. "Truth and Method: Feminist Standpoint Theory Revisited." *Signs* 22.2 (1997): 341–65. Print.

Hendrix, Dannie M., and Orval Holcomb, eds. *Psychology of Memory*. New York: Nova, 2012. Print.

Henschen, Doug, Ben Werther, and Scott Gnau. "Big Data Debate: End Near for Data Warehousing?" *InformationWeek*. UBM Tech, 19 Nov. 2012. Web. 8 Oct. 2013.

Henz, Martin. *Objects for Concurrent Constraint Programming*. New York: Springer, 1998. Print.

Hernandez, Michael J. *Database Design for Mere Mortals: A Hands-On Guide to Relational Database Design*. Upper Saddle River: Addison-Wesley, 2013. Print.

Hernstein, Richard J. "Rational Choice Theory: Necessary but Not Sufficient." *American Psychologist* 45.3 (1990): 356–67. Print.

Herrman, John. "How to Get Started: 3D Modeling and Printing." *Popular Mechanics*. Hearst Digital Media, 15 Mar. 2012. Web. 31 Jan. 2016.

Hey, Tony, and Gyuri Pápay. *The Computing Universe: A Journey through a Revolution*. New York: Cambridge UP, 2015. Print.

Hider, Philip. *Information Resource Description: Creating and Managing Metadata*. Chicago: Amer. Lib. Assoc., 2012. Print.

Higgins, Julian P. T., and Sally Green. *Cochrane Handbook for Systematic Reviews of Interventions*. Hoboken: Wiley, 2008. Print.

Highfield, Roger. "Fast Forward to Cartoon Reality." *Telegraph*. Telegraph Media Group, 13 June 2006. Web. 31 Jan. 2016.

Highhouse, Scott. "The Influence of Douglas McGregor." *TIP: The Industrial-Organizational Psychologist* 49.2 (Oct. 2011): 105–7. Print.

Hill, Janette R., Llyan Song, and Richard E. West. "Social Learning Theory and Web-based Learning Environments: A Review of Research and Discussion Implications." American Journal of Distance Education 23.2 (2009): 88–103. Print.

Hillmann, Diane I., and Elaine L. Westbrooks. *Metadata in Practice*. Chicago: Amer. Lib. Assn., 2004. Print.

"History and Timeline." *Open Group*. Open Group, n.d. Web. 28 Feb. 2016.

History of Cryptography: An Easy to Understand History of Cryptography. N.p.: Thawte, 2013. *Thawte*. Web. 4 Feb. 2016.

"History of Java Technology." *Oracle.com*. Oracle, n.d. Web. 22 Aug. 2013.

"The History of the Integrated Circuit." *Nobelprize.org*. Nobel Media, 2014. Web. 31 Mar. 2016.

Hoehle, Hartmut, Eusebio Scornavacca, and Sid Huff. "Three Decades of Research on Consumer Adoption and Utilization of Electronic Banking

Channels: A Literature Analysis." *Decision Support Systems* 54.1 (2012): 122–32. Print.

Hoffstein, Jeffrey, Jill Pipher, and Joseph H. Silverman. *An Introduction to Mathematical Cryptography.* 2nd ed. New York: Springer, 2014. Print.

Hofmann, Markus, and Leland R. Beaumont. "Content Transfer." *Content Networking: Architecture, Protocols, and Practice.* San Francisco: Elsevier, 2005. 25–52. Print.

Hofstedt, Petra. *Multiparadigm Constraint Programming Languages.* New York: Springer, 2013. Print.

Holcombe, Jane, and Charles Holcombe. *Survey of Operating Systems.* New York: McGraw, 2015. Print.

Holt, Thomas J., Adam M. Bossler, and Kathryn C. Seigfried-Spellar. *Cybercrime and Digital Forensics: An Introduction.* New York: Routledge, 2015. Print.

Horenstein, Henry. *Digital Photography: A Basic Manual.* Boston: Little, 2011. Print.

Horibe, Frances. *Managing Knowledge Workers: New Skills and Attitudes to Unlock the Intellectual Capital in your Organization.* New York: Wiley, 1999. Print.

Horton, Sarah, and Whitney Quesenbery. *Universal Design for Web Accessibility.* New York: Rosenfeld Media, 2013. Print.

"How Firewalls Work." *Boston University Information Services and Technology.* Boston U, n.d. Web. 28 Feb. 2016.

Howard, Jennifer. "Publishers See Online Mega-Courses as Opportunity to Sell Textbooks." *Chronicle of Higher Education.* Chronicle of Higher Ed., Sept. 2010. Web. 5 Aug. 2013.

Howe, Jeff. *Crowdsourcing: Why the Power of the Crowd Is Driving the Future of Business.* New York: Crown, 2009. Print.

Hoyle, Rick H., ed. *Handbook of Structural Equation Modeling.* New York: Guilford, 2012. Print.

Hricko, Mary. *Design and Implementation of Web-Enabled Teaching Tools.* Hershey: Idea Group, 2002. Print.

Hu, Helen. "MOOCs Changing the Way We Think about Higher Education." *Diverse Issues in Higher Education.* Diverse Issues in Higher Ed., April 2013. Web. 29 July 2013.

Hu, Li-tze, and Peter M. Bentler. "Cutoff Criteria for Fit Indexes in Covariance Structure Analysis: Conventional Criteria versus New Alternatives." *Structural Equation Modeling* 6.1 (1999): 1–55. Print.

Human Brain Project. Human Brain Project, 2013. Web. 16 Feb. 2016.

Humphrey, Neil. *Social and Emotional Learning: A Critical Appraisal.* Thousand Oaks: Sage, 2013. Print.

Hunt, Morton. *How Science Takes Stock: The Story of Meta-analysis.* New York: Russell Sage Foundation, 1997.

Hutchinson, Lee. "Home 3D Printers Take Us on a Maddening Journey into Another Dimension." *Ars Technica.* Condé Nast, 27 Aug. 2013. Web. 6 Jan. 2016.

Huth, Alexa, and James Cebula. *The Basics of Cloud Computing.* N.p.: Carnegie Mellon U and US Computer Emergency Readiness Team, 2011. PDF file.

Hyde, Randall. *Write Great Code: Understanding the Machine.* Vol. 1. San Francisco: No Starch, 2005. Print.

Ingham, Kenneth, and Stephanie Forrest. *A History and Survey of Network Firewalls.* Albuquerque: U of New Mexico, 2002. PDF file.

Iniewski, Krzysztof. *Embedded Systems: Hardware, Design, and Implementation.* Hoboken: Wiley, 2013. Print.

Inmon, William H. *Building the Data Warehouse.* 4th ed. Indianapolis: Wiley, 2011. Print.

Innovation in Produce Design: From CAD to Virtual Prototyping. Eds. Monica Bordegoni and Caterina Rizze. London: Springer, 2001. Print.

"Intro to Algorithms." *Khan Academy.* Khan Acad., 2015. Web. 19 Jan. 2016.

"Introduction to Image Files Tutorial." *Boston University Information Services and Technology.* Boston U, n.d. Web. 11 Feb. 2016.

"iOS: A Visual History." *Verge.* Vox Media, 16 Sept. 2013. Web. 24 Feb. 2016.

ITL Education Solutions. *Introduction to Information Technology.* 2nd ed. Delhi: Pearson, 2012. Print.

Iversen, Jakob, and Michael Eierman. *Learning Mobile App Development: A Hands-On Guide to Building Apps with iOS and Android.* Upper Saddle River: Addison-Wesley, 2014. Print.

Jackson, Charles L. "Wireless Efficiency Versus Net Neutrality." *Federal Communications Law Journal* 63.2 (2011): 445–80. Print.

Jacobs, Deborah L. "The Trouble with Crowdfunding." *Forbes.com.* Forbes.com LLC, 17 Apr. 2013. Web. 7 Aug. 2013.

Jacobson, Douglas, and Joseph Idziorek. *Computer Security Literacy: Staying Safe in a Digital World.* Boca Raton: CRC, 2013. Print.

James, Mike. "The Triumph of Deep Learning." *I Programmer*. I-programmer.info, 14 Dec. 2012. Web. 27 Sept. 2013.

Janert, Philipp K. *Feedback Control for Computer Systems*. Sebastopol: O'Reilly, 2014. Print.

"Java Programming Style Guidelines." *GeoSoft*, Geotechnical Software Services, Apr. 2015, geosoft.no/development/javastyle.html. Accessed 14 Feb. 2017.

Javary, Michèle. "Evolving Technologies and Market Structures: Schumpeterian Gales of Creative Destruction and the United Kingdom Internet Service Providers' Market." *Journal of Economic Issues* 38.3 (2004): 629–57. Print.

Jayaswal, Bijay K., and Peter C. Patton. *Design for Trustworthy Software: Tools, Techniques, and Methodology of Developing Robust Software*. Prentice Hall, 2007.

Jeannot, Emmanuel, and J. Žilinskas. *High Performance Computing on Complex Environments*. Hoboken: Wiley, 2014. Print.

Jeffery, Clinton, et al. "Integrating Regular Expressions and SNOBOL Patterns into String Scanning: A Unifying Approach." *Proceedings of the 31st Annual ACM Symposium on Applied Computing*, 2016, pp. 1974–79, doi:10.1145/2851613.2851752.

_____, et al. "Pattern Matching in Unicon." *Unicon Project*, 16 Jan. 2017, unicon.org/utr/utr18.pdf. Accessed 4 May. 2017.

Jemielniak, Dariusz. *The New Knowledge Workers*. Cheltenham: Elgar, 2012. Print.

Jeng, Judy. "What Is Usability in the Context of the Digital Library and How Can It Be Measured?" *Information Technology and Libraries* 24.2 (2005): 47–56. Web. 21 Aug. 2013.

Jennings, Tom. "An Annotated History of Some Character Codes." *World Power Systems*. Tom Jennings, 29 Oct. 2004. Web. 16 Feb. 2016.

Jex, Steve M., and Thomas W. Britt. *Organizational Psychology: A Scientist-Practitioner Approach*. 3rd ed. Malden: Wiley, 2014. Print.

Johnson, Burke, and Sharlene Nagy Hesse-Biber. *The Oxford Handbook Of Multimethod And Mixed Methods Research Inquiry*. Oxford: Oxford UP, 2015. *eBook Collection (EBSCOhost)*. Web. 30 June 2015.

Johnson, Jeff. *Designing with the Mind in Mind*. 2nd ed. Waltham: Morgan, 2014. Print.

Johnson, R. Burke, and Anthony Onwuegbuzie. "Mixed Methods Research: A Research Paradigm Whose Time Has Come." *Educational Researcher* 33.7 (2004): 14–26. Print.

_____, Anthony Onwuegbuzie, and Lisa A. Turner. "Toward a Definition of Mixed Methods Research." *Journal of Mixed Methods Research* 1.2 (2007): 112–33. Print.

Joiner, Richard, et al. "Comparing First and Second Generation Digital Natives' Internet Use, Internet Anxiety, and Internet Identification." *CyberPsychology, Behavior & Social Networking* 16.7 (2013): 549–52. Print.

Jones, Chris. "A New Generation of Learners? The Net Generation and Digital Natives." *Learning, Media & Technology* 35.4 (2010): 365–68. Print.

Jones, Kristopher B. *SEO: Your Visual Blueprint for Effective Internet Marketing*. 3rd ed. Indianapolis: Wiley, 2013. Print.

Jussim, Lee. *Social Perception and Social Reality: Why Accuracy Dominates Bias and Self-Fulfilling Prophecy*. New York: Oxford UP, 2012. Print.

Kahn, David. *The Codebreakers: The Story of Secret Writing*. New York: Macmillan, 1967.

Kale, Vivek. *Guide to Cloud Computing for Business and Technology Managers*. Boca Raton: CRC, 2015. Print.

Kalla, Siddharth. "Calculate Standard Deviation." *Explorable*. Explorable.com, 27 Sept. 2009. Web. 4 Oct. 2013.

Kalyanam, Kirthi, and Monte Zweben. "The Perfect Message at the Perfect Moment." *Harvard Business Review* 83.11 (2005): 112–20. Print.

Kapp, Karl M. *The Gamification of Learning and Instruction: Game-Based Methods and Strategies for Training and Education*. San Francisco: Pfeiffer, 2012. Print.

Karthikeyan, K., and M. Sangeetha. "Page Rank Based Design and Implementation of Search Engine Optimization." *International Journal of Computer Applications* 40.4 (2012): 13–18. Print.

Kass, Robert E. "Statistical Inference: The Big Picture." *Statistical Science* 26.1 (2011): 1–9. Print.

Katz, Jonathan, and Yehuda Lindell. *Introduction to Modern Cryptography*. 2nd ed. Boca Raton: CRC, 2015. Print.

Kearns, Kate. *Semantics*. 2nd ed. New York: Palgrave, 2011. Print.

Keizer, Garret. *Privacy*. New York: Picador, 2012. Print.

Kelby, Scott. *The Digital Photography Book: Part One*. 2nd ed. San Francisco: Peachpit, 2013. Print.

Kellner, Douglas. "New Media and New Literacies: Reconstructing Education for the New Millennium."The Handbook of New Media: Social Shaping and Consequences of ICTs. Eds. Leah Lievrouw and Sonia Livingstone. Thousand Oaks: Sage, 2002. Print.

Kelly, Gordon, "Apple iOS 9: 11 Important New Features." *Forbes*. Forbes.com, 16 Sept. 2015. Web. 28 Feb. 2016.

Kemper, Bitsy. *The Right to Privacy: Interpreting the Constitution*. New York: Rosen, 2015. Print.

Kenny, David A., Deborah A. Kashy, and William L. Cook. *Dyadic Data Analysis*. New York: Guilford, 2006. Print.

Kerdvilbulvech, Chutisant. "A New Method for Web Development Using Search Engine Optimization." *International Journal of Computer Science and Business Informatics* 3.1 (2013). Web. 26 Aug. 2013.

Khan, Gul N., and Krzysztof Iniewski, eds. *Embedded and Networking Systems: Design, Software, and Implementation*. Boca Raton: CRC, 2014. Print.

Khurana, Atika, et al. "Stronger Working Memory Reduces Sexual Risk Taking in Adolescents, Even After Controlling for Parental Influences." *Child Development* (2015). n. pag. Web. 24 Jun. 2015.

Kim, Chang-Hun, et al. *Real-Time Visual Effects for Game Programming*. Singapore: Springer, 2015. Print.

Kim, Sung Yeon, and Kwangsu Cho. "Usability and Design Guidelines of Smart Canes for Users with Visual Impairments." *International Journal of Design* 7.1 (2013): 99–110. Print.

Kingsley-Hughes, Adrian, and Kathie Kingsley-Hughes. *Beginning Programming*. Wiley Publishing, 2005.

Kizza, Joseph Migga. *Ethical and Social Issues in the Information Age*. 5th ed. London: Springer, 2013. Print.

Kline, Rex B. *Principles and Practice of Structural Equation Modeling*. 3rd ed. New York: Guilford, 2011. Print.

Knobloch, Leanne K., and Laura E. Miller. "Uncertainty and Relationship Initiation." *Handbook of Relationship Initiation*. Ed. Susan Sprecher, Amy Wenzel, and John Harvey. New York: Psychology P, 2008. 121–34. Print.

"Knowledge Base: Technologies in 3D Printing." *DesignTech*. DesignTech Systems, n.d. Web. 6 Jan. 2016.

Koblin, Aaron. "Q&A: Aaron Koblin: The Data Visualizer." Interview by Jascha Hoffman. *Nature* 486.7401 (2012): 33. Print.

Koenig, Andrew. "The Snocone Programming Language." *Snobol4*, www.snobol4.com/report.htm. Accessed 4 May 2017.

Kojić, Miloš, et al. *Computer Modeling in Bioengineering: Theoretical Background, Examples and Software*. Hoboken: Wiley, 2008. Print.

Kolowich, Steve. "The Online Student." *Inside Higher Ed*. Inside Higher Ed., 25 July 2012. Web. 14 Sept. 2014.

Koricheva, Julia, Jessica Gurevitch, and Kerrie Mengersen, eds. *Handbook of Meta-analysis in Ecology and Evolution*. Princeton: Princeton UP, 2013. Print.

Korpela, Jukka K. *Unicode Explained*. Sebastopol: O'Reilly Media, 2006. Print.

Kosky, Philip, et al. *Exploring Engineering: An Introduction to Engineering and Design*. 4th ed. Waltham: Academic, 2016. Print.

Krar, Steve, Arthur Gill, and Peter Smid. *Computer Numerical Control Simplified*. New York: Industrial, 2001. Print.

Krill, Paul. "JavaScript Creator Ponders Past, Future." *InfoWorld*. Infoworld, 23 June 2008. Web. 5 Aug. 2013.

Krimsky, Sheldon, and Tania Simoncelli. *Genetic Justice: DNA Data Banks, Criminal Investigations, and Civil Liberties*. New York: Columbia UP, 2011. Print.

——. "Do Financial Conflicts of Interest Bias Research? An Inquiry into the 'Funding Effect' Hypothesis." *Science, Technology, & Human Values*, vol. 38, no. 4, 2013, pp. 566–87.

Kristol, David M. "HTTP Cookies: Standards, Privacy, and Politics." *ArXiv.org*. Cornell U Lib., 9 May 2001. Web. 25 Sept. 2013.

Krug, Steve. Don't Make Me Think: A Common Sense Approach to Web Usability. 3rd ed. Berkeley: New Riders, 2014. Print.

Kruk, Robert. "Public, Private and Hybrid Clouds: What's the Difference?" *Techopedia*. Techopedia, 18 May 2012. Web. 21 Jan. 2016.

Kulisch, Ulrich. *Computer Arithmetic and Validity: Theory, Implementation, and Applications*. 2nd ed. Boston: De Gruyter, 2013. Print.

Kuo, Sen M., Bob H. Lee, and Wenshun Tian. *Real-Time Digital Signal Processing: Fundamentals, Implementations and Applications*. 3rd ed. Hoboken: Wiley, 2013. Print.

Laberge, Robert. *The Data Warehouse Mentor: Practical Data Warehouse and Business Intelligence Insights.* New York: McGraw, 2011. Print.

Lackey, Ella Deon, et al. "Introduction to Public-Key Cryptography." *Mozilla Developer Network.* Mozilla, 21 Mar. 2015. Web. 4 Feb. 2016.

Lakhtakia, A., and R. J. Martín-Palma. *Engineered Biomimicry.* Amsterdam: Elsevier, 2013. Print.

Lalanda, Philippe, Julie A. McCann, and Ada Diaconescu, eds. *Autonomic Computing: Principles, Design and Implementation.* London: Springer, 2013. Print.

Lande, Daniel R. "Development of the Binary Number System and the Foundations of Computer Science." *Mathematics Enthusiast* 1 Dec. 2014: 513–40. Print.

Lane, David M. "Measures of Variability." *Online Statistics Education: An Interactive Multimedia Course of Study.* Lane, n.d. Web. 4 Oct. 2013.

Lane, Jonathan, et al. "JavaScript Primer." *Foundation Website Creation with HTML5, CSS3, and JavaScript.* New York: Apress, 2012. Print.

Lathi, B. P. *Linear Systems and Signals.* 2nd rev. ed. New York: Oxford UP, 2010. Print.

Lau, Albert. "Data Sets: Narrative Visualization." *Many Eyes.* IBM, 18 May 2011. Web. 24 Feb. 2014.

Law, Averill M. *Simulation Modeling and Analysis.* 5th ed. New York: McGraw, 2015. Print.

Lawton, Kevin, and Dan Marom. *The Crowdfunding Revolution: How to Raise Venture Capital Using Social Media.* New York: McGraw. 2012. Print.

Lee, Kent D., and Steve Hubbard. *Data Structures and Algorithms with Python.* Springer, 2015.

Lee, M. C. "Factors Influencing the Adoption of Internet Banking: An Integration of TAM and TPB with Perceived Risk and Perceived Benefit." *Electronic Commerce Research and Applications* 8.3 (2009): 130–41. Print.

Lee, Roger Y., ed. *Applied Computing and Information Technology.* New York: Springer, 2014. Print.

Lee, Silvia Wen-Yu, and Chin-Chung Tsai. "Technology-Supported Learning in Secondary and Undergraduate Biological Education: Observations from Literature Review." *Journal of Science Education and Technology* 22.2 (2013): 226–33. Print.

Leech, Nancy L., and Anthony Onwuegbuzie. "A Typology of Mixed Methods Research Designs." *Quality & Quantity* 43.2 (2009): 265–75. Print.

Lesk, Michael. *Understanding Digital Libraries.* 2nd ed. San Francisco: Morgan, 2004. Print.

Lessig, Lawrence. *Remix: Making Art and Commerce Thrive in the Hybrid Economy.* New York: Penguin, 2008. Print.

Letherby, Gayle, et al. *Objectivity and Subjectivity in Social Research.* Sage Publications, 2013.

Levin, Eden S., ed. *Working Memory: Capacity, Developments and Improvement Techniques.* New York: Nova, 2011. Print.

Levine, Timothy R., Kim Sang-Yeon, and Merissa Ferrara. "Social Exchange, Uncertainty, and Communication Content as Factors Impacting the Relational Outcomes of Betrayals." *Human Communication* 13.4 (2010): 303–18. Print.

Levy, Steven. *In the Plex: How Google Thinks, Works, and Shapes Our Lives.* New York: Simon, 2011.

Lewand, Robert. *Cryptological Mathematics.* Washington, DC: Mathematical Association of America, 2000.

Li, Han-Xiong, and XinJiang Lu. *System Design and Control Integration for Advanced Manufacturing.* Hoboken: Wiley, 2015. Print.

Li, Juanzi. *Semantic Web and Web Science.* New York: Springer, 2013. Digital file.

Liberson, W. T., et al. "Functional Electrotherapy: Stimulation of the Peroneal Nerve Synchronized with the Swing Phase of the Gait of Hemiplegic Patients." *Archives of Physical Medicine and Rehabilitation* 42 (1961): 101–5. Print.

Lichtblau, Eric. "In Secret, Court Vastly Broadens Powers of N.S.A." *New York Times* 7 July 2013, New York ed.: A1. Print.

Lien, Tracey. "Virtual Reality Isn't Just for Video Games." *Los Angeles Times.* Tribune, 8 Jan. 2015. Web. 23 Mar. 2016.

Light, Jennifer. "Rethinking the Digital Divide." *Harvard Educational Review* 71.4 (2001): 709–33. Print.

Lima, Manuel. *Visual Complexity: Mapping Patterns of Information.* Princeton: Princeton Architectural, 2013. Print.

Linder-Pelz, Susie. *NLP Coaching: An Evidence-Based Approach for Coaches, Leaders and Individuals.* Philadelphia: Kogan, 2010. Print.

Lipiansky, Ed. *Electrical, Electronics, and Digital Hardware Essentials for Scientists and Engineers.* Hoboken: Wiley, 2013. Print.

Littell, Julia H., Jacqueline Corcoran, and Vijayan Pillai. *Systematic Reviews and Meta-analysis.* New York: Oxford UP, 2008. Print.

Liu, Alan. *The Laws of Cool: Knowledge Work and the Culture of Information.* Chicago: U of Chicago P, 2004. Print.

Liu, Shih-Chii, Tobi Delbruck, Giacomo Indiveri, Adrian Whatley, and Rodney Douglas. *Event-Based Neuromorphic Systems.* Chichester: Wiley, 2015. Print.

Livingston, Steven, and Gregor Walter-Drop, eds. *Bits and Atoms: Information and Communication Technology in Areas of Limited Statehood.* New York: Oxford UP, 2014. Print.

Löbner, Sebastian. *Understanding Semantics.* New York: Routledge, 2013. Digital file.

Loehlin, John C. *Latent Variable Models: An Introduction to Factor, Path, and Structural Equation Analysis.* 4th ed. New York: Routledge, 2011. Print.

Lohr, Steve. "Humanizing Technology: A History of Human-Computer Interaction." *New York Times: Bits.* New York Times, 7 Sept. 2015. Web. 31 Jan. 2016.

Longino, Helen E. "Feminist Standpoint Theory and the Problems of Knowledge." *Signs* 19.1 (1993): 201–12. Print.

Loo, Alfred Waising, ed. *Distributed Computing Innovations for Business, Engineering, and Science.* Hershey: Information Science Reference, 2013. Print.

Looker, Dianne, and Victor Thiessen. "The Digital Divide in Canadian Schools: Factors Affecting Student Access to and Use of Information Technology." *Statistics Canada.* Statistics Canada, 2003. PDF file.

Loshin, Peter. *IPv6: Theory, Protocol, and Practice.* 2nd ed. San Francisco: Morgan Kaufmann, 2004. Print.

Luckham, David C. *Event Processing for Business.* Hoboken: Wiley, 2012. Print.

Lunenberg, Fred C. "Expectancy Theory of Motivation: Motivating by Altering Expectations." *International Journal of Management, Business and Administration* 15.1 (2011): n.p. Print.

MacLennan, Bruce J. *Principles of Programming Languages: Design, Evaluation, and Implementation.* 3rd ed., Oxford UP, 1999.

Maddux, Cleborne D., and D. Lamont Johnson. *Technology in Education: A Twenty-Year Retrospective.* Hoboken: Taylor, 2013. Print.

Madhav, Sanjay. *Game Programming Algorithms and Techniques: A Platform-Agnostic Approach.* Upper Saddle River: Addison, 2014. Print.

Magsamen-Conrad, Kate, et al. "Bridging the Divide: Using UTAUT to Predict Multigenerational Tablet Adoption Practices." *Computers in Human Behavior* 50 (2015): 186–196. Print.

Mahowald, Mary B. "On Treatment of Myopia: Feminist Standpoint Theory and Bioethics." *Feminism and Bioethics: Beyond Reproduction.* Ed. Susan M. Worf. New York: Oxford UP, 1996. Print.

Maiwald, Eric. *Network Security: A Beginner's Guide.* 3rd ed. New York: McGraw, 2013. Print.

Mallick, Pradeep Kumar, ed. *Research Advances in the Integration of Big Data and Smart Computing.* Hershey: Information Science Reference, 2016. Print.

Malvik, Callie. "Graphic Design vs. Web Design: Which Career Is Right for You?" *Rasmussen College.* Rasmussen Coll., 25 July 2013. Web. 16 Mar. 2016.

Manjoo, Farhad. "Planet Android's Shaky Orbit." *New York Times* 28 May 2015: B1. Print.

Marchant, Ben. "Game Programming in C and C++." *Cprogramming.com.* Cprogramming.com, 2011. Web. 16 Mar. 2016.

Marchewka, Jack T. *Information Technology Project Management.* 5th ed. Hoboken: Wiley, 2015. Print.

Marsland, Stephen. *Machine Learning: An Algorithmic Perspective.* Boca Raton: Taylor, 2009. Print.

Mason, Paul. *Understanding Computer Search and Research.* Chicago: Heinemann, 2015. Print.

Matulka, Rebecca. "How 3D Printers Work." *Energy.gov.* Dept. of Energy, 19 June 2014. Web. 6 Jan. 2016.

May, Amy, and Kelly E. Tenzek. "Seeking Mrs. Right: Uncertainty Reduction in Online Surrogacy Ads." *Qualitative Research Reports in Communication* 12.1 (2011): 27–33. Print.

Mayer, Richard E. *The Cambridge Handbook for Multimedia Learning.* New York: Cambridge UP, 2005. Print.

McCauley, Renée, et al. "Debugging: A Review of the Literature from an Educational Perspective." *Computer Science Education* 18.2 (2008): 67–92. Print.

McClure, Stuart, Joel Scambray, and George Kurtz. *Hacking Exposed: Network Security Secrets & Solutions.* 7th ed. New York: McGraw, 2012. Print.

McConnell, Robert, James Haynes, and Richard Warren. "Understanding ASCII Codes." *NADCOMM.* NADCOMM, 14 May 2011. Web. 16 Feb. 2016.

McCormick, Ty. "Gamification: A Short History." *Foreign Policy.* FP Group, 24 June 2013. Web. 20 Aug. 2013.

McCracken, Harry. "Ten Momentous Moments in DOS History." *PCWorld.* IDG Consumer, n.d. Web. 31 Jan. 2016.

McDonald, Nicholas G. "Past, Present, and Future Methods of Cryptography and Data Encryption." *SpaceStation.* U of Utah, 2009. Web. 4 Feb. 2016.

McFedries, Paul. *Fixing Your Computer: Absolute Beginner's Guide.* Indianapolis: Que, 2014. Print.

McGonigal, Jane. *Reality Is Broken: Why Games Make Us Better and How They Can Change the World.* New York: Penguin, 2011. Print.

McGregor, Douglas. *The Human Side of Enterprise.* New York: McGraw-Hill, 1960. Print.

McKinnon, Alan, et al., eds. *Green Logistics: Improving the Environmental Sustainability of Logistics.* 3rd ed. Philadelphia: Kogan, 2015. Print.

McMillan, Charles. "Global Logistics and International Supply Chain Management." *Supply Chain Management, Marketing and Advertising, and Global Management.* Ed. Hossein Bidgoli. Hoboken: Wiley, 2010. 68–88. Print. Vol. 2 of *The Handbook of Technology Management.* 3 vols.

McMillan, Robert. "IBM Bets $3B That the Silicon Microchip Is Becoming Obsolete." *Wired.* Condé Nast, 9 July 2014. Web. 10 Mar. 2016.

McMurdo, Thomas, and Birdie MacLennan. "The Vermont Digital Newspaper Project and the National Digital Newspaper Program: Cooperative Efforts in Long-Term Digital Newspaper Access and Preservation." *Library Resources & Technical Services* 57.3 (2013): 148–63. Web. 21 Aug. 2013.

Menezes, Alfred J., Paul C. van Oorschot, and Scott A. Vanstone. *Handbook of Applied Cryptography.* Boca Raton: CRC, 1996. Print.

Meszaros, Gerard. *xUnit Test Patterns: Refactoring Test Code.* Pearson Education, 2009.

Metz, J. R. *The Drone Wars: Uncovering the Dynamics and Scope of United States Drone Strikes.* Diss. Wesleyan U, 2013. Print.

Metz, Sandi. *Practical Object-Oriented Design in Ruby: An Agile Primer.* Upper Saddle River: Addison, 2012. Print.

"Microprocessors." *MIT Technology Review.* MIT Technology Review, 2016. Web. 11 Mar. 2016.

"Microprocessors: Explore the Curriculum." *Intel.* Intel Corp., 2015. Web. 11 Mar. 2016.

Miles, Jeffrey A., ed. *New Directions in Management and Organizations Theory.* Newcastle upon Tyne: Cambridge Scholars, 2014. Print.

Miller, Charles, and Aaron Doering. *The New Landscape of Mobile Learning: Redesigning Education in an App-Based World.* New York: Routledge, 2014. Print.

Miller, Greg. "Making Memories." *Smithsonian* May 2010: 38–45. Print.

Miller, Luke. *The Practitioner's Guide to User Experience Design.* New York: Grand Central, 2015. Print.

Miller, Michael J. "The Rise of DOS: How Microsoft Got the IBM PC OS Contract." *PCMag.com.* PCMag Digital Group, 10 Aug. 2011. Web. 31 Jan. 2016.

Miller, Michelle D. *Minds Online: Teaching Effectively with Technology.* Cambridge: Harvard UP, 2014. Print.

Min, Seong-Jae. "From the Digital Divide to the Democratic Divide: Internet Skills, Political Interest, and the Second-Level Digital Divide in Political Internet Use." *Journal of Information Technology & Politics* 7 (2010): 22–35. Taylor & Francis Online. PDF file.

"The Mind-Blowing Possibilities of Quantum Computing." *TechRadar.* Future, 17 Jan. 2010. Web. 26 Mar. 2016.

Mitchell, Tom M. *Machine Learning.* New York: McGraw, 1997. Print.

———. *The Discipline of Machine Learning.* Pittsburgh: Carnegie Mellon U, 2006. PDF file.

Mohamed, Arif. "A History of Cloud Computing." *Computer Weekly.* TechTarget, 2000–2015. Web. 8 June 2015.

Morin, Amanda. "5 Ways Kids Use Working Memory to Learn." *Understood.* Understood.org, 16 Dec. 2013. Web. 24 Jun. 2015.

Morin, Christophe. "Neuromarketing: The New Science of Consumer Behavior." *Society* 48.2 (2011): 131–35. Print.

Morreale, Patricia, and Kornel Terplan, eds. *The CRC Handbook of Modern Telecommunications.* 2nd ed. Boca Raton: CRC, 2009. Print.

Morrison, Foster. *The Art of Modeling Dynamic Systems: Forecasting for Chaos, Randomness, and Determinism.* 1991. Mineola: Dover, 2008. Print.

"MS-DOS: A Brief Introduction." *Linux Information Project.* Linux Information Project, 30 Sept. 2006. Web. 31 Jan. 2016.

Mueller, Scott. *Upgrading and Repairing PCs.* 22nd ed. Indianapolis: Que, 2015. Print.

Mullins, Craig S. *Database Administration: The Complete Guide to DBA Practices and Procedures.* 2nd ed., Addison-Wesley, 2013.

Myers, Courtney Boyd, ed. *The AI Report.* Forbes. Forbes.com, 22 June 2009. Web. 18 Dec. 2015.

Myers, Glenford J., Tom Badgett, and Corey Sandler. *The Art of Software Testing.* Hoboken: Wiley, 2012. Print.

Nair, Mahendhiran, Mudiarasan Kuppusamy, and Ron Davison. "A Longitudinal Study on the Global Digital Divide Problem: Strategies to Close Cross-Country Digital Gap." *The Business Review* Cambridge 4.1 (2005): 315–26. PDF file.

Nasri, Wadie, and Lanouar Charfeddine. "Motivating Salespeople to Contribute to Marketing Intelligence Activities: An Expectancy Theory Approach." *International Journal of Marketing Studies* 4.1 (2012): 168. Print.

National Information Standards Organization. *Understanding Metadata.* Bethesda: NISO P, 2004. Digital file.

National Telecommunications and Information Administration. "Digital Nation: Expanding Internet Usage." *National Telecommunications and Information Administration.* US Dept. of Commerce, 2011. PDF file.

Nayeem, Sk. Md. Abu, Jyotirmoy Mukhopadhyay, and S. B. Rao, eds. *Mathematics and Computing: Current Research and Developments.* New Delhi: Narosa, 2013. Print.

Neapolitan, Richard E. *Foundations of Algorithms.* 5th ed. Burlington: Jones, 2015. Print.

Neiderreiter, Harald, and Chaoping Xing. *Algebraic Geometry in Coding Theory and Cryptography.* Princeton: Princeton UP, 2009. Print.

Neo, Emily, and Philip J. Calvert. "Facebook and the Diffusion of Innovation in New Zealand Public Libraries." *Journal of Librarianship and Information Science* 44.4 (2012): 227–37. Print.

Neuliep, James W. "The Influence of Theory X and Y Management Style on the Perception of Ethical Behavior in Organizations." *Journal of Social Behavior & Personality* 11.2 (June 1996): 301–11. Print.

New York Times Editorial Board. "Ruling Drones, Before They Rule Us." *New York Times* (January 10, 2016): SR10. http://www.nytimes.com.

Newman, Jared. "Android Laptops: The $200 Price Is Right, but the OS May Not Be." *PCWorld.* IDG Consumer & SMB, 26 Apr. 2013. Web. 27 Jan. 2016.

———. "With Android Lollipop, Mobile Multitasking Takes a Great Leap Forward." *Fast Company.* Mansueto Ventures, 6 Nov. 2014. Web. 27 Jan. 2016.

Nielson, Hanne Riis, and Flemming Nielson. *Semantics with Applications: A Formal Introduction.* New York: Wiley, 1992. Print.

Nixon, Robin. *Learning PHP, MySQL, JavaScript, CSS & HTML5: A Step-by-Step Guide to Creating Dynamic Websites.* 3rd ed., O'Reilly, 2014.

Nobel, Carmen. "Neuromarketing: Tapping into the 'Pleasure Center' of Consumers." *Forbes.* Forbes.com, 1 Feb. 2013. Web. 29 June 2015.

Noergaard, Tammy. *Embedded Systems Architecture: A Comprehensive Guide for Engineers and Programmers.* 2nd ed. Boston: Elsevier, 2012. Print.

Norstrom, Per. "Engineers' Non-Scientific Models in Technology Education." *International Journal of Technology and Design Education* 23.2 (2013): 377–90. Print.

Northrup, Tony. "Firewalls." *TechNet.* Microsoft, n.d. Web. 28 Feb. 2016.

Nystrom, Robert. *Game Programming Patterns.* N.p.: Author, 2009–14. Web. 16 Mar. 2016.

O'Brien, James, and George Marakas. *Management Information Systems.* 10th ed. New York: McGraw-Hill, 2010. Pint.

O'Connor, Joseph, and John Seymour. *Introducing NLP: Psychological Skills for Understanding and Influencing People.* San Francisco: Conary P, 2011. Print.

O'Toole, Greg. *Sustainable Web Ecosystem Design.* New York: Springer, 2013. Print.

Oki, Eiji, et al. *Advanced Internet Protocols, Services, and Applications.* Hoboken: Wiley, 2012. Print.

Olley, Allan. "Can Machines Think Yet? A Brief History of the Turing Test." *Bubble Chamber.* U of Toronto's Science Policy Working Group, 23 June 2014. Web. 18 Dec. 2015.

Oloruntoba, Samuel. "SOLID: The First 5 Principles of Object Oriented Design." *Scotch*. Scotch.io, 18 Mar. 2015. Web. 1 Feb. 2016.

Oppy, Graham, and David Dowe. "The Turing Test." *Stanford Encyclopedia of Philosophy (Spring 2011 Edition)*. Ed. Edward N. Zalta. Stanford U, 26 Jan. 2011. Web. 18 Dec. 2015.

Orris, J. B. "A Visual Model for the Variance and Standard Deviation." *Teaching Statistics* 33.2 (2011): 43–45. Print.

Orwick, Penny, and Guy Smith. *Developing Drivers with the Windows Driver Foundation*. Redmond: Microsoft P, 2007. Print.

Orzan, G., I. A. Zara, and V. L. Purcarea. "Neuromarketing Techniques in Pharmaceutical Drugs Advertising: A Discussion and Agenda for Future Research." *Journal of Medicine and Life* 5.4 (2012): 428–32. Print.

"Our Story." *Pixar*. Pixar, 2016. Web. 31 Jan. 2016.

Owen, Mark. *Practical Signal Processing*. New York: Cambridge UP, 2012. Print.

Paar, Christof, and Jan Pelzi. *Understanding Cryptography: A Textbook for Students and Practitioners*. Heidelberg: Springer, 2010. Print.

Paine, Jocelyn. "Programs that Transform Their Own Source Code; or: the Snobol Foot Joke." *Dr. Dobb's*, UBM, 17 Jan. 2010, www.drdobbs.com/architecture-and-design/programs-that-transform-their-own-source/228701469. Accessed 4 May 2017.

Pandolfi, Luciano. *Distributed Systems with Persistent Memory: Control and Moment Problems*. New York: Springer, 2014. Print.

Parashar, Manish, and Salim Hariri, eds. *Autonomic Computing: Concepts, Infrastructure, and Applications*. Boca Raton: CRC, 2007. Print.

Parent, Rick. *Computer Animation: Algorithms and Techniques*. Waltham: Elsevier, 2012. Print.

Parisi, Tony. *Learning Virtual Reality: Developing Immersive Experiences and Applications for Desktop, Web, and Mobile*. Sebastopol: O'Reilly, 2015. Print.

Park, Jung-ran. *Metadata Best Practices and Guidelines*. London: Routledge, 2011. Print.

Parker, Jason, "The Continuing Evolution of iOS." *CNET*. CBS Interactive, 7 May 2014. Web. 26 Feb. 2016.

Parrinello, Michael. C. "Prevention of Metabolic Syndrome from Atypical Antipsychotic Medications: Applying Rogers' Diffusion of Innovations Model in Clinical Practice." *Journal of Psychosocial Nursing & Mental Health Services* 50.12 (2012): 36–44. Print.

"Part Two: Communicating with Computers—The Operating System." *Computer Programming for Scientists*. Oregon State U, 2006. Web. 31 Jan. 2016.

Patel, Ruchika, and Parth Bhatt. "A Review Paper on Digital Watermarking and Its Techniques." *International Journal of Computer Applications* 110.1 (2015): 10–13. Web. 14 Mar. 2016.

Patterson, David A., and John L. Hennessy. *Computer Organization and Design: The Hardware/Software Interface*. 5th ed. Waltham: Morgan, 2013. Print.

Pelleau, Marie, and Narendra Jussien. *Abstract Domains in Constraint Programming*. London: ISTE, 2015. Print.

Peng, Chih-Wei, et al. "Review: Clinical Benefits of Functional Electrical Stimulation Cycling Exercise for Subjects with Central Neurological Impairments." *Journal of Medical and Biological Engineering* 31.1 (2011): 1–11. Print.

Pentecost, Kathryn. "Imagined Communities in Cyberspace." *Social Alternatives* 30.2 (2011): 44–47. Print.

Pentland, Alex. "The Data-Driven Society." *Scientific American* 309.4 (2013): 78–83. Print.

Pérez, André. *Mobile Networks Architecture*. Hoboken: Wiley, 2012. Print.

Perfetti, Christine, and Jared M. Spool. "Macromedia Flash: A New Hope for Web Applications." *UIE.com*. User Interface Engineering, 2002. Web. 5 Aug. 2013.

Periathiruvadi, Sita, and Anne N. Rinn. "Technology in Gifted Education: A Review of Best Practices and Empirical Research." *Journal of Research on Technology in Education* 45.2 (2012–2013): 153–69. Print.

Peters, Thomas A., and Lori Bell, eds. *The Handheld Library: Mobile Technology and the Librarian*. Santa Barbara: ABC-CLIO, 2013. Print.

Piatetsky-Shapiro, Gregory. "Knowledge Discovery in Real Databases: A Workshop Report." *AI Magazine* 11, no. 5 (January 1991).

Pigott, Terri D. *Advances in Meta-analysis*. New York: Springer, 2012. Print.

Piore, Adam. "Mind in the Machine." *Discover* June 2013: 52-59. Print.

Pollitt, Mark. "A History of Digital Forensics." *Advances in Digital Forensics VI*. Ed. Kam-Pui Chow and Sujeet Shenoi. Berlin: Springer, 2010. 3–15. Print.

Portmore, Douglas W. "Imperfect Reasons and Rational Options." *Noûs* 46.1 (2012): 24–60. Print.

Portner, Paul, Klaus von Heusinger, and Claudia Maienborn. *Semantics: An International Handbook of Natural Language Meaning*. Berlin: De Gruyter Mouton, 2013. Digital file.

Potter, Michael. "Loyalism, Women and Standpoint Theory." *Irish Political Studies* (2012): 1–17. Web. 22 July 2013.

Powell, Juliette. *33 Million People in the Room: How to Create, Influence, and Run a Successful Business with Social Networking*. Upper Saddle River: FT, 2009. Print.

Prandoni, Paolo, and Martin Vetterli. *Signal Processing for Communications*. Boca Raton: CRC, 2008. Print.

Prescott, Bonnie. "Improved Method of Electrical Stimulation Could Help Treat Damaged Nerves." *Beth Israel Deaconess Medical Center*. Beth Israel Deaconess Med. Ctr., 21 Nov. 2011. Web. 7 Oct. 2013.

"The Printed World." *Economist*. Economist Newspaper, 10 Feb. 2011. Web. 6 Jan. 2016.

Proakis, John G., and Dimitris G. Manolakis. *Digital Signal Processing: Principles, Algorithms, and Applications*. 4th ed. Upper Saddle River: Prentice, 2007. Print.

Prokopenko, Mikhail. *Advances in Applied Self-Organizing Systems*. London: Springer, 2013. Print.

Pugliese, Joseph. *State Violence and the Execution of Law: Biopolitical Caesurae of Torture, Black Sites, Drones*. New York: Routledge, 2013. Print.

Puryear, Martin. "Programming Trends to Look for This Year." *TechCrunch*. AOL, 13 Jan. 2016. Web. 7 Feb. 2016.

"A Quantum Leap in Computing." *NOVA*. WGBH/PBS Online, 21 July 2011. Web. 24 Mar. 2016.

Quian, Quiroga R., and Stefano Panzeri. *Principles of Neural Coding*. Boca Raton: CRC, 2013. Print.

Ragnarsson, K. T. "Functional Electrical Stimulation after Spinal Cord Injury: Current Use, Therapeutic Effects and Future Directions." *Spinal Cord* 46.4 (2008): 255–74. Print.

Rainie, Lee. "The State of Digital Divides." *Pew Research Internet Project*. Pew Research Center, 5 Nov. 2013. Web. 11 Nov. 2014.

Rajasekaran, Sanguthevar. *Multicore Computing: Algorithms, Architectures, and Applications*. Boca Raton: CRC, 2013. Print.

Ramirez, Artemio. "The Effect of Interactivity on Initial Interactions: The Influence of Information Seeking Role on Computer-Mediated Interaction." *Western Journal of Communication* 73.3 (2009): 300–25. Print.

Ransdell, Sarah, Brianna Kent, Sandrine Gaillard-Kenney, and John Long. "Digital Immigrants Fare Better than Digital Natives Due to Social Reliance." *British Journal of Educational Technology* 42.6 (2011): 931–38. Print.

Rao, P. N. CAD/CAM: Principles and Applications. 3rd Ed. India: Tata McGraw, 2010. Print.

Raymond, Eric S. "Origins and History of Unix, 1969–1995." *The Art of UNIX Programming*. Boston: Pearson Education, 2004. Print.

Redd, Steven P., and Alex Mintz. "Policy Perspectives on National Security and Foreign Policy Decision Making." *Policy Studies Journal* 41.5 (2013): 511–37. Print.

Reese, Terry, Jr., and Kyle Banerjee. *Building Digital Libraries: A How-to-Do-It Manual*. New York: Neal, 2008. Print.

Reimer, Jeremy. "A History of the GUI." *Ars Technica*. Condé Nast, 5 May 2005. Web. 31 Jan. 2016.

Reinhardt, W., et al. "Knowledge Worker Roles and Actions—Results of Two Empirical Studies." *Knowledge and Process Management* 18.3 (2011): 150–174. Print.

Renda, Andrea. "Competition-Regulation Interface in Telecommunications: What's Left of the Essential Facility Doctrine." *Telecommunications Policy* 34.1 (2010): 23–35. Print.

Renko, Maija, K. Galen Kroeck, and Amanda Bullough. "Expectancy Theory and Nascent Entrepreneurship." *Small Business Economics* 39.3 (2012): 667–84. Print.

Reynolds, Mark C. "Modeling the Java Bytecode Verifier." *Science of Computer Programming* 78.3 (2013): 327–42. Print.

Reza, Juan Rolando. "Java Supervenience." *Computer Languages, Systems & Structures* 38.1 (2012): 73–97. Print.

Rice, Daniel M. *Calculus of Thought: Neuromorphic Logistic Regression in Cognitive Machines*. Waltham: Academic, 2014. Print.

Richards, Jack C., and Theodore S. Rodgers. *Approaches and Methods in Language Teaching*. New York: Cambridge UP, 2001. Print.

Ringquist, Evan J. *Meta-analysis for Public Management and Policy*. San Francisco: Jossey, 2013. Print.

Ripley, Amanda. "Playing Defense Against the Drones." *Atlantic* (November, 2015). http://www.theatlantic.com/magazine/archive/2015/11/playing-defense-against-the-drones/407851/. This article discusses interactions with and reactions to drones in various places.

Roberts, Dan, and Spencer Ackerman. "NSA Mass Phone Surveillance Revealed by Edward Snowden Ruled Illegal." *Guardian*. Guardian News and Media, 7 May 2015. Web. 29 June 2015.

Roberts, Fred S., and Barry Tesman. *Applied Combinatorics*. 2nd ed. Boca Raton: Chapman, 2012. Print.

Roberts, Richard M. *Computer Service and Repair*. 4th ed. Tinley Park: Goodheart, 2015. Print.

Roblyer, M. D., and Aaron H. Doering. *Integrating Educational Technology into Teaching*. 6th ed. Boston: Pearson, 2013. Print.

Rogers, Everett M. *Diffusion of Innovations*. 5th ed. New York: Free Press, 2003. Print.

Rose, Nathan S., and Craik, Fergus I. M. "A Processing Approach to the Working Memory/Long-Term Memory Distinction: Evidence from the Levels-of-Processing Span Task." *Journal of Experimental Psychology: Learning, Memory, and Cognition* 38.4 (2012): 1019–029. Print.

Rosenberg, Marc J. *E-Learning: Strategies for Delivering Knowledge in the Digital Age*. New York: McGraw, 2001. Print.

Rosenthal, Robert, and Ralph L. Rosnow. *Artifacts in Behavioral Research*. New York: Oxford UP, 2009. Print.

Rountree, Derrick, and Ileana Castrillo. *The Basics of Cloud Computing*. Waltham: Elsevier, 2014. Print.

Ruiz, Rebecca R. "F.C.C. Sets Net Neutrality Rules." New York Times. New York Times, 12 Mar. 2015. Web. 2 Apr. 2015.

Rushton, Andrew. *VHDL for Logic Synthesis*. 3rd ed. Hoboken: Wiley, 2011. Print.

Ryan, Janel. "Five Basic Things You Should Know about Cloud Computing." *Forbes*. Forbes.com, 30 Oct. 2013. Web. 30 Oct. 2013.

Sadler, Ian. *Logistics and Supply Chain Integration*. Thousand Oaks: Sage, 2007. Print.

Saeed, John I. *Semantics*. 4th ed. Hoboken: Wiley, 2015. Print.

Sahin, Ismail I. "Detailed Review of Rogers' Diffusion of Innovations Theory and Educational Technology-related Studies Based on Rogers' Theory." *The Turkish Online Journal of Educational Technology* 5.2 (2006): 14–23. PDF file.

Salmon, Gilly. *E-Moderating: The Key to Teaching and Learning Online*. London: Kogan. 2003. Print.

Salz, Peggy Anne, and Jennifer Moranz. *The Everything Guide to Mobile Apps: A Practical Guide to Affordable Mobile App Development for Your Business*. Avon: Adams Media, 2013. Print.

Sammons, John. *The Basics of Digital Forensics: The Primer for Getting Started in Digital Forensics*. Waltham: Syngress, 2012. Print.

Sandberg, Bobbi. *Networking: The Complete Reference*. 3rd ed. New York: McGraw, 2015. Print.

Sanders, James. "Hybrid Cloud: What It Is, Why It Matters." *ZDNet*. CBS Interactive, 1 July 2014. Web. 10 Jan. 2016.

Sanders, Laura. "Brain Separates Working Memory: Hemispheres Independent in Mental Version of RAM." *Science News* 30 July 2011: 10–11. Print.

Santora, Marc. "In Hours, Thieves Took $45 Million in A.T.M. Scheme." *New York Times* 9 May 2013, New York ed.: A1. Print.

Santos, Raul Aquino, and Arthur Edwards Block, eds. *Embedded Systems and Wireless Technology*. Boca Raton: CRC, 2012. Print.

Sarkar, Advait. "The Impact of Syntax Colouring on Program Comprehension." *Psychology of Programming Interest Group*, 2015, www.ppig.org/sites/default/files/2015-PPIG-26th-Sarkar.pdf. Accessed 17 Feb. 2017.

Sarkar, Jayanta. *Computer Aided Design: A Conceptual Approach*. Boca Raton: CRC, 2015. Print.

Sasao, Tsutomu. *Switching Theory for Logic Synthesis*. New York: Springer, 1999. Print.

Sauter, Martin. *From GSM to LTE: An Introduction to Mobile Networks and Mobile Broadbands*. Hoboken: Wiley, 2011. Print.

Savage, Terry Michael, and Karla E. Vogel. *An Introduction to Digital Multimedia*. 2nd ed. Burlington: Jones, 2014. 256–58. Print.

Schensul, Stephen L., Jean J. Schensul, and Margaret D. Le Compte. *Initiating Ethnographic Research: A Mixed Methods Approach*. Lanham: AltaMira, 2012. Print.

Schewe, Jeff. *The Digital Negative: Raw Image Processing in Lightroom, Camera Raw, and Photoshop*. San Francisco: Peachpit, 2012. Print.

_____. *The Digital Print: Preparing Images in Lightroom and Photoshop for Printing*. San Francisco: Peachpit, 2013. Print.

Schildt, Herbert. *Java: The Complete Reference*. 8th ed. New York: McGraw, 2011. Print.

Schlinger, Henry D. "The Myth of Intelligence." *The Psychological Record* 53.1 (2012): 2. Print.

Schmidt, Frank L., and John E. Hunter. *Methods of Meta-Analysis*. 3rd ed. Los Angeles: Sage, 2015. Print.

Schmitt, Christopher. *Designing Web & Mobile Graphics: Fundamental Concepts for Web and Interactive Projects*. Berkeley: New Riders, 2013. Print.

Schneider, David I. *An Introduction to Programming Using Visual Basic*. 10th ed., Pearson, 2017.

Schönsleben, Paul. *Integral Logistics Management: Operations and Supply Chain Management in Comprehensive Value-Added Networks*. 3rd ed. Boca Raton: Auerbach, 2007. Print.

Schou, Corey, and Steven Hernandez. *Information Assurance Handbook: Effective Computer Security and Risk Management Strategies*. New York: McGraw, 2015. Print.

Schultz, Kenneth F., and David A. Grimes. "Blinding in Randomised Trials: Hiding Who Got What." *Lancet* 359.9307 (2002): 696–700. *RHL: The WHO Reproductive Health Library*. Web. 11 June 2015.

Schwartz, Bennett L. *Memory: Foundations and Applications*. Thousand Oaks: Sage, 2011. Print.

Scott, David Meerman. *The New Rules of Marketing and PR*. 4th ed. Hoboken: Wiley, 2013. Print.

Scott, John. "Rational Choice Theory." *Understanding Contemporary Society: Theories of the Present*. Ed. Gary Browning, Abigail Halcli, and Frank Webster. London: Sage, 2000. 126–37. Print.

Scott, Michael L. *Programming Language Pragmatics*. 4th ed., Elsevier, 2016.

Segall, Richard S., Jeffrey S. Cook, and Qingyu Zhang, eds. *Research and Applications in Global Supercomputing*. Hershey: Information Science Reference, 2015. Print.

Seidl, Martina, et al. *UML@Classroom: An Introduction to Object-Oriented Modeling*. Cham: Springer, 2015. Print.

Semela, Tesfaye, Thorsten Bohl, and Marc Kleinknecht. "Civic Education in Ethiopian Schools: Adopted Paradigms, Instructional Technology, and Democratic Citizenship in a Multicultural Context." *International Journal of Educational Development* 33.2 (2013): 156–64. Print.

Severance, Charles. "JavaScript: Designing a Language in 10 Days." *Computer* 45.2 (2012): 7–8. Print.

"Shadow RAM Basics." *Microsoft Support*. Microsoft, 4 Dec. 2015. Web. 10 Mar. 2016.

Shahzad, Farhat. "Forging the Nation as an Imagined Community." *Nations and Nationalism* 18.1 (Jan. 2012): 21–38. Print.

Sharples, Mike, et al., "Innovating Pedagogy 2012: Exploring New Forms of Teaching, Learning, and Assessment, to Guide Educators and Policy Makers." *The Open University*. Open U, 2012. Web. 5 Aug. 2013.

Shaw, Hollie. "Advertisers Are Looking inside Your Brain: Neuromarketing Is Here and It Knows What You Want." *Financial Post*. Natl. Post, 17 Apr. 2015. Web. 29 June 2015.

Shenoi, Belle A. *Introduction to Digital Signal Processing and Filter Design*. Hoboken: Wiley, 2006. Print.

Sheppard, Rob. *Digital Photography: Top One Hundred Simplified Tips and Tricks*. Indianapolis, IN: Wiley, 2010. Print.

Shin, Dong-Hee. "A Comparative Analysis of Net Neutrality: Insights Gained by Juxtaposing the US and Korea." *Telecommunications Policy* 38.11 (2014): 1117–133. Print.

Shin, Tacksoo, Mark L. Davison, and Jeffrey D. Long. "Effects of Missing Data Methods in Structural Equation Modeling with Nonnormal Longitudinal Data." *Structural Equation Modeling* 16.1 (2009): 70–98. Print.

Shinder, Deb. "So You Want to Be a Computer Forensics Expert." *TechRepublic*. CBS Interactive, 27 Dec. 2010. Web. 2 Feb. 2016.

Shinohara, Kazunori, and Hiroshi Okuda. "Dynamic Innovation Diffusion Modeling." *Computational Economics* 35.1 (2010): 51–62. Print.

Shivde, Geeta, and Michael C. Anderson. "On the Existence of Semantic Working Memory: Evidence for Direct Semantic Maintenance." *Journal of Experimental Psychology: Learning, Memory, and Cognition* 37.6 (2011): 1342–70. Print.

Shustek, Len. "Microsoft MS-DOS Early Source Code." *Computer History Museum*. Computer History Museum, 2013. Web. 31 Jan. 2016.

Shweiki, Ehyal, et al. "Applying Expectancy Theory to Residency Training: Proposing Opportunities to Understand Resident Motivation and Enhance Residency Training." *Advances in Medical Education and Practice* 6 (2015): 339–46. Print.

Silberschatz, Abraham, Peter B. Galvin, and Greg Gagne. *Operating Systems Concepts*. 9th ed. Hoboken: Wiley, 2012. Print.

Silicon Valley Historical Association. "Cisco Systems." *Silicon Valley Historical Association.* Silicon Valley Historical Assoc., 2013. Web. 29 July 2013.

Silver, H. Ward. "Digital Code Basics." *Qst* 98.8 (2014): 58–59. PDF file.

Silver, Nate. *The Signal and the Noise: Why So Many Predictions Fail—but Some Don't.* New York: Penguin, 2012. Print.

Simpson, Kyle. *JavaScript and HTML5 Now.* Sebastopol: O'Reilly Media, 2012. Kindle file.

Singel, Ryan. "Google Busted with Hand in Safari-Browser Cookie Jar." *Wired.* Condé Nast, 17 Feb. 2012. Web. 25 Sept. 2013.

———. "You Deleted Your Cookies? Think Again." *Wired.* Condé Nast, 10 Aug. 2009. Web. 25 Sept. 2013.

Singer, Natasha. "Making Ads That Whisper to the Brain." *New York Times.* New York Times, 13 Nov. 2010. Web. 7 Oct. 2013.

Singh, Ajit, D. C. Upadhyay, and Hemant Yadav. "The Analytical Data Warehouse: A Sustainable Approach for Empowering Institutional Decision Making." *International Journal of Engineering Science and Technology* 3.7 (2011): 6049–57. PDF file.

Sito, Tom. *Moving Innovation: A History of Computer Animation.* Cambridge: MIT P, 2013. Print.

Sklar, Joel. *Principles of Web Design.* 6th ed. Boston: Cengage, 2015. Print.

Sloane, Paul, ed. *A Guide to Open Innovation and Crowdsourcing: Advice from Leading Experts.* London: Kogan, 2012. Print.

Smiraglia, Richard P., ed. *Metadata: A Cataloger's Primer.* London: Routledge, 2012. Print.

Smith, Laurence. *Cryptography: The Science of Secret Writing.* New York: Dover Publications, 1971.

Smith, Mary Lee, and Gene V. Glass. "Meta-analysis of Psychotherapy Outcome Studies." *American Psychologist* 32 (1977): 752–60. Print.

Snape, Paul, and Wendy Fox-Turnbull. "Perspectives of Authenticity: Implementation in Technology Education." *International Journal of Technology and Design Education* 23.1 (2013): 51–68. Print.

Snoke, David W. *Electronics: A Physical Approach.* Boston: Addison, 2014. Print.

Soffer, Tal, Rafi Nachmias, and Judith Ram. "Diffusion of Web Supported Instruction in Higher Education—the Case of Tel-Aviv University." *Journal of Educational Technology & Society* 13.3 (2010): 212. Print.

Solnon, Christine. *Ant Colony Optimization and Constraint Programming.* Hoboken: Wiley, 2010. Print.

Sommer, Jeff. "What the Net Neutrality Rules Say." *New York Times.* New York Times, 12 Mar. 2015. Web. 2 Apr. 2015.

Song, Dong-Ping. *Optimal Control and Optimization of Stochastic Supply Chain Systems.* London: Springer, 2013. Print.

Soto, María, André Rossi, Marc Sevaux, and Johann Laurent. *Memory Allocation Problems in Embedded Systems: Optimization Methods.* Hoboken: Wiley, 2013. Print.

Sozański, Krzysztof. *Digital Signal Processing in Power Electronics Control Circuits.* New York: Springer, 2013. Print.

Spence, Ewan. "New Android Malware Strikes at Millions of Smartphones." *Forbes.* Forbes.com, 4 Feb. 2015. Web. 11 Mar. 2016.

"Spyware." *Secure Purdue.* Purdue U, 2010. Web. 11 Mar 2016.

St. Clair-Thompson, Helen, ed. *Working Memory: Developmental Differences, Component Processes and Improvement Mechanisms.* New York: Nova, 2013. Print.

St. Germain, H. James de. "Debugging Programs." *University of Utah.* U of Utah, n.d. Web. 31 Jan. 2016.

Stalling, William. *Network Security Essentials: Applications and Standards.* 5th ed. Upper Saddle River: Prentice, 2014. Print.

———. *Operating Systems: Internals and Design Principles.* Boston: Pearson, 2015. Print.

———, and Lawrie Brown. *Computer Security: Principles and Practice.* 3rd ed. Boston: Pearson, 2015. Print.

Steck, Andreas J., and Barbara Steck. *Brain and Mind: Subjective Experience and Scientific Objectivity.* Springer, 2016.

Steier, Sandy. "To Cloud or Not to Cloud: Where Does Your Data Warehouse Belong?" *Wired.* Condé Nast, 29 May 2013. Web. 9 Oct. 2013.

Steinberg, Don. *The Kickstart Handbook: Real-life Crowdfunding Success Stories.* Philadelphia: Quirk, 2012. Print.

Steinberg, Scott. "Amanda Palmer on Crowdfunding and the Rebirth of the Working Musician." *Rollingstone.com.* Rolling Stone, 29 Aug. 2013. Web. 7 Aug. 2013.

Steinhauer, Jennifer, and Jonathan Weisman. "U.S. Surveillance in Place since 9/11 Is Sharply Limited." *New York Times*. New York Times, 2 June 2015. Web. 29 June 2015.

Stephens, Rod. *Beginning Database Design Solutions*. Indianapolis: John Wiley & Sons, 2009. Print.

Stewart, Matthew. "Theories X and Y, Revisited." *Oxford Leadership Journal* 1.3 (June 2010): 1–5. Print.

Stoetzler, Marcel, and Nira Yuval-Davis. "Standpoint Theory, Situated Knowledge and the Situated Imagination." *Feminist Theory* 3.3 (2002): 315–33. Print.

Stojkovic, Zlatar. *Computer-Aided Design in Power Engineering: Application of Software Tools*. New York: Springer, 2012. Print.

Stokes, Jon. *Inside the Machine: An Illustrated Introduction to Microprocessors and Computer Architecture*. San Francisco: No Starch, 2015. Print.

———. "RAM Guide Part I: DRAM and SDRAM Basics." *Ars Technica*. Condé Nast, 18 July 2000. Web. 10 Mar. 2016.

Stokes, Jon. *Inside the Machine: An Illustrated Introduction to Microprocessors and Computer Architecture*. San Francisco: No Starch, 2015. Print.

Stonebank, M. "UNIX Introduction." *University of Surrey*. U of Surrey, 2000. Web. 28 Feb. 2016.

"Storage vs. Memory." *Computer Desktop Encyclopedia*. Computer Lang., 1981–2016. Web. 10 Mar. 2016.

Strahilevitz, Lior Jacob. "Toward a Positive Theory of Privacy Law." *Harvard Law Review* 126.7 (2013): 2010–42. Print.

Stuart, Allison. "File Formats Explained: PDF, PNG and More." *99Designs*. 99Designs, 21 May 2015. Web. 11 Feb. 2016.

Stummer, Christian, et al. "Innovation Diffusion of Repeat Purchase Products in a Competitive Market: An Agent-Based Simulation Approach." *European Journal of Operational Research* 245.1 (2015): 157–167. Print.

Stuntebeck, Vince, and Trevor Jones. "B2B Event-Driven Marketing: Triggers Your Analytics Shouldn't Miss." *MarketingProfs*. MarketingProfs, 22 Apr. 2013. Web. 7 Oct. 2013.

Suebsaeng, Asawin. "Drones: Everything You Ever Wanted to Know but Were Always Afraid to Ask." *Mother Jones* (March, 2013). http://www.motherjones.com/politics/2013/03/drones-explained. This question-and-answer article provides an overview of drones.

Suie, Daniel, Sara Elwood, and Michael Goodchild, eds. *Crowdsourcing Geographic Knowledge: Volunteered Geographic Information (VGI) in Theory and Practice*. New York: Spring, 2013. Print.

Sun, Jiming, Vincent Zimmer, Marc Jones, and Stefan Reinauer. *Embedded Firmware Solutions: Development Best Practices for the Internet of Things*. Berkeley: ApressOpen, 2015. Print.

Supino, Phyllis G. "Fundamental Issues in Evaluating the Impact of Interventions: Sources and Control of Bias." *Principles of Research Methodology: A Guide for Clinical Investigators*. Ed. Supino and Jeffrey S. Borer. New York: Springer, 2012. 79–110. Print.

Surowiecki, James. *The Wisdom of Crowds: Why the Many Are Smarter than the Few and How Collective Wisdom Shapes Business, Economies, Societies, and Nations*. New York: Doubleday, 2004. Print.

Surprenant, Aimée M., and Ian Neath. *Principles of Memory*. New York: Taylor, 2009. Print.

Sutherland, Ivan. Sketchpad: A Man-Machine Graphical Communication System. Diss. MIT. Cambridge: MIT, 1963. Print.

Swigonski, Mary E. "The Logic of Feminist Standpoint Theory for Social Work Research." *Social Work* 39.4 (1994): 387–93. Print.

Takahashi, Dean. "The App Economy Could Double to $101 Billion by 2020." *VB*. Venture Beat, 10 Feb. 2016. Web. 11 Mar. 2016.

Talbot, James, and Justin McLean. *Learning Android Application Programming: A Hands-On Guide to Building Android Applications*. Upper Saddle River: Addison, 2014. Print.

Tan, Li, and Jean Jiang. *Digital Signal Processing: Fundamentals and Applications*. 2nd ed. Boston: Academic, 2013. Print.

Tanenbaum, Andrew S., and Herbert Bos. *Modern Operating Systems*. 4th ed. Boston: Pearson, 2014. Print.

Tansey, Keith, et al. "Restorative Neurology of Motor Control after Spinal Cord Injury." *Restorative Neurology of Spinal Cord Injury*. Ed. Milan R. Dimitrijevic et al. Oxford: Oxford UP, 2012. 43–64. Print.

Tatalović, Mićo. "How Mobile Phones Increased the Digital Divide." *SciDev.Net*. SciDev.Net, 26 Feb. 2014. Web. 11 Nov. 2014.

Tatnall, Arthur, and Bill Davey, eds. *Reflections on the History of Computers in Education: Early Use of Computers and Teaching about Computing in Schools*. Heidelberg: Springer, 2014. Print.

Taylor, Jeremy J. "Confusing Stats Terms Explained: Standard Deviation." *Stats Make Me Cry.* Taylor, 1 Aug. 2010. Web. 4 Oct. 2013.

Teddlie, Charles, and Abbas Tashakkori. *Foundations of Mixed Methods Research: Integrating Quantitative and Qualitative Approaches in the Social and Behavioral Sciences.* Thousand Oaks: Sage, 2008, Print.

———. "Mixed Methods Research: Contemporary Issues in an Emerging Field." *The SAGE Handbook of Qualitative Research.* 4th ed. Eds. Norman K. Denzin and Yvonna S. Norman. Thousand Oaks: Sage, 2011. 285–300. Print.

Teo, Timothy. "An Initial Development and Validation of a Digital Natives Assessment Scale (DNAS)." *Computers & Education* 67 (2013): 51–57. Print.

Thakur, A. A. G. Banerjee, and S. K. Gupta. "A Survey of CAD Model Simplification Techniques for Physics-Based Simulation Applications." *CAD Computer Aided Design* 41.2 (2009): 65–80. Print.

Theodorelis-Rigas, Haris. "From 'Imagined' to 'Virtual Communities': Greek–Turkish Encounters in Cyberspace." *Studies in Ethnicity and Nationalism* 13.1 (Apr. 2013): 2–19. Print.

"Theories X and Y." *Economist.* Economist Newspaper, 6 Oct. 2008. Web. 17 Sept. 2013.

Thompson, Penny. "The Digital Natives as Learners: Technology Use Patterns and Approaches to Learning." *Computers & Education* 65 (2013): 12–33. Print.

Thompson. A. "Applying Rational Choice Theory to International Law: The Promise and Pitfalls." *Journal of Legal Studies* 31.1, part 2 (2002): n. pag. Print.

"Timeline of Computer History: 1963." *Computer History Museum.* Computer History Museum, 1 May 2015. Web. 23 Feb. 2016.

Toal, Ray. "Algorithms and Data Structures." *Ray Toal.* Loyola Marymount U, n.d. Web. 19 Jan. 2016.

Tomei, Lawrence A., ed. *Encyclopedia of Information Technology Curriculum Integration.* 2 vols. Hershey: Information Science Reference, 2008. Print.

Tooley, Mike. *Electronic Circuits: Fundamentals and Applications.* 4th ed. New York: Routledge, 2015. Print.

Toomey, Warren. "The Strange Birth and Long Life of Unix." *IEEE Spectrum.* IEEE, 28 Nov. 2011. Web. 28. Feb. 2016.

Toplak, Cirila, and Irena Šumi. "Europe(an Union): Imagined Community in the Making." *Journal of Contemporary European Studies* 20.1 (Mar. 2012): 7–28. Print.

Trigonis, John T. *Crowdfunding for Filmmakers: The Way to a Successful Film Campaign.* Studio City: Michael Wiese Productions, 2013. Print.

Tripathy, B. K., and D. P. Acharjya, eds. *Global Trends in Intelligent Computing Research and Development.* Hershey: Information Science Reference, 2014. Print.

Tsai, Wei-Tek, et al., eds. *Crowdsourcing : Cloud-Based Software Development.* Heidelberg: Springer, 2015. *eBook Collection (EBSCOhost).* Web. 19 June 2015.

Tu, Chih-Hsiung, and Marina McIsaac. "The Relationship of Social Presence and Interaction in Online Classes." *American Journal of Distance Education* 16.3 (2002): 131–50. Print.

Tulving, Endel. *Elements of Episodic Memory.* 1983. Oxford: Oxford UP, 2008. Print.

———. "Episodic and Semantic Memory." *Organization of Memory.* Ed. Tulving and Wayne Donaldson. New York: Academic, 1972. 381–03. Print.

———. Episodic Memory: From Mind to Brain." *Annual Review of Psychology* 53 (2002): 1–5. Print.

"Turing Test." *Encyclopædia Britannica.* Encyclopædia Britannica, 23 Sept. 2013. Web. 18 Dec. 2015.

Turing, Alan M. "Computing Machinery and Intelligence." *Mind* 59.236 (1950): 433–60. Web. 23 Dec. 2015.

Turner, L. H., and R. West. *Introducing Communication Theory.* 4th ed. New York: McGraw, 2010. 147–65. Print.

Turse, Nick, and Tom Engelhardt. *Terminator Planet: The First History of Drone Warfare, 2001–2050.* TomDispatch, 2012. Kindle file.

Tutorial on Data Representation *NTU.edu.* Nanyang Technological U, Jan. 2014. Web. 20 Feb. 2016.

Ugah, Akobundu, and Uche Arua. "Expectancy Theory, Maslow's Hierarchy of Needs, and Cataloguing Departments." *Library Philosophy and Practice* 1 (2011): 51. Print.

"Unicode 8.0.0." *Unicode.org.* Unicode, 17 June 2015. Web. 3 Mar. 2016.

"The Unicode Standard: A Technical Introduction." *Unicode.org.* Unicode, 25 June 2015. Web. 3 Mar. 2016.

United States Congress. House Committee on Oversight and Government Reform. Subcommittee on

National Security and Foreign Affairs. *Rise of the Drones: Hearing before the Subcommittee on National Security and Foreign Affairs of the Committee on Oversight and Government Reform, House of Representatives, One Hundred Eleventh Congress, second session*. Washington: GPO, 2011. Print.

United States. Federal Communications Commission. *Preserving the Open Internet: Report and Order*. Washington: GPO, 2010. Print.

"Unretouched by Human Hand." *Economist*. Economist Newspaper, 12 Dec. 2002. Web. 14 Mar. 2016.

Urdan, Timothy C. *Statistics in Plain English*. 3rd ed. New York: Routledge, 2010. Print.

User Interface Design Basics "User Interface Design Basics." *Usability*. US Dept. of Health and Human Services, 2 Feb. 2016. Web. 2 Feb. 2016.

Vacca, John R. *Computer and Information Security Handbook*. Amsterdam: Kaufmann, 2013. Print.

———, ed. *Network and System Security*. 2nd ed. Waltham: Elsevier, 2014. Print.

Van Bel, Egbert J., Ed Sander, and Alan Weber. *Follow That Customer! The Event-Driven Marketing Handbook*. Chicago: Racom, 2011. Print.

van Deursen, Alexander, and Jan van Dijk. "Internet Skills and the Digital Divide." *New Media and Society* 13.6 (2010): 893–911. PDF file.

Van Roy, Peter, and Seif Haridi. *Concepts, Techniques, and Models of Computer Programming*. MIT P, 2004.

———. *Concepts, Techniques, and Models of Computer Programming*. Cambridge: MIT P, 2004. Print.

Van Schuppen, Jan H., and Tiziano Villa, eds. *Coordination Control of Distributed Systems*. Cham: Springer, 2015. Print.

Vetterli, Martin, Jelena Kovačević, and Vivek K. Goyal. *Foundations of Signal Processing*. Cambridge: Cambridge UP, 2014. Print.

Vrbová, Gerta, Olga Hudlicka, and Kristin Schaefer Centofanti. *Application of Muscle/Nerve Stimulation in Health and Disease*. Dordrecht: Springer, 2008. Print. Advances in Muscle Research 4.

Wagner, Carl. "Choice, Chance, and Inference." *Math.UTK.edu*. U of Tennessee, Knoxville, 2015. Web. 10 Feb. 2016.

Wagner, Sean. "'Functional Electrical Stimulation' Treatment Improves Walking Ability of Parkinson's Patients." *Medical News Today*. MediLexicon Intl., 3 June 2008. Web. 7 Oct. 2013.

Wake, Lisa, *The Role of Brief Therapy in Attachment Disorders*, London: Karnac, 2010. Print.

Walker, Helen M., and Joseph Lev. *Statistical Inference*. New York: Holt, 1953. Print.

Walker, Tim. "Gamification in the Classroom: The Right or Wrong Way to Motivate Students?" *NEA Today*. Natl. Educ. Assn., 23 June 2014. Web. 28 July 2015.

Wang, C., D. M. Shannon, and M. E. Ross. "Students' Characteristics, Self-Regulated Learning, Technology Self-Efficacy, and Course Outcomes in Online Learning." *Distance Education* 34.3 (2013): 302–23. Print.

Watt, David A. *Programming Language Design Concepts*. West Sussex: Wiley, 2004. Print.

Way, Jeffrey. "9 Confusing Naming Conventions for Beginners." *Envato Tuts+*, Envato, 22 Oct. 2010, code.tutsplus.com/articles/9-confusing-naming-conventions-for-beginners–net-15584. Accessed 14 Feb. 2017.

Weisberg, Robert W., and Lauretta M. Reeves. *Cognition: From Memory to Creativity*. Hoboken: Wiley, 2013. Print.

Weisfeld, Matt. *The Object-Oriented Thought Process*. 4th ed. Upper Saddle River: Addison, 2013. Print.

Weisstein, Eric W. "Standard Deviation." *Wolfram MathWorld*. Wolfram Research, n.d. Web. 4 Oct. 2013.

Werbach, Kevin, and Dan Hunter. *For the Win: How Game Thinking Can Revolutionize Your Business*. Philadelphia: Wharton, 2012. Print.

"What Is a Driver?" *Microsoft Developer Network*. Microsoft, n.d. Web. 10 Mar. 2016.

"What Is an Event?" *Eventricity*. Eventricity Ltd, n.d. Web. 7 Oct. 2013.

"What Is Unicode?" *Unicode.org*. Unicode, 1 Dec. 2015. Web. 3 Mar. 2016.

"What Is Unix?" *Knowledge Base*. Indiana U, 2015. Web. 28 Feb. 2016.

"What Went Wrong? Finding and Fixing Errors through Debugging." *Microsoft Developer Network Library*. Microsoft, 2016. Web. 31 Jan. 2016.

Wheeler, Leigh Ann. *How Sex Became a Civil Liberty*. New York: Oxford UP, 2013. Print.

White, Ron. *How Computers Work: The Evolution of Technology*. Illus. Tim Downs. 10th ed. Indianapolis: Que, 2015. Print.

Whitson, G. M. "Artifical Intelligence." *Applied Science*, edited by Donald R. Franceschetti, Salem

Press, a Division of EBSCO Information Services, Inc., 2012, pp. 121–127.

Wilken, Rowan, and Gerard Goggin, eds. *Mobile Technology and Place*. New York: Routledge, 2012. Print.

Williams, Brad, David Damstra, and Hal Stern. *Professional WordPress: Design and Development*. 3rd ed. Indianapolis: Wiley, 2015. Print.

Williams, Christine B., and Jane Fedorowicz. "Rational Choice and Institutional Factors Underpinning State-Level Interagency Collaboration Initiatives." *Transforming Government: People, Process and Policy* 6.1 (2012): 13–26. Print.

Williams, P. John. "Research in Technology Education: Looking Back to Move Forward." *International Journal of Technology and Design Education* 23.1 (2013): 1–9. Print.

Williams, Paul. "A Short History of Data Warehousing." *Dataversity*. Dataversity Educ., 23 Aug. 2012. Web. 8 Oct. 2013.

Williams, Rhiannon, "iOS 9: Should You Upgrade?" *Telegraph*. Telegraph Media Group, 16 Sept. 2015. Web. 25 Feb. 2016.

———. "Apple iOS: A Brief History." *Telegraph*. Telegraph Media Group, 17 Sept. 2015. Web. 25 Feb. 2016.

Williams, Richard N. *Internet Security Made Easy: Take Control of Your Computer*. London: Flame Tree, 2015. Print.

Wilson, Peter. *The Circuit Designer's Companion*. 3rd ed. Waltham: Newnes, 2012. Print.

Winder, Catherine, and Zahra Dowlatabadi. *Producing Animation*. Waltham: Focal, 2011. Print.

Winograd, Morley. *Millennial Momentum: How a New Generation Is Remaking America*. New Brunswick: Rutgers UP, 2011. Print.

Winston, Patrick H. *Artificial Intelligence*. Reading, Mass. Addison-Wesley, 1999. Print.

Wolf, Marilyn. *High Performance Embedded Computing: Architectures, Applications, and Methodologies*. Amsterdam: Elsevier, 2014. Print.

Wood, David. *Interface Design: An Introduction to Visual Communication in UI Design*. New York: Fairchild, 2014. Print.

Wood, Lamont. "The 8080 Chip at 40: What's Next for the Mighty Microprocessor?" *Computerworld*. Computerworld, 8 Jan. 2015. Web. 12 Mar. 2016.

Woodill, Gary. *The Mobile Learning Edge: Tools and Technologies for Developing Your Teams*. New York: McGraw, 2011. Print.

Woods, Dan. "Why Adopting the Declarative Programming Practices Will Improve Your Return from Technology." *Forbes*. Forbes.com, 17 Apr. 2013. Web. 2 Mar. 2016.

Worstall, Tim. "Is Unix Now the Most Successful Operating System of All Time?" *Forbes*. Forbes.com, 7 May 2013. Web. 7 Mar. 2016.

Wright, Matthew. "Diversity and the Imagined Community: Immigrant Diversity and Conceptions of National Identity." *Political Psychology* 32.5 (Oct. 2011): 837–62. Print.

Wright, Oliver. "Hacking Scandal: Is This Britain's Watergate?" *Independent* 9 July 2011, Independent.co.uk. Web. 1 Aug. 2013.

Wright, Tim. *Learning JavaScript: A Hands-On Guide to the Fundamentals of Modern JavaScript*. Boston: Addison, 2012. Print.

Wysocki, Robert K. *Effective Project Management: Traditional, Agile, Extreme*. 7th ed., John Wiley & Sons, 2014.

Yamamoto, Jazon. *The Black Art of Multiplatform Game Programming*. Boston: Cengage, 2015. Print.

Yetiv, Steve A. *National Security through a Cockeyed Lens: How Cognitive Bias Impacts US Foreign Policy*. Johns Hopkins UP, 2013.

Yi, Esther. "Inside the Quest to Put the World's Libraries Online." *Atlantic*. Atlantic Monthly Group, 26 July 2012. Web. 21 Aug. 2013.

Yoo, Christopher S. "Protocol Layering and Internet Policy." *University of Pennsylvania Law Review* 161.6 (2013): 1707–71. Print.

Young, Thomas Elliott. *The Everything Guide to Crowdfunding: Learn How to Use Social Media for Small-Business Funding*. Avon: F+W Media, 2013. Print.

Yu, Cheng, and Song Jia. *Computer Aided Design : Technology, Types, and Practical Applications*. New York: Nova, 2012. eBook Collection (EBSCOhost). Web. 5 June 2015.

Yule, Daniel, and Jamie Blustein. "Of Hoverboards and Hypertext." *Design, User Experience, and Usability. Design Philosophy, Methods, and Tools*. Berlin: Springer, 2013. Print.

Zakas, Nicholas C. "Cookies and Security." *NCZOnline*. Zakas, 12 May 2009. Web. 25 Sept. 2013.

———. "HTTP Cookies Explained." *NCZOnline*. Zakas, 5 May 2009. Web. 25 Sept. 2013.

———. *The Principles of Object-Oriented JavaScript*. No Starch Press, 2014.

Zelkowitz, Marvin V., ed. *The Internet and Mobile Technology*. Burlington: Academic, 2011. Print.

Zeller, Andreas. *Why Programs Fail: A Guide to Systematic Debugging*. Burlington: Kaufmann, 2009. Print.

Zemsky, Robert. "With a MOOC MOOC Here and a MOOC MOOC There, Here a MOOC, There a MOOC, Everywhere a MOOC MOOC." *Jour. of General Ed.* 63.4 (2014): 237–243. Print.

Zetter, Kim, and Brian Barrett. "Apple to FBI: You Can't Force us to Hack the San Bernardino iPhone." *Wired*. Condé Nast, 25 Feb. 2016. Web. 15 Apr. 2016.

Zhang, Xiaojun, et al. "Using Diffusion of Innovation Theory to Understand the Factors Impacting Patient Acceptance and Use of Consumer E-Health Innovations: A Case Study in a Primary Care Clinic." *BMC Health Services Research* 15.1 (2015): 1–15. PDF file.

Zichermann, Gabe, and Christopher Cunningham. *Gamification by Design: Implementing Game Mechanics in Web and Mobile Apps*. Sebastopol: O'Reilly, 2011. Print.

———, and Joselin Linder. *The Gamification Revolution: How Leaders Leverage Game Mechanics to Crush the Competition*. New York: McGraw, 2013. Print.

Zimmermann, Thomas Ede, and Wolfgang Sternefeld. *Introduction to Semantics: An Essential Guide to the Composition of Meaning*. Boston: De Gruyter Mouton, 2013. Print.

Zoccai, Giuseppe Biondi. *Network Meta-Analysis: Evidence Synthesis with Mixed Treatment Comparison*. Hauppage: Nova, 2014. Print.

INDEX

Note: Locators in bold at the beginning of entries indicate the actual description of those terms.

2D
   animation, 44–46
   line drawings, 56
   vector-based graphics, 53
   VR (virtual reality), 18
3D printing, **1–3**
   advantages over subtractive manufacturing, 1, 3
   alternative methods, 3
   extrusion printing, 3
   liquid, 1–3
   powder-based, 3
   printers, 3, 57
   software and modeling, 1
   uses, 3
   See also fused deposition modeling (FDM); selective layer sintering (SLS); stereolithography (SLA)
3D rendering, 44, 45, 46
4D BIM (building information modeling), 55, 57

A-0, 295
abstractions, 156, 222–224
additive manufacturing (AM). *See* 3D printing
Adleman, Leonard, 34
Adobe
   Creative Cloud, 11, 31, 304
   Flash, 46, 145, 165, 285, 288
   Photoshop, 11, 31, 99, 141, 301
agile software development, 66–67, 126, 260
al-Khwarizmi, Muhammad ibn Musa, 4
algorithm(s), **4–6**, 37–38, 48, 50
   applications, 5–6
   cipher, 32
   conditional processes, 4–5
   data-mining, 79
   deterministic, 4, 5
   distributed, 4, 5
   elegance, 4, 5
   evolutionary, 5
   function, 4
   goodness, 4, 5
   iterative, 5
   machine-learning, 173
   nondeterministic, 5
   quantum, 228, 229
   quicksort, 5
   recursive, 4, 5
   state, 4
   types, 5
   See also combinatorics
American Standard Code for Information Interchange (ASCII), **6–8**, 274, 275, 276, 296
   character encoding, 6–7
   codes, 7
   standard, 7, 21
   uses, 8
American Standards Association (ASA), 7
analytic combinatorics, 37, 38
Anderson, Benedict, 147
Android OS, **8–10**
   application program interface (API), 8, 9
   "cloud rot," 10
   features, 9–10
   "forking," 10
   future of, 10
   merging with Chrome, 10
   open-source development, 9–10
   origin and installation, 9
   popularity of, 9, 10
   pros and cons of, 10
   swype keyboard, 9
animation variables (avars), 16, 17, 44–45
annotation, 35, 39, 40, 243
Apple, 276
   iPhone, 9, 11, 158–160, 165, 221, 305
   iOS, 9, 10, 158–160, 278
   iTunes, 303
   OS X, 278
   Pay, 305
   See also iOS
application program interface (API), 8, 9
application suite, 11, 13

Index

application(s), **11–13**
  careers in, 13
  cross-platform, 11
  evolution of, 11–13
  security implications, 13
  software, 11–13
  suite, 11, 13
  types, 13
  web, 11, 13, 30, 165, 226, 227
  See also mobile apps
app(s). See application(s)
arrays, 41, 42, 43, 144, 253, 268
artificial intelligence (AI), 14, 16, 78, 182, 269, 270
Atkinson, Richard C., 289
Atkinson, William, 301
attacking/hacking, 13, 51–53, 94
  See also malware
autonomic components (ACs), 14, 15, 16
autonomic computing, **14–16**
  design of, 14–16
  feedback control system, 15
  multi-agent system, 14, 15–16
  peer-to-peer (P2P) systems, 16
  promise of, 16
  self-management, 14–16
avatars, **16–17**
  animation of, 17
  See also virtual reality (VR)

Backus, John, 295
Baddeley, Alan, 289
Bandler, Richard, 207
Barricelli, Nils, 78
BASIC (Beginner's All-purpose Symbolic Instruction Code), 223, 297
basic input/output system (BIOS), 132, 195, 232
Bayes, Thomas, 173, 174
Beauty and the Beast, 44
Berger, Charles, 272, 273
Betist, Pim, 71
binary hexadecimal representations, **19–22**
  binary number system (base-2 system), 19–22
  decimal number system (base-10 system), 19–20, 22
  hexadecimal number system (base-16 system), 19, 22
binary number system (base-2 system), 19–22
  disadvantage, 22
  importance, 20

binder jetting, 1, 3
biometrics, 51, 101
bit width, 6, 7
bit(s), 7, 19, 20
BitTorrent, 303
Boole, George, 169, 170
Boolean expressions, 23–24, 25, 42, 60, 61, 143
Boolean logic, 25, 43, 169–170
Boolean operators, **23–24**, 41, 42, 43, 60, 142, 143
  conditional statements, 23, 24
    if, 23, 24, 143, 253, 254, 265
    if-else, 23, 24, 43
    if-then-else, 25, 26–27, 42, 59–61
  control flow, 23
  precedence, 23
  short-circuiting, 23
  uses, 23–24
bootstrapping, 14, 16
branching logic, **25–27**, 61
  drawbacks of, 26
  importance of, 27
  uses of, 26, 27
Bush, George W., 221
byte, 19, 20, 152

C, 243, 277, 291, 298
C++, 136, 139, 163, 215, 223, 243, 285, 300
Caesar, Julius, 32, 76–77
  Caesar cipher, 76
Calebrese, Richard, 272, 273
Cars, 45
Cascading Style Sheets (CSS), 284, 288
Catmull, Ed, 44
central processing unit (CPU), 68–70, 87–88, 184, 193–196, 231
  See also CPU design
character encoding, 6–8, 274
  See also American Standard Code for Information Interchange (ASCII); Morse code; Unicode
characters, **28–29**, 6–8, 279, 280
  special, 29, 274
  uses, 29
  See also character encoding
Chomsky, Noam, 239
Chrome, 9, 10
cipher algorithm, 32
class, 35, 39, 200, 215, 217, 218
class-based inheritance, 215–216

cloud computing, **30–31**, 81
   advantages, 31
   cloud networks
      design, 30
      types, 30, 31
   disadvantages, 31
   as service
      infrastructure (IaaS), 30–31
      platform (PaaS), 30, 31
      software (SaaS), 30, 31
   third-party data center, 30
"cloud rot," 10
COBOL (Common Business-Oriented Language), 218, 223, 296
coding, **32–34**
   rules, 32–33
   theory, 37
   See also encryption
color coding, **35–37**
   approaches to, 35–36
   for clarity, 35
   commonly colored elements, 36
   Finnish experiment, 35
   goal of, 35
   semantic highlighting, 35, 36
   syntax highlighting, 35, 36
   uses, 36–37
combinatorics, **37–39**
   analytic, 37, 38
   applications of
      algorithm design, 38–39
      coding, 39
      graph theory, 37, 38
      Sudoku puzzles, 38
   basics of, 37–38
   combinatorial design, 37, 38
   enumerative, 37, 38, 39
   Konigsberg bridge problem, 38
   permutations, 38, 39
   See also algorithms
command, 28, 29, 35, 39, 40, 86, 240, 243
command line/command-line interface, 11, 104, 105, 138, 139
command-line interpreter, 276, 278
comment programming, **39–41**
   comments
      code annotation, 40
      creation, 40
      "header," 40
   software development, 40–41
   useful, 40, 41
Communications Act (1934), 204
comparison operators, **41–43**
   power of comparison, 43
   uses, 42–43
Compatible Time-Sharing System (CTSS), 296
computer-aided design (CAD), **53–57**, 1
   applications, 53–54, 56
   process
      drafting, 56
      solid modeling, 55, 56
      wire-frame model, 56
   SketchUp, 54
   See also Computer-aided design and computer-aided manufacturing software (CAD/CAM)
Computer-aided design and computer-aided manufacturing software (CAD/CAM), **55–57**
   applications, 55–56, 57
      four-dimensional building information modeling (4D BIM), 55, 57
      rapid prototyping, 55, 57
   origin, 55
   process, 56–57
      computer numerical control (CNC), 56–57
      drafting, 56
      solid modeling, 55, 56
      wire-frame model, 56
   rasters, 55, 56
   vectors, 55, 56
computer-aided manufacturing (CAM), 56–57
   applications, 56, 57
   process
      computer numerical control (CNC), 56–57
   See also Computer-aided design and computer-aided manufacturing software (CAD/CAM)
computer animation, **44–46**
   animation variables (avars), 17, 44–45
   in films, 44, 45
   history of, 44
   process, 44, 45–46
   storyboards, 45
   three-dimensional (3D), 45
   tools, 46
   two-dimensional (2D), 45–46

Index

computer-assisted instruction (CAI), **57–59**
  adaptive instruction, 59
  advantages, 58–59
    speech recognition, 58
    word prediction, 57, 58
  disadvantages, 59
  distance education, 59
  learner-controlled programs, 57, 58
  learning
    strategy, 57
    style, 57
  pedagogy, 57
  traditional instruction vs., 58
computer memory, **46–48**
  flash, 46, 48
  nonvolatile, 46, 47
  random access (RAM), 46, 47
  solid state disks (SSDs), 48
  virtual, 46, 47
  volatile, 46, 47
computer modeling/computer models, **48–50**
  creation, 50
  importance, 50
  Manhattan Project, 48
  mathematical models, 49
  simulations, 48–49
  types
    continuous, 49
    dynamic, 49
    farming, 49–50
    static, 49
computer security, **51–53**
  goal of, 51, 53
  hardware security
    biometrics, 51
    device fingerprinting, 51
    intrusion detection system, 51
    trusted platform module (TPM), 51
  network security, 51–52
    backdoors, 51, 52, 133
    phishing, 51, 52, 206
    principle of least privilege, 51, 52
  software security, 52
Computer Security Institute (CSI), 205
conditional expressions, 41, 42, 59–61, 142, 143
  See also conditional operators
conditional operators, **59–61**
  applications, 60–61
  benefits of simplified code, 61

bool, 59, 60, 61
if-then-else construct, 59–61
implementation, 60
operand, 60–61
ternary operator, 60
uses, 60
See also conditional expressions
constraint programming, **62–63**
  constraints, 62, 63
  domain, 62, 63
  feasibility vs. optimization, 63
  functional programming, 62
  imperative programming, 62
  map coloring, 62–63
  models
    perturbation, 62
    refinement, 62
control characters, 6, 274
control systems, **64–66**
  applications, 65, 66
  autonomous agent, 64
  fault detection, 64
  linear, 65
  performance, 64–65
    actuators, 64, 65
    automatic sequential control system, 64, 65
  system agility, 64
  system identification, 64
  types
    closed-loop, 64, 65
    open-loop, 64
cookies, 145–146
  first-party, 145
  persistent, 145
  session, 145
  third-party, 145–146
  See also HTTP (Hypertext Transfer Protocol) cookie
Cormier, Dave, 177
cowboy coding, **66–68**
  agile development, 66–67
  controversies, 67–68
  process, 66–67
  uses, 67
  waterfall development, 66
CP/M, 298
CPU (central processing unit) design, **68–70**
  goals, 69–70
    control unit design, 68, 69
    datapath design, 68, 69

logic implementation, 68, 69
protocol processors, 68, 69–70
instruction sets
microcode, 70
random logic, 70
microcontroller, 68, 69
Moore's law, 70
peripherals, 68, 70
reduced instruction set computer (RISC) chips, 70
Craik, Fergus, 168
crowdfunding, **71–72**, 73
donation-based, 71
investment, 71–72
uses, 72
websites, 72
crowdsourcing, **72–74**
cloud-based, 74
crowd creation, 73
crowd voting, 73
Internet resources for, 74
process, 73
uses, 74
"What's on the Menu," 74
cryptography, **74–77**
asymmetric/public-key, 32, 75, 76
RSA public-key cipher, 34
cipher machines/rotor machines
Enigma machine, 75
ciphertext, 32–34, 75, 117–118
computer-age, 75–76
digital signature, 76, 119
decryption, 32, 34, 75, 76–77, 117, 118
hash function, 74, 75, 76
importance of, 77
origin, 32, 75
substitution ciphers, 32–33, 34, 75, 76
symmetric/secret-key, 32, 75, 76
techniques, 75
transposition ciphers, 32, 34, 75, 76
See also coding; encryption
customer relationship management (CRM), 79, 122
cybercrimes, 93, 94, 129, 131, 174

Dahl, Ole-John, 297
data mining, **78–80**
algorithms, 79
applications

customer relationship management (CRM), 79
detection of fraud and terrorists, 80
direct marketing, 80
genomics and proteomics, 80
healthcare, 80
risk and collection analytics, 80
boosting, 79
bootstrap aggregation, 79
controversies
Health Insurance Portability and Accountability Act (HIPAA) (1996), 80
history, 78–79
"knowledge discovery in databases" (KDD), 78
machine learning vs., 173–174
problems addressed
predictive modeling, 79
segmentation (data clustering), 79
summarization, 79
visualization, 79
data source, 48, 50
data warehouse, **80–81**
data marts, 81
database design, **82–83**
database structuring conventions, **83–85**
attributes, 83–84
foreign key, 83, 84
primary key, 83, 84
entities, 83–84
optimization by normalization, 84–85
triggers, 83, 84
Davidson, Donald, 239, 240
debugging, **85–87**, 13, 35, 40, 119, 120, 160, 162, 243, 244, 260
bugs, 85, 86
compilation errors, 86
identification and addressal of, 86–87
logic errors, 86
run-time errors, 86
Y2K, 86
delta, 86, 87
in-circuit emulator, 86, 87
integration testing, 86, 87
memory dumps, 86
software patches, 86
decimal number system (base-10 system), 19–20, 22
decryption, 32, 34, 75, 76–77, 117, 118
See also cryptography; encryption
Dennard, Robert, 230

deterministic algorithm, 4, 5
device drivers, **87–89**
   device managers, 87, 89
   input/output instructions, 87–88
   interface, 87, 88, 89
   making of, 89
   peripherals, 88
   virtual, 88, 89
   working mechanism, 88
Devlin, Barry A.
   "Architecture for a Business and Information System, An," 80
diffusion of innovations (DOI), **89–90**
   adoption/adopters, 90
   concerns-based adoption model (CBAM), 90
   stages, 90
   theory, 90
digital divide, **91–92**, 97
digital forensics, **93–95**
   cybercrimes, 93, 94, 129, 131, 174
   Electronic Communications Privacy Act (ECPA) (1986), 93, 94
   evolution of, 93–94
   future challenges, 94–95
   metadata, 93, 94
   policy, 94
   Scientific Working Group on Digital Evidence (SWGDE), 93, 94–95
   techniques
      logical copying, 93, 94
      physical copying, 94
digital libraries, **95–96**
digital native, **96–97**
   digital divide, 97
   digital immigrants, 96, 97
   generations
      first vs. second (2DNs), 97
digital photography, **98–99**
   digital cameras
      advantages over film cameras, 99
      origin, 98
      working mechanism, 98–99
digital signal processors (DSPs), **99–101**
   applications, 100–101
   biometric scanning, 101
   working mechanism, 100
      fixed-point arithmetic, 99, 100
      floating-point arithmetic, 99, 100
      Harvard architecture, 99, 100
      multiplier-accumulator, 99–100
      pipelined architecture, 100
      semiconductor intellectual property (SIP) block, 100
digital watermarking, **101–103**
   applications, 102, 103
   carrier signal, 101 102
   classification
      1-bit vs. multibit, 101, 102
      robust vs. fragile, 102, 103
   crippleware, 101, 103
   noise-tolerant signal, 101
   qualities of, 102–103
   reversible data hiding, 101, 102–103
   technique, 102
Dijkstra, Edsger, 297
Dingli, Alexiei
   New Digital Natives: Cutting the Chord, The, 97
directed energy deposition (DED), 1, 3
disk operating system (DOS), **103–105**
   graphical user interface (GUI), 104, 105
   MS-, 103, 257, 299
      auxiliary utility programs, 104
      command line, 104, 105
      development, 104
      impact of, 105
      nongraphical systems, 104, 105
      parts, 104
      shell, 104
      use, 105
   QDOS (Quick and Dirty Operating System), 103, 104, 105
distributed algorithm, 4, 5
DRAKON chart, 4
drone warfare, **105–106**
   criticism of drone strikes, 106
   Vietnam War, 106
   See also drones
drones, **107–108**
   history of, 107
   Israeli-developed, 106
   problems with, 108
   rules for, 108
   unmanned aerial vehicles (UAVs), 105, 106, 107
   uses, 105–106, 107–108
   See also drone warfare
Dube, Chetan, 16

Eastman, George, 98
e-banking, **110**
Eich, Brendan, 164, 165
e-learning, **111–112**
   See also massive open online courses (MOOCs)
electronic circuits, **113–115**
   applications of, 114–115
   counter, 113, 114
   electric vs., 113
   integrated circuits (IC), 113–114
   logic gates, 114
      BCD-to-seven-segment decoder/driver, 113, 114
      inverter, 113, 114
      negative-AND (NAND) gate, 113, 114
   programmable oscillator, 113, 114
   retriggerable single shot, 113, 114
electronic communication software, **115–116**
   asynchronous communication, 115–116
      e-mail, 115
      multicast, 115, 116
      push technology, 115
      Short Message Service (SMS), 115, 116
   privacy concerns, 116
   synchronous communication, 116
      multicast, 115, 116
      video calling, 116
      voice over Internet Protocol (VoIP), 115, 116
Electronic Communications Privacy Act (ECPA) (1986), 93, 94, 115, 116
Electronic Recording Machine, Accounting (ERMA), 296
encryption, **32–34, 117–119**, 31, 75–76, 77
   authentication and security, 117, 118–119
   ciphertext, 32–34, 75, 117–118
   hardware vs. software, 119
   hashing algorithm, 117, 118
   Pretty Good Privacy (PGP), 117, 118–119, 302
   rules, 32–33
   substitution, 32–33, 34, 75, 76
   systems in practice, 118–119
   transposition, 32, 34, 75, 76
   types
      asymmetric/public-key, 32, 34, 75, 76, 117–118
      symmetric/secret-key, 32, 75, 76, 117
   See also coding; decryption
enumerative combinatorics, 37, 38, 39
episodic memory, 238–239

error handling, **119–121**
   in computer programs, 119–120
      logic errors, 86, 87, 119, 120
      syntax errors, 86, 87, 119, 120
   debugging, 119, 120
   error handler, 119, 120, 121
   in practice, 121
   for software improvement, 121
   structured, 120–121
      try/catch blocks, 120, 121
   unstructured, 120
event-based marketing. *See* event-driven marketing (EDM)
event-driven marketing (EDM), **121–122**
   National Australia Bank (NAB), 122
   Teradata, 122
evolutionary algorithm, 5
expectancy theory, **123–124**
   applications, 124
   components, 123
   theoretical implications, 123
experimenter's bias, **124–125**
   interpreter bias, 124, 125
   observer bias, 124, 125
   self-fulfilling prophecy, 124, 125
      Pygmalion effect, 125
extreme programming (XP), **126–127**, 247
   activities, 126–127
   drawbacks, 127
   flexible development strategy, 127
   iterations, 126, 127
   pair programming, 126, 127
   in real-world scenarios, 127
   unit tests, 126, 127
   values of, 126

Farber, David, 252
Federal Aviation Administration (FAA), 108
Federal Communications Commission (FCC), 204
firewalls, **129–131**, 31, 52, 205, 206
   and computer security, 131
   history of, 129
   proxy server, 129
   types, 129–131
      application-level, 129, 130
      host-based, 129, 130
      network, 129
      packet filters, 129, 131
      stateful filters, 129, 131

firmware, **131–133**
   in automobile manufacturing, 133
   basic input/output system (BIOS), 132
   embedded systems, 131, 132
   flashing, 131, 132, 133
   free software, 131, 132
   in gaming consoles, 132, 133
   homebrew, 131–133
   modification and replacement, 132–133
   nonvolatile memory, 131, 132
   vulnerability of, 133
      "backdoors," 133
flash memory, 46, 48
flowchart, 4, 5, 58
"forking," 10
FORTRAN, 218, 223, 295
FreeBSD, 302
functional electrical stimulation (FES), **133–134**
   applications, 134
   development, 133
functional neuromuscular stimulation (FNS).
        See functional electrical stimulation (FES)
fused deposition modeling (FDM), 2, 3
Futureworld, 44

Gabriel, Richard P., 291
game programming, **135–136**
   as a career, 136
   game loop, 135
   homebrew, 135
   object-oriented programming (OOP), 135
   process, 135–136
   pseudocode, 135, 136
   source code, 135–136
   tools, 136
gamification, **137–138**
Gardner, Harvard Howard, 261–262
   Frames of Mind: The Theory of Multiple Intelligences, 261
GNU (for Gnu's Not Unix), 299
Google, 5, 8, 258, 263
   Android OS, 8–10
   App Engine, 31
   Chrome, 9, 10
graph theory, 37, 38
Graphical user interface (GUI), **138–140**
   application-specific, 138, 140
   direct manipulation interfaces, 138, 139
   elements of, 139

   future of, 140
   interface metaphors, 138, 139
   object-oriented user interfaces (OOUIs), 138, 139
   origin of, 139
   text interface vs., 138–139
   user-centered design, 138, 139
   WIMP (windows, icons, menus, and pointer objects), 140
graphics formats, **140–142**
   color models
      CMYK, 141, 142
      RGB, 140, 141, 142
   compressed data, 140
   digital imaging, 140–141
   GIF (Graphics Interchange Format), 142
   image compression, 141–142
      lossless, 140, 141, 142
      lossy, 140, 142
      LZW, 140, 141, 142
   JPEG, 142
   SVG (Scalable Vector Graphics), 142
Grinder, John, 207
Griswold, Ralph, 252, 254
guard close, **142–144**
   Boolean operators, 143
   conditional expression, 143
   creation of, 143–144
   replacement of nested conditional statements, 143
   uses, 143, 144

hacking/attacking, 13, 51–53, 94
   See also malware
Hadoop, 303
Hall, Gene, 90
hamming distance, 6, 7
Hartsock, Nancy, 250
Hegel, Georg Wilhelm Friedrich, 249
hexadecimal number system (base-16 system), 19, 22, 274, 275, 276
Hitch, Graham, 289
Hopper, Grace, 295
HTTP (Hypertext Transfer Protocol) cookie, **145–146**
Hull, Chuck, 1
hybrid clouds, 30, 31
HyperCard, 301
hypertext markup language (HTML), 164, 165, 183, 284, 285, 287, 288, 305

IBM (International Business Machines
    Corporation), 7, 8, 14, 31, 104, 230, 253,
    295–301, 303, 304
  CICS (Customer Information Control System), 298
  FORTRAN project, 295
  SABRE, 297
if-else statements, 23, 24, 43
if-statements, 23, 24, 143, 253, 254, 265
if-then-else statements, 25, 26–27, 42, 59–61
imagined communities, **147–148**
immersive mode, 8, 9
"inbetweening". *See* keyframing
incremental development, **149–151**
  advantages, 149–151
  disadvantages, 150
  in practice, 150
  software development life cycle (SDLC), 149
  waterfall development, 149–151
Information and Communication Technology
    (ICT), 91–92
Information technology (IT), **151–152**
  data production growth, 152
  dot-com bubble, 152
  history of, 151–152
  telecom equipment, 151, 152
information visualization, **153–154**
Inmon, William H.
  Building the Data Warehouse, 80
integrated development environments (IDEs),
    35, 37
Internet Protocol (IP), **154–155**
  address identification, 155
  datagrams, 155
  IPv4, 154–155
  IPv6, 155
  Transmission Control Protocol (TCP), 155
Internet Protocol Interoperability and Collaboration
    System (IPICS), 204
interpolation, 28, 29
inversion of control, **155–157**
  benefits over traditional programming, 156–157
  coupling, 156
  decoupling, 156–157
  drawbacks, 157
  "Hollywood principle," 156
  object-oriented programming, 156
  process, 156
  run-time binding, 156, 157
  uses, 157

iOS, **158–160**
  capacitive touch screen system, 158
  jailbreaking, 158–159
  multitouch gestures, 158
  new innovations, 159–160
  3D Touch, 158, 160
  version updates, 159–160
  widgets, 158, 159
iterative algorithm, 5
iterative constructs, **160–162**
  loops
    considerations with, 161
    event, 160, 161
    for, 160, 161
    infinite, 161
    in practice, 161–162
    in software development, 162
    while, 160, 161
Iverson, Kenneth
  Programming Language, A, 296

Java programming language, **163–164**
  comments in, 243–244
  development, 163–164
  Java Virtual Machine (JVM), 164
  1.0, 302
JavaScript, **164–165**, 42, 161, 223, 243, 284, 285,
    288, 301, 302

Kapor, Mitch, 299
Karem, Abraham, 106
Kemeny, John, 297
keyframing, 17, 44, 46
Kilby, Jack, 114
Kildall, Gary, 298
Kimball, Ralph, 81
knowledge worker, **166**
Kurtz, Thomas, 297

Labyrinth, 44
levels of processing theory, **168**
  elaborative rehearsal, 168
  "multi-level" theory vs., 168
Liberson, W. T., 133
Linux, 9, 11, 89
  kernel, 301
Lockhart, Robert, 168
LOGO, 297
logic synthesis, **169–170**

logistics, **170–172**
   in business sectors, 171–172
      disposal, 172
      distribution, 171–172
      emergency, 172
      green, 172
      inbound, 171
      outbound, 171
      procurement, 171
      production, 171, 172
      reverse, 172
   management processes, 170–172
Lotus 1-2-3, 299

Mac OS, 11, 138
   X, 303
machine learning, **173–174**, 14, 16, 78, 79
   algorithms, 173
      categories of, 173
      decision-tree, 173
      perceptron, 173
      support vector machines (SVMs), 173
   data mining vs., 173–174
machining. *See* subtractive manufacturing
malware, **174–176**
   addressing threat from, 176
   Heartbleed bug, 305
   history of, 175–176
   malicious programming, 175
   types
      adware, 174, 175, 176
      ransomware, 175, 176
      scareware, 175, 176
      spyware, 175, 176
      viruses, 175
      worms, 175
   Stuxnet virus, 175, 304
   zombie computers, 175, 176
Martin, Robert Cecil, 216, 264, 265
massive open online courses (MOOCs), **177–179**, 59, 112
   criticism of, 178
   features of, 177
   providers, 177
   traditional schooling vs., 177–178
Material Design, 8, 10
material extrusion, 1, 3
material jetting, 1, 3
Mathematica, 301

MATH-MATIC, 295
Matlab, 300
McGregor, Douglas, 262, 263
   Human Side of Enterprise, The, 262
meta-analysis, **179–180**
metacomputing, **181–182**
   domain-dependent complexity, 181
   meta-complexity, 181
   middle computing, 181
   mimicking the brain, 182
   necessity for, 181
   by networking, 181
   supercomputers, 181, 185, 301
   ubiquitous computing, 181, 182
metadata, **183–184**, 93, 94, 96, 98, 155
   categories
      administrative, 183
      structural, 183
   privacy concerns, 183
   regulations on collection of, 183–184
   uses of, 183
microprocessors, **184–186**
   advancement of, 186
   capacity of, 185
      million instructions per second (MIPS), 184, 185
   characteristics of, 185–186
      clock speed, 184, 186
      data width, 184, 186
   central processing unit (CPU), 184, 185
   functions of, 184–185
   history of, 185
   transistors, 184, 185
Microsoft (MS), 10, 276
   -DOS, 103–105, 257, 299
      auxiliary utility programs, 104
      command line, 104, 105
      development, 104
      impact of, 105
      nongraphical systems, 104, 105
      parts, 104
      shell, 104
      use, 105
   Office, 11, 257
      Excel, 11, 301
      365, 304
      Word, 11, 176, 299, 301
   Windows
      95, 105
      NT, 302

3.0, 301
XP, 303
Visual Studio, 302
mixed methods research (MMR), **186–187**
   challenges of, 187
   paradigm pluralism, 187
   triangulation, 187
mobile apps, **188–190**
   emulators, 188, 190
   market for, 190
   mobile website, 188, 189
   platform lock, 190
   and social change, 190
   types of mobile software, 188–190
   utility programs, 188, 189
mobile computing, 8, 9, 158, 192
mobile technology, **191–192**
   applications, 191–192
   devices, 191–192
   networks, 191, 192
   operating systems, 192
   See also smartphones
Moe, J. H., 133
Montague, Richard, 239, 240
Moore, Gordon, 70, 186
   Moore's law, 186
Morse, Samuel, 6
   Morse code, 6–7
motherboards, **193–195**
   BIOS, 195
   core voltage, 193, 194
   crosstalk, 193, 194
   designing of, 194
   evolution of, 193
   forms of, 194
   functions of, 193
   overclocking, 194
   printed circuit board (PCB), 193, 194
   trace impedance, 193, 194
   tuning, 193, 194
multi-agent system, 14, 15–16
multiprocessing operating system (OS), **195–197**
   communication architecture, 195
   coupling, 195, 196–197
      loose, 196
      tight, 196–197
   multitasking, 197
   parallel processing, 195
   processor symmetry, 195, 197
   symmetric multiprocessing (SMP), 197
   single processor OS vs., 195–196
multitasking, 9, 97, 158, 197, 277, 278
multitenancy, 30, 31
multitouch gestures, 9, 10, 158
multi-user operating system (OS), **197–199**, 276, 277, 278
   cost effectiveness, 199
   development of, 198
   multiterminal configuration, 197, 198
   resource allocation, 198, 199
   shared computing, 198–199
   time-sharing, 198–199
   See also UNIX
Murphy, Paul T.
   "Architecture for a Business and Information System, An," 80

naming conventions, **200–203**
   common elements in, 202
   factors for consideration, 200, 202
   importance of, 202–203
   necessity for, 202
National Security Agency (NSA), 34, 183, 184, 221
net neutrality, **203–204**
   opponents of, 204
   proponents of, 203
   regulations for, 204
network security, **205–206**, 51–52
   authorized user password, 205
   intrusion detection and prevention systems, 205
   retinal scan, 205
   security breaches, 206
   See also firewalls
neuro-linguistic programming (NLP), **207–208**
   cross-over mirroring, 207
   mirroring, 207
   reframing, 207
neuromarketing, **208–210**
   criticism of, 210
   ethical concerns, 210
   medical technologies used for
      electroencephalography (EEG), 208, 209
      eye tracking, 208, 209
      functional magnetic resonance imaging (fMRI), 208, 209
      galvanic skin response, 208
      steady-state topography (SST), 208, 209
   process of, 208–210

neuromorphic chips, **210–212**
   for complex tasks, 212
   designing of, 211–212
   functioning of, 211
   Human Brain Project (HBP), 210, 211–212
   memristor, 211, 212
   neuroplasticity, 211, 212
   parallel processing, 212
   traditional microprocessor chips vs., 211, 212
neuromuscular electrical stimulation (NMES). *See* functional electrical stimulation (FES)
nibble, 19, 20, 22
nondeterministic algorithm, 5
nonvolatile memory, 46, 47, 131, 132
normalization, 84–85, 274, 275, 276
Nygaard, Kristen, 297

Obama, Barack, 106, 184, 204
objectivity, **213–214**, 249
   cognitive bias, 213
   cultural bias, 213, 214
   maintenance of
      double-blind studies, 213, 214
      peer review, 214
      randomization, 214
      replication of research 214
      single-blind studies, 214
   in qualitative research, 213–214
   in real-world research, 214
   research ethics, 214
   subjectivity vs., 213, 214, 249
object-oriented analysis (OOA), 215, 216
object-oriented design (OOD), **215–217**
   applications, 216–217
      automatic teller machine (ATM), 216
   class-based inheritance, 215, 216
   drawbacks, 217
   object-oriented analysis (OOA), 215, 216
   principles
      dependency inversion, 216
      interface segregation, 216
      Liskov substitution, 216
      open-closed, 216
      single responsibility, 216
   processes
      object design, 216
      system design, 215–216
   prototypal inheritance, 215, 216

object-oriented programming (OOP), **217–219**, 35, 39, 42, 135, 155, 156, 163, 200, 259–260
   hybrid languages, 218, 219
   for improving software development, 218–219
   inheritance technique, 218
   object-oriented programming languages (OOPLs), 217, 218, 219
   process, 217–218
open-source software, 9–10, 105
operating system (OS)
   Android, 8–10
   Linux, 9, 11, 89, 301
   See also disk operating system (DOS)
operator(s), 23
   See also Boolean operators

PageMaker, 300
parallel processing, 195, 212, 254
parameters, 17, 25, 26, 48, 50, 65, 115, 120, 144, 259
Pascal, 298
peer-to-peer (P2P) systems, 16, 203
Perl (Practical Extraction and Report Language), 300
Personal Home Page (PHP), 288
Petrofsky, Jerrold, 134
photofabrication. *See* stereolithography (SLA)
photo-solidification. *See* stereolithography (SLA)
Pixar, 44, 45, 46
Plankalkül (Plan Calculus), 295
platforms, 11, 30, 188–190
   Google, 8–10, 189
   iOS, 158, 189
Polonsky, Ivan, 252
Post, H. W., 133
powder bed fusion, 1, 3
Prensky, Marc, 96, 97
   On the Horizon, 96
Pretty Good Privacy (PGP), 117, 118–119, 302
printable characters, 6
Privacy Act (1974), 221
privacy rights, **220–221**
   Board of Education v. Earls, 220
   Bowers v. Hardwick, 220, 221
   of celebrities vs. freedom of the press
      death of Diana (Princess of Wales), 221
      phone tapping of celebrities in UK, 221
   Griswold v. Connecticut, 220
   of individuals vs. law enforcement needs

Automatic License Plate Readers (LPRs)
    use, 221
  DNA collection of non-convicted individuals, 221
  FBI vs. Apple dispute, 221
  Lawrence v. Texas, 221
  Paparazzi Reform Initiative, 221
  Privacy Act (1974), 221
  Roe v. Wade, 220
  USA PATRIOT Act (2001), 183, 221
private clouds, 30, 31
procedural programming, 155, 156, 217, 218
programming languages, **222–224**
  abstraction, 222–224
  declarative language, 222
    Structured Query Language (SQL), 223
  future of, 224
  imperative language, 222
    COBOL, 223
  paradigms/styles, 223
  pseudocode use, 224
  semantics, 222–224
  structuring of, 223–224
  turing complete, 222, 223
  types
    first-generation, 222
    second-generation, 222–223
    third-generation, 223
  uses, 222
prototyping, **225–227**
  advantages, 226
  approaches to
    evolutionary, 226
    extreme, 226
    throw-away (rapid/close-ended), 225–226
    incremental, 226
  categories
    horizontal, 225
    vertical, 225
  disadvantages, 226
  linear development methodologies vs., 226
  software
    for web applications development, 226–227
  steps involved
    prototype development, 225
    prototype revision, 225
    requirements identification, 225
    user evaluation, 225
  user interaction/user involvement, 226, 227

pseudocode, 61, 135, 136, 224, 280
public clouds, 30, 31

quantum computing, **228–229**
  challenges, 229
  entanglement, 228, 229
  potential of
    Shor's algorithm, 229
  quantum algorithms, 228, 229
  quantum bit (qubit), 228, 229
  quantum computers design, 229
  quantum logic gate, 228, 229
  quantum physics, 228, 229
  subatomic computation theories, 228
  superposition, 228–229
  wave-particle duality, 228
quicksort algorithm, 5

Rand, Sperry, 295
random access memory (RAM), **230–232**, 46, 47, 193
  direct-access storage, 230
  history, 230–231
  types
    dynamic random-access memory (DRAM), 230–231
    nonvolatile random-access memory (NVRAM), 230, 231
    shadow RAM, 230, 231–232
    static random-access memory (SRAM), 230, 231
  uses, 231
rapid application development (RAD), **232–234**, 150, 226, 247, 283, 284
  applications of, 233–234
    development of solutions, 233–234
  development cycle
    application generation, 233
    business modeling, 232
    data modeling, 232
    process modeling, 232–233
    testing and turnover, 233
  features of, 233
  integration, 232, 233
  iteration, 232, 233
rapid prototyping, 3, 55, 57, 232
rasters, 55, 56, 140–142
rational choice theory, **234–236**
read only memory (ROM), 47, 195, 230–232

recursive algorithm, 4, 5
   "divide and conquer," 5
render farms, 17, 44, 46
rendering, 17, 44, 45, 46, 274, 275
resource distribution, 14, 16
Ritchie, Dennis, 298
Rivest, Ronald, 34
robotics, 1, 14, 16, 270
Rogers, Everett
   Diffusions of Innovations, 89–90
Rosenblatt, Frank, 173
Rosenthal, Robert, 124, 125
RS-232-C standard, 298

Scientific Working Group on Digital Evidence (SWGDE), 93, 94–95
Scratch, 303
Search Engine Optimization (SEO), **237–238**
   crawling, 237
   in web design, 286
   in web graphic design, 287, 288
selective layer sintering (SLS), 2, 3
self-star properties, 14–16
   See also autonomic computing
semantic memory, **238–239**
   Alzheimer's disease, 239
   episodic memory vs., 238–239
   semantic dementia, 239
semantics, **239–240**, 222, 223–224, 285
Seychell, Dylan
   New Digital Natives: Cutting the Chord, The, 97
Shamir, Adi, 34
Shannon, Claude, 32, 295
   Mathematical Theory of Communication, The, 295
sheet lamination, 1, 3
Shiffrin, Richard M., 289
Shor, Peter, 229
signal processing, **240–242**
   analog signal, 240, 241–242
   applications, 242
   digital signal, 241–242
   digital signal processor (DSP)
      fixed-point, 240
      floating-point, 241, 242
   information extraction from signals, 242
   linear predictive coding, 241, 242
   mechanism, 242
      Fourier transform, 241, 242
Simula, 297

simulation, **16–18**, 48–49
   educational applications, 18
   See also virtual reality (VR)
smartphones, 97, 203, 222, 257, 270, 284, 286, 301
   apps, 11, 110, 150, 182, 188–192
   cameras, 183
   crowd searching using, 73
   malware, 176
   popularity of, 9, 10
Snowden, Edward, 221
social media
   Instagram, 304
   Facebook, 9, 176, 263, 304
   Twitter, 9, 190
source code, 243, 9, 13, 28, 29, 30, 35, 36, 39, 120, 135–136, 276, 278, 298, 302
source code comments, **243–245**
   annotation, 243
   creation of effective, 244–245
   delimiters, 243–244
   syntactical rules, 243
   uses, 244
   value of, 245
special characters, 29, 274
Speedcode, 295
spiral development, **245–247**, 150
   benefits, 247
   drawbacks, 247
   as hybrid methodology, 247
   in large projects, 247
   phases, 245–247
      engineering, 246
      evaluation, 246–247
      risk analysis, 246
      risk planning, 246
   risk management, 247
Stallman, Richard, 299
standard deviation, **248–249**
   empirical rule, 249
   population, 248–249
   sample, 248–249
   square root of the variance, 248
standpoint theory, **249–250**
   criticism of
      dualism or double thinking, 250
   essentialism, 249
   feminist standpoint theory, 250

statistical inference, **251–252**
    structural equation modelling (SEM), 252
stereolithography (SLA), 1–3
string-oriented symbolic language (SNOBOL), **252–254**
    data types, 253
    development, 252–253
    disadvantages, 253
    modern usage, 254
    strengths, 253
    string manipulation, 253–254
string(s), **28–29**, 7, 23, 42, 279, 280
    concatenation, 29
    interpolation, 28, 29
    literals, 28, 29
    special characters in, 29
    uses, 29
structural equation modeling (SEM), **254–255**, 252
    latent variables, 255
    mediators, 254, 255
substitution ciphers, 32–33, 34, 75, 76
    Caesar cipher, 76
    Playfair Square cipher, 33
subtractive manufacturing, 1, 3
supercomputers, 181, 185, 301
surface tessellation language (STL), 1
Sutherland, Ivan, 53
    Sketchpad, 296–297
syntax, 28, 29, 39, 40, 41, 60, 163, 222–224, 243, 253, 285
    errors, 86, 87, 119–120
    highlighting, 35, 36
system software, 11, 188–189

technology in education, **256–258**
    80s, 257
    present, 257–258
    70s, 256–257
Telecommunications Act (1996), 204
test doubles, **259–260**
    pros and cons, 259–260
    in real-world scenarios, 260
    roles in testing, 259
        dummy object, 259
        fake object, 259, 260
        mock object, 259
        test spy, 259
        test stub, 259, 260
theory of multiple intelligences (MI), **261–262**

Theory X, **262–263**
Theory Y, **262–263**
Thompson, Kenneth, 298
Tolman, Edward, 123
Torvalds, Linus, 9, 301
touch-screen technology, 9, 10, 140, 158
Toy Story, 17, 44
traditional manufacturing. *See* subtractive manufacturing
Transformation Priority Premise (TPP), **264–266**
    application of, 265–266
    considerations with, 265
    core principle, 265
    power of transformations, 266
    recursion, 264, 265
    test-driven development (TDD), 264, 265, 266
    unconditional statement, 264, 265
transposition ciphers, 32, 34, 75, 76
tree structures, **266–268**
    advantages, 268
    binary, 266, 267, 268
    importance of, 268
    median, 266, 268
    nodes, 266–268
        child, 266–268
        parent, 267
        root, 267, 268
        sibling, 267
    uses, 268
    value in programming, 268
trigger-based marketing. *See* event-driven marketing (EDM)
Tulving, Endel, 238–239
    Elements of Episodic Memory, 239
    Organization of Memory, 238
Turing, Alan, 269, 270
    "Computing Machinery and Intelligence," 270
turing complete, 222, 223
turing test, **269–270**
    artificial intelligence, 269, 270
    automaton, 269
    chatterbot, 269, 270
    criticism of, 270
    imitation game, 269, 270
    legacy of, 270
    natural language processing (NLP), 270
"tweening". *See* keyframing

uncertainty reduction theory (URT), **272–273**
   assumptions of, 273
   axioms of, 273
   behavioral uncertainty, 272
   cognitive uncertainty, 272
   strategies to defeat uncertainty, 273
Unicode, **274–276**, 8, 29
   character encoding, 274
   development of, 274
   glyph, 274, 275
   grapheme, 274, 275, 276
   hexadecimal, 274, 275, 276
   normalization, 274, 275, 276
   process, 275–276
   rendering, 274, 275
   special characters, 274
   standard, 275
   uses, 276
unit test, 126–127, 259–260, 264–266, 282
UNIX, **276–278**, 198, 298
   characteristics of, 277–278
      command-line interpreter, 276, 278
      hierarchical file system, 276, 277, 278
      kernel, 277, 278
   multitasking system, 277, 278
   multiuser capabilities, 277, 278
   origin of, 277
   uses, 278
   variants
      "certified UNIX operating systems,"278
      "UNIX-like operating systems," 278
unmanned aerial vehicles (UAVs), 105, 106, 107
   See also drones
USA Freedom Act (2015), 184
USA PATRIOT Act (2001), 183, 221
utility programs, 11, 104, 188, 189

variables, 23–25, 28, 29, 35–37, 39, 48, 50
   animation (avars), 17, 44–45
   applications of, 280–281
   characters, 279, 280
   considerations with
      lifetime, 280
      scope, 280
   string, 279, 280
   See also variables and values
variables and values, **279–281**
   importance of, 281
   storing and accessing data, 279–280

vat photopolymerization, 1, 3
vectors, 5, 55, 56, 140–142
Verilog, 300
virtual memory, 46, 47
virtual reality (VR), 16–18, 44
   in 3D, 18
   virtual worlds, 17
   See also avatars; simulation
Visicalc, 299
volatile memory, 46, 47
Vroom, Victor, 123

Wall, Larry, 300
waterfall development, **282–284**, 66, 149–151, 225, 245, 247
   methodology, 282–283
   phases
      implementation, 282
      maintenance, 283
      requirements analysis, 282
      system design, 282
      unit testing, 282–283
   in practice, 283–284
   pros and cons of, 283
   rapid application development (RAD), 283, 284
   for software development, 284
web applications, 11, 13, 30, 165, 226, 227
web design, **284–286**
   computer languages used, 284, 285
   elements in, 286
   Flash, 285
   history of, 284–285
   impact of, 285–286
   responsive, 286
   search engine optimization (SEO) in, 286
web graphic design, **287–289**
   functionality, 288
   logotype, 287, 288
   search engine optimization (SEO) in, 287, 288
   tableless, 287
   techniques for, 288
   typography, 287, 288
   user-generated, 288–289
   wireframe, 287, 289
widgets, 9, 158, 159
Wolfram, Stephen, 301
WordStar, 299
working memory, **289–290**
   brain parts contributing to, 290

impairment of, 290
short-term memory vs., 289
subsystems
   phonological loop, 289, 290
   visual-spatial "sketch pad"/"mind's eye," 289–290
worse-is-better (software development approach), **291–292**
   applications, 292
   characteristics, 291–292
      completeness, 292
      consistency, 292
      correctness, 292
      simplicity, 291–292
   strengths and weaknesses, 292

Year 2000 (Y2K) bug, 86, 302–303

Zadeh, Lotfi, 78
Zuse, Konrad, 295